Next Generation Nanoelectrochemistry

Online virtual event

29 November – 1 December 2021

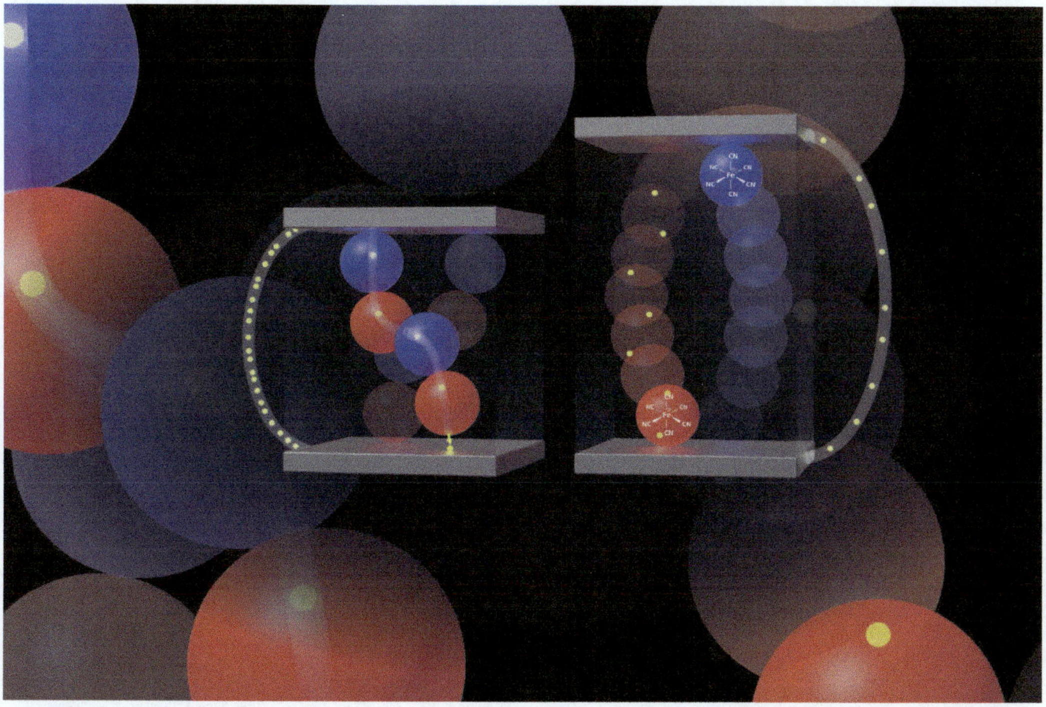

FARADAY DISCUSSIONS

Volume 233, 2022

ROYAL SOCIETY OF **CHEMISTRY**

The **Faraday Division** of the Royal Society of Chemistry, previously the Faraday Society, was founded in 1903 to promote the study of sciences lying between chemistry, physics and biology.

Faraday Discussions (Print ISSN 1359-6640, Electronic ISSN 1364-5498) is published 8 times a year by the Royal Society of Chemistry, Thomas Graham House, Science Park, Milton Road, Cambridge, UK CB4 0WF.

Volume 233 ISBN 978-1-83916-428-6

2022 annual subscription price: print+electronic £1133 US $1994; electronic only £1079, US $1899.
Customers in Canada will be subject to a surcharge to cover GST. Customers in the EU subscribing to the electronic version only will be charged VAT.

All orders, with cheques made payable to the Royal Society of Chemistry, should be sent to the Royal Society of Chemistry Order Department, Royal Society of Chemistry, Thomas Graham House, Science Park, Milton Road, Cambridge, CB4 0WF, UK
Tel +44 (0)1223 432398; E-mail **orders@rsc.org**

If you take an institutional subscription to any Royal Society of Chemistry journal you are entitled to free, site-wide web access to that journal. You can arrange access via Internet Protocol (IP) address at **www.rsc.org/ip**

Customers should make payments by cheque in sterling payable on a UK clearing bank or in US dollars payable on a US clearing bank.

Printed in the UK

FSC MIX Paper from responsible sources FSC® C013604 www.fsc.org

ROYAL SOCIETY OF CHEMISTRY

Faraday Discussions

Faraday Discussions are unique international discussion meetings that focus on rapidly developing areas of chemistry and its interfaces with other scientific disciplines.

Information for Authors

Next Generation Nanoelectrochemistry

Faraday Discussions

www.rsc.org/faraday_d

A General Discussion on Next Generation Nanoelectrochemistry was held virtually on the 29th and 30th November and 1st December 2021.

The Royal Society of Chemistry is the world's leading chemistry community. Through our high impact journals and publications we connect the world with the chemical sciences and invest the profits back into the chemistry community.

CONTENTS

ISSN 1359-6640; ISBN 978-1-83916-428-6

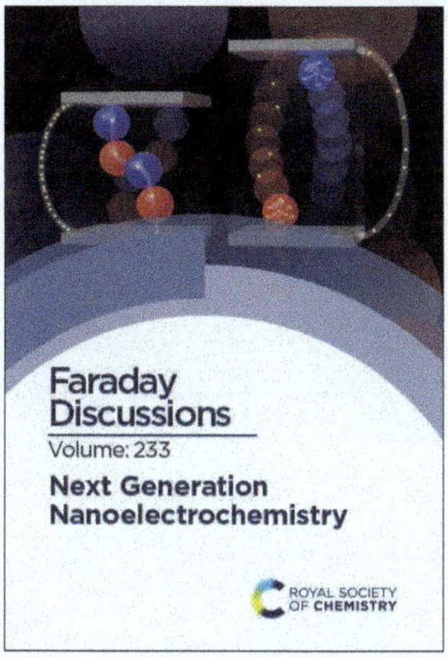

Cover
See Kurihara *et al.*, *Faraday Discuss.*, 2022, **233**, 206–221.

Nano-confined electrochemical reaction studied by electrochemical surface forces apparatus.

Image reproduced by permission of Kazue Kurihara from *Faraday Discuss.*, 2022, **233**, 206.

PREFACE

INTRODUCTORY LECTURE

PAPERS AND DISCUSSIONS

POSTER SPONSORSHIP

ROYAL SOCIETY OF CHEMISTRY | FARADAY DIVISION

Faraday Discussions

CONCLUDING REMARKS

ADDITIONAL INFORMATION

ROYAL SOCIETY OF CHEMISTRY

EDITORIAL

Preface

Yi-Tao Long * and Hong-Yuan Chen

DOI: 10.1039/d2fd90005j

The years of 2020 and 2021 have brought many uncertainties and challenges to the global scientific community. This *Faraday Discussion* on Next Generation Nanoelectrochemistry was proposed in the fall of 2018, and follows on from the *Faraday Discussion* on Single Entity Electrochemistry in York in September 2016 and Electrochemistry at nano-interfaces in Bath in June 2018. We planned an in-person scientific meeting as a continuing event after a fruitful 2018 Discussion for the growing nanoelectrochemistry community. Due to the COVID-19 pandemic, the committee was faced with the hard decision to transition this *Faraday Discussion* into the virtual meeting format. To overcome the limitations of online meetings, the Royal Society of Chemistry offered a creative arrangement which allowed the best accessibility across different time zones around the world. This special online meeting format not only brought researchers together, but served as a form of close community building around a topic of current interest in nanoelectrochemistry.

This *Faraday Discussion* talked about the challenges in both fundamental and applied nanoelectrochemistry, where new concepts and new knowledge play key roles. This field is rapidly developing and requires interactions with other scientific disciplines. Following the 2016, 2018, and 2021 *Faraday Discussion* on nanoelectrochemistry, some discussion points and talks have become landmarks in this field. We hope that "nanoelectrochemistry" can be the topic of a meeting series to promote future developments in this exciting research area. With strong supports from the Royal Society of Chemistry, a follow-up *Faraday Discussion* on "New Horizons in Nanoelectrochemistry" will be held in the year 2024. Looking forward, we welcome you to join us face-to-face in Nanjing in 2024. Hopefully, that will be a great opportunity to physically connect with the nanoelectrochemistry community and move electrochemical science forward.

Nanjing University, Nanjing, China. E-mail: yitaolong@nju.edu.cn

Faraday Discussions

ROYAL SOCIETY OF CHEMISTRY

PAPER

Spiers Memorial Lecture. Next generation nanoelectrochemistry: the fundamental advances needed for applications

Yanfang Wu, [a] Sina Jamali, [a] Richard D. Tilley [b] and J. Justin Gooding *[a]

Received 12th November 2021, Accepted 18th November 2021
DOI: 10.1039/d1fd00088h

Nanoelectrochemistry, where electrochemical processes are controlled and investigated with nanoscale resolution, is gaining more and more attention because of the many potential applications in energy and sensing and the fact that there is much to learn about fundamental electrochemical processes when we explore them at the nanoscale. The development of instrumental methods that can explore the heterogeneity of electrochemistry occurring across an electrode surface, monitoring single molecules or many single nanoparticles on a surface simultaneously, have been pivotal in giving us new insights into nanoscale electrochemistry. Equally important has been the ability to synthesise or fabricate nanoscale entities with a high degree of control that allows us to develop nanoscale devices. Central to the latter has been the incredible advances in nanomaterial synthesis where electrode materials with atomic control over electrochemically active sites can be achieved. After introducing nanoelectrochemistry, this paper focuses on recent developments in two major application areas of nanoelectrochemistry; electrocatalysis and using single entities in sensing. Discussion of the developments in these two application fields highlights some of the advances in the fundamental understanding of nanoelectrochemical systems really driving these applications forward. Looking into our nanocrystal ball, this paper then highlights: the need to understand the impact of nanoconfinement on electrochemical processes, the need to measure many single entities, the need to develop more sophisticated ways of treating the potentially large data sets from measuring such many single entities, the need for more new methods for characterising nanoelectrochemical systems as they operate and the need for material synthesis to become more reproducible as well as possess more nanoscale control.

[a]*School of Chemistry and Australian Centre for NanoMedicine, The University of New South Wales, Sydney, New South Wales 2052, Australia. E-mail: justin.gooding@unsw.edu.au*
[b]*School of Chemistry and Electron Microscope Unit, Mark Wainwright Analytical Centre, The University of New South Wales, Sydney, New South Wales 2052, Australia*

Introduction

Nanoelectrochemistry in the broadest sense is where electrochemical properties and/or electrode materials are investigated in the nanoscale regime.[1] Integral to this definition is the word "investigated" which implies more than just occurring at the nanoscale. After all, the magnitude of the electrical double layer means that virtually all electrochemical processes occur at the nanoscale. *Investigated* implies that we understand how properties that we observe arise from the nanoscale control. That is, adding the science to the technology. To really understand electrochemistry at the nanoscale requires the development of new measurement tools that can probe some of the myriad of basic phenomena that occur at the nanoscale.[2] It also requires reproducible nanoscale, or even atomic, control, over the materials used to make nanoelectrochemical devices.[3]

Nanoscale electrochemistry itself is not new, but the interest in nanoelectrochemistry is accelerating.[3–5] This interest is driven by the development of incredible new measurement tools for understanding electrochemistry at the nanoscale and the astonishing advances in nanomaterials synthesis that allow us to make so many nanoscale electrochemical systems.[2] Together, the measurement tools and the materials give researchers the ability to rationally design electrode interfaces at the nanoscale. The speed of developments in nanoscale electrochemistry is demonstrated by the small-focused conferences that provide a snapshot of cutting-edge thoughts at a specific point in time. *Faraday Discussion* meetings are ideal for revealing the zeitgeist of a field as the published proceedings are as much about the discussion of papers presented in the field as the papers themselves.[6]

In nanoscale electrochemistry, there has been a series of these meetings. The first in 2016 was on "Single Entity Electrochemistry".[7] This meeting was focused on the common questions and methods involved in the electrochemistry of single entities whether they be molecules, particles or cells (Fig. 1).[8] The meeting effectively defined a field even though studying single entities electrochemically dated back to the Coulter counter in 1949,[9] to electrophysiology and the patch clamp amplifiers that measure individual ion-channels since 1976,[10] and single cell electrochemistry dating back to 1982.[11] What was new, was a plethora of new techniques that were developed in the early 2010's that allowed us to start looking at a much larger range of single entity phenomena. In fact the paper from the first lecture by the great N. J. Tao, was focused entirely on measurement methods for exploring the electrochemistry of single entities.[2] That is not to say that in 2016 the science was focused entirely on how to measure single entities. There were papers on fundamental nanoscale phenomena such as nanobubbles,[12] how ion size influenced double layer charge screening more dramatically with decreasing size of electrodes[13] and how water structure in nanoscale pores influenced proton transport.[14] But the majority of the papers were focused on the new types of information these new techniques gave us from investigating single entities.

The focus on new measurement tools was perfectly understandable because the rise of single entity electrochemistry was initiated by the development of new instruments to investigate single entities. The small size of the signals arising from such methods saw two approaches to obtain the electrochemical information. The first approach was to use high quality electronics and to keep the

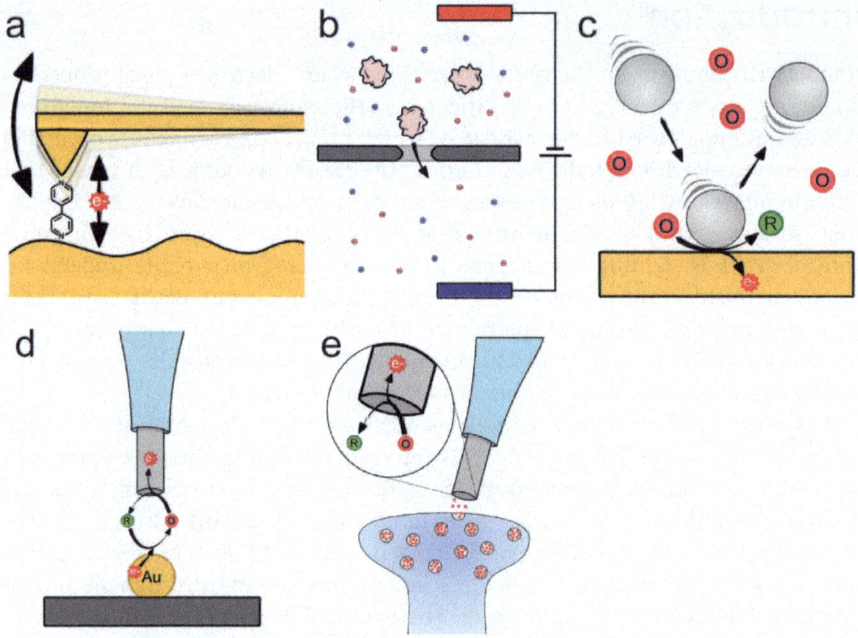

Fig. 1 The family of methods that make up the field of single entity electrochemistry. (a) Scanning tunneling microscope to explore electron transport processes at the single molecule level. (b) Nanopores for detecting single molecules. (c) The electrochemistry of single nanoparticles as they collide with a microelectrode. (d) Scanning electrochemical microscope (SECM) measurement of a single nanoscale entity on a surface. (e) Electrochemistry of single entities released from, or within, cell membranes. Adapted with permission from ref. 8. Copyright 2016, John Wiley & Sons, Inc.

electrons transferred in the electrochemical processes as electrons.[15–18] The second strategy was to convert electrons transferred into photons.[19–21] The advantage of the first approach is that the only processes occurring are electrochemical but the capability of instruments, in terms of response time and sensitivity, is limited by the electronics. Advances in electronics are reducing that limitation (DOI: 10.1039/D1FD00055A). The advantage of the latter approach is that through the incredible developments in microscopy, single photon, and hence single electron, detectors are now well-established. The limitation is, as stated by White and colleagues, "a mechanism of electron-transfer-to-photon-transduction is required" for some types of fundamental questions to be answered.[22]

An implication of panels d and e in Fig. 1 is that single entity electrochemistry could involve looking at single entities within a much larger interface. So just two years after the first meeting, a second *Faraday Discussions* meeting focused on the broader field of "Electrochemistry at Nano-Interfaces". Although some elegant new methods were being developed to understand nanoscale electrochemistry,[23–25] by this time the field had already moved forward significantly. Many more papers presented at the meeting reported on obtaining fundamental understanding of nanoelectrochemistry using some of the new experimental methods. Discussion of applications of nanoelectrochemistry also came to the fore. The paper aligned with the opening lecture of that meeting reflected the move towards fundamental understanding of nanoscale electrochemistry with White and colleagues placing nanoscale electrochemistry in a more historical context and future trajectory.[22] The paper really highlighted the motivations behind the emergence of interest before venturing towards understanding electrochemical dynamics and stochastic electrochemistry. Another important concept that was well articulated in that meeting was from the final paper by Bohn with a discussion of length scales.[26] The variance in length scales in nanoelectrochemical systems is diverse. Length scales vary from electrons transferring over typically less than a nanometre, molecular interactions being on the scale of a few nanometres,[27] diffusion commonly being on the length scale of microns, whilst for analytical measurements sample sizes will be millimetres to centimetres.[27] Being cognizant of these different length scales can be vitally important in understanding nanoelectrochemical systems as we will see below.

Thus far we have focused more on the amazing new instruments that have been developed for investigating electrochemical processes. We would argue that the intense interest in nanoelectrochemistry has also arisen because of the concurrent developments in nanomaterials synthesis.[28] That is, we need something at the nanoscale to investigate. Nanoparticles can now be synthesised, to present specific crystal facets, with multiple different materials precisely located in a nanoparticle to synergistically facilitate electrochemical reactions,[29] to give a perfectly ordered distribution of atoms of materials in an alloy[30] and even precisely locate single catalytic atoms on a substrate.[31] These new materials have given us the ability to design electrode interfaces at the nanoscale, or even molecular level,[32] in ways that were never previously possible. These new nanomaterials presented electrochemistry with a myriad of exotic electrode materials and architectures, that have revealed a range of new electrochemistry, and required new instrumentation and understanding of the electrochemical processes occurring.[2,33]

The paper herein serves as the introduction to the third of these meetings, being held in 2021, entitled "Next Generation Nanoelectrochemistry". So where are we now and how has the field of nanoelectrochemistry changed in these 3 years since the last meeting? These are of course 3 years affected by a global pandemic that has changed our lives, but it does not seem to have slowed the pace of development in nanoelectrochemistry. As we can see from some of the papers to be presented in this meeting, the shift to application has progressed, new measurement methods (DOI: 10.1039/D1FD00063B) continue to be developed and advances in fundamental knowledge using some of the exquisite methods for probing nanoelectrochemical systems are being made. So in this paper, aligned with the title of the meeting, we will look forward, guided by the needs of the applications and the fundamental science that must follow to advance these applications. We will focus on materials and devices and how they drive the applications in nanoelectrochemistry forward. We will frame this around electrocatalysis (DOI: 10.1039/D1FD00043H) and nanoelectrochemical sensing (DOI: 10.1039/D1FD00057H, DOI: 10.1039/D1FD00049G, DOI: 10.1039/D1FD00042J). From these applications we will discuss some of the fundamental science developments requiring attention to drive our understanding of nanoelectrochemistry forward and hence allowing us to realise more of the technologies.

Electrocatalysis

Nowhere has it been more evident that the rise in nanoelectrochemistry has been driven by developments in materials chemistry than in electrocatalysis (energy generation in terms of fuel cells) and energy storage devices such as batteries (Fig. 2). The driver in the materials development in electrocatalysis has been to obtain more active catalysts and catalysts made from cheaper materials. The adsorption of reactants onto electrode materials to reduce the energy required to make a reaction proceed, and the desorption of products, are central to catalyst performance. As such, the dominant mode of thought in materials design for catalysts is the Sabatier principle,[34] where the adsorption energy of the rate limiting step must be just right such that the reactants bind and the products leave. This leads to the classical volcano plots where the catalytic activity rises with adsorption energy which promotes the reaction of substances at the electrode surface until the optimal binding energy is reached, whereupon the reaction rate declines again as products do not desorb from the electrode surface sufficiently rapidly. The adsorption–desorption steps are linked by linear scaling relationships that arise when the bond between the electrode and the reactant is the same as the product and the electrode.[35,36] That is, increasing the adsorption

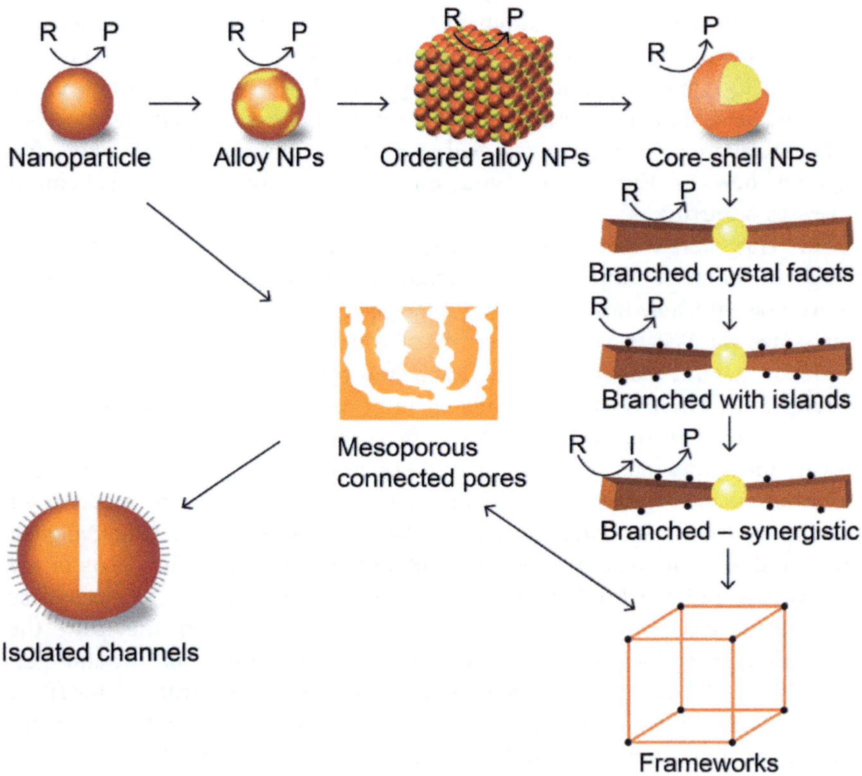

Fig. 2 The progression in sophistication of nanoparticle (NP) synthesis for electrocatalysis. R, I, and P represent reactants, intermediates, and products of a given electrocatalytic reaction, respectively.

energy of the reactant to facilitate the reaction makes it harder for the product to leave. As such the peak of the volcano plot represents the peak catalytic activity achievable based on optimising the adsorption–desorption energies alone.

Initially the attraction of nanoparticles was the wide range of materials that can be used to explore more catalytically active materials. Their high surface area to volume ratio can mean more active sites can be presented to the species in solution, and that lower amounts of expensive catalytically active material can be used. Nanoparticle design however, has become progressively more sophisticated to go beyond simply optimising the chemical composition of the particle or alloying the particle, to controlling the distribution of atoms in alloys[30] or nanoparticle structure to further enhance activity (Fig. 2). For example there has been a trend towards nanoparticles being designed with specific crystal facets dominating, that have superior catalytic activity[37,38] or stability.[39] Furthermore, nanoparticles with three-dimensional shapes to maximise the presentation of certain crystal facets have been synthesised.[28] Taking the sophistication and nanoscale control further, building on the ability to control crystal facets and nanoparticle shape, multicomponent electrocatalytic nanoparticles can be designed. For example we have synthesised branched nickel nanoparticles upon which very small islands of platinum were grown using a slow growth method using a Fisher–Porter bottle for the hydrogen evolution reaction.[40] The slow growth allowed epitaxial growth of Pt onto a surface where the spacing between Ni atoms is defined by the Ni(0002) such that there is a 8–9% difference in the atomic spacing. This means considerable compressive strain was imparted on the Pt if the islands were small enough. This strain alters the adsorption–desorption energies of the reactants–products to the catalytic material.[41,42] What was observed was the smaller the platinum island, the higher the specific and mass activity. As such, 1.9 nm Pt islands had $\sim$1.4 times the compressive strain than 2.7 nm islands and as a result 2.3 times the specific activity.

With the growth of small amounts of catalytic materials on another nanoparticle that serves as the substrate, this substrate could be passive or contribute synergistically in the electrocatalytic activity. We show this in a later paper on the hydrogen evolution reaction at Pt islands on amorphous Ni where there is no epitaxial growth of the Pt and hence no strain. However, the Ni substrate facilitates the cleaving of the H–OH bond in the rate-limiting water dissociation step, providing more H* intermediates that can recombine at the Pt sites to form H_2.[43] As such, it was shown that by increasing the number of interfaces between the Pt islands and the Ni support that superior catalytic activity could be achieved.

Beyond small islands of catalytic materials, the ultimate synergistic effect between catalytic centres and a facilitating substrate can be achieved with single atom electrocatalysts. Not only do single atom catalysts use a minimal amount of expensive catalytic material and potentially have higher catalytic activity, but the number of interfaces between a catalytic site and a facilitating substrate would be maximised with single atom catalysts. Such a possibility has recently been shown by Ciucci *et al.*[31] who synthesised single Pt atom catalysts on a ruthenium oxide surface for methanol oxidation reaction (see single atom reactive sites in the electron microscopy images in Fig. 3). The single Pt atom catalysts on RuO_2 showed 15.3 times higher catalytic activity than commercial platinum on carbon whilst single Pt atoms on carbon showed no activity. These results were

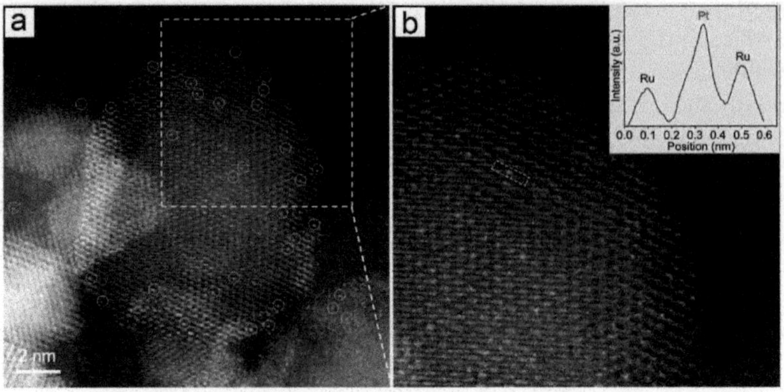

Fig. 3 (a and b) Representative electron microscope images of Pt_1/RuO_2. The inset in (b) shows the intensity profile along with the dashed rectangle. Adapted from ref. 31 from Springer Nature under the terms of the CC-BY Creative Commons Attribution 4.0 International license (https://creativecommons.org/licenses/by4.0/).

interpreted as the RuO_2 being highly tolerant to CO poisoning, an intermediate in the reaction, whilst carbon was not.

The other major trend in materials synthesis for electrocatalysis has been to make porous catalytic electrode materials. The initial motivation of highly porous electrodes was simply to increase the surface area. Potentially highly tortuous pore spaces could also introduce strain to give high catalytic activity. These porous structures however, can also present a different solution environment to the bulk solution,[44] where the change in solution conditions in the pore spaces could also influence the catalytic activity. This was elegantly shown by Erlebacher and co-workers who explored the utility of etched foils of a Pt_3Ni alloy, to give a mesoporous film with recessed active sites, for oxygen reduction reaction.[45] The mesoporous structure of the films resulted in increased specific activity relative to the non-etched foils. The mesoporosity of the Pt_3Ni was described as a way of controlling diffusion rates to the active sites. In a later paper[46] they also showed that the solution conditions inside the nanoporous environment, which could be controlled to be different to the bulk solution, could influence the reaction outcomes. Usually in reduction reactions, such as the oxygen reduction reaction, there is competition from the hydrogen evolution reaction; a reaction which is pH dependent. They showed using unbuffered solutions, that the pH increases during the hydrogen evolution reaction, the depletion of the H^+ saw the hydrogen evolution reaction suppressed, and hence provided better catalytic activity.[46]

In mesoporous films such as those described above, the pores are nanoscale in size and therefore relate to mass transport and nanoconfinement effects.[44] However, most mesoporous films have a broad range of pore sizes and the pores are typically interconnected, which makes it challenging to draw unambiguous conclusions about the impact of nanoconfinement. To help elucidate the impact of nanoconfinement on electrocatalysis, we recently started exploring isolated nanoscale channels in nanoparticles to understand the impact of nanoconfinement on electrocatalytic reactions. An array of isolated channels is very different to interconnected nanopores because with isolated channels, each individual channel is a separate diffusional system, distinct from a layer of interconnected pores which is a single diffusional system like a partition membrane.[47] To form isolated channels in a Pt–Ni catalyst, Pt_3Ni nanoparticles were prepared followed by the Ni domains being etched from the nanoparticle to give Pt based nanoparticles with isolated channels which are between 1.5 – 3 nm in diameter. The exterior of the nanoparticles was passivated with a surfactant such that the reaction can only occur within the channel.[48] The nanoparticles were called nanozymes as their three-dimensional geometry resembled enzymes, and showed three times higher specific activity than the best performing PtNi oxygen reduction catalysts reported to that point in time.[49] In later work where we compared three different channel diameters, we showed the narrower the channel the higher the specific activity (Fig. 4). A continuum model suggested that part of the reason for the high specific activity was due to the active transport of protons into the isolated channel by migration. This migration was facilitated by the fact that with these small substrate channels, $\sim$1.5 nm in diameter, the electrical double layers overlapped such that the charge imbalance from the potential applied to the channel walls could not be compensated for by the counter ions.[50]

Returning to the linear scaling laws discussed above, and how they represent an effective limit to catalyst improvement from optimisation of the adsorption–desorption energies at an electrocatalyst surface, the nanoconfinement examples above suggest that this is one strategy where the linear scaling laws could be broken. This is in line with the suggestion by Montoya *et al.*,[51] that nanoconfinement may be one of the strategies to get performance beyond the scaling laws. Other strategies they suggested involving the nanoscale catalytic interface architecture include using alloys that have dual functionality, incorporating ligands to bind to intermediates and designing interfacial sites between materials, whilst electrolyte engineering and adding promotors are two solution-based strategies.[51] We have incorporated more than one of these strategies for the reduction of carbon dioxide at a silver core-mesoporous shell nanoparticle. In this nanoparticle design, the silver converts the CO_2 to CO at much lower potentials

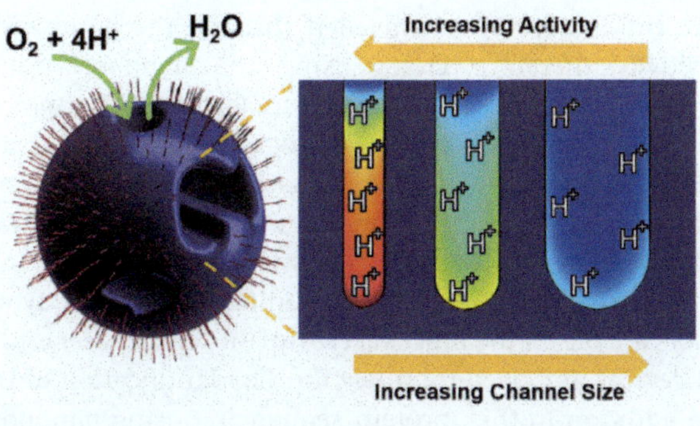

Fig. 4 The impact of nanoscale confinement on electrocatalytic performance. At low overpotentials where electron transfer kinetics dominate, the narrower the channel diameter, the higher the specific activity of the electrocatalyst for the oxygen reduction reaction. Reproduced from ref. 50 with permission from the Royal Society of Chemistry under the terms of the CC-BY Creative Commons Attribution 3.0 Unported license (https://creativecommons.org/licenses/by/3.0/).

than on copper, but the conversion of the CO to higher order organic molecules occurs at the copper. This cascade reaction allows C3 organic molecules to be produced at much lower potentials than previously reported. This is attributed to not just the spill over mechanism at the interface of the two metals, but also that the nanoconfinement effectively traps lots of CO near the copper, this changes the solution conditions where the reactions proceed.[29]

Sensing

The use of nanomaterials in combination with nanoelectrochemistry is ubiquitous in modern sensing. The fabrication of nanoelectrodes has advanced significantly in recent years to the point where we are not only able to perform electrochemical measurements inside cells, but also deliver reagents as well (DOI: 10.1039/D1FD00057H). Similarly, microfabrication has allowed the development of electrochemical sensors that can measure laterally along lipid bilayers (DOI: 10.1039/D1FD00045D). Many of the principles of nanoparticle design, where different components could work synergistically to achieve the desired reaction could be exploited to allow selective detection of the analyte using the well-established electroanalytical approach of modified electrodes.[52] However, here we will focus on nanoscale sensors for monitoring single entities. Crucial to the performance of nanoscale electrochemical sensors is the difference in length scales as discussed in the introduction. This is because time constants for diffusion have a squared dependency on the distance something must diffuse, so in analytical applications where the sample size must be large, on the order of millilitres or even litres, and low concentrations of analyte must find a nanometre sized sensor, response times can be inordinately long.[53]

Nanopore sensors exhibit the incredible power of nanoscale sensors and also the challenges with regards to length scales they face. The classical nanopore sensor, first described in the 1980s,[54] comprises a nanoscale orifice in an otherwise insulating membrane. An electrode is placed on either side of the membrane and the conductance due to ions flowing through the nanopore is measured. When a single molecule translocates through the nanopore, there is a drop in the conductance, represented as a resistive spike, with the current restored to baseline levels once the molecule has travelled through the nanopore. The single molecule sensitivity therefore arises using the nanopore to confine the measurement to a few femtolitres so a single molecule takes appreciable volume in the nanopore such that the ion flow through the nanopore is altered by the molecule. Although most of the initial success with nanopore sensors have been using biological nanopores such as α-hemolysin ion-channels,[55] gramicidin ion-channels[56] and aerolysin ion-channels;[57] in more recent times more and more solid-state nanopores[58] have been used including nanopipette type nanopores. Nanopore sensors are one of the most successful biosensors ever and are now sold as commercial devices and used routinely for the identification of DNA and RNA sequences.[59] In addition to this, protein sequencing using nanopores has been emerging rapidly.[60,61] For example, Long and co-workers demonstrated the utility of a biological nanopore to differentiate single peptide molecules based on their characteristics in ionic current recording during the passage of the peptide molecule through the nanopore.[62] The work showed that three kinds of peptide molecules with similar sequences but varied lengths can be distinguished effectively, and stated the potential in using nanopores to sequence peptides with structural information about molecules of interest.

Selectivity can be a challenge for nanopore sensors as other species in the sample for analysis that can translocate through the nanopore may also give a signal. Sometimes the different species that translocate through the nanopore can be differentiated by the characteristics of the signal[63] but to use nanopores in complex biological fluids has only been demonstrated in a few instances.[64,65] Selectivity in nanopores can be enhanced by adding selective binding ligands to the interior of the nanopore using protein engineering for the biological nanopores or surface modification for solid-state nanopores. A nice example of this is by Minteer and co-workers (DOI: 10.1039/D1FD00042J) where the internal walls of a glass nanopore were modified with putidaredoxin which could bind with its redox partner $P450_{cam}$ monooxygenase, which entered the nanopore from the bulk solution. The work showed that the resistive pulses observed when the putidaredoxin was present where much longer than if the binding ligand on the nanopore walls did not bind to $P450_{cam}$ (Fig. 5). Such a strategy would certainly facilitate selectivity in a nanopore experiment. The much longer residence time is attributed to the binding between the two proteins. Note the apparent lifetime of the binding is facilitated by the nanoconfinement in the nanopore and the amount of binding ligands in the nanopores. This is because the nanoconfinement ensures a high probability of collisions of the $P450_{cam}$ with the walls of the nanopores and the high density of binding partners biases the equilibrium towards a binding event. Therefore, a $P450_{cam}$ molecule that is dissociated from its putidaredoxin binding partner has a high probability of being captured again. Nanoconfinement and a high density of binding ligands is a common strategy used to extend the lifetime of binding partners in single molecule sensing.[66]

The nanoconfinement inside nanopores can also be used to facilitate other reactions and monitor them over extraordinarily long lifetimes. For example, White, Burrows and colleagues have shown that single molecule reactions can be

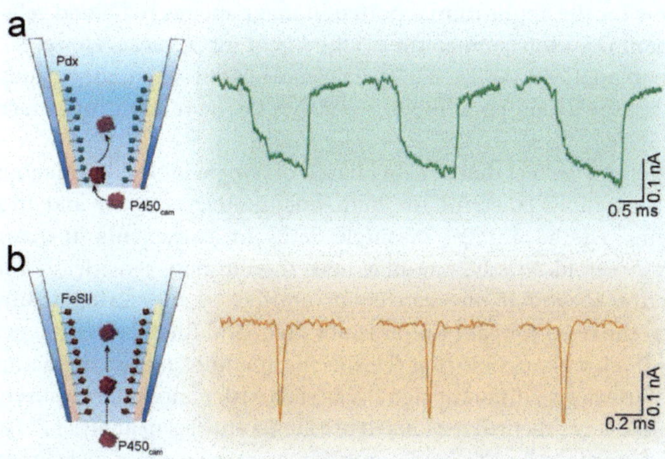

Fig. 5 Schematic of $P450_{cam}$ translocation and representative resistive pulse signals for translocation events using (a) Pdx- and (b) FeSII-modified glass nanopores. Reproduced from DOI: 10.1039/D1FD00042J with permission from the Royal Society of Chemistry.

followed inside an α-hemolysin nanopore.[67] They showed that an acid–base titration of cytosine bases in a single DNA molecule containing a C:C mismatch could be repeatedly monitored over a period of 3 hours.

Despite the single molecule sensitivity, and their incredible success as sensors for molecular identification, nanopore sensors have not been as successful for quantitative analysis.[68] This is not a signal problem, but a mass transport problem related to the different length scales. The sequential nature of the single molecule measurement means that at low concentrations of analyte the interval between resistive spikes becomes inordinately long. For example, at 1 fM the time between spikes is estimated to be approximately 10 minutes.[53] As such nanopore sensors when used in quantitative analysis typically have detection limits in the μM to nM range. Attempts to improve the detection limit have focused on using electrical fields to speed up the mass transport of the molecules to the nanopore.[69,70] We have addressed the issue for detecting the protein prostate specific antigen (PSA) using antibody-modified magnetic nanoparticles that are dispersed throughout the sample to capture the PSA.[65] Upon application of a magnetic field the nanoparticles were brought rapidly to the nanopore where they then blocked the nanopore. This innovation allows aM detection limits in an array of 9 nanopores.[71] More nanopores would be preferable so the analysis could be performed in parallel rather than sequentially, but this would require each nanopore to be independently addressable.

The nanoparticle collision experiments shown in Fig. 1c have also been proposed for analysis. In fact, examples have been reported for detecting cells,[72] nanoparticles,[73] proteins,[74] and viruses,[75] and to prove host–guest interactions (DOI: 10.1039/D1FD00054C). Such a strategy is again essentially a sequential measurement and as such will be challenged at low concentrations with the frequency of events. This can be seen in a study by Compton and co-workers where the collisions of red blood cells with an electrode are used for cell counting.[72] The study showed the ability to detect pM levels of red blood cells based on the number of collisions in 300 s. At 0.5 pM there was on average only 1 collision in that 300 s timeframe which would be insufficient for quantitative analysis. This is not an issue for the application reported in that paper as red blood cells are far more abundant than this, and so the method is fit for purpose. However, it does highlight that applications of nanoparticle collisions need to be judiciously selected as we expect nanoparticle collisions may not be suitable for ultrasensitive sensing.[76]

Taken together, we feel that sensing based on single molecules, nanoparticles or cells, is incredibly powerful for both quantitative analysis and molecular understanding. The advantages of single entity measurements in quantitative analysis are threefold. Firstly, one may have the ultimate sensitivity of a single molecule/entity, second if one can count entities by measuring many single entities, then the need for calibration is obviated, and finally when we get to the single entity level, it is possible that the specific and nonspecific interactions with the sensing surface give different signals, as seen with nanopore measurements.[76] Hence there are very practical reasons to use single entities in analysis.[68] However, for these advantages to be achieved requires many single entity measurements to be performed. Therefore, the serial nature of the measurement methods described above may be a challenge.

The advances needed to apply nanoscale electrochemistry

Beyond the two applications discussed above in electrocatalysis and sensing, nanoscale electrochemistry plays a role in modern battery technology, *in vivo* electrochemical sensing technology, supercapacitors, dye sensitized solar cells and displays.[1] To fully realise the potential of these devices we need to understand these technologies in detail. With that in mind there are a number of themes that have more than one application and we will discuss the challenges related to each of these below.

The need to understand nanoconfinement

With nanoscale structures on an electrode surface, commonly there will be a region where nanoconfinement may play a role in the electrochemical response. By nanoconfinement we mean a nanosized region where the solution is confined in spaces of less than 100 nm in dimension. There are some recent excellent reviews on this topic.[77,78] Nanoconfinement was a significant element of the discussion above in electrocatalysis, especially with the nanoporous systems mentioned, and sensing with the nanopore sensors. However, it is far more ubiquitous than those specific examples being relevant to some degree for nearly all nanoelectrochemical systems. This is because one can envisage there is an aspect of nanoconfinement in systems as simple as a nanoparticle sitting on a surface,[79] to two-dimensional lamellar structures,[80] and to isolated channels in a membrane.[81] Importantly as we go from the nanoparticles on surfaces to isolated channels, the degree of nanoconfinement, based on the ability of bulk solution to access the nanoconfined volume increases, or the degree of freedom decreases. Therefore, the impact of nanoconfinement will, in principle, also increase.

The unanswered question in most cases is how much of a role does nanoconfinement really play? Take for example our study on nanozymes depicted in Fig. 4. The enhanced catalytic activity was attributed to nanoconfinement causing a local high concentration of protons, a necessary reactant in the oxygen reduction reaction. The concentration of protons inside the substrate channel was further enhanced by the active transport of protons into the substrate channel. This is because at the concentration of electrolyte used, and with substrate channels of 2 nm in diameter or less, the electrical double layers overlapped and as such charge compensation was not completely achieved. With the protons being the only counter ion in the system, mass transport of the reactant into the substrate channel was enhanced. This conclusion was supported by a classical continuum model of the electrochemical system. As the simplest possible model, a continuum model was chosen, especially as it gave conclusions consistent with the data, and the simplification in the modelling relative to the actual experimental system is justified. However, one could argue that because of the small diameter of the nanopores, strain generated by the curvature could be significant, so how can one rule out that strain is not the main reason for the excellent electrocatalytic performance? We attempted to address this by making mesoporous particles with similar pore diameters but the nanopores were interconnected. In that case the specific activity was significantly lower, thus pointing out the impact of nanoconfinement on the specific activity.

What the nascent studies into nanoconfinement make clear is that it can play a major role in nanoelectrochemical systems. The examples we presented herein have centred on the confinement ensuring higher amounts of reactants maintained close to each other.[82,83] The small size of the channels also provides the potential to use such materials in sensing and exploiting their ability to offer size selectivity to enhance the specificity of the sensor.[84–86] Nanoconfinement can also be used as a means to alter solvent properties[87,88] and molecular adsorption behaviour.[89]

Questions related to how significant a role nanoconfinement plays is one of the reasons we feel a fundamental understanding of nanoconfinement, and the exploitation of nanoconfinement, are among the next frontiers in electrochemistry. To enhance our understanding of the processes occurring in nanoconfined electrochemical systems, we feel experimental systems with isolated channels, that experience the maximum amount of nanoconfinement, are particularly useful for teasing out the impact of nanoconfinement on a reaction outcome. This is why we have focused on our nanozyme construct with isolated channels in nanoparticles. However, even then the heterogeneity in the size of the channels, as reflected by the fact that we quote the proportion of channels above or below 2 nm,[50] the limited ability to tune the channel diameter limits the quality of information we can acquire. Therefore, more controlled materials are needed, and we are exploring membrane-based approaches to forming nanoconfined systems – such as those by Martin and colleagues, and the surfactant based templating approaches, popularized by Bartlett and Attard[90] in the early 2000s for templating metals, and more recently by Walcarius and colleagues for templating silica[81] – are particularly attractive systems. This is because not only does one get many nanoconfined channels in one electrode that have low polydispersity but also there is control over the size of the channels with subnanometre control.[81,90–92]

Even if we can make better defined materials to observe enhanced performance in electrochemical devices due to nanoconfinement, we also need to understand why. We highlighted the justification of the use of a continuum model in understanding the enhanced performance in an electrocatalytic system.[50] The continuum model considers the ions in the diffuse double layer as point charges and did not consider the structure of water. However, in that case the length scale we were exploring – the micron length scales of mass transport –are compatible with a continuum model, but once we get inside the channel more atomistic models may be required. This is because within the nanoconfined channel the ion size is significant in terms of forming the electrical double layer,[13] or even entering the nanoconfined space. Perhaps even more complicated than this are questions that arise regarding what is the water structure like inside the channels? How is it affected by the ion composition? And what role does it have on the transport of reactants within the channel? Some of these questions are being answered in non-electrochemical systems as we see with Rébiscoul *et al.*;[93] having shown for confinement in mesoporous silica that below 2.6 nm the water dynamics are slowed by the confinement, but at 6.6 nm the concentration and kosmotropic properties of the ions have a greater influence on the water dynamics than the confinement. However, to our knowledge they have not begun to be addressed in systems where the species being transported are reacting electrochemically within the channels. In addition to this, we feel that modelling on the gas–liquid–solid interface may be needed as well. This is because bubbles often exist in liquids, not to mention the situations where gases play a role as reactants or products involved in electrochemical reactions for energy conversion. Thus, attention should be drawn to a better understanding of the gas–liquid–solid interface and its influence on charge transfer and mass transport behaviours for gas-involved electrochemical reactions which would shed light on the future design of electrode materials and structures, surface physicochemical properties, and reaction conditions.

The need to detect many single entities

We feel methods to measure many single entities in parallel need to be developed. The challenge with electrochemistry is the requirement for individually addressable electrodes in an electrochemical array that can make many parallel measurements connected to a circuit, and is definitely a limitation of electrochemistry. That is not to say it cannot be done as we have seen with arrays of 512 nanopores used for DNA sequencing.[94] There is still a limitation with the density of elements in an array due to the space on a chip taken by leads and bonding pads, although strategies to limit this have been developed.[95,96] We have shown on semi-conducting electrodes how to address individual areas on a monolithic surface using light, but this is a serial method again.[97] And even though we can interrogate different regions incredibly rapidly using a projector-based approach,[98] the stochastic nature of single entity measurements may make this impractical.

As such, we are of the view that for performing massively parallel measurements, optical methods have distinct advantages over electrochemical methods. That is because CMOS cameras and other optical detectors can easily monitor many single-molecule emissions at the same time as long as they are spatially or temporally separated. This is after all the principle behind super-resolution light microscopy. And of course, microscopy approaches can do this over wide fields of view meaning a lot of data can be obtained simultaneously. This does not mean that optical detection cannot involve electrochemistry. In the introduction we mentioned the two schools of thought from the first *Faraday Discussions* on "Single Entity Electrochemistry" on whether to keep the electron as an electron, or turn the electron into a photon. We have electrochemical methods like electrochemiluminescence (which has now been shown to go to the single molecule level[99]), the plasmonic imaging of current,[21] and the electrochemical modulation of fluorescence, which we have shown can be monitored at the single molecule level.[100] We feel that optical imaging of electrochemical processes, and/or applying electrochemistry to optics, will experience renewed interest in the electrochemistry community.

The question of whether to keep an electron as an electron or convert it to a photon however is not a question of one approach being better than another. Rather, certain scientific questions or applications may be better suited to one approach over another. So, we see the two approaches as completely complementary.

From the applications discussed above, the need to detect many single entities may seem one that only applies to sensing applications as there we articulated the need for many single entity measurements. However, we see a strong need in many areas of nanoscale and single entity electrochemistry; especially when we want to obtain fundamental information. The stochastics of single molecules and nanoparticles inevitably means many have to be measured before common behaviours become evident over normal variability in behaviours. That is, many measurements are needed so that the signals rise above the noise. This is what we see with molecular electronic measurements using the scanning tunnelling microscopy approach[15] and with nanoparticle collisions.[101] We also feel, that with the variability in materials being used in nanoscale electrochemistry, in some cases for the exquisite information we are obtaining from techniques that relate nanoscale information to the bulk behaviour, many measurements are needed.

How to get the most out of the data

Making many measurements, often from systems that are fundamentally stochastic, means huge amounts of data that to the untrained eye may look like noise. Electrochemists have been excellent at extracting fundamental information from individual pieces of data. Great examples of this are the turning of individual current profiles into information about individual exocytosis events from individual cells[102] or the conversion of single-molecule trajectories through a nanopore into single-molecule ionic spectra.[103] However, as we have seen with molecular electronics measurements using the STM break junction, we also need to acquire and analyse large volumes of data before a definitive picture of the complex molecular system can be painted.

So how do we turn these large data sets into information? In some cases, like with the STM break junctions, the simplest approach is to plot histograms of individual events to determine the most frequent events. To perform such analysis, does of course rely on some assumptions about what one would expect to see in a given experiment, something that has worked well but could mean important information is lost as it is unexpected. Methods to look at such data with no *a priori* information could rectify this. Albrecht and colleagues have addressed this issue for the STM break junction experiments using a multi-parameter vector-based classification process that allows unsupervised analysis and classification of data.[104] The break junction experiment is of course very highly controlled. Other single entity experiments can be less controlled as we have seen in single cell impedance measurements[105] where the cells were exposed to anti-cancer drugs. When doxorubicin was added to media containing breast cancer cells on a silicon electrode the impedance signals suddenly became very noisy, which was related to ion-flux through the cell membranes due to the opening of ion-channels. We examined the data with a simple frequency domain transformation of data by a mathematical algorithm such as fast Fourier transformation to identify a characteristic frequency where the magnitude of this signal related to the amount of drug that entered the cytoplasm.[105] Other simple approaches could be the Hilbert–Huang transformation[106] to extract frequency events and associated intensities and draw correlation between changes in signature electrochemical functions and ion transport through cellular ion channels. But one does wonder what more information could be extracted with the emerging machine learning techniques or when there is more complicated electrochemical data.[107–109]

The intrinsically stochastic nature of events, *i.e.*, heterogeneous electron transfer, particle collision, cell growth and division, generates a random fluctuation of electrochemical functions that needs to be treated, analysed and understood. This becomes challenging as the data is amalgamated with the instrumental and environmental noise that is generated by the data acquisition instrument, circuitry, interfaces, electrical and electromagnetic sources of noise. Deep learning algorithms are required to filter out extrinsic noise so that the "clean" intrinsic data can be treated and analysed independently. Further development of low noise instrumentation capable of high bandwidth (>100 kHz) electrochemical measurement with low signal-to-noise ratio is addressed by Ying and co-workers (DOI: 10.1039/D1FD00055A). Using solely model and theoretically driven data to train deep learning algorithms can confuse the trained architecture when it is used to analyse real-life data that may not completely obey the model or may include contributions from sources that are not accounted for in the theoretical model. Prerequisite considerations and ways of preventing erroneous analytics are discussed by Zhang and co-workers (DOI: 10.1039/D1FD00050K).

The need for more characterisation methods

Single entity electrochemistry is a branch of electrochemistry that really arose from the development of a number of instrumental methods that could measure electrochemical information from single nanoparticles or molecules. The most common strategy to achieving signals from a single entity that are discernable from all the other entities that might exist in a system, is to reduce the measured area to a very small area or volume such that there is only a single entity in the measured volume. A classic example of this is scanning electrochemical cell microscopy (SECCM) where the scanning probe is an electrolyte filled tip that

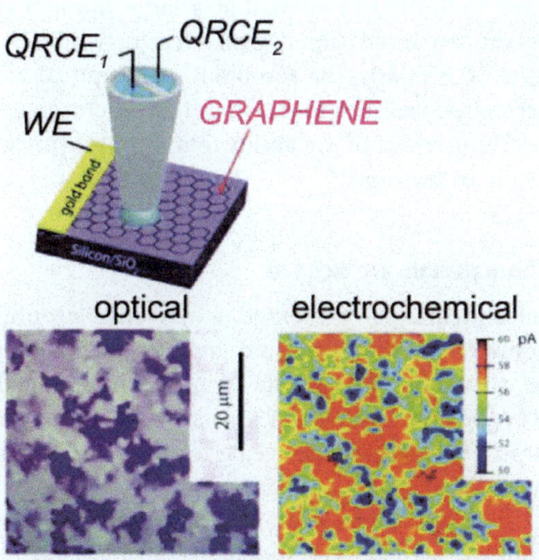

Fig. 6 Mapping of the heterogeneous electron transfer activity of graphene using scanning electrochemical cell microscopy (SECCM). Adapted with permission from ref. 110. Copyright 2012 American Chemical Society.

forms a complete electrochemical cell with the substrate being characterised such that the electrochemistry is confined to ~500 nm in diameter as defined by the tip. An example is shown in Fig. 6 where electron transfer at graphene sheets is measured at a highly localised level and correlated with the structure of the chemically vapour deposited graphene such that higher electron transfer kinetics were increased with the number of graphene layers, but there were subtle effects on the electron transfer kinetics based on how the sheets stacked.[110] An important part of the conclusions arrived at in that study was the correlation of the electrochemical measurements with optical microscopy, shown in Fig. 6, as well as Raman spectroscopy and atomic force microscopy.

We feel that developing techniques that provide correlative measurement is exceedingly powerful. We certainly think that new measurement methods for characterising electrochemical reactions *in situ* or even *operando* will be developed. The strategy of reducing the measurement volume to one where there is typically one or no entity in the measurement volume is analogous to the one used for single molecule sensing.[68] The strength of such techniques is the richness of information they provide. The potential limitation is that to measure many regions is a serial process and hence is limited by the speed at which the measurement can be performed and the number of measurements required.

Light microscope methods do possess some advantages with regards to throughput, but it does depend on the scientific questions being asked. Certainly, the plasmonic imaging approach by Tao and colleagues can allow us to observe variance in electrocatalytic responses from many individual nanoparticles as the electrochemical reactions are proceeding. Even greater resolution has been achieved from using super resolution light microscopy-based image analysis methods when exploring the change in electrochemistry of fluorescent molecules.[111] Using such an approach the variation in the electrocatalytic performance at different domains on a single nanoparticle can be identified.[112] Such strategies can potentially be used to look at many nanoparticles simultaneously.[113] Furthermore, we are beginning to show that a large number of common fluorescent molecules can be interrogated with electrochemical processes at the single molecule level.[100] Similarly, the combination of optical methods for *in situ* monitoring of electrochemical reactions using light microscopy is not limited to plasmonics with a recent report of the ability to monitor a single electrochemical luminescent molecule in solution.[99]

Better defined nanomaterials are needed

The section on electrocatalysis presented a narrative around the incredible advances in nanomaterials synthesis and our ability to control the electrochemical interface, through the synthesis of nanomaterials with almost atomic level precision. The positive impact of the synthesis of sophisticated nanomaterials with purposely designed nanoscale features for electrocatalytic performance was also made clear.

The need for better defined nanomaterials is two-fold. First, it is obvious that the greater the control, with higher spatial resolution, the more tailored the electrode design needs to be to satisfy the requirements of the applications. The second reason is less obvious, but equally important. This relates to knowing what electrode design is required for a given application. To know what the optimal electrode design is, we need to understand the fundamental electrochemistry of a system. To fully understand the impact of design elements on response at the nanoscale, requires characterisation tools that can monitor reaction processes at the nanoscale as well. As we have seen above, many of these characterisation methods have been developed, but because of the heterogeneity of electrode surfaces, many different regions of an electrode must be characterised before one can be statistically be sure of the conclusions one can make. This is naturally time consuming at best and potentially limits the ability for us to derive the fundamental information we seek if the material itself changes over time as alluded to above. Saying that, the number of measurements required to elucidate trends above uncertainty is directly related to the heterogeneity of the material itself. The more homogeneous, the less measurements are required to get statistically robust information.

The synthesis of new and more highly defined nanomaterials is also reaching an exciting level of sophistication. This is highlighted in bi-metallic nanomaterials where precise atomic arrangement is now possible with examples ranging from ordered and disordered alloy nanoparticles, nanocages and single atom catalysts. This atomic level of synthesis gives rise to new and unique nanoscale environments for nanoelectrochemistry. These sites can now be fully characterised by aberration corrected transmission electron microscopes that enable a detailed understanding of the relationship between the atomic site and nanoelectrochemistry performance.

Conclusions

In this paper we have sought to give an overview of where nanoelectrochemistry is currently and where it might go in the next few years. We predict that the application side of this field will become more and more the focus of research. However, we fear that if the development of nanoscale architectures on electrodes is not accompanied with a fundamental understanding of how these architectures influence the electrochemical behaviour, the field of nanoelectrochemistry will not progress as rapidly as it could. The development of the new measurement techniques continuing apace is important herein, especially with the development of methods that allow the correlation of the electrochemistry with other measurement methods. Our view is that understanding the impact of nanoconfinement on electrochemistry is a field of study that is an emerging frontier of electrochemistry and will play a big role on these systems. We also feel that high throughput approaches to measuring single entities electrochemically is important for both fundamental understanding in nanoelectrochemistry and for the application of single entities for analytical applications.

Conflicts of interest

There are no conflicts to declare.

Acknowledgements

We acknowledge funding from the Australian Research Council for the Discovery Projects (DP210102698, JJG and DP190102659, RDT).

Notes and references

1 M. V. Mirkin and S. Amemiya, *Nanoelectrochemistry*, CRC Press, Boca Raton, FL, 2015.

2 Y. Wang, X. Shan and N. Tao, *Faraday Discuss.*, 2016, **193**, 9–39.

3 R. W. Murray, *Chem. Rev.*, 2008, **108**, 2688–2720.

4 S. M. Oja, M. Wood and B. Zhang, *Anal. Chem.*, 2013, **85**, 473–486.

5 S. M. Oja, Y. Fan, C. M. Armstrong, P. Defnet and B. Zhang, *Anal. Chem.*, 2016, **88**, 414–430.

6 *The website for Faraday Discussions meetings*, https://www.rsc.org/events?searchtext=faraday&eventtype=3&rsc=1&membernetworkoptions=116&pagetitle=178.

7 R. M. Crooks, *Faraday Discuss.*, 2016, **193**, 533–547.

8 J. J. Gooding, *Angew. Chem., Int. Ed.*, 2016, **55**, 12956–12958; *Angew. Chem.*, 2016, **128**, 13148–13150.

9 W. H. Coulter, *USA Pat.*, US2656508A, 1949.

10 E. Neher and B. Sakmann, *Nature*, 1976, **260**, 799–802.

11 E. Neher and A. Marty, *Proc. Natl. Acad. Sci. U. S. A.*, 1982, **79**, 6712–6716.

12 S. R. German, M. A. Edwards, Q. Chen, Y. Liu, L. Luo and H. S. White, *Faraday Discuss.*, 2016, **193**, 223–240.

13 Y. Gao, Y. Liu and S. Chen, *Faraday Discuss.*, 2016, **193**, 251–263.

14 J. Huang, J. Zhang and M. Eikerling, *Faraday Discuss.*, 2016, **193**, 427–446.

15 B. Xu and N. J. Tao, *Science*, 2003, **301**, 1221–1223.

16 A. J. Bard, F. R. F. Fan, J. Kwak and O. Lev, *Anal. Chem.*, 1989, **61**, 132–138.

17 R. C. Engstrom and C. M. Pharr, *Anal. Chem.*, 1989, **61**, 1099A–1104A.

18 D. Hellberg, F. Scholz, F. Schauer and W. Weitschies, *Electrochem. Commun.*, 2002, **4**, 305–309.

19 V. Brasiliense, A. N. Patel, A. Martinez-Marrades, J. Shi, Y. Chen, C. Combellas, G. Tessier and F. Kanoufi, *J. Am. Chem. Soc.*, 2016, **138**, 3478–3483.

20 D. Han, G. M. Crouch, K. Fu, L. P. Zaino Iii and P. W. Bohn, *Chem. Sci.*, 2017, **8**, 5345–5355.

21 X. Shan, U. Patel, S. Wang, R. Iglesias and N. Tao, *Science*, 2010, **327**, 1363–1366.

22 M. A. Edwards, D. A. Robinson, H. Ren, C. G. Cheyne, C. S. Tan and H. S. White, *Faraday Discuss.*, 2018, **210**, 9–28.

23 A. R. Hillman, R. Barker, R. M. Dalgliesh, V. C. Ferreira, E. J. R. Palin, R. M. Sapstead, E. L. Smith, N.-J. Steinke, K. S. Ryder and A. D. Ballantyne, *Faraday Discuss.*, 2018, **210**, 429–449.

24 T. Sun, D. Wang and M. V. Mirkin, *Faraday Discuss.*, 2018, **210**, 173–188.

25 J.-F. Lemineur, J.-M. Noël, C. Combellas, D. Ausserré and F. Kanoufi, *Faraday Discuss.*, 2018, **210**, 381–395.

26 P. W. Bohn, *Faraday Discuss.*, 2018, **210**, 481–493.

27 S. O. Kelley, C. A. Mirkin, D. R. Walt, R. F. Ismagilov, M. Toner and E. H. Sargent, *Nat. Nanotechnol.*, 2014, **9**, 969–980.

28 A. R. Poerwoprajitno, L. Gloag, S. Cheong, J. J. Gooding and R. D. Tilley, *Nanoscale*, 2019, **11**, 18995–19011.

29 P. B. O'Mara, P. Wilde, T. M. Benedetti, C. Andronescu, S. Cheong, J. J. Gooding, R. D. Tilley and W. Schuhmann, *J. Am. Chem. Soc.*, 2019, **141**, 14093–14097.

30 H.-S. Chen, T. M. Benedetti, V. R. Gonçales, N. M. Bedford, R. W. J. Scott, R. F. Webster, S. Cheong, J. J. Gooding and R. D. Tilley, *J. Am. Chem. Soc.*, 2020, **142**, 3231–3239.

31 Z. Zhang, J. Liu, J. Wang, Q. Wang, Y. Wang, K. Wang, Z. Wang, M. Gu, Z. Tang, J. Lim, T. Zhao and F. Ciucci, *Nat. Commun.*, 2021, **12**, 5235.

32 J. J. Gooding and S. Ciampi, *Chem. Soc. Rev.*, 2011, **40**, 2704–2718.

33 L. A. Baker, *J. Am. Chem. Soc.*, 2018, **140**, 15549–15559.

34 P. Sabatier, *La catalyse en chimie organique*, 1920.

35 A. Kulkarni, S. Siahrostami, A. Patel and J. K. Nørskov, *Chem. Rev.*, 2018, **118**, 2302–2312.

36 A. Vojvodic and J. K. Nørskov, *Natl. Sci. Rev.*, 2015, **2**, 140–143.

37 K. Zhou and Y. Li, *Angew. Chem., Int. Ed.*, 2012, **51**, 602–613; *Angew. Chem.*, 2012, **124**, 622–635.

38 M. T. M. Koper, *Nanoscale*, 2011, **3**, 2054–2073.

39 L. Gloag, T. M. Benedetti, S. Cheong, Y. Li, X.-H. Chan, L.-M. Lacroix, S. L. Y. Chang, R. Arenal, I. Florea, H. Barron, A. S. Barnard, A. M. Henning, C. Zhao, W. Schuhmann, J. J. Gooding and R. D. Tilley, *Angew. Chem., Int. Ed.*, 2018, **57**, 10241–10245; *Angew. Chem.*, 2018, **130**, 10398–10402.

40 A. Alinezhad, L. Gloag, T. M. Benedetti, S. Cheong, R. F. Webster, M. Roelsgaard, B. B. Iversen, W. Schuhmann, J. J. Gooding and R. D. Tilley, *J. Am. Chem. Soc.*, 2019, **141**, 16202–16207.

41 P. Strasser, S. Koh, T. Anniyev, J. Greeley, K. More, C. Yu, Z. Liu, S. Kaya, D. Nordlund, H. Ogasawara, M. F. Toney and A. Nilsson, *Nat. Chem.*, 2010, **2**, 454–460.

42 M. Luo and S. Guo, *Nat. Rev. Mater.*, 2017, **2**, 17059.

43 A. Alinezhad, T. M. Benedetti, J. Lian, V. R. Gonçales, J. J. Gooding and R. D. Tilley, *Chem. Commun.*, 2021, **57**, 2788–2791.

44 M. Seo and T. D. Chung, *Curr. Opin. Electrochem.*, 2019, **13**, 47–54.

45 J. Snyder, T. Fujita, M. W. Chen and J. Erlebacher, *Nat. Mater.*, 2010, **9**, 904–907.

46 E. E. Benn, B. Gaskey and J. D. Erlebacher, *J. Am. Chem. Soc.*, 2017, **139**, 3663–3668.

47 C. Andronescu, J. Masa, R. D. Tilley, J. J. Gooding and W. Schuhmann, *Curr. Opin. Electrochem.*, 2021, **25**, 100644.

48 T. M. Benedetti, C. Andronescu, S. Cheong, P. Wilde, J. Wordsworth, M. Kientz, R. D. Tilley, W. Schuhmann and J. J. Gooding, *J. Am. Chem. Soc.*, 2018, **140**, 13449–13455.

49 R. Chattot, T. Asset, P. Bordet, J. Drnec, L. Dubau and F. Maillard, *ACS Catal.*, 2017, **7**, 398–408.

50 J. Wordsworth, T. M. Benedetti, A. Alinezhad, R. D. Tilley, M. A. Edwards, W. Schuhmann and J. J. Gooding, *Chem. Sci.*, 2020, **11**, 1233–1240.

51 J. H. Montoya, L. C. Seitz, P. Chakthranont, A. Vojvodic, T. F. Jaramillo and J. K. Nørskov, *Nat. Mater.*, 2017, **16**, 70–81.

52 E. Katz, I. Willner and J. Wang, *Electroanalysis*, 2004, **16**, 19–44.

53 L. Höfler and R. E. Gyurcsányi, *Anal. Chim. Acta*, 2012, **722**, 119–126.

54 M. Sugawara, K. Kojima, H. Sazawa and Y. Umezawa, *Anal. Chem.*, 1987, **59**, 2842–2846.

55 O. Braha, B. Walker, S. Cheley, J. J. Kasianowicz, L. Song, J. E. Gouaux and H. Bayley, *Chem. Biol.*, 1997, **4**, 497–505.

56 B. A. Cornell, V. L. B. Braach-Maksvytis, L. G. King, P. D. J. Osman, B. Raguse, L. Wieczorek and R. J. Pace, *Nature*, 1997, **387**, 580–583.

57 R. Stefureac, Y.-T. Long, H.-B. Kraatz, P. Howard and J. S. Lee, *Biochemistry*, 2006, **45**, 9172–9179.

58 L. Xue, H. Yamazaki, R. Ren, M. Wanunu, A. P. Ivanov and J. B. Edel, *Nat. Rev. Mater.*, 2020, **5**, 931–951.

59 M. Jain, S. Koren, K. H. Miga, J. Quick, A. C. Rand, T. A. Sasani, J. R. Tyson, A. D. Beggs, A. T. Dilthey, I. T. Fiddes, S. Malla, H. Marriott, T. Nieto, J. O'Grady, H. E. Olsen, B. S. Pedersen, A. Rhie, H. Richardson, A. R. Quinlan, T. P. Snutch, L. Tee, B. Paten, A. M. Phillippy, J. T. Simpson, N. J. Loman and M. Loose, *Nat. Biotechnol.*, 2018, **36**, 338–345.

60 L. Restrepo-Pérez, C. Joo and C. Dekker, *Nat. Nanotechnol.*, 2018, **13**, 786–796.

61 J. A. Alfaro, P. Bohländer, M.-J. Dai, M. Filius, C. J. Howard, X. F. van Kooten, S. Ohayon, A. Pomorski, S. Schmid, A. Aksimentiev, E. V. Anslyn, G. Bedran, C. Cao, M. Chinappi, E. Coyaud, C. Dekker, G. Dittmar, N. Drachman, R. Eelkema, D. Goodlett, S. Hentz, U. Kalathiya, N. L. Kelleher, R. T. Kelly, Z. Kelman, S. H. Kim, B. Kuster, D. Rodriguez-Larrea, S. Lindsay, G. Maglia, E. M. Marcotte, J. P. Marino, C. Masselon, M. Mayer, P. Samaras, K. Sarthak, L. Sepiashvili, D. Stein, M. Wanunu, M. Wilhelm, P. Yin and A. Meller, *Nat. Methods*, 2021, **18**, 604–617.

62 Y.-T. Long and M.-N. Zhang, *Sci. China, Ser. B: Chem.*, 2009, **52**, 731–733.

63 O. Braha, L.-Q. Gu, L. Zhou, X. Lu, S. Cheley and H. Bayley, *Nat. Biotechnol.*, 2000, **18**, 1005–1007.

64 Y. Lin, Y.-L. Ying, X. Shi, S.-C. Liu and Y.-T. Long, *Chem. Commun.*, 2017, **53**, 11564–11567.

65 K. Chuah, Y. Wu, S. R. C. Vivekchand, K. Gaus, P. J. Reece, A. P. Micolich and J. J. Gooding, *Nat. Commun.*, 2019, **10**, 2109.

66 Y. Wu, D. Bennett, R. D. Tilley and J. J. Gooding, *Adv. Mater.*, 2020, **32**, 1904339.

67 H. Ren, C. G. Cheyne, A. M. Fleming, C. J. Burrows and H. S. White, *J. Am. Chem. Soc.*, 2018, **140**, 5153–5160.

68 J. J. Gooding and K. Gaus, *Angew. Chem., Int. Ed.*, 2016, **55**, 11354–11366; *Angew. Chem.*, 2016, **128**, 11526–11539.

69 M. Wanunu, W. Morrison, Y. Rabin, A. Y. Grosberg and A. Meller, *Nat. Nanotechnol.*, 2010, **5**, 160–165.

70 K. J. Freedman, L. M. Otto, A. P. Ivanov, A. Barik, S.-H. Oh and J. B. Edel, *Nat. Commun.*, 2016, **7**, 10217.

71 Y. Wu, Y. Yao, S. Cheong, R. D. Tilley and J. J. Gooding, *Chem. Sci.*, 2020, **11**, 12570–12579.

72 L. Sepunaru, S. V. Sokolov, J. Holter, N. P. Young and R. G. Compton, *Angew. Chem., Int. Ed.*, 2016, **55**, 9768–9771; *Angew. Chem.*, 2016, **128**, 9920–9923.

73 E. J. E. Stuart, Y.-G. Zhou, N. V. Rees and R. G. Compton, *RSC Adv.*, 2012, **2**, 6879–6884.

74 J. E. Dick, C. Renault and A. J. Bard, *J. Am. Chem. Soc.*, 2015, **137**, 8376–8379.

75 J. E. Dick, A. T. Hilterbrand, A. Boika, J. W. Upton and A. J. Bard, *Proc. Natl. Acad. Sci. U. S. A.*, 2015, **112**, 5303–5308.
76 Y. Wu, R. D. Tilley and J. J. Gooding, *J. Am. Chem. Soc.*, 2019, **141**, 1162–1170.
77 K.-Y. Fu, S.-R. Kwon, D. Han and P. W. Bohn, *Acc. Chem. Res.*, 2020, **53**, 719–728.
78 S.-M. Lu, Y.-Y. Peng, Y.-L. Ying and Y.-T. Long, *Anal. Chem.*, 2020, **92**, 5621–5644.
79 Y. Zheng, A. H. Soeriyadi, L. Rosa, S. H. Ng, U. Bach and J. Justin Gooding, *Nat. Commun.*, 2015, **6**, 8797.
80 Y. Zhang, X. Weng, H. Li, H. Li, M. Wei, J. Xiao, Z. Liu, M. Chen, Q. Fu and X. Bao, *Nano Lett.*, 2015, **15**, 3616–3623.
81 A. Walcarius, E. Sibottier, M. Etienne and J. Ghanbaja, *Nat. Mater.*, 2007, **6**, 602–608.
82 M. Tagliazucchi and I. Szleifer, *J. Am. Chem. Soc.*, 2015, **137**, 12539–12551.
83 F. M. Gilles, M. Tagliazucchi, O. Azzaroni and I. Szleifer, *J. Phys. Chem. C*, 2016, **120**, 4789–4798.
84 T. Nasir, G. Herzog, M. Hébrant, C. Despas, L. Liu and A. Walcarius, *ACS Sens.*, 2018, **3**, 484–493.
85 A. Walcarius, *Chem. Soc. Rev.*, 2013, **42**, 4098–4140.
86 A. Walcarius, *Curr. Opin. Electrochem.*, 2018, **10**, 88–97.
87 L. Fumagalli, A. Esfandiar, R. Fabregas, S. Hu, P. Ares, A. Janardanan, Q. Yang, B. Radha, T. Taniguchi, K. Watanabe, G. Gomila, K. S. Novoselov and A. K. Geim, *Science*, 2018, **360**, 1339–1342.
88 Z. Li, G. Jeanmairet, T. Méndez-Morales, M. Burbano, M. Haefele and M. Salanne, *J. Phys. Chem. Lett.*, 2017, **8**, 1925–1931.
89 J. M. Martínez de la Hoz and P. B. Balbuena, *Phys. Chem. Chem. Phys.*, 2013, **15**, 1647–1654.
90 G. S. Attard, P. N. Bartlett, N. R. B. Coleman, J. M. Elliott, J. R. Owen and J. H. Wang, *Science*, 1997, **278**, 838–840.
91 L. Eliad, G. Salitra, A. Soffer and D. Aurbach, *J. Phys. Chem. B*, 2001, **105**, 6880–6887.
92 J. M. Elliott, L. M. Cabuché and P. N. Bartlett, *Anal. Chem.*, 2001, **73**, 2855–2861.
93 M. Baum, F. Rieutord, F. Juranyi, C. Rey and D. Rébiscoul, *Langmuir*, 2019, **35**, 10780–10794.
94 E.-A. Raiber, R. Hardisty, P. van Delft and S. Balasubramanian, *Nat. Rev. Chem.*, 2017, **1**, 0069.
95 B. Lam, J. Das, R. D. Holmes, L. Live, A. Sage, E. H. Sargent and S. O. Kelley, *Nat. Commun.*, 2013, **4**, 2001.
96 Z. Lin, Y. Takahashi, Y. Kitagawa, T. Umemura, H. Shiku and T. Matsue, *Anal. Chem.*, 2008, **80**, 6830–6833.
97 M. H. Choudhury, S. Ciampi, Y. Yang, R. Tavallaie, Y. Zhu, L. Zarei, V. R. Gonçales and J. J. Gooding, *Chem. Sci.*, 2015, **6**, 6769–6776.
98 Y. B. Vogel, V. R. Gonçales, L. Al-Obaidi, J. J. Gooding, N. Darwish and S. Ciampi, *Adv. Funct. Mater.*, 2018, **28**, 1804791.
99 J. Dong, Y. Lu, Y. Xu, F. Chen, J. Yang, Y. Chen and J. Feng, *Nature*, 2021, **596**, 244–249.
100 S. Fan, J. E. A. Webb, Y. Yang, D. J. Nieves, V. R. Gonçales, J. Tran, G. Hilzenrat, M. Kahram, R. D. Tilley, K. Gaus and J. J. Gooding, *Angew.*

Chem., Int. Ed., 2019, **58**, 14495–14498; *Angew. Chem.*, 2019, **131**, 14637–14640.
101 R. Dasari, D. A. Robinson and K. J. Stevenson, *J. Am. Chem. Soc.*, 2013, **135**, 570–573.
102 C. Amatore, S. Arbault, I. Bonifas, M. Guille, F. Lemaître and Y. Verchier, *Biophys. Chem.*, 2007, **129**, 181–189.
103 S.-C. Liu, M.-X. Li, M.-Y. Li, Y.-Q. Wang, Y.-L. Ying, Y.-J. Wan and Y.-T. Long, *Faraday Discuss.*, 2018, **210**, 87–99.
104 M. Lemmer, M. S. Inkpen, K. Kornysheva, N. J. Long and T. Albrecht, *Nat. Commun.*, 2016, **7**, 12922.
105 Y. Yang, F. M. Mansfeld, M. Kavallaris, K. Gaus, R. D. Tilley and J. J. Gooding, *Chem. Sci.*, 2021, **12**, 2558–2566.
106 X. Li, Y.-L. Ying, X.-X. Fu, Y.-J. Wan and Y.-T. Long, *Angew. Chem., Int. Ed.*, 2021, **60**, 24582–24587; *Angew. Chem.*, 2021, **133**, 24787–24792.
107 T. Albrecht, G. Slabaugh, E. Alonso and S. M. M. R. Al-Arif, *Nanotechnology*, 2017, **28**, 423001.
108 A. Mistry, A. A. Franco, S. J. Cooper, S. A. Roberts and V. Viswanathan, *ACS Energy Lett.*, 2021, **6**, 1422–1431.
109 P. Puthongkham, S. Wirojsaengthong and A. Suea-Ngam, *Analyst*, 2021, **146**, 6351–6364.
110 A. G. Güell, N. Ebejer, M. E. Snowden, J. V. Macpherson and P. R. Unwin, *J. Am. Chem. Soc.*, 2012, **134**, 7258–7261.
111 X. Zhou, N. M. Andoy, G. Liu, E. Choudhary, K.-S. Han, H. Shen and P. Chen, *Nat. Nanotechnol.*, 2012, **7**, 237–241.
112 J. B. Sambur, T.-Y. Chen, E. Choudhary, G. Chen, E. J. Nissen, E. M. Thomas, N. Zou and P. Chen, *Nature*, 2016, **530**, 77–80.
113 X. Zhou, E. Choudhary, N. M. Andoy, N. Zou and P. Chen, *ACS Catal.*, 2013, **3**, 1448–1453.

Faraday Discussions

ROYAL SOCIETY OF CHEMISTRY

PAPER

An ultra-low noise amplifier array system for high throughput single entity analysis†

Cheng-Bing Zhong,[a] Hui Ma,[a] Jia-Jun Wang,[a] Lin-Lin Zhang,[a] Yi-Lun Ying,*[ab] Rong Wang,*[c] Yong-Jing Wan[c] and Yi-Tao Long[a]

Received 17th August 2021, Accepted 13th September 2021
DOI: 10.1039/d1fd00055a

Electrochemical measurements at the single entity level provide ultra-sensitive tools for the precise diagnosis and understanding of basic biological and chemical processes. By decoding current signatures, single-entity electrochemistry provides abundant information on charges, sizes, shapes, catalytic performances and compositions. The accuracy of single-entity electrochemistry highly relies on advanced instrumentation to achieve the amperometric resolution at the sub-picoampere level and the temporal resolution at the sub-microsecond level. Currently, it is still a challenge for paralleling amplifiers to allow low-noise and high bandwidth single-entity electrochemical measurements. Herein, we developed a low-noise four-channel electrochemical instrumentation that integrates an Au electrode array with amplifiers in the circuit board. With this amplifier array, we achieved a high bandwidth (>100 kHz) electrochemical measurement. The further practical experiments proved the capability of this amplifier array system in acquiring transient signals from both single-molecule detection with an aerolysin nanopore and single Pt nanoparticle catalysis during the dynamic collision process. Paired with appropriate microfluidic array systems, our instrumentation will enable an extraordinarily high-throughput feature for single-entity sensing.

1. Introduction

The thirst for understanding the heterogeneous and stochastic behaviors of single cells, single particles and single molecules has pushed the rapid development of single-entity electrochemistry.[1–5] This implies that an electrochemical detection method should have the ability to measure a large quantity of entities individually to uncover their subtle differences. To achieve this goal, electrochemical sensors need to satisfy the high throughput single-molecule

[a]*State Key Laboratory of Analytical Chemistry for Life Science, School of Chemistry and Chemical Engineering, Nanjing University, Nanjing, 210023, P. R. China. E-mail: yilunying@nju.edu.cn*

[b]*Chemistry and Biomedicine Innovation Center, Nanjing University, Nanjing, 210023, P. R. China*

[c]*School of Information Science and Engineering, East China University of Science and Technology, Shanghai 200237, P. R. China. E-mail: rongwang@ecust.edu.cn*

† Electronic supplementary information (ESI) available. See DOI: 10.1039/d1fd00055a

measurements. However, the dynamic single-entity process yields a short duration time (microsecond to millisecond) and a small current amplitude (sub-picoampere to nanoampere).[6–8] The requirements of the sufficient signal-to-noise ratio (SNR) and temporal resolution challenge the design of low-noise amplifier arrays for single-entity electrochemistry.

Generally, the current response from the single-entity detection is converted to an electric current at an electrode interface, then is transformed into a voltage output by a transimpedance amplifier (TIA). This signal is further modulated by the regulation circuit if needed and the following low-pass filter to improve the SNR. After digitization at a high sampling rate, the signal is recorded and stochastically analyzed for extracting single-entity features. The temporal resolution of such a measurement is typically dependent on the maximum bandwidth, which is limited by the design of the amplifier, the properties of the electrode and the capacitance of the confined sensing interface[6] (*e.g.* a nanopore in single-molecule detection) (Fig. 1a–c). During the recording, the frequency of the low-pass filter is usually fixed to 5–10 kHz in order to meet the satisfactory SNR. In single-entity electrochemical detection, the SNR is defined as the ratio of the signal current amplitude (ΔI) to the peak-to-peak noise of the baseline current (I_{noise}), which is $\text{SNR} = \dfrac{\Delta I}{I_{\text{noise}}}$ (Fig. 1c). Usually, the value of the SNR should exceed 3 for the following identification of the current amplitude and duration of each signal. Although a series of developed algorithms have been applied to recover the short signals from the noisy data,[9–12] their performance still highly relies on the SNR of the recorded raw data. The SNR can be improved either by reducing the

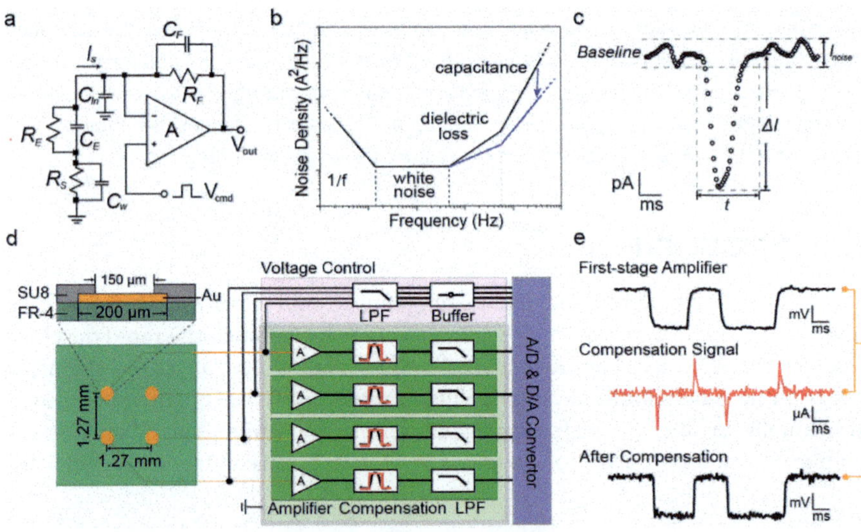

Fig. 1 Schematic of an amplifier array for single-entity measurements. (a) A circuit model of the electrode and the amplifier; (b) an example of noise power spectral density for ultra-low current detection, illustrated as a function of frequency; (c) a representative transient current signal from single-entity measurements; (d) illustration of the amplifier array and electrode array design; (e) illustration of output signal responses before and after the compensation circuit modulation.

noise (I_{noise}) or increasing the current response of ΔI. Various experimental strategies have been developed to increase both the amplitude and the duration of the current signal,[13–16] but the reduction of I_{noise} is full of difficulties. There needs to be continuous efforts in amplifier design and the improvement of low-noise electrochemical sensing interfaces.

To lower I_{noise}, the noise contribution of I_{noise} needs to be understood. The equivalent model is given in Fig. 1a. An electrode can be equivalent to a resistor R_E, along with a capacitance from the electrode interface (C_E) in parallel. In addition, the capacitance of the recording electronics includes the contributions from the electrode connecting wire (C_W), the amplifier input (C_{In}) and the amplifier parasitic capacitance (C_F) of a feedback resistor (R_F) (Fig. 1a). The power spectral density (PSD) spectrum for the single-entity measurement can be divided into four regions (Fig. 1b) which are the flick noise (or $1/f$ noise), thermal and shot noise (or white noise), dielectric noise and capacitance noise. The $1/f$ noise in the low frequency regime is mainly attributed to the current fluctuations from the charge carriers binding/unbinding with the sensing interface.[17,18] The moderate frequency region is associated with the white noise from the sensing interface or circuit electronics. At the high frequency regime, the major noise contribution is from the amplifier's voltage noise together with the total capacitance of the input. In the whole regime, the root mean square of I_{noise} could be described by:[19]

$$I_{\mathrm{RMS}}(\mathrm{B}) = \frac{2\pi}{\sqrt{3}} B^{\frac{3}{2}} (C_{\mathrm{E}} + C_{\mathrm{w}} + C_{\mathrm{In}} + C_{\mathrm{F}}) v_{\mathrm{n}} \tag{1}$$

where B is the measurement system bandwidth and v_n is the voltage noise density ($V^2\ Hz^{-1}$). Therefore, to meet the requirements of a high temporal resolution (high bandwidth) and a high SNR, the amplifier array system requires the elegant design of the electronics and the maximum reduction of the parasitic capacitance.

In this work, a two-stage amplifier with a low input capacitance is designed for a low-noise four-channel electrochemical instrumentation which reaches a high bandwidth of >100 kHz. The performance of this four-channel array device is verified by parallelly recording the current response from both single-molecule detection using a biological nanopore and single-entity collisions.

2. Experiments

2.1 Current amplifier array

The four-channel amplifier array consists of four-channel current amplifiers, a four-channel compensation circuit, four-channel filters, a four-channel 5 kHz low-pass filter and bias voltage control units (Fig. 1d). An Axon Digidata 1440 (Molecular Devices, USA) was used to control the analog-to-digital conversion and transfer the data to the computer.

Usually, a large feedback resistor (R_F) is adopted in TIA in order to obtain the high current gain up to 10^9 for ultra-low current amplification. However, the large resistance leads to a high parasitic capacitance, which generates the dominant noise source in the high frequency regime (eqn (1)). Moreover, the crosstalk among the multi-channel amplifier circuits increases the noise and distorts the transient signals from the single-entity measurements. To suppress noise and improve the bandwidth of the multi-channel amplifier, we adopted a two-stage amplifier configuration, together with a frequency compensation circuit, for the design of the multi-channel amplifier array (Fig. 1e). Herein, the first and second stage amplifiers provided gains of 160 dB and 20 dB, respectively. Therefore, the ultra-low current at the picoampere level can be amplified to the millivolt level with a 10^9 current gain. The final output voltage of the two-stage current amplifier was fine-tuned by the following frequency composition circuit (Fig. 2a). The performance of this circuit design was evaluated using the simulation software LTspice (Analog Device, USA).

For reducing the noise from the bias voltage, the voltage buffer units were used to separate the noise interaction between the voltage generator and amplifiers. A two-order Bessel filter with a cut-off frequency of 1 kHz was employed to suppress the voltage noise. A home-designed Faraday cage comprising two separated cages isolated the whole amplifier array and the electrode array for suppressing the possible noise (Fig. 2b and c).

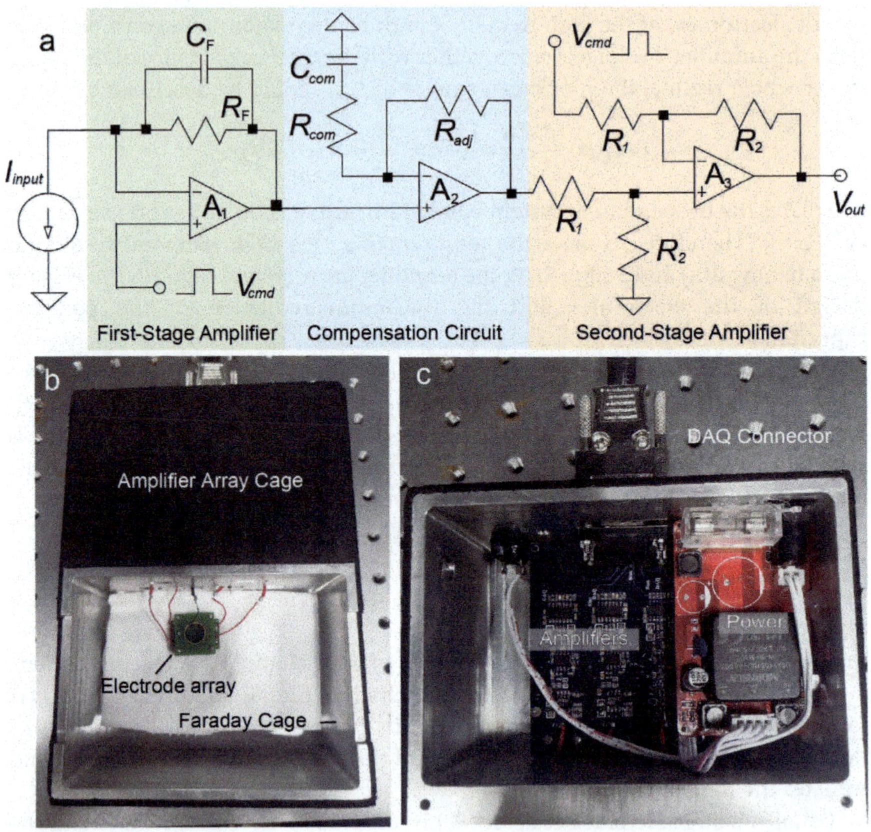

Fig. 2 (a) The circuit of the single-channel amplifier. The first-stage amplifier, compensation circuit and second-stage amplifier are shown in the green, blue and yellow boxes, respectively; (b) a photograph of the experimental setup. An Au electrode array is put in a Faraday cage and linked to the amplifier inside the back cage. (c) A photograph of the four-channel amplifier array in the back cage of the home-designed Faraday cage. The amplifier array is connected to analog-to-digital (A/D) convertor by a data acquisition (DAQ) connector.

2.2 Software-based simulation

The equivalent circuit model was simulated using LTspice (Analog Device, USA) for optimizing the bandwidth of the amplifier array. The simulation was implemented at the transient mode for 1 ms and the AC mode at frequencies from 10 Hz to 1 MHz.

2.3 Fabrication of the cavity array

As shown in Fig. 1d, the Au electrode (~200 μm in diameter), positioned with a 1.27 mm spacing in a 2 × 2 array chip (fabricated by the Shanghai Fast-PCB Circuit Technology Company, China), was used for single-particle collision measurements. The whole surface of the chip was coated by the insulating polymer SU-8 (MicroChem, USA). Micrometer-scale cavities were fabricated through the SU-8 polymer by photo exposure.[20] As a result, the SU-8 polymer with 25 μm thickness support had 150 μm micro-cavities in between. Therefore, the final diameter of the active Au electrode was 150 μm. The ground Ag/AgCl electrode was placed outside for offering the reference potential. The characterization of the electrode array is shown in Fig. S1.†

2.4 Single-entity measurements

2.4.1 Chemicals. Trypsin–EDTA, decane (anhydrous, ≥99%), KCl, HCl and Pt nanoparticles (Pt NPs, $d = 3$ nm, <5 nm) were purchased from Sigma-Aldrich Co., Ltd., USA. $HClO_4$ was purchased from Shanghai Lingfeng Chemical Reagent Co. Ltd., China. 1,2-Diphytanoyl-*sn*-glycero-3-phosphocholine (in chloroform, ≥99%) was purchased from Avanti Polar Lipids, Inc., USA. The poly(dA)$_4$ (5′-AAAA-3′) sample was synthesized and HPLC-purified by Sangon Biotech Co., Ltd., China. Glycerinum was purchased from Amresco, Inc., USA. The reagents and materials used were all at analytical grade. All solutions were prepared using Milli-Q ultrapure water with a resistance of 18.2 MΩ cm^{-1} at 25 °C (EMD Millipore, USA).

2.4.2 Single-molecule detection with an aerolysin nanopore. The wild-type proaerolysin was purified and activated by following the previous procedure.[16,21] The single-molecule detection experiments were performed according to our previous papers.[16,21] The lipid bilayer was formed on the Delrin perfusion bilayer cuvette containing an aperture which was ~50 μm in diameter (Warner Instruments, USA). Then, a single aerolysin nanopore was inserted into the lipid bilayer. Two compartments of the chamber were filled with an electrolyte solution of 1.0 M KCl, 10 mM Tris and 1.0 mM EDTA at pH 8.0. A pair of Ag/AgCl electrodes was employed to apply the potential across the aerolysin nanopore. The ionic current was recorded at a sampling rate of 100 kHz with a low-pass Bessel filter of 5 kHz. The four-channel amplifier and experimental chambers were placed inside a custom-designed Faraday cage (Fig. 2b and c).

2.4.3 Single Pt NP collisions with the Au electrode array. An Au electrode array was printed on the circuit board and connected with the amplifier array (Fig. 1d and 2b). Before the collision experiments, the electrode array was cleaned with 500 mM H_2SO_4 using cyclic voltammetry (CV) at a scan rate of 4.00 V s^{-1}. Then, the electrode array was characterized by performing cyclic voltammetry in an aqueous solution containing 0.9 mM ferrocenemethanol and 100 mM KNO_3 at a scan rate of 0.01 V s^{-1} (Fig. S2†). The time-resolved current traces from the single Pt NP collisions were acquired in 10 mM $HClO_4$ at an applied potential of −0.30 V *vs.* Ag/AgCl in the presence of 10 nM Pt NPs. The data was recorded for the first 5 min at a sampling rate of 100 kHz with a low-pass Bessel filter of 5 kHz.

3. Results and discussions

3.1 Performance assessments of the current amplifier array

A schematic of the circuit for the proposed two-stage current amplifier architecture is illustrated in Fig. 2a. A transient current response generated from the electrode array is amplified and converted to a readable voltage for the following calibration and digitization. The amplification circuit consists of a first-stage amplifier (R_F and A_1), a second-stage amplifier (R_1, R_2 and A_3) and a frequency compensation circuit (R_{com}, C_{com}, R_{adj} and A_2). The compensation circuit is connected between the output of the first stage amplifier and the input of the second stage amplifier to achieve the bandwidth compensation. The first-stage amplifier employs a trans-impendence amplifier circuit for providing a larger gain of ionic current. The gain of the first-stage amplifier is determined by R_F and the parasitic capacitance C_F in R_F, which inevitably leads to the signal distortion. The gain (G_1) of the first-stage amplifier is given by eqn (2):

$$G_1(\mathrm{dB}) = 20\ \mathrm{lg}\left(\sqrt{\frac{{R_F}^2}{1+4\pi^2 f^2 {C_F}^2 {R_F}^2}}\right) \tag{2}$$

where f is the frequency. This compensation circuit calibrates the distortion of the signal after the first-stage amplifier by virtue of combining the non-inverting amplifier and differential circuit. The working principle of the compensation circuit is shown in Fig. 1e. Theoretically, the compensation circuit provides gain to calibrate the signal distortion at a high frequency. The gain (G_{com}) of the compensation circuit is given by eqn (3):

$$G_{com}(\mathrm{dB}) = 20\ \mathrm{lg}\left(\sqrt{\frac{1+4\pi^2 f^2 {C_{com}}^2 (R_{com}+R_{adj})^2}{1+4\pi^2 f^2 {C_{com}}^2 {R_{com}}^2}}\right) \tag{3}$$

The second-stage amplifier incorporates a trans-impendence amplifier circuit and subtraction circuit to offset the bias voltage. The gain of the second-stage amplifier is determined by the ratio of R_2/R_1. As a result, the total gain (G) of the amplifier is given by eqn (4):

$$G(\mathrm{dB}) = 20\ \mathrm{lg}\left(R_F \frac{R_2}{R_1}\sqrt{\frac{1+4\pi^2 f^2 {C_{com}}^2 (R_{com}+R_{adj})^2}{(1-4\pi^2 f^2 C_F R_F R_{com} R_{adj})^2 + 4\pi^2 f^2 (C_F R_F + C_{com} R_{com})^2}}\right) \tag{4}$$

The two-stage amplifier reaches a gain of 10^9 when $R_F = 100\ \mathrm{M\Omega}$, $R_1 = 10\ \mathrm{k\Omega}$ and $R_2 = 100\ \mathrm{k\Omega}$. The bandwidth of TIA is usually defined by the −3 dB bandwidth. This is the frequency of the −3 dB bandwidth at which the output voltage is reduced to 70% of the lower frequency (or DC) output levels.[22] To determine the bandwidth of the two-stage amplifier, a series of current inputs were generated with an amplitude of 100 pA at varying frequencies from 10 kHz to 200 kHz (Fig. 3a). Below 100 kHz, the amplitude of the output signal is barely changed and the signal shape does not show any distortions. However, the amplitude of the output signal decreases to 70% at 100 kHz and to 40% at 200 kHz (Fig. 3b). After the compensation circuit, the −3 dB bandwidth significantly increases from 1 kHz to over 100 kHz with the R_{com}, C_{com} and R_{adj} set to 20 Ω, 50 nF and 150 Ω, respectively (Fig. 3c). During the measurements, the parasitic capacitance on the board would be increased due to the crosstalk from the channels and the structure design of the board. To simulate the bandwidth reduction in practical electrochemical measurements, the parasitic capacitance was further increased (Fig. 3d). When the parasite capacitance is set 0.5 pF, the bandwidth is decreased to 90 kHz, but still meets the requirement for single-entity measurement.

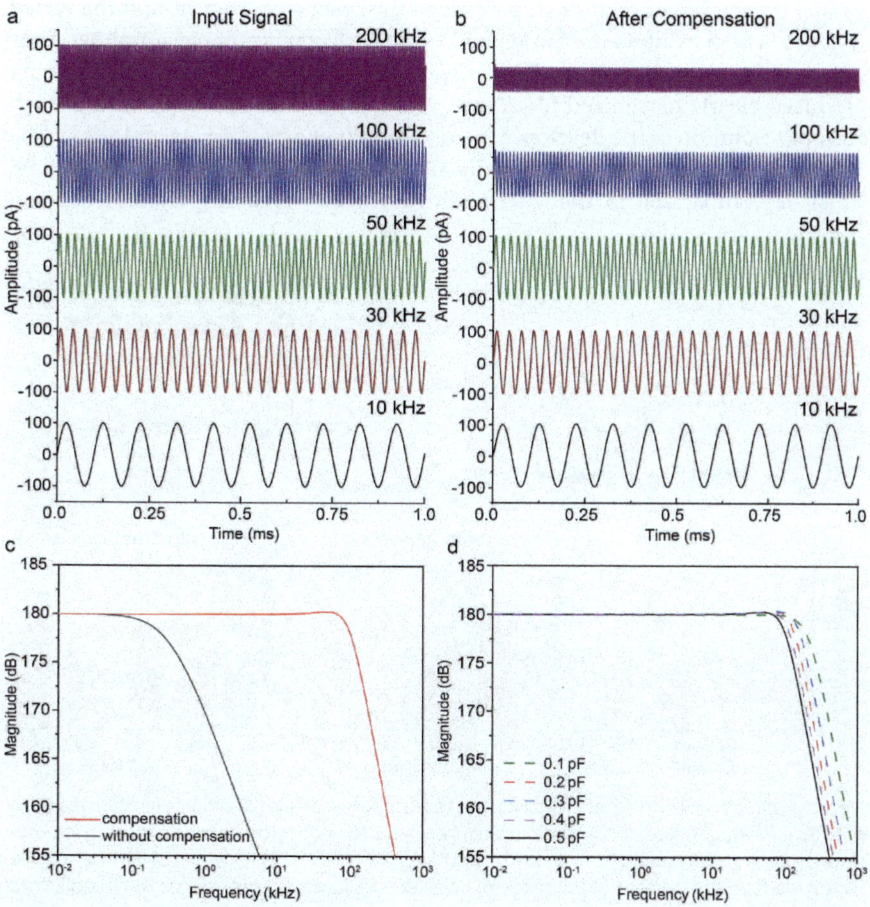

Fig. 3 Circuit simulation of the amplifier array performance. (a) Input simulation signals; (b) output simulation signals after compensation. (c) Bandwidth of the four-channel amplifiers without compensation (black) and with compensation (red); (d) bandwidth of the four-channel amplifiers with compensation under different parasitic capacitances.

3.2 Single-entity measurements with the current amplifier array

To test the performance of the developed current amplifier array in a practical single-entity experiments, it was used for both single-molecule detection with an aerolysin nanopore and single-particle collision measurements with an Au electrode array (Fig. 4a and b). The four-channel single-molecule detection was operated in four individual nanopore chambers, while the single-particle collisions were measured from an integrated Au electrode array with four Au electrodes. The four-channel current amplifier array was connected to the Au electrode array to ensure the independent recording of each Au electrode. After 5 kHz filtering, the current noise (I_{RMS}) of the amplifier array was 1.0 ± 0.1 pA under the open circuit condition. Here, the standard deviation describes the noise difference from the four channels. The results in Fig. 4c demonstrate the uniform noise performance among the channels. By carrying out single-molecule measurements with aerolysin biological nanopores, four channels show a similar noise level of 2.0 ± 0.2 pA after the formation of the aerolysin nanopore on the lipid bilayer (Fig. 4d). The *I*–*V* curves of the aerolysin nanopore demonstrate a strong linear relationship between the open pore current and the varied voltages, which exhibits the fairly good noise performance of our amplifier array at various voltages (Fig. S3†). The addition of poly(dA)$_4$ into the *cis* side chamber produced clearly recognized blockages. The raw current data supports the high current resolution of the developed current amplifier array (Fig. 4a and S4a†). The SNR value for a translocation event is 15 ± 1 under a bias voltage of +100 mV, which is comparable to that of the single-channel recording system of Axon

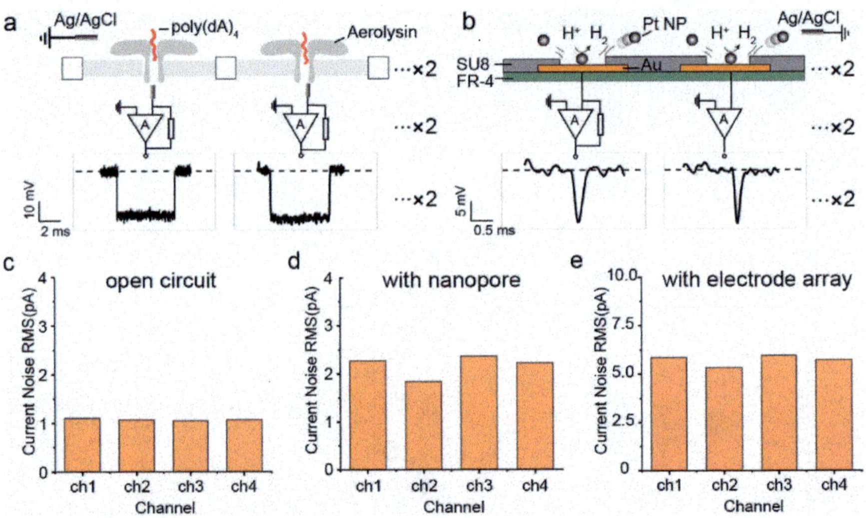

Fig. 4 Evaluation of the performance of the amplifier array for single-entity measurements. (a) Parallel electrochemical measurements for nanopore-based single-molecule detection. A model analyte, poly(dA)$_4$, is driven into the wild-type aerolysin nanopore, which leads to the current blockages. (b) Parallel measurements for single Pt NP collisions with the integration of the electrode array and four-channel amplifier array. The low-pass filter is set to 5 kHz. The current noise evaluation of the four-channel amplifier system without load (c), with the wild-type aerolysin nanopore in 1.0 M KCl, 10 mM Tris and 1.0 mM EDTA at pH 8.0 and +100 mV (d) and with the Au electrode array in 10 mM $HClO_4$ under −300 mV *vs.* Ag/AgCl, filtered by a 5 kHz low-pass filter (e).

Digidata 1440 and Cube D0 (ref. 23) (developed by our group). Moreover, the blockage events with a current amplitude as low as 5 pA (SNR = 1) could be clearly identified, which meets the on-demand requirements of the designed amplifier array.

To achieve high throughput single-particle collision measurements, we designed a four-channel Au electrode array (Fig. 4b). The baseline I_{RMS} values from four independent electrodes showed comparable noise levels before and after the addition of Pt NPs. Similar to the previous studies,[24–27] the Pt NPs electrochemically catalyze proton reduction to hydrogen during the collision time. Using the presented amplifier array, the low-noise faradaic current recording from single Pt NP catalysis events were achieved (Fig. 4e and S4b†).

4. Conclusions and outlook

In summary, this study demonstrates a new design for multi-channel amplifier arrays to achieve the high-bandwidth acquisition of ultra-low current responses from single-entity measurements. The simulation reveals that the bandwidth of the amplifier array is greater than 100 kHz. The experimental results show that the four channels exhibit comparable low noise levels for single-molecule analysis and single-particle collisions. To further suppress the noise and enhance the SNR, the diameter of the Au electrode in the array needs to be reduced for lowering the C_E contribution to the noise. Similarly, in the biologic nanopore experiment, the size of the lipid bilayer support should be reduced so as to decrease the noise from membrane capacitance. These two types of optimizations could be achieved by fabricating nanostructures using optical lithography or electron/ion beam lithography.[28–31] Our design could be expanded to more channels for the high-throughput electrochemical analysis of single entities. But the large-scale electrochemical measurement could hardly be achieved with current circuit designs using commercialized amplifier chips. It is an urgent requirement for the elegant CMOS (complementary metal oxide semiconductor) design to be used to fabricate ultrasensitive amplifiers for achieving high gains at relatively low noise. Although various CMOS amplifier systems have been proposed,[32–36] future efforts are required in order to simultaneously achieve large-scale parallelization and high bandwidth recording. The upscaling of the integrated nanopores or micro-/nano-electrodes with CMOS amplifier arrays needs to further overcome the cross-talk interference among the detection units. Future studies will require multi-disciplinary knowledge to combine electrochemical technology with integrated circuits for boosting the development of single-entity electrochemistry measurements.

Conflicts of interest

There are no conflicts to declare.

Acknowledgements

This research was supported by the National Natural Science Foundation of China (22027806, 22090050 and 61871183) and the Fundamental Research Funds for the Central Universities (2020102025S, 0205-14380258).

References

1 J. J. Gooding and K. Gaus, *Angew. Chem., Int. Ed.*, 2016, **55**, 11354–11366.
2 L. A. Baker, *J. Am. Chem. Soc.*, 2018, **140**, 15549–15559.
3 S. M. Lu, Y. Y. Peng, Y. L. Ying and Y. T. Long, *Anal. Chem.*, 2020, **92**, 5621–5644.
4 H. Ren and M. A. Edwards, *Curr. Opin. Electrochem.*, 2021, **25**, 100632.
5 Y. Wang, X. Shan and N. Tao, *Faraday Discuss.*, 2016, **193**, 9–39.
6 Z. Gu, Y. L. Ying and Y. T. Long, *Sci. China: Chem.*, 2018, **61**, 1483–1485.
7 W. Ma, H. Ma, J. F. Chen, Y. Y. Peng, Z. Y. Yang, H. F. Wang, Y. L. Ying, H. Tian and Y. T. Long, *Chem. Sci.*, 2017, **8**, 1854–1861.
8 J. Ustarroz, M. Kang, E. Bullions and P. R. Unwin, *Chem. Sci.*, 2017, **8**, 1841–1853.
9 Z. Gu, Y. L. Ying, C. Cao, P. He and Y. T. Long, *Anal. Chem.*, 2015, **87**, 907–913.
10 Z. X. Wei, Y. L. Ying, M. Y. Li, J. Yang, J. L. Zhou, H. F. Wang, B. Y. Yan and Y. T. Long, *Anal. Chem.*, 2019, **91**, 10033–10039.
11 J. Zhang, X. Liu, Y. L. Ying, Z. Gu, F. N. Meng and Y. T. Long, *Nanoscale*, 2017, **9**, 3458–3465.
12 X. J. Sui, M. Y. Li, Y. L. Ying, B. Y. Yan, H. F. Wang, J. L. Zhou, Z. Gu and Y. T. Long, *Journal of Analysis and Testing*, 2019, **3**, 134–139.
13 K. Fu, S. R. Kwon, D. Han and P. W. Bohn, *Acc. Chem. Res.*, 2020, **53**, 719–728.
14 H. Ma, J. F. Chen, H. F. Wang, P. J. Hu, W. Ma and Y. T. Long, *Nat. Commun.*, 2020, **11**, 2307.
15 H. Ma, W. Ma, J. F. Chen, X. Y. Liu, Y. Y. Peng, Z. Y. Yang, H. Tian and Y. T. Long, *J. Am. Chem. Soc.*, 2018, **140**, 5272–5279.
16 Y. Q. Wang, M. Y. Li, H. Qiu, C. Cao, M. B. Wang, X. Y. Wu, J. Huang, Y. L. Ying and Y. T. Long, *Anal. Chem.*, 2018, **90**, 7790–7794.
17 S. M. Bezrukov and J. J. Kasianowicz, *Phys. Rev. Lett.*, 1993, **70**, 2352–2355.
18 R. M. M. Smeets, U. F. Keyser, N. H. Dekker and C. Dekker, *Proc. Natl. Acad. Sci. U. S. A.*, 2008, **105**, 417–421.
19 J. K. Rosenstein, M. Wanunu, C. A. Merchant, M. Drndic and K. L. Shepard, *Nat. Methods*, 2012, **9**, 487–492.
20 S. C. Saha, F. Thei, M. R. R. De Planque and H. Morgan, *Sens. Actuators, B*, 2014, **199**, 76–82.
21 C. Cao, D. F. Liao, J. Yu, H. Tian and Y. T. Long, *Nat. Protoc.*, 2017, **12**, 1901–1911.
22 C. A. Bunge, M. Beckers and B. Lustermann, *Polymer Optical Fibres*, Woodhead Publishing, 2017, ch. 3, pp. 47–118.
23 Z. Gu, H. F. Wang, Y. L. Ying and Y. T. Long, *Sci. Bull.*, 2017, **62**, 1245–1250.
24 X. Xiao and A. J. Bard, *J. Am. Chem. Soc.*, 2007, **129**, 9610–9612.
25 P. A. Defnet, C. Han and B. Zhang, *Anal. Chem.*, 2019, **91**, 4023–4030.
26 Z. P. Xiang, H. Q. Deng, P. Peljo, Z. Y. Fu, S. L. Wang, D. Mandler, G. Q. Sun and Z. X. Liang, *Angew. Chem., Int. Ed.*, 2018, **57**, 3464–3468.
27 M. Chen, S.-M. Lu, Y.-Y. Peng, Z. Ding and Y.-T. Long, *Chem.–Eur. J.*, 2021, **27**, 11799–11803.
28 D. W. M. Arrigan, *Analyst*, 2004, **129**, 1157–1165.
29 Y. Liu, X. Li, J. Chen and C. Yuan, *Front. Chem.*, 2020, **8**, 1102–1111.
30 B. Lu, G. Kocharyan and J. J. Schmidt, *Biotechnol. J.*, 2014, **9**, 446–451.

31 R. Liu, R. Chen, A. T. Elthakeb, S. H. Lee, S. Hinckley, M. L. Khraiche, J. Scott, D. Pre, Y. Hwang, A. Tanaka, Y. G. Ro, A. K. Matsushita, X. Dai, C. Soci, S. Biesmans, A. James, J. Nogan, K. L. Jungjohann, D. V. Pete, D. B. Webb, Y. Zou, A. G. Bang and S. A. Dayeh, *Nano Lett.*, 2017, **17**, 2757–2764.

32 J. K. Rosenstein, S. Ramakrishnan, J. Roseman and K. L. Shepard, *Nano Lett.*, 2013, **13**, 2682–2686.

33 A. Uddin, S. Yemenicioglu, C. H. Chen, E. Corigliano, K. Milaninia and L. Theogarajan, *Nanotechnology*, 2013, **24**, 155501.

34 G. Ferrari, F. Gozzini, A. Molari and M. Sampietro, *IEEE J. Solid-State Circuits*, 2009, **44**, 1609–1616.

35 J. Kim, R. Maitra, K. D. Pedrotti and W. B. Dunbar, *IEEE Transactions on Biomedical Circuits and Systems*, 2013, **7**, 285–295.

36 S. Magierowski, S. Z. Islam, Y. Huang and E. Ghafar-Zadeh, *2014 36th Annual International Conference of the IEEE Engineering in Medicine and Biology Society*, 2014, pp. 1564–1569.

Faraday Discussions

PAPER

Inclusion of multiple cycling of potential in the deep neural network classification of voltammetric reaction mechanisms

Luke Gundry, Gareth Kennedy, Alan M. Bond * and Jie Zhang *

Received 12th August 2021, Accepted 5th October 2021
DOI: 10.1039/d1fd00050k

The use of deep neural networks (DNNs) for the classification of electrochemical mechanisms using simulated voltammograms with one cycle of potential for training has previously been reported. In this paper, it is shown how valuable additional patterns for mechanism distinction become available when a new DNN is trained simultaneously on images obtained from three cycles of potential using tensor inputs. Significant improvements, relative to the single cycle training, in achieving the correct classification of E, $EC_{1^{st}}$ and $EC_{2^{nd}}$ mechanisms (E = electron transfer step and $C_{1^{st}}$ and $C_{2^{nd}}$ are first and second order follow up chemical reactions, respectively) are demonstrated with noisy simulated data for conditions where all mechanisms are close to chemically reversible and hence difficult to distinguish, even by an experienced electrochemist. Challenges anticipated in applying the new DNN to the classification of experimental data are highlighted. Directions for future development are also discussed.

1. Introduction

The development of machine learning methods, such as deep neural networks (DNNs), for the classification of electrochemical reaction mechanisms using cyclic voltammetry has become viable and of significant interest in recent years thanks to the rapid increase in computing power.[1–7] Earlier computationally supported mechanistic classifications in voltammetry employed pattern recognition.[8–12] The DNN tool has a neural network architecture which consists of a series of layered nodes connected in a network array. When analysing an input image, the DNN applies a filter for each convolutional layer. This generates a series of filters of $(n \times n)$ kernel size, where each filter consists of n^2 independent weights. These kernel sizes represent the pixel size of the filters and extract or repress information in the image by relating pixels to the patterns in the data for each feature. Automatic variation of the parameters for the multiple filters allows the DNN to be trained on specific patterns present in the images, which are

School of Chemistry, Monash University, Clayton, Vic. 3800, Australia. E-mail: Alan.Bond@monash.edu; Jie.Zhang@monash.edu

then used to identify the mechanism represented by the image. A key feature of artificial neural networks is that the weights are optimised to achieve the correct classification by continuously applying a DNN architecture to a very large database. This approach is referred to as supervised learning.[3]

Recent developments in computer science have provided open source software libraries for machine learning, such as Tensorflow[13] and PyTorch.[14] Their widespread availability has facilitated the development of machine learning algorithms that can be implemented with modern high speed advanced parallel computing. Previously, a DNN machine learning method was developed in our laboratories for the mechanistic classification of simple electrochemical processes.[3] The method was trained using extensive simulations of cyclic voltammograms achieved by use of the publicly available MECSim software.[15] Successful training was achieved on solution based electrochemical processes that encompassed heterogeneous one- and multi-electron transfer reactions that could be coupled with a homogeneous first order chemical reaction.[3] The DNN method used in this earlier study stores cyclic voltammetric data for the initial cycle of potential as a 100×100 dimensional image and passes them through three two dimensional convolution neural network layers with max pooling in what is referred to as single cycle DNN (SCDNN). However, the DNN architecture allows multiple cycles of potential to be incorporated into the neural network *via* multiple channel inputs.

In this paper, the previously used DNN method has been extended to accommodate training for three cycles of potential. It is well known to electrochemists that the examination of multiple rather than just the initial cycle of potential significantly enhances the mechanistic insights available from cyclic voltammetry. Emphasis in this paper is given to machine learning based distinctions between mechanisms associated solely with electron transfer (E) and those coupled to irreversible first ($EC_{1^{st}}$) or second order ($EC_{2^{nd}}$) homogeneous chemical reactions, described in eqn (1) and (2) with the corresponding rate constants $k_f^{EC_1}$ and $k_f^{EC_2}$.

Cyclic voltammograms for E, $EC_{1^{st}}$ and $EC_{2^{nd}}$ reactions, particularly when close to the chemically reversible limit, have many common features and are therefore difficult to discriminate experimentally. This scenario is emphasised in initial classification studies in this paper. Fig. 1 shows an example where simulated voltammograms for the initial cycle of potential for $EC_{1^{st}}$ and $EC_{2^{nd}}$ reactions are essentially indistinguishable. However, as also shown in Fig. 1, differences are significantly enhanced in the second and subsequent cycles of potential with the modeled parameters of reversible potential (E^0), analyte concentration (C), diffusion coefficient (D), uncompensated resistance (R_u) and capacitance coefficient (C_{dl}). To better distinguish between $EC_{1^{st}}$ and $EC_{2^{nd}}$ mechanisms, voltammetric experiments also can be undertaken over a range of scan rates and/or analyte concentrations.[16,17] Nevertheless, in principle, analysis of voltammograms with multiple cycles of potential by DNN also could be used to help discriminate between these two reaction mechanisms.

$$\begin{aligned} &A + e^- \rightleftharpoons B \\ &B \xrightarrow{k_f^{EC_1}} C \end{aligned} \qquad (1)$$

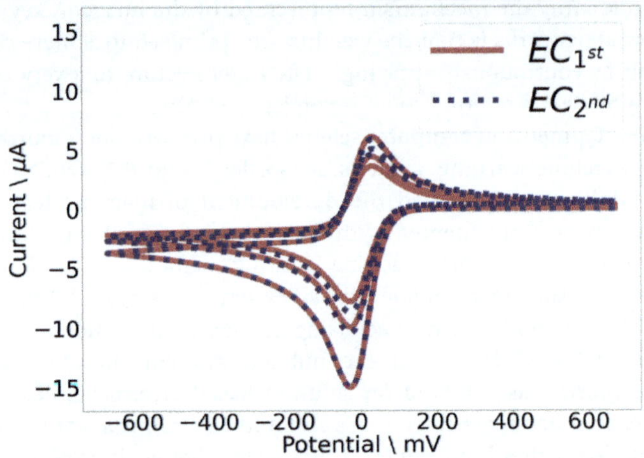

Fig. 1 Simulated initial three cycles of potential in cyclic voltammograms for $EC_{1^{st}}$ ($k_f^{EC_1} = 0.1\ s^{-1}$) and $EC_{2^{nd}}$ ($k_f^{EC_2} = 10^3\ M^{-1}\ s^{-1}$) mechanisms where the electron transfer (E) step is reversible. $E^0 = 0.000$ mV, scan rate = 100 mV s^{-1}, $C = 1.0$ mM, $D = 5.0 \times 10^{-6}$ $cm^2\ s^{-1}$, $R_u = 0\ \Omega$ and $C_{dl} = 0$ F cm^{-2}.

$$\begin{gathered} A + e^- \rightleftharpoons B \\ 2B \xrightarrow{k_f^{EC_2}} C \end{gathered} \tag{2}$$

2. Neural network architecture

The important extension to the method used in previous work,[3] is that images of multiple cycles of potential can now be analysed simultaneously by introducing multiple DNN channels into the initial layer. This approach is commonly used for the classification of colour images by splitting the image into red, green and blue channels.[18–21] In the present DNN mechanism classification study, three sequential images derived from the first three cycles of potential are passed to

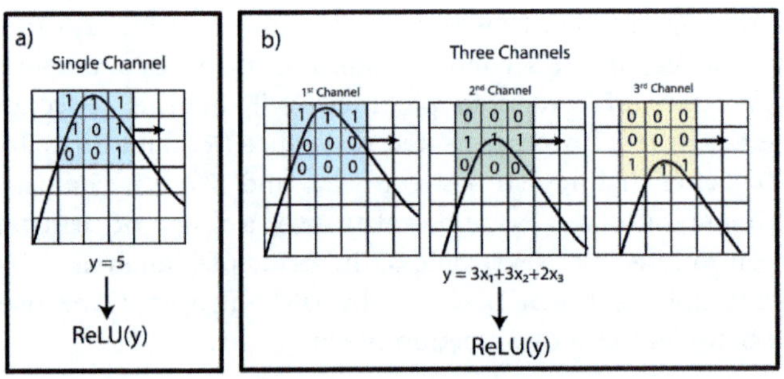

Fig. 2 Initial layer of SCDNN (a) and MCDNN (b) based on Kennedy *et al.*[3] For single cycle, a filter (blue) scans along the image, summing the overlap for each point before passing to the ReLU activation function. In the MCDNN, the filters are a 3 dimensional tensor that sums along each channel and multiplies the convolutional sum by a scalar (x_1, x_2 and x_3 for channels 1, 2 and 3 respectively) before passing to the activation function.

three channels of a DNN and treated as a tensor for each feature filter, as shown in Fig. 2. The neural network takes the multiple input images of the cyclic voltammogram and converts them into a single 100 × 100 × 3 pixel density tensor as the input data for the DNN. The channels are computed as a 3D convolution in the initial layer, with each channel having a weighted value. In this study, sequential cycles of potential are used. However, any form of images could be used that, for example accommodate variable analyte concentration or scan rate.

In applying the new DNN to three cycles of potential, the neural network used previously with just the initial cycle of potential (now referred to as the SCDNN) is expanded by the addition of multiple channels into the initial convolutional layer. The effect of this extension is shown in Fig. 2, with each channel seeking the horizontal line marked in the figure at relatively low points in the image. Thus, there are now three channels used as an input, with each representing a cycle of potential (now referred to as multiple cycle MCDNN). Each cycle of potential is normalised to the value of max(|current|) and then each cycle is passed consecutively to the neural network as a separate channel.

3. Synthetic training data

A large number of inputs are required to "train" the DNN to optimise weights that are used to classify the input image. Training could be based on access to

Table 1 Parameter ranges used in simulations of cyclic voltammetric data for the classification of E, $EC_{1^{st}}$ and $EC_{2^{nd}}$ mechanisms. In addition, the potential range was fixed at 1.34 V, the scan rate at 100 mV s^{-1}, the working electrode area at 0.079 cm^2, the temperature at 20 °C and the analyte concentration at 1.0 mM. Note that * means that this parameter is randomly sampled on the log scale

Mechanism	Parameter/unit	Low	High
E	*R_u/Ω	1	300
	D/cm^2 s^{-1}	3×10^{-6}	9.5×10^{-6}
	C_{dl}/F cm^{-2}	0	4.0×10^{-5}
	E^0/mV	−250	250
	*k^0/cm s^{-1}	0.001	10.0
	α	0.45	0.55
$EC_{1^{st}}$	*R_u/Ω	1	300
	D/cm^2 s^{-1}	1×10^{-6}	5.5×10^{-6}
	C_{dl}/F cm^{-2}	0	4.0×10^{-5}
	E^0/mV	−250	250
	*k^0/cm s^{-1}	0.001	1.0
	α	0.45	0.55
	*$k_f^{EC_1}$/s^{-1}	1×10^{-2}	4×10^{-2}
	Note that k_b was set to 0		
$EC_{2^{nd}}$	*R_u/Ω	1	300
	D/cm^2 s^{-1}	3×10^{-6}	9.5×10^{-6}
	C_{dl}/F cm^{-2}	0	4.0×10^{-5}
	E^0/mV	−250	250
	*k^0/cm s^{-1}	0.001	1.0
	α	0.45	0.55
	*$k_f^{EC_2}$/M^{-1} s^{-1}	10	30
	Note that k_b was set to 0		

a comprehensive collection of experimental data. However, since this possibility was not available, cyclic voltammograms with three cycles of potential (5004 per reaction mechanism) were generated from simulations based on linear diffusion and Butler–Volmer kinetics consisting of an electron transfer rate (k^0) and a charge transfer coefficient (α) using the input parameters given in Table 1. Both SCDNN and MCDNN were trained on 5004 simulations per reaction mechanism, since more training data did not lead to improved prediction accuracy. Five epochs was also found to be the point where further epochs did not lead to an increase of the convergence of the DNN training accuracy and loss. Simulations were undertaken for three cycles of potential. The data for the first and all three cycles of potential were then used in the SCDNN and MCDNN, respectively. In this manner, the training data was essentially the same for both DNNs, which minimised possible bias arising from the choice of training data. 80% of these simulations were then used for training and 20% for validation. A deficiency of this training process is that the patterns contained in the training data are derived from idealised theoretical models that do not incorporate nuances, such as noise, that are present in real experimental data.[22,23] Accordingly, the DNNs were designed to identify patterns that exist in simulated data which should be prominent enough to be identifiable in "noisy" experimental data.

The parameter ranges used in this study were selected so that the synthetic data should adequately reveal characteristic patterns required to allow the distinction of E (electron transfer only), $EC_{1^{st}}$ (electron transfer with follow up first order chemical reaction) and $EC_{2^{nd}}$ (electron transfer with follow up second order chemical reaction) reactions. Where necessary, parameters were varied logarithmically (denoted by *). The simplest case of the (E) mechanism was simulated by assuming that the electron transfer process is modelled by the Butler–Volmer relationship with mass transport governed by planar diffusion. A double layer model with a potential independent capacitance (C_{dl}) was assumed in simulating the background current. The simulation also took into account the effect of the IR_u drop (I = current, R_u = uncompensated resistance). Accordingly, DNN training was based on cyclic voltammetry with variation in the reversible formal potential (E^0), heterogeneous electron transfer rate constant (k^0), charge transfer coefficient (α), diffusion coefficients (D), which were assumed to be the same for all species, C_{dl} and R_u. In the case of the $EC_{1^{st}}$ and $EC_{2^{nd}}$ mechanisms, the homogeneous chemical reactions accompanying a one electron charge transfer E process were simulated using standard kinetic models and by assuming that the chemical reaction was fully irreversible.[24,25]

The DNNs were designed to accept and analyse patterns available in data obtained at a scan rate of 100 mV s^{-1}. In the initial study, the chemical rate constants selected only induced small differences in the voltammetry relative to the pure E mechanism (equal reduction and oxidation peak current magnitudes). In generating simulated data used for training, the rate constants for the backward chemical reactions (k_b) were set to zero. The E process limits for the $EC_{1^{st}}$ and $EC_{2^{nd}}$ reactions were reached if $k_f^{EC_1}$ was less than about 1 × 10^{-2} s^{-1} and $k_f^{EC_2}$ less than about 1 × 10^{1} M^{-1} s^{-1}, respectively. Conversely, if $k_f^{EC_1}$ was greater than about 300 s^{-1} and $k_f^{EC_2}$ greater than about 3 × 10^{5} M^{-1} s^{-1}, then full chemical irreversibility in cyclic voltammograms applied with no peak current evident in the data obtained on reversing the potential scan direction. Broader based training for the $EC_{1^{st}}$ and $EC_{2^{nd}}$ processes was undertaken below using simulated data which had $k_f^{EC_1}$ and $k_f^{EC_2}$ values within the limits of full chemical reversibility and full irreversibility.

Since the k_f values selected for initial training with the $EC_{1^{st}}$ and $EC_{2^{nd}}$ mechanisms were carefully chosen so that the voltammograms approach the chemically reversible limit in both cases, this represents a scenario where even an experienced electrochemist would find it difficult to distinguish between E, $EC_{1^{st}}$ or $EC_{2^{nd}}$ mechanisms by visual inspection of the cyclic voltammograms. The impact of the k_f values in this regime is shown in Fig. 3. In the $EC_{1^{st}}$ (Fig. 3b and more clearly in Fig. 3c), a larger decrease in the peak current with each cycle of potential is evident, relative to a simpler E process. In comparison, the $EC_{2^{nd}}$ (Fig. 3d and e) process displays less decay of the peak current with each cycle of potential.

If the electron transfer process is reversible, the simulated peak current for an EC process is slightly larger than for the E mechanism.[26] To make the distinction between E, $EC_{1^{st}}$ and $EC_{2^{nd}}$ even more difficult for the DNN training step, the range of possible values for the diffusion coefficient was slightly decreased for the $EC_{1^{st}}$ case so that the peak currents for all three mechanisms were very similar.

4. Utilised software and hardware

Software was written in Python using the Tensorflow[13] and Keras[27] machine learning libraries for development of the DNN algorithm and training steps, with simulated data being stored in a local PostgreSQL database for computational ease. All calculations were undertaken on a desktop running an Ubuntu operating system with software run on an Intel® generation 8 *i*7 CPU, containing 6 cores and 12 threads and a clock speed of 3.2 GHz. Neural network training used a NVIDIA Quadro® P620 GPU, which contained 2 GB of graphics RAM.

The DNN training used nine processor threads running in parallel on the CPU to collect the simulated data from the PostgreSQL database. The current data were then transformed into 100×100 or $100 \times 100 \times 3$ images to create the SCDNN and MCDNN, respectively. This process occurred simultaneously with GPU based DNN training and acted as a database pipeline for GPU calculations. The image generation was completed in batches of 42 images at a time. The training on 15 012 (5004 per reaction mechanism) simulations required 24 minutes for the SCDNN and 47 minutes for the MCDNN, using 5 training epochs. The generation of simulations also employed parallel processing on 6 threads to speed up the computation times, with individual simulations taking 0.5 to 2 seconds. However, as the simulations only needed to be generated once before storage in the database, this latter exercise required relatively little computational time.

5. Results and discussion

The two neural networks (SCDNN and MCDNN) were trained initially for each of the E, $EC_{1^{st}}$ and $EC_{2^{nd}}$ processes using bounds in the parameter space specified in Table 1.

5.1 Accuracy of classification using the simulated data

The simulated data used for the initial training were used to provide a limited form of validation and comparison of both the SCDNN and MCDNN methods

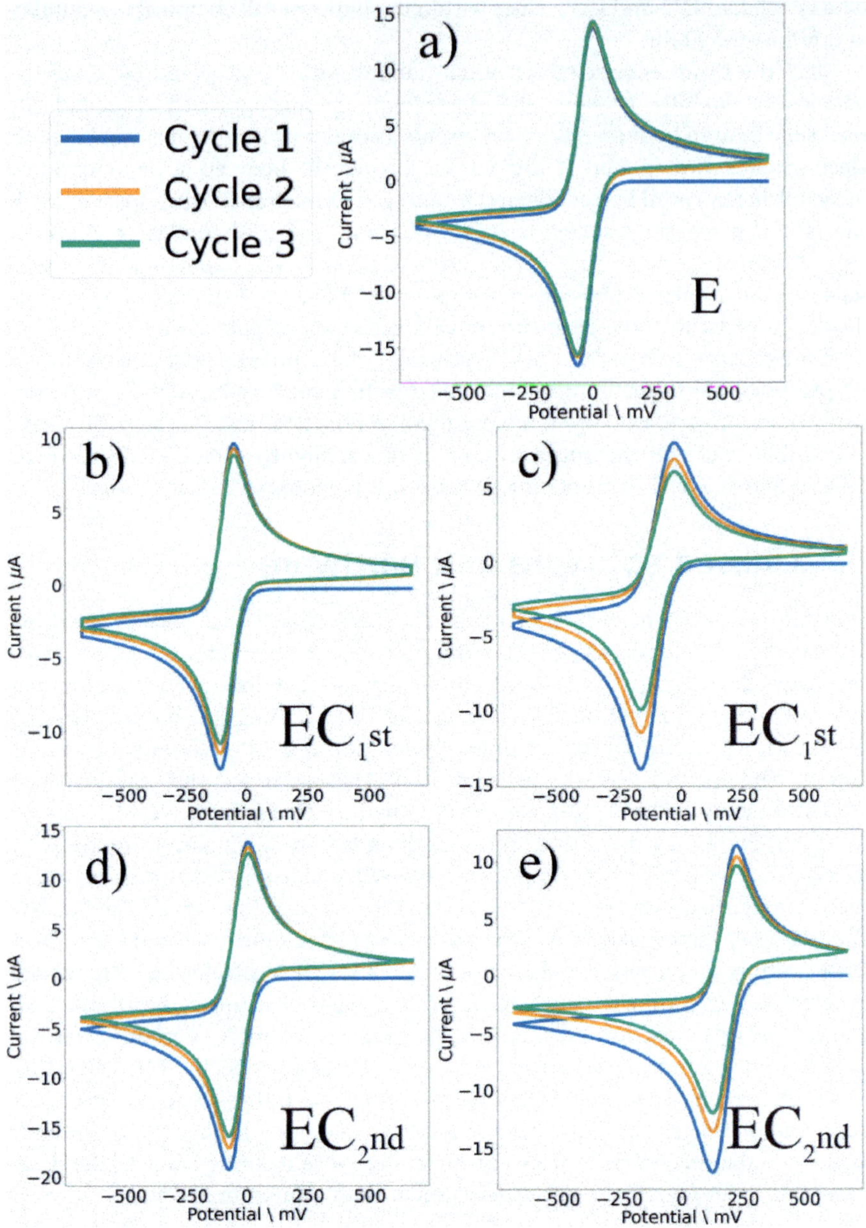

Fig. 3 Simulated cyclic voltammograms (three cycles of potential) obtained at a scan rate of 100 mV for an E mechanism with $k^0 = 0.8$ cm s^{-1} (a), an $EC_{1^{st}}$ mechanism (b and c, with $k_f^{EC_1}$ being 0.00105 s^{-1} and 0.0342 s^{-1} respectively) and an $EC_{2^{nd}}$ mechanism (d and e, with $k_f^{EC_2}$ being 15.1 M^{-1} s^{-1} and 23.1 M^{-1} s^{-1} respectively). All other parameters were randomly generated from within the ranges specified in Table 1.

(repeat training on same data set). In this exercise, the fully trained DNNs were applied to 300 randomly sampled simulations generated using parameter values within the ranges provided in Table 1. Examples of the cyclic voltammograms generated in this manner are provided in Fig. 3.

The high level of success achieved in the classification of all three mechanisms by analysis of the randomly selected simulated data using parameters specified in Table 1 confirms that both DNNs are able to successfully exploit different patterns available for the identification of each of the E, $EC_{1^{st}}$ and $EC_{2^{nd}}$ processes in the absence of experimental uncertainty. Significantly, for the $EC_{1^{st}}$ and $EC_{2^{nd}}$ mechanisms, the MCDNN is found to exhibit a superior capability to correctly classify the simulated reaction mechanisms than the SCDNN, as shown in Fig. 4. This is attributed to the addition of the multiple cycles of potential providing more information, enabling the MCDNN to identify patterns which are not present or less prominent in the single cycle data. Fig. 4c and d include a label that identifies examples where the DNN is unable to identify the reaction mechanism with greater then 65% confidence. In this initial validation exercise using simulated data with $EC_{1^{st}}$ and $EC_{2^{nd}}$ mechanisms close to the E limit, the SCDNN is unable to correctly identify over 40% of the EC type reaction mechanisms. In contrast, the use of three cycles adds sufficient new pattern information to allow the correct identification of 84% of reaction mechanisms using MCDNN.

The sample output from an $EC_{1^{st}}$ reaction displayed in Fig. 5 shows the feature maps of the initial layer and final convolutional layer. The DNN creates an emphasis on the patterns associated with the brighter regions. The features being amplified are those in the mass transport controlled regions of the voltammogram, which constitute the largest region of the image. A significantly greater

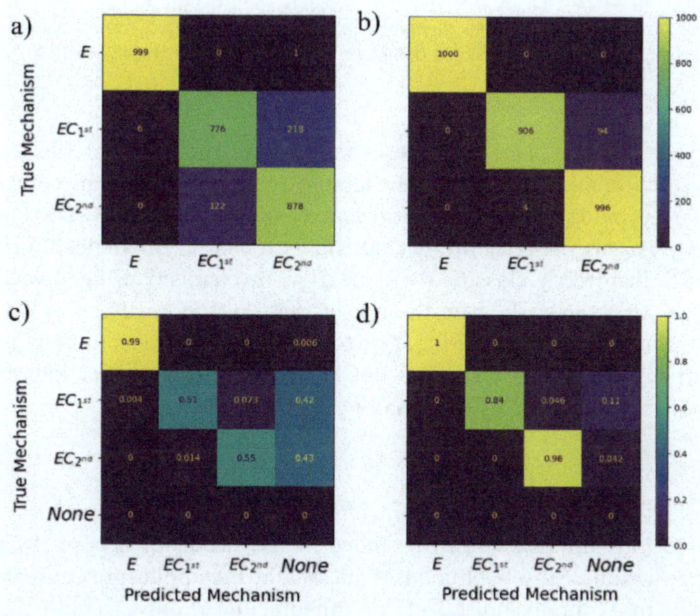

Fig. 4 Confusion matrices for the identification of the 3000 randomly sampled simulated data sets generated from Table 1. The confusion matrices show the most probable reaction mechanisms predicted by (a) single cycle SCDNN and (b) three cycle MCDNN. The confusion matrices in (c) and (d) represent normalised fits where only the mechanisms with 65% or more confidence are treated as the known reaction mechanisms, otherwise the reaction mechanism is classified as unknown; (c) is for single cycle SCDNN and (d) is for three cycle MCDNN.

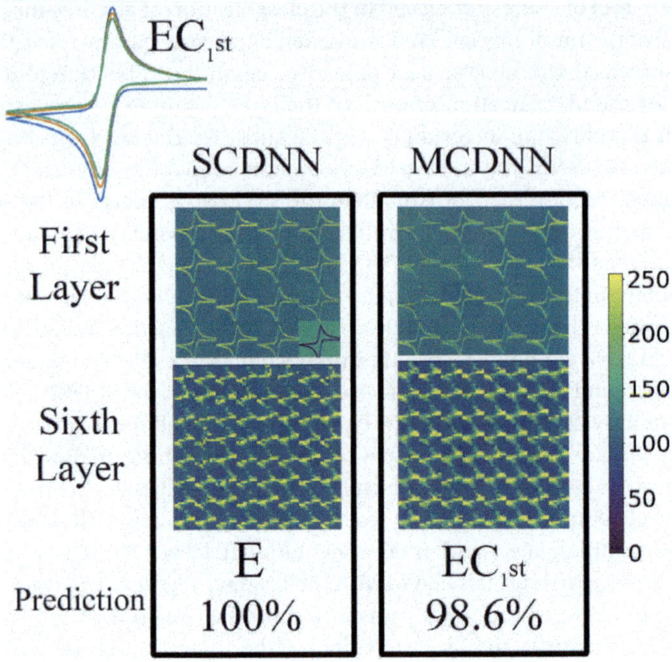

Fig. 5 Feature maps for the first and sixth layer of the SCDNN and MCDNN derived for a simulated $EC_{1^{st}}$ process simulated with $k^0 = 0.01$ cm s^{-1}, $k_f^{EC_1} = 0.015$ s^{-1}, $D = 5.0 \times 10^{-6}$ cm^2 s^{-1} and R_u = 200 Ω, as displayed in top left of the figure. The other simulated parameters lie within the ranges given in Table 1. The classified mechanisms and the corresponding confidence levels are provided at the bottom of the figure.

current decay in these potential regions upon cycling the potential is expected for the $EC_{1^{st}}$ reaction relative to that of the simple E reaction, due to the contribution from the irreversible chemical reaction step. Consequently, the MCDNN was able to correctly classify the reaction mechanism with 98.6% confidence. In contrast, the SCDNN incorrectly classified the reaction mechanism to be E with 100% confidence. After the sixth layer, the images were further analysed by two dense layers to obtain the final prediction. Due to the difficulty of visually displaying the images in these dense layers it has not been possible to further identify what voltammetric features are used in making the final prediction.

5.2 Accuracy of classification using noisy synthetic data

The conclusions inferred from the above metric need not necessarily provide a good prediction of the likelihood of success in the application to real experimental data that contains noise and experimental uncertainty. To more rigorously test their robustness, the SCDNN and MCDNN based methods were also more broadly trained on 15 012 simulations (5004 per reaction mechanism) generated from logarithmically scaled $k_f^{EC_1}$ values for the $EC_{1^{st}}$ (0.01–300 s^{-1}) and $k_f^{EC_2}$ values for $EC_{2^{nd}}$ (10–3 × 10^5 M^{-1} s^{-1}) mechanisms. This training now encompasses the regime from minimal influence of the chemical step, as in the initial exercise, to almost complete chemical irreversibility, where the heterogeneous electron transfer step for the $EC_{1^{st}}$ and $EC_{2^{nd}}$ processes was treated as fully reversible with k^0 being set as a constant of 1000 cm s^{-1}. In the case of the E mechanism, simulations were undertaken with k^0 varied over the range $k^0 = 0.01$–1 cm s^{-1}; the changes in the training data were added to focus on the classification ability of the DNN when only one kinetic parameter (k^0, $k_f^{EC_1}$ and $k_f^{EC_2}$) was varied for each reaction mechanism. Other values set as constants for the training data used in this section were $D = 5.0 \times 10^{-6}$ cm^2 s^{-1}, $R_u = 0\ \Omega$, $C_{dl} = 0.00$ F cm^{-2}, $E^0 = 0.00$ mV and $\alpha = 0.5$. The other experimental constants were the same as those of the training data shown in Table 1. Once trained, these DNNs were applied to the classification of simulated data sets containing known levels of Gaussian noise with a mean of zero and standard deviations calculated as a percentage of the maximum current. As shown in Fig. 6a, both the SCDNN and MCDNN methods are able to correctly identify the mechanisms over 80% of the time when the Gaussian noise was 0.5%. As the noise increased to 2%, the MCDNN performs better in most cases. However, with larger than experimentally expected noise levels above 2%, the single cycle based SCDNN method performs slightly better.

In the second form of validation with noise added to the simulated cyclic voltammetric data, while both DNNs perform acceptably well in classification when the Gaussian noise level is below 1%, the MCDNN performance is superior because it can access the additional patterns present in multiple cycles of potential. With 1% noise, the MCDNN achieves an almost perfect success rate of 99% for the $EC_{1^{st}}$ mechanism, whereas the SCDNN only achieves a success rate of 75%. While performing slightly better than the MCDNN, the SCDNN also becomes highly problematic with 2% noise, with the majority of examples being classified as $EC_{2^{nd}}$ (Fig. 6b). This represents a significant shift to predictions with high noise levels being incorrectly labelled as having an $EC_{2^{nd}}$ reaction mechanism, as seen in the confusion matrix in Fig. 6b. A related shift occurs in the SCDNN labelling with respect to the identification of E, EE and EC undertaken by Kennedy *et al.*[3] In this study, high noise levels led the SCDNN to classify the E

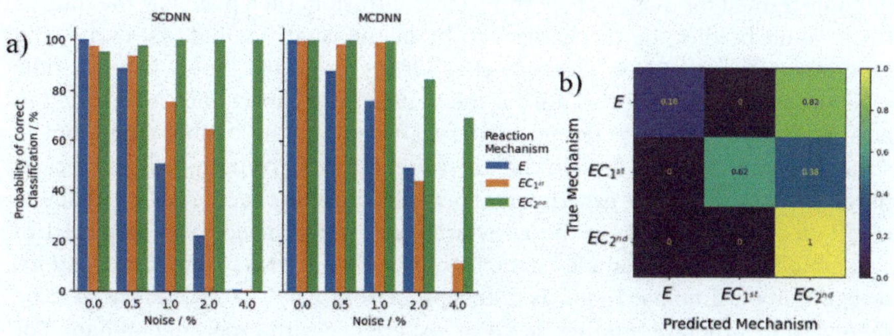

Fig. 6 (a) Bar graph showing the success probability in mechanism classification for 500 randomly sampled simulations per mechanism. The logarithmically scaled parameter ranges available for the simulated data are as follows: $EC_{1^{st}} = 0.01$–300 s^{-1} and $EC_{2^{nd}} = 10$–3×10^5 M^{-1} s^{-1}, and for the E mechanism the range of k^0 is 0.01–1 cm s^{-1}. (b) Normalised confusion matrix for the classification of 3000 randomly sampled reaction mechanisms using the SCDNN, when the simulated data contains a Gaussian distribution of noise with a standard deviation of 2% of the maximum current. The correctly classified mechanisms are shown in the diagonal of the confusion matrix.

reaction mechanism predominately as an EE reaction mechanism. A possible cause of this misclassification may be the over expression of the patterns associated with the noise by the softmax activation layer on the final DNN layer, leading to bias in the data analytics. In principle, a more robust DNN could be developed to analyse very noisy data. However, it would be preferable to redo a very noisy experiment so that high quality data can be used for analysis.

In summary, MCDNN training achieves significantly enhanced success in correct classification, relative to that of the SCDNN, when applied to simulated data without noise and when noise is present at the level of one percent or less when both DNNs are trained in limited regions of k_f for the $EC_{1^{st}}$ and $EC_{2^{nd}}$ mechanisms. It is reasonable to assume that the MCDNN multiple cycle method also would outperform the single cycle method if training sets covered a wider range of parameter space or contained a wider range of mechanisms in the classification selection, since exploitable additional information content is always available in the multiple cycles of potential images.

5.3 Issues that may arise in applications to experimental data

In future work, when applications to experimental data are undertaken, the additional issues listed below are expected to be encountered.

(1) A major difference in undertaking classification of mechanisms with experimental data is that, while the DNNs will still be trained on simulated data derived from exact models, real experimental data will introduce issues associated with model uncertainty. For example, the Butler–Volmer relationship employed in the present simulations may not strictly apply to the experimental electron transfer reaction. Furthermore, if subtle mechanistic nuances, such as a small amount of convection, radial diffusion or adsorption, contribute to the experimental data but is not included in the modelling, then the experimental data is likely to confuse the MCDNN. For example, in an extreme case where a very small microdisk electrode is used, radial diffusion could provide a substantial steady state type of contribution to the voltammetry. If a DNN was given this data but only trained for mass transport by linear diffusion, then plausibly the trained DNN would predict the mechanism to be of the catalytic kind. As always, the experimenter using DNNs needs to check if the predicted model is chemically realistic. Confusion may also arise if the double layer capacitance is assumed to be independent of the potential in the training data, as in this study, but is strongly potential dependent in the experimental data. Training should be carefully chosen to reflect the experimental non-idealities, so that the DNN will not be readily fooled or biased, or the DNN architecture will need to be designed to accommodate "out of sample" data.[28–30] Accordingly, DNNs trained for pattern recognition on simulated data, as in this proof of principle study, may need to be refined to incorporate training based on experience with experimental data. For example, instrumental distortion introduced by an insufficiently fast potentiostat rise time[31,32] may erroneously modify regions which are important in the DNN layers' feature maps, shown in Fig. 5. Thus, measurement limitations need to be considered when designing DNNs that need to generalise well in actual experimental settings.

(2) Training undertaken with one specific set of parameters may need to be modified to accommodate changes dictated by experimental requirements. For

example, a DNN trained with data simulated at 20 °C may not be able to correctly classify experimental data obtained at a temperature of −70 °C. Variations in other parameter values, such as the scan rate, concentration and potential range, in training and experimental data also may cause confusion to DNNs. Some of these problems could in principle be addressed by introducing training based on simulating voltammograms with dimensionless rather than dimensioned forms of the parameters used in this study. Even a difference in the initial scan direction used in training and in experiments may cause confusion.

(3) Parameter ranges chosen for the simulations used in training may need to be chosen carefully in the mechanistic classification of experimental data by DNNs to minimise the possibility that the patterns identified by the DNN are not a result of overfitting by the DNN at the training stage. This could be accommodated by automated unsupervised learning, such as a clustering technique, to automatically identify poorly defined sets of simulations.

(4) An important issue of concern when using experimental data may arise from the fact that the output probability in DNNs is largely defined by the training data and how the weights used in the DNNs layers are optimised. This optimisation, when applied to experimental data containing inherent uncertainty, may not always lead to a singular convergence of the projected probability. Consequently, two DNNs trained on the same simulated voltammetric data could generate differences in the probability of classification when applied to experimental data. The validity of different predictions would need to be confirmed *via* the application of the DNN to a wider range of experiments or expert analysis of experiments that have been specifically designed to test the DNNs and confirm their accuracy and robustness when applied to a particular set of experimental data.

In future work, as well as reporting the outcomes of using DNNs in classification studies derived from extensive experimental data, the DNN method will be expanded to cover a wider range of reaction mechanisms. The use of more pattern sensitive artificial neural networks, including fully connected 1D convolutional DNNs[4] or incorporation of a more parameter based double-layer feedforward artificial neural network,[2] as well as techniques in addition to cyclic voltammetry, also may need to be considered as a means of addressing deficiencies identified from experience with applications of machine learning to experimental data.

6. Conclusion

In this study, a SCDNN reaction mechanism classifier trained with cyclic voltammetric images derived from simulations of the first cycle of potential has been extended to a MCDNN method that uses the initial three cycles of potential. Training with both classes of neural networks was undertaken with simulated data for E, $EC_{1^{st}}$ and $EC_{2^{nd}}$ mechanisms. The training for the $EC_{1^{st}}$ and $EC_{2^{nd}}$ mechanisms included employing parameters for the C step that generated images in a regime where only minimal perturbation is induced relative to the E case. Thus, classification in this regime is based on analysis of a problem in a regime where all three mechanisms provide similar cyclic voltammetric data, making classification on the basis of the visual inspection of the cyclic voltammograms very difficult, even for an experienced electrochemist.

Despite the significant improvement in classification success achieved using noisy synthetic simulated data with one specific training set, it is suggested that additional broad based training will be needed to facilitate more robust mechanistic classification with experimental data applied to a wider range of mechanisms. For example, training with simulations based on dimensionless rather than dimensioned parameters, as used in this study, could prove valuable, noting that simulations of cyclic voltammograms are usually based on dimensionless parameter inputs.[33]

Limitations of the DNN method applied to voltammetric classification include possible overfitting to patterns that do not exist in experimental data and the difficulty in the generalisation of training to accommodate large possible variations in conditions and techniques employed to obtain the experimental data. The former may be addressed by using more pattern sensitive DNNs, such as fully connected 1D convolutional DNNs.[4] The use of training derived from simulations based on imperfect models can also be problematic when applied to experimental data.

Conflicts of interest

There are no conflicts of interests to declare.

Acknowledgements

Jie Zhang and Alan M. Bond gratefully acknowledge the Australian Research Council for the financial support provided by the award of the Discovery Grants DP170101535 and DP210100606.

Notes and references

1 P. K. Kundu, P. Panchariya and M. Kundu, *ISA Trans.*, 2011, **50**, 487–495.
2 P. J. Grahn, G. W. Mallory, O. U. Khurram, B. M. Berry, J. T. Hachmann, A. J. Bieber, K. E. Bennet, H.-K. Min, S.-Y. Chang, K. H. Lee and J. L. Lujan, *Front. Neurosci.*, 2014, **8**, 169.
3 G. F. Kennedy, J. Zhang and A. M. Bond, *Anal. Chem.*, 2019, **91**, 12220–12227.
4 S. N. Dean, L. C. Shriver-Lake, D. A. Stenger, J. S. Erickson, J. P. Golden and S. A. Trammell, *Sensors*, 2019, **19**, 2392.
5 J.-J. Ye, C.-H. Lin and X.-J. Huang, *J. Electroanal. Chem.*, 2020, **872**, 113934.
6 A. M. Bond, *J. Solid State Electrochem.*, 2020, **24**, 2041–2050.
7 L. Gundry, S.-X. Guo, G. Kennedy, J. Keith, M. Robinson, D. Gavaghan, A. M. Bond and J. Zhang, *Chem. Commun.*, 2021, **57**, 1855–1870.
8 R. A. DePalma and S. P. Perone, *Anal. Chem.*, 1979, **51**, 825–828.
9 R. A. DePalma and S. P. Perone, *Anal. Chem.*, 1979, **51**, 829–832.
10 L. K. Bieniasz, *J. Electroanal. Chem.*, 1996, **406**, 45–52.
11 E. P. Sapozhnikova, M. Bogdan, B. Speiser and W. Rosenstiel, *J. Electroanal. Chem.*, 2006, **588**, 15–26.
12 L. K. Bieniasz and H. Rabitz, *Anal. Chem.*, 2006, **78**, 8430–8437.
13 M. Abadi, A. Agarwal, P. Barham, E. Brevdo, Z. Chen, C. Citro, G. S. Corrado, A. Davis, J. Dean, M. Devin, S. Ghemawat, I. Goodfellow, A. Harp, G. Irving, M. Isard, Y. Jia, R. Jozefowicz, L. Kaiser, M. Kudlur, J. Levenberg, D. Mané,

R. Monga, S. Moore, D. Murray, C. Olah, M. Schuster, J. Shlens, B. Steiner, I. Sutskever, K. Talwar, P. Tucker, V. Vanhoucke, V. Vasudevan, F. Viégas, O. Vinyals, P. Warden, M. Wattenberg, M. Wicke, Y. Yu and X. Zheng, *TensorFlow: Large-Scale Machine Learning on Heterogeneous Systems*, 2015, software available fromhttps://www.tensorflow.org/.

14 A. Paszke, S. Gross, S. Chintala, G. Chanan, E. Yang, Z. DeVito, Z. Lin, A. Desmaison, L. Antiga and A. Lerer, *NIPS 2017 Workshop on Autodiff*, NIPS, 2017.

15 *MECSim Software Package*, http://www.garethkennedy.net/MECSim.html, accessed September 2021.

16 R. G. Compton, J. C. Eklund, L. Nei, A. M. Bond, R. Colton and Y. A. Mah, *J. Electroanal. Chem.*, 1995, **385**, 249–255.

17 J. A. Alden, A. M. Bond, R. Colton, R. G. Compton, J. C. Eklund, Y. A. Mah, P. J. Mahon and V. Tedesco, *J. Electroanal. Chem.*, 1998, **447**, 155–171.

18 K. He, X. Zhang, S. Ren and J. Sun, *2016 IEEE Conference on Computer Vision and Pattern Recognition (CVPR)*, IEEE, 2016.

19 A. Krizhevsky, I. Sutskever and G. E. Hinton, *Advances in Neural Information Processing Systems*, 2012.

20 K. He, X. Zhang, S. Ren and J. Sun, *Proceedings of the IEEE Conference on Computer Vision and Pattern Recognition (CVPR)*, 2016.

21 W. Rawat and Z. Wang, *Neural Comput.*, 2017, **29**, 2352–2449.

22 M. Robinson, A. N. Simonov, J. Zhang, A. M. Bond and D. Gavaghan, *Anal. Chem.*, 2019, **91**, 1944–1953.

23 L. Gundry, G. Kennedy, J. Keith, M. Robinson, D. Gavaghan, A. M. Bond and J. Zhang, *ChemElectroChem*, 2021, **8**, 2238–2258.

24 A. W. Bott, S. W. Feldberg and M. Rudolph, *Curr. Sep.*, 1996, **15**, 67–71.

25 A. W. Bott, *Curr. Sep.*, 1999, **18**, 9–16.

26 J.-M. Savéant and C. Costentin, *Elements of Molecular and Biomolecular Electrochemistry*, Wiley, 2019.

27 F. Chollet *et al.*, *Keras*, 2015, https://keras.io, accessed September 2021.

28 H. Noh, T. You, J. Mun and B. Han, *Proceedings of the 31st International Conference on Neural Information Processing Systems*, NIPS, 2017.

29 S.-M. Moosavi-Dezfooli, A. Fawzi, O. Fawzi and P. Frossard, *Proceedings of the IEEE Conference on Computer Vision and Pattern Recognition*, 2017, pp. 1765–1773.

30 X. Liu, M. Cheng, H. Zhang and C.-J. Hsieh, *Proceedings of the European Conference on Computer Vision (ECCV)*, 2018.

31 A. W. Colburn, K. J. Levey, D. OHare and J. V. Macpherson, *Phys. Chem. Chem. Phys.*, 2021, **23**, 8100–8117.

32 P. Kissinger, *Laboratory Techniques in Electroanalytical Chemistry*, Marcel Dekker, Inc., New York, 1996.

33 R. G. Compton and C. E. Banks, *Understanding Voltammetry*, Imperial College Press, 2010.

Faraday Discussions

ROYAL SOCIETY OF CHEMISTRY

DISCUSSIONS

Electrochemical data mining: from information to knowledge: general discussion

Tim Albrecht, Xiangkun Elvis Cao, Dongfei Chen, Manuel Corva, Martin A. Edwards, Andrew Ewing, Stefano Fornasaro, J. Justin Gooding, Luke Gundry, Ayumi Hirano-Iwata, Grant Jeffcoat, Ali Reza Kamali, Frédéric Kanoufi, Serge G. Lemay, Ndrina Limani, Steven Linfield, Xu Liu, Si-Min Lu, Gabriel N. Meloni, Zhongqun Tian, Kristina Tschulik, Swathi Naidu Vakamulla Raghu, Hui Wei and Yi-Lun Ying

DOI: 10.1039/d2fd90001g

Yi-Lun Ying opened discussion of the introductory lecture by Justin Gooding: The nanointerface needs more comprehensive models to describe the interaction networks among water and ions. Would you please comment on how these models and understanding guide the applications of nanoelectrochemistry?

Justin Gooding answered: This is a very difficult question which I don't think we know enough yet to give you a clear answer. What we do know is that with surfaces inside the substrate channels of our enzyme-like nanoparticles we have two to three layers of ice like water. Inside the channels that might mean that all the water is in an ice-like state. How this affects transport in the channel and the electrical double layer in electrochemical systems are big questions for us. Certainly, regarding transport, there is evidence for proton migration inside these channels and possibly sodium and potassium if they are present. But in our systems our materials are not well defined enough to tease out such effects. This was what my point was about regarding needing better defined materials to really understand the nature of the nanointerface in nanoparticle systems. There is however a lot of information we can extract from studies performed in the transport of species through nanochannels. I don't think we yet have an understanding though for case that I described where the nanochannels themselves are also reactive.[1,2]

1 T. M. Benedetti, C. Andronescu, S. Cheong, P. Wilde, J. Wordsworth, M. Kientz, R. D. Tilley, W. Schuhmann and J. J. Gooding, Electrocatalytic nanoparticles that mimic the three dimensional geometric architecture of enzymes: nanozymes, *J. Am. Chem. Soc.*, 2018, **140**, 13449–13455, DOI: 10.1021/jacs.8b08664.

2 J. Wordsworth, T. M. Benedetti, A. Alinezhad, R. D. Tilley, M. A. Edwards, W. Schuhmann and J. J. Gooding, The importance of nanoscale confinement to electrocatalytic performance, *Chem. Sci.*, 2020, **11**, 1233–1240, DOI: 10.1039/c9sc05611d.

Si-Min Lu enquired: I have a question on the COMSOL modeling in your study. You mentioned that the migration plays a important role in the electrocatalysis. I am wondering if there is any possibility that the electric convection could enhance the mass transport in the confined channel as the electroosmotic flow may exist in the channel where the double layer is large enough. The clever marriage between finite element modeling and molecular dynamics (MD) simulation may be a good way for unveiling the double layer structure in the nanoconfinement.

Justin Gooding replied: This is another very good question that I don't think we have a definitive answer to yet. What we believe our COMSOL modelling shows is there is some active transport of protons into the channels but also there is a possibility that the water structure inside the channels is influencing proton transport. All we can say with the COMSOL modelling, as it is a continuum model, is the increase in catalytic activity we see is consistent with the active transport of protons going into the channel. In our model however, it approximates the surface potential as charge on the channel walls and the diffuse double layer is composed of point charges. So we do need molecular dynamics simulations inside the channels to really understand what the relative effects might be. So I agree that a combination of COMSOL modelling with MD simulation would be a good way to start to understand such systems, although we also need better defined experimental systems and we do need MD models to include the consumption of the protons within the channel.

Swathi Naidu Vakamulla Raghu said: It was mentioned that in array-like structures each channel will have its own micro-environment of sorts. Is this the case say for a homogeneous structure such as with metal oxide nanotubes?

Justin Gooding responded: So why I make this statement is from our work on porous silicon,[1] so not from electrochemical systems but optical devices. In the case of porous silicon you have well-defined columnar pores, so each nanopore is an isolated pore. There what we see is the materials act like a sponge, if we have low concentrations of analyte outside the structure, because each nanopore is a separate diffusional system, we get high concentrations of analyte inside the nanoporous structure. This is because each open channel is in equilibrium with the exterior solution so if a molecule goes into the channel, the other channels nearby do not feel the effect of that molecule entering the structure. However, in a mesoporous structure where each pore is connected with the others, when a molecule enters one pore this is felt by the other pores as they are all interconnected within the nanostructure.[2]

1 K. A. Kilian, T. Böcking, K. Gaus, J. King-Lacroix, M. Gal and J. J. Gooding, Hybrid lipid bilayers in nanostructured silicon: a biomimetic mesoporous scaffold for optical detection of cholera toxin, *Chem. Commun.*, 2007, 1936–1938, DOI: 10.1039/b702762a.
2 J. Wordsworth, T. M. Benedetti, A. Alinezhad, R. D. Tilley, M. A. Edwards, W. Schuhmann and J. J. Gooding, The importance of nanoscale confinement to electrocatalytic performance, *Chem. Sci.*, 2020, **11**, 1233–1240, DOI: 10.1039/c9sc05611d.

Hui Wei asked: You showed nice examples of how to design enzymatic channels in nanozymes. I am wondering whether you have considered the design of more flexible (softer) channels in (inorganic) nanomaterials? We know natural enzymes are usually softer and more flexible.

Another question is: you demonstrated the nanoconfinement effect in electrochemical catalysis systems. Would the thermally driven catalysis have similar effects as shown in electrochemistry driven catalysis?

My third question is: By using external magnetic force and magnetic particles, you can improve the detection sensitivity. For the analytes of ultralow concentration, how would one enhance the interaction between analytes and the antibodies (or other recognition elements) on a magnetic particle?

Justin Gooding replied: With the rigid structures that we have made one could use the small size of the channels to enhance selectivity *via* size exclusion. The idea of flexible channels to mimic allosteric enzymes is not something we have considered yet. What we would like to do is make nanoparticles where the reaction only occurs at the end of the channels, rather than also reacting along the channel walls, to better mimic enzymes. We feel this would deconvolute some of the uncertainties related to what is occurring in our system. We don't currently know how to make the materials flexible anyway. We are actually not trying to mimic every aspect of an enzyme but rather with that work we are trying to just mimic how an enzyme controls the solution properties where the reaction proceeds to facilitate that reaction. That is what we would like to know how to do because we think that could give us significant advantages in electrocatalysis. Regarding the second question, in many ways I think similar effects can occur in non-electrochemically driven systems, but in our system we can use potential to control the reaction which is an option other approaches may not provide to the same extent. Regarding the third question and facilitating interactions between the analyte and antibodies in magnetic nanoparticles. We still rely on the affinity between the antibody and the analyte but by having many antibodies on the nanoparticle we actually enhance the chances of an analyte to remain bound to the nanoparticle as we have discussed previously.[1]

1 Y. F. Wu, D. Bennett, R. D. Tilley and J. J. Gooding, How nanoparticles transform single molecule measurements into quantitative sensors, *Adv. Mater.*, 2020, **32**, 1904339.

Ndrina Limani enquired: In the 'Electrocatalysis' part of your presentation (DOI: 10.1039/d1fd00088h), some ORR and OER electrocatalytic activities were presented, mainly to illustrate the performance of novel nanomaterials. However, there are several methods that could be used to assess the electrocatalytic activity (SECM, RRDE…) and I was wondering how was it assessed in those cases?

Justin Gooding answered: To determine the electrocatalytic activity we usually used rotating disc electrodes but of course also had to determine the electroactive area. Most frequently for Pt surfaces we used Cu underpotential deposition (UPD). The reasoning for using Cu UPD rather than H UPD was for two different reasons. The first reason is that because of the high curvature on some of the nanoparticle systems we deal with there is a possibility of strain effects that will influence the H UPD as we have shown in ref 1. The second reason is for the nanozyme-style particles that mimic the 3D geometry of enzymes the hydrogen can also pass through the surfactant layer around the nanoparticles, but the copper does not, so we get the area of the substrate channel only. This is discussed in ref. 2.

1 A. Alinezhad, L. Gloag, T. M. Benedetti, S. Cheong, R. F. Webster, M. Roelsgaard, B. B. Iversen, W. Schuhmann, J. J. Gooding and R. D. Tilley, Direct growth of highly strained Pt islands on branched Ni nanoparticles for improved hydrogen evolution reaction activity, *J. Am. Chem. Soc.*, 2019, **141**, 16202–16207.
2 T. M. Benedetti, C. Andronescu, S. Cheong, P. Wilde, J. Wordsworth, M. Kientz, R. D. Tilley, W. Schuhmann and J. J. Gooding, Electrocatalytic nanoparticles that mimic the three-dimensional geometric architecture of enzymes: nanozymes, *J. Am. Chem. Soc.*, 2018, **140**, 13449–13455.

Frédéric Kanoufi communicated: This was a very inspiring talk showing a wide spectrum of single entity inspection and the challenges to face. The inspection of nanoporous nanoparticle activity is really inspiring. You particularly mentioned the effect of migration within single nanoparticle nanopores. It is a major field of research in water and electrolyte transport in single nanostructures (*e.g.* nanotubes, see ref. 1). In this field the many works suggest osmotic transport of water within nanoconfined regions. Do you think this might happen in your system (it would increase the electrocatalysis mass transport)? It is definitely an interesting aspect to unveil by electrochemistry or to combine with electrocatalysis.

1 K. Falk, F. Sedlmeier, L. Joly, R. R. Netz and L. Bocquet, *Nano Lett.*, 2010, **10**, 4067.

Justin Gooding communicated in reply: Yes I agree that osmotic transport could be occurring in our nanoparticles with nanopores and I mentioned studies in nanochannels and nanotubes. When I was referring to migration, I think it is more accurate for us to say that our results are consistent with active transport of protons into the channel without us having the control over the experimental system to tease out which of the various possibilities of that active transport could be the dominant effect.

Ali Reza Kamali opened discussion of the paper by Yi-Lun Ying: Very interesting work! I am wondering whether you have studied the performance of the device in non-aqueous phases too?

Yi-Lun Ying answered: We have not studied the performance of our amplifier system in non-aqueous phases yet. Our amplifier array could be used to work in other conditions such as organic phase and solid phase if it is adopted to a suitable electrode array.

Si-Min Lu enquired: The instrumental advances are pivotal to single entity electrochemistry at the nanoscale as the electrochemical instrumentation with high bandwidth can resolve more accurate information on mass transport and electron transfer of single entities at the confined interface. This study enables parallel nanoelectrochemical measurements. I am interested in the chip architecture, what is the distances between each electrode in the array? Why? Have you optimized the distance by simulation?

Yi-Lun Ying replied: We are currently working on optimizing the distance between electrodes. The interference between electrodes would be diminished when the distance is set to >0.5 mm according to previous work.[1] Here, the distance set into 1.27 mm which was accessible to the fabrication of the array chip.

1 M. Akhtari, D. Emin, B. M. Ellingson, D. Woodworth, A. Frew and G. W. Mathern, *J. Appl. Phys.*, 2016, **119**, 064701.

Zhongqun Tian asked: In the electrochemical environment, a single entity is not isolated and its behavior may be influenced by many factors. So its characteristic current including noise may provide some information about the nanopore/nanoelectrode such as edge, shape, soft and hard electrode materials, dynamical change of modified layer and biomolecules, *etc.* The noise analysis may be important for deep understanding of the mechanism and getting a full physico-chemical picture of the single entity. So I wonder how much useful information on the sub-entity level and/or molecular/atomic level one can extract from the so called noise? And how to technically distinguish the real useful noise from thermal and artificial noises generated by electronics and instrument? If possible could you summarize or predict the methods and approaches of the noise analysis in this field?

Yi-Lun Ying responded: Indeed, much information could be extracted from the noise on the sub-entity level and/or molecular/atomic level. The algorithm for big data analysis is useful to calculate dynamic physico-chemical properties from signals. We summarized our developed algorithm into the following aspects:

(i) Locating and recovering the attenuated signal. The duration time of the filter distorts the blockages. This attenuation could be calibrated by using a second-order-differential-based calibration. Then, the current amplitude of the blockages was recovered with an integrated method.[1,2] The results showed that the relative error for evaluating the blockages with a duration as short as 0.01 ms was significantly decreased to <20% which was lower than the conventional method (>800%). In order to automatically locate the short blockages from the highly noisy data, we developed a Modified Hidden Markov Model (MHMM) which includes the Fuzzy c-Means to initialize the HMM parameters and Viterbi training algorithm to optimize the HMM. This algorithm shows strong tolerance to the noise which enables detection of nanopore blockages at the highest bandwidth of the amplifier.[3]

(ii) Identification of similar blockades from the mixture. The analytes that differ with only a subgroup (*e.g.* methylcytosine and cytosine) often exhibit similar blockage currents or/and durations. To improve the sensing ability of nanopores, a shapelet-based machine learning approach was used to discriminate mixed analytes that exhibit nearly identical blockage current amplitude and durations.[4] This algorithm would facilitate the application of nanopore techniques in complex real samples. Moreover, the adaBoost model improved the capacity of the nanopore in efficient identification of single nucleotide differences.[5]

(iii) Tracing time-series single-molecule characteristics. The transition conformation of a single molecule usually generates time-series spike-like current fluctuations which could be described by using a four-stage Markov chain model.[6] This method built discrete-state stochastic models for describing long-term statistical dynamics. Therefore, the combination of nanopores with the Markov chain model clarified the transition pathways of the β-hairpin peptide. For single nanoparticles collision experiments, the envelope algorithm was developed to identify multi-peaks and gave the integrated charges calculation.[7]

(iv) Mining ion-interaction networks from the "noise". The transport of single analyte and ions through a biological nanopore is governed by interaction networks among the confined ions, analyte and residues at the inner surface of the nanopore. However, the ion interaction networks only generate very weak ionic fluctuations which are usually buried in the blockage noise. In order to achieve qualitative and spectral analysis of such weak fluctuations, we developed an advanced frequency method based on the Hilbert–Huang transform.[8] This method quantitatively bridges the gap between frequency characteristics and ion interaction networks inside the nanopore. Especially, the peak frequency (f_m) and its amplitude (a_m) were used as a frequency fingerprint to describe and predict dynamic equilibriums for ion interaction networks. We anticipate that the frequency characteristics could be extended to study interaction networks for various analytes, allowing the nanopore-based biopolymer sequencing.

1 Z. Gu, Y.-L. Ying, C. Cao, P. He and Y.-T. Long, *Anal. Chem.*, 2015, **87**, 907.
2 Z. Gu, Y.-L. Ying, C. Cao, P. He and Y.-T. Long, *Anal. Chem.*, 2015, **87**, 10653.
3 J. Zhang, X. Liu, Y.-L. Ying, Z. Gu, F.-N. Meng and Y.-T. Long, *Nanoscale*, 2017, **9**, 3458, DOI: 10.1039/c6nr09135k.
4 Z.-X. Wei, Y.-L. Ying, M.-Y. Li, J. Yang, J.-L. Zhou, H.-F. Wang, B.-Y. Yan and Y.-T. Long, *Anal. Chem.*, 2019, **91**, 10033.
5 X.-J. Sui, M.-Y. Li, Y.-L. Ying, B.-Y. Yan, H.-F. Wang, J.-L. Zhou, Z. Gu and Y.-T. Long, *J. Anal. Test.*, 2019, **3**, 134.
6 S.-C. Liu, Y.-L. Ying, W.-H. Li, Y.-J. Wan and Y.-T. Long, *Chem. Sci.*, 2021, **12**, 3282, DOI: 10.1039/d0sc06106a.
7 C.-Y. Xu, R.-J. Yu, X. Ni, S.-W. Xu, X.-X. Fu, Y.-J. Wan, Y.-L. Ying and Y.-T. Long, *Chin. J. Chem.*, 2021, **39**, 1936.
8 X. Li, Y.-L. Ying, X.-X. Fu, Y.-J. Wan and Y.-T. Long, *Angew. Chem., Int. Ed.*, 2021, **60**, 24582.

Ayumi Hirano-Iwata commented: The current noise level is also dependent on the sampling frequency of recordings. If you use a higher sampling frequency, the noise will also increase. Because the sampling frequency is dependent on how fast the target reactions are, it would be helpful if you could show us the time resolution and noise level that are required for your target reactions.

Yi-Lun Ying responded: The noise of the chemical system and circuit system is mainly affected by three factors of the instrument: bandwidth of amplifiers, cut-off frequency of the low-pass filter and the sampling frequency. Moreover, the capacitance from the electrode (C_E) and connection wires (C_W) also contribute to the noise. To achieve a satisfactory signal-to-noise ratio (SNR) in the electrochemical measurement, a low-pass filter was adopted, which gave the cut-off frequency of 5 kHz. The noise level of the single-entity electrochemical measurement in our system at 5 kHz filter was about 4–5 pA RMS. The time resolution was about 0.2 ms. In order to avoid the signal distortion, the value of the sampling rate usually should be set to 2.56 times higher than that of the cut-off frequency of the filter according to the Nyquist sampling rate. Therefore, the sampling rate in our system should be set above 12.8 kHz. Here, we set the sampling rate to as high as 100 kHz which is sufficient enough to avoid the signal distortion from analog-to-digital conversion.

Martin A. Edwards asked: In your system, do you know how much of the noise is coming from the amplifiers *vs.* the electrodes and chemistry of the system? You present results from relatively large microelectrodes (200 µm). I expect the noise from the electrodes and their capacitance to scale with their area (radius squared) – thus there might be great opportunities to drastically lower the capacitance/noise. Do you anticipate that this would be useful?

Yi-Lun Ying answered: That would be useful. In our system, the noise mainly originates from the electrodes and chemistry of system. The electrode interface capacitance (C_E) mainly contributes to the noise. The capacitance is proportional to the area of the electrode. Therefore, the reduction of the area of the electrode will be very useful to lower the noise.

Andrew Ewing addressed Martin A. Edwards and Yi-Lun Ying: Further to the previous question by Martin A. Edwards: that implies interference noise, right? Need a noise spectrum analysis? 1 pA is good for a 200 µm electrode, but not for a smaller electrode.

Yi-Lun Ying replied: For the experiment, we should compromise between the bandwidth in the whole recording system and the required signal-to-noise ratio.

Martin A. Edwards replied: Yes, noise spectrum is what I was thinking. Interference (background processes) was what I was imagining, but there are likely also electrical components due to the double-layer capacitance that might have some impact. Yes, 1 pA seems good for a 200 µm electrode, although as you are well aware, we should also consider the frequency we are assessing it at to make truly meaningful comparisons.

Serge Lemay enquired: The compensation stage also amplifies noise from the first amplifier stage in the additional frequency window, correct? Could you please quantify the impact that this has on the overall performance of the circuit?

Yi-Lun Ying answered: The compensation circuit indeed amplifies the input noise and noise from the first amplifier stage. But the compensation circuit only amplifies the input signal at the high frequency region of >1 kHz, see Fig. 1 here. In this region, the noise is from dielectric loss and capacitance (Fig. 1b in the paper; DOI: 10.1039/d1fd00055a). In other words, the compensation circuit compensates the gain loss of the first-stage amplifier at the high frequency region. As shown in Fig. 3c in the paper (DOI: 10.1039/d1fd00055a), the −3 db bandwidth was increased from 1 kHz to over 100 kHz. In order to further suppress the noise, the low-pass filter could be used. For example, in our single-molecule detection assay, a 5 kHz low-pass filter was employed to further reduce the noise in the high frequency region.

Andrew Ewing asked: Have you analysed the noise types using a frequency spectrum analyser? Just off the top of the head, the success of the 2-stage amplifier I think suggests you are limited by shot noise, but 1 pA is not really to that limit. What does the frequency spectrum look like if you put in high frequencies?

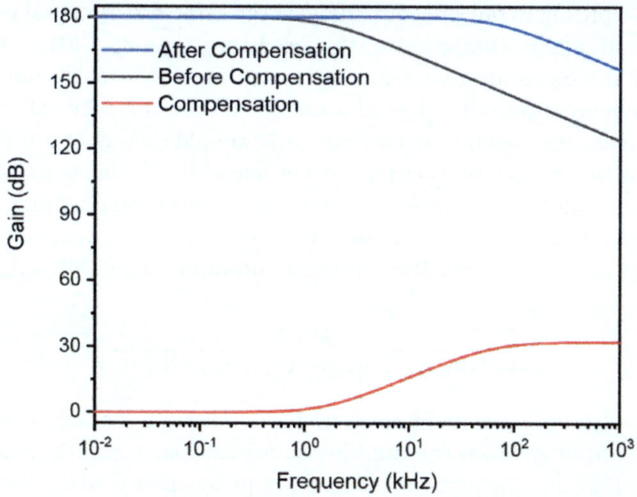

Fig. 1 Bandwidth simulation for the amplifier circuit performance. Bandwidth of four-channel amplifier circuits without compensation (black), the compensation circuit (red) and after compensation (blue).

Yi-Lun Ying responded: The noise in the experiments depends not only on the amplifier but also on the electrode array and the connection wire. Currently, we adopt an electrode array with a relatively large electrode of diameter of 150 μm (Fig. 2b in the paper; DOI: 10.1039/d1fd00055a). This electrode array caused additional noise in the measurement. Therefore, we adopted a 5 kHz Bessel low-pass filter to remove the noise at high frequencies. The further incorporation of a nanoelectrode array would be beneficial for the noise reduction.

Kristina Tschulik said: Thank you for sharing your work on an amplifier for high bandwidth, low current applications. Developments in this direction will be essential to overcome existing limitations in single entity electrochemistry. Could you please specify what is the maximum filter frequency you have been able to use for filtering in nanoparticle impact studies to get appreciable signal-to-noise ratios for the example systems you present in your paper (DOI: 10.1039/d1fd00055a)?

Yi-Lun Ying replied: This is a really good question. We are exploring the potential highest bandwidth for recording single-nanoparticle collisions at the electrode. The whole bandwidth of the single-nanoparticle collisions depends not only on the amplifier but also on the electrode. The double layer capacitance of the electrode is proportional to the electrode area. Consider the electrode capacitance (C_E) as a simple parallel plate capacitor given by

$$C_E = \frac{\varepsilon A}{d}$$

where A is the area of the electrode and d is the thickness of the double layer. Therefore, the greatly reduced surface area of the micro-/nanoelectrode results in a small C. According to our previous study, at the same experimental conditions, the root mean square (RMS) noise decreased to 2.7 pA at 5 kHz filter as the diameter of the electrode reduced to 480 nm.[1] Moreover, a 30 nm close-type wireless nanopore electrode achieved a RMS of 1.3 pA in AgNPs collision experiments.[2] Usually, the signal-to-noise ratio (SNR) should exceed 3 for the following identification of each signal from the current recording. To achieve a satisfactory SNR, the bandwidth of the whole system is expected to reach >5 kHz if a nanoelectrode with low capacitance is adopted. Note that the defined maximum bandwidth (B_{max}) at which the SNR exceeds a minimum value (SNR_{min}) is given by

$$B_{max} \approx \left(\frac{\Delta I \sqrt{3}}{SNR_{min} \times 2\pi (C_E + C_W + C_{In} + C_F) v_n} \right)^{2/3}$$

where C_W is capacitance contributions from the electrode connecting wire, C_{In} is the amplifier input and C_F is the amplifier parasitic capacitance (C_F) of a feedback resistor (R_F). If C_E and C_W could be reduced to 1 pF together with the low C_F of this designed amplifier (0.1 pF), it may be possible for the bandwidth to reach a maximum value of 100 kHz. However, such a high bandwidth needs continuous efforts in amplifier design, chip connection and electrode array manufacturing. Also, the mass transport of the redox molecule should be considered when we perform the high bandwidth recording.

1 Y. Zheyao, M. Wei, Y. Yilun and L. Yitao, *Acta Chim. Sin.*, 2017, **75**, 671.
2 R.-J. Yu, S.-W. Xu, S. Paul, Y.-L. Ying, L.-F. Cui, H. Daiguji, W.-L. Hsu and Y.-T. Long, *ACS Sens.*, 2021, **6**, 335.

Kristina Tschulik noted: For readers who want to test the bandwidth of their lab equipment (which is not always transparent *e.g.* for commercial potentiostats), I would like to suggest a hands-on guide for electrochemists we've developed.[1]

1 K. Kanokkanchana, E. N. Saw and K. Tschulik, *ChemElectroChem*, 2018, **5**, 3000–3005, DOI: 10.1002/celc.201800738.

Yi-Lun Ying added: The temporal resolution of the single-entity electrochemical measurement is inversely related to the bandwidth in the recording system but also is limited by the required signal-to-noise ratio (SNR). Various research studies have discussed the impact of bandwidth, including but not limited to ref. 1 and 2. Our group has also elaborated the bandwidth impact in perspective and review papers.[3–5]

1 E. Kätelhön, E. E. L. Tanner, C. Batchelor-McAuley and R. G. Compton, Destructive nano-impacts: What information can be extracted from spike shapes?, *Electrochim. Acta*, 2016, **199**, 297.
2 R. Gao, M. A. Edwards, J. M. Harris and H. S. White, Shot noise sets the limit of quantification in electrochemical measurements, *Curr. Opin. Electrochem.*, 2020, **22**, 170.
3 J. Wang, Y.-L. Ying, C.-B. Zhong, L.-M. Zhang, F. Yan and Y.-T. Long, Instrumentational implementation for parallelized nanopore electrochemical measurements, *Analyst*, 2021, **146**, 4111, DOI: 10.1039/d1an00471a.
4 Z. Gu, Y.-L. Ying and Y.-T. Long, Nanopore sensing system for high-throughput single molecular analysis, *Sci. China: Chem.*, 2018, **61**, 1483.
5 Y.-L. Ying, Z. Ding, D. Zhan and Y.-T. Long, Advanced electroanalytical chemistry at nanoelectrodes, *Chem. Sci.*, 2017, **8**, 3338, DOI: 10.1039/c7sc00433h.

Gabriel Meloni questioned: In Fig. 4b of the paper (DOI: 10.1039/d1fd00055a) it looks like there is still some distortion in the transient (looks a bit like a discharge) while you are still away from the frequency limitation of your amplifier. Is that due to an under compensation of the compensating circuit? And how do you set that compensation? Also, in the simulations you used a top-end parasitic capacitance of 0.5 pF, isn't that too small for a real system, considering the length of the wires you have in Fig. 2 of the paper? Final point, any comments on power supply stability? For me not many people think about it.

Yi-Lun Ying replied: (i) About the signal distortion. In order to achieve a high SNR, we set the low-pass filter at 5 kHz. Therefore, the maximum bandwidth of the whole measurement system is 5 kHz. The possible distortion originates from the low-pass filter. However, the current shape of the recorded signal is similar to the previous single-nanoparticle collision for H_2 evolution reactions using a commercial amplifier at 5 kHz.[1] Therefore, each channel of our amplifier array showed comparable performance to that from a commercially available single-channel amplifier (*e.g.* Axon 200B).

(ii) About compensation. By adjusting the parameters of R_{com}, C_{com} and R_{adj} in the compensation circuit, the −3 dB bandwidth of the amplifier could reach as high as over 100 kHz (Fig. 3c of the paper; DOI: 10.1039/d1fd00055a). The final values of R_{com}, C_{com}, R_{adj} were set to 20 Ω, 50 nF and 150 Ω, respectively.

(iii) About parasitic capacitance. Fig. 3d in the paper simulates the bandwidth reduction from crosstalk of the multi-channels and the structure design of the board. It is reasonable to set the parasitic capacitance of feedback resistance to 0.5 pF based on our circuit board design. This value is not the total capacitance for the whole measurement system in single-particle collisions.

(iv) About power supply stability. The power supply was stabilized by a low dropout regulator (LDO) which is an essential power management circuit. The voltage noise is suppressed to <20 μV RMS.

1 M. Chen, S.-M. Lu, Y.-Y. Peng, Z. Ding and Y.-T. Long, *Chem.–Eur. J.*, 2021, **27**, 11799.

Manuel Corva asked: I am interested in the Pt nanoimpacts you present in the paper (DOI: 10.1039/d1fd00055a). Why is no HER steady-state current observed despite working in 10 mM $HClO_4$?

Yi-Lun Ying responded: The rapid current spike is due to the initial hydrogen adsorption process (the Volmer step, $H^+ + e^- \rightarrow H_{ads}$), while the slow decay is due to the hydrogen evolution process. At potentials ranging from 0 to −400 mV, the overpotential is insufficient to drive the evolution reaction, and therefore the hydrogen adsorption will proceed until the entire Pt surface is covered by a monolayer of H_{ads} (Fig. 4B of the paper; DOI: 10.1039/d1fd00055a). The fact that the current spike returns to the baseline also suggests that it is likely due to a self-limiting process, which supports this hypothesis.

Frédéric Kanoufi enquired: You showed a very promising instrumental improvement for increasing the frequency range of single entity electrochemistry studies. It seems in the bandwidth spectrum curve presented in Fig. 2 of the paper (DOI: 10.1039/d1fd00055a) that at frequencies in the 10^5 Hz range (10^2 kHz) the gain is (slightly but still by >10%?) higher than at lower frequencies. Does it mean that in this frequency range, because of higher gain, the signal is overestimated compared to the other frequency ranges? If yes, and even if it is likely a minor effect, could you tell how much error can be made in terms of charge or current for quantitative single entity NP analysis?

Yi-Lun Ying replied: The slight increase in gain at high frequency in Fig. 3c of the paper (DOI: 10.1039/d1fd00055a) is caused by a little overshoot in the compensation circuit. We can solve this problem by adjusting the parameters of the compensation circuit. But it hardly causes an impact on the detection of the single entity in our experiments since the low-pass filter is set to 5 kHz.

Xu Liu communicated: The low-noise four-channel electrochemical instrumentation as described in your paper (DOI: 10.1039/d1fd00055a) looks promising. You talked about the challenges in achieving large-scale electrochemical measurement due to the current circuit designs with commercialized amplifier chips. Can you include more discussions on other factors (in addition to the circuit design) that could compromise the large-scale electrochemical measurement? Also, would you be able to comment on the potential of large-scale electrochemical measurement for other methods in the literature?

Yi-Lun Ying communicated in reply: To achieve large-scale electrochemical measurement, a parallelized working electrode is a must. The direct integration of the nanoelectrode array with the amplifier array is necessary to further reduce the capacitance from the electrode surface and connection wires. The high throughput measurement brings a massive dataset. Therefore, an automatic and intelligent data analysis algorithm is required to real-time and accurately read single-entity characteristics from the massive current data. In addition, during the process of electron transfer measurement at each working electrode, the reference/ground electrode combines with the unique one. Therefore, the reaction at the reference/ground electrode would cause the cross-talk. This factor should be considered for the further improvement.

Xiangkun Elvis Cao communicated: A vital component of the multi-channel amplifier arrays, as shown in your study, is the ability to reduce noise. The paper (DOI: 10.1039/d1fd00055a) studied some approaches for noise reduction (*e.g.*, by reducing the diameter of the Au electrode in the array). Can you also comment on some other potential methods to suppress the noise?

Yi-Lun Ying communicated in response: As discussed in our paper (DOI: 10.1039/d1fd00055a), the noise contribution of the ultra-low current measurement is mainly from the electrode interface, electrode connecting wire, amplifier input and amplifier parasitic capacitance of the feedback resistor. Therefore, the noise reduction methods should suppress the noise contributions from the above aspects. First, the small size of the electrode array (*e.g.* nanoelectrode array) certainly could reduce the capacitance from the electrode surface leading to a low current noise level. Moreover, the direct integration of the amplifier array with the micro/nanoelectrode would significantly minimize the connection distance, thereby, reducing the noise. Utilizing the multi-channel amplifiers designed using CMOS (Complementary Metal Oxide Semiconductor) technology could further suppress the noise. In addition, the current fluctuations from the charge carriers binding/unbinding with the sensing interface give the noise at the low frequency region. Therefore, the inhibition of the unwanted charge fluctuations (*e.g.* wetting and de-wetting, protonation and deprotonation) would benefit the noise reduction. The possible methods include the hydrophilic treatment (*e.g.* O_2 plasma) of the nanoscale sensing interface and the surface modification of hydrophilic compounds for the solution assay.

Serge Lemay opened discussion of the paper by Luke Gundry: This is a rather open-ended question, but thinking of a single-variable, quasi-continuous function $I(t)$ in terms of a 2D array of independent pixels takes some getting used to! To what extent does the coarse-graining process in two dimensions impact the information content of the CVs? Intuitively one expects that a fine enough grid will capture all of the information (our digitally recorded CVs are also discretized, after all), but I find it difficult to shake the feeling that *something* must be lost during this transformation.

Luke Gundry replied: Hey thanks for the question, its quite a technical one so my apologies in advance. What helped me conceptualize the 2D array was saving the 100 × 100 image to an Excel sheet where each pixel contains the value of the array. To imagine this simply think that the deep neural network (DNN) is "looking" at the voltammogram through squinted eyes. As for using a larger voltammogram 2D array, there are two main issues. The first issue is that the increase of array size drastically increases computational complexity (due to the n^2 relationship for a $n \times n$ image), the second issue is that the DNN is "looking" at the image with a kernel that scans across the 2D array for each layer. Where this kernel would have to be increased such that the DNN looks at the same amount of information in the comparative layers. Each layer also has a max-pooling after it which decreases the size of the array for the next layer, so the expansion would only matter for the first few layers anyhow.

Ultimately, a range of things were tested in this study which were not published, such as a 224 × 224 2D array with a more complex DNN layer design using the single and multiple channels. This approach is the more common in most image based DNNs, the biggest issue we had to contend with for these expanded DNNs was over fitting in the training process leading to poor generalization of the DNN, which could be corrected using more dropout layers but it was very challenging to get right for all reaction mechanisms we were originally trying. It was found that the simpler DNN with the inclusion of the 3 channels leads to the largest increase in improvement and didn't have the issue with over fitting.

Overall, as much as there is definitely information lost in the coarse-graining process there is still enough information for the DNN to make an estimated prediction on what the reaction mechanism is for these reaction mechanisms.

Steven Linfield enquired: Can the DNN incorporate the mischaracterisations observed during testing with simulated data? For example, you demonstrated that the MCDNN could correctly assess 98.6% of the EC_1 simulations. Can the voltammogram features (or even simulation parameters) of the 1.4% of mischaracterised simulations be analysed and fed back into the DNN to improve the training or even shorten the training time? If this is possible, is there a limit to repeating this with less certain assessments (*i.e.* 84%) until probability of correct characterisation is high enough for real data?

Luke Gundry responded: This question was answered in the talk but I will add the plots used in the training of the DNN. In training, the parameters in the DNN are modified to maximise accuracy (a ratio of correct predicted mechanism for the training data) in the prediction and minimise the loss (a function to measure the error in the predictions) of the DNN predictions on the training data. Once this process has been completed for all the training data the DNN will then test itself on the remaining validation data. The full process is referred to as an epoch, where a DNN can complete multiple epochs in a row incorporating the errors in the prediction; the DNN should do as many epochs as is required for the accuracy to converge to a maximum and loss to converge to a minimum.

Fig. 2 here displays the model loss and accuracy plots for the multichannel DNN to show how the two values converge to a stable solution after 5 epochs.

Steven Linfield asked: You highlighted the difficulties of applying DNNs to the analysis of real voltammetry. Are there any downsides to including more simulation parameters (you mentioned for example convection, radial diffusion, and adsorption) that might model real experiments more closely and train the DNN for better characterisation of real voltammograms?

Luke Gundry answered: The main downside with including more parameters and increasing the complexity in the model is time. As in the computational time required to run the simulation and the manual time required to set up the mathematics in the simulation. For the simulation package used, convection can be modelled so including this would only lead to an increase of time to generate the simulation. However, since ten thousand simulations are required for each reaction mechanism a small increase in simulation time adds up quickly. Inclusion of say radial diffusion would lead to a massive increase in the mathematical modelling complexity and therefore time. This is due to the change from linear to radial coordinates so simulation of this would probably be unfeasible unless access to large computer infrastructure was available.

Stefano Fornasaro enquired: Could you please comment about the validation of the DNN models. How do you select the 80% of the data for training? I was wondering if you consider methods to check the instability coming from this choice.

Luke Gundry responded: The 80% for training is randomly selected from the 5004 for each reaction mechanism. Generally the training data needs to be enough samples to cover all possible cases which is why we have to simulate the large number of simulations for training. The other 20% of data is used for prediction validation and is the data not selected for training. Besides the optimisation process done in training which is discussed in another question, the instability of the prediction is checked to not be an effect by increasing the amount of training data used and generating a few DNN models using the same process to make sure it has converged to a constant rate of predictions and the

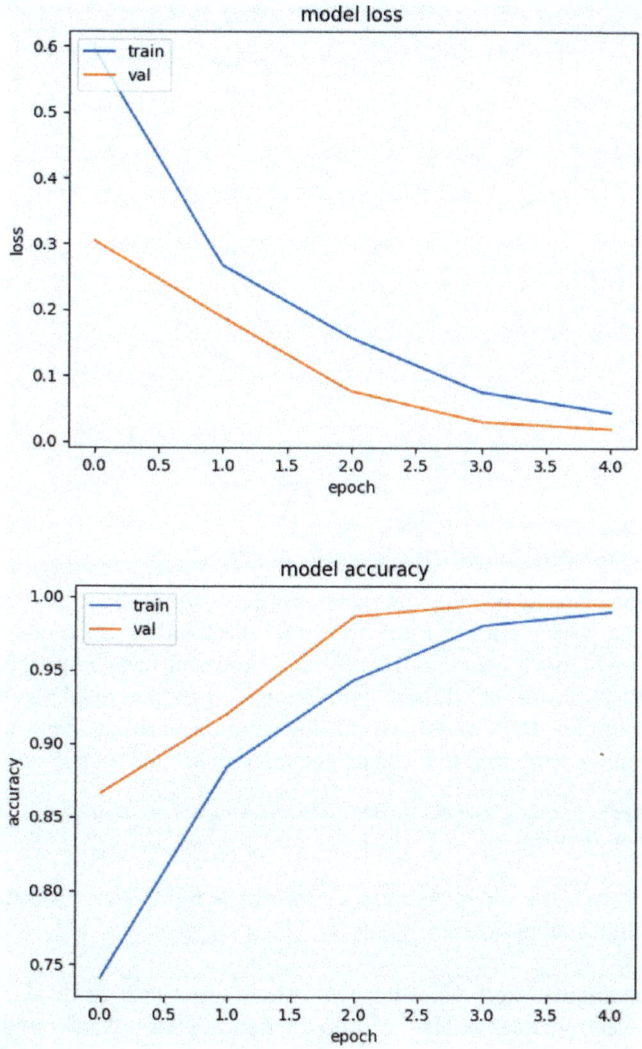

Fig. 2 Model loss and accuracy plots for the multichannel DNN.

predictions have converged to a stable answer independent of the randomly selected training data.

Tim Albrecht said: Following on from Fornasaro's question, in your work, you do not seem to use an "unseen" testset to test how well the network has learned. Did you optimise any hyperparameters during the validation phase? If not, it would seem that the validation set has been used as a testset. Is this correct?

Luke Gundry replied: For the DNNs published in this paper (DOI: 10.1039/d1fd00050k) we did not do hyperparameter tuning, the main reason for this was it was done under the study published by a coauthor of this paper.[1] Generally, I did try out a range of other hyperparameter values and none really led to any

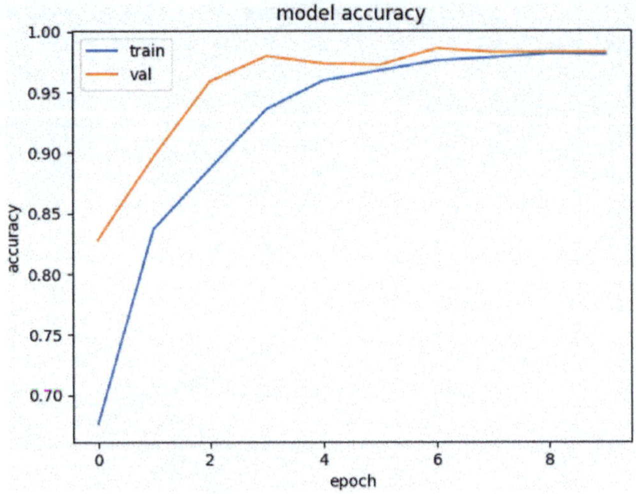

Fig. 3 Accuracy plot for a multicycle DNN.

improvements. For the training data themselves, we simulated more then enough data then split 80% of it for training and 20% for validation which was completed the end of each epoch, where 5–10 epochs were found to be required. For each epoch the DNN "looks" at all these simulations. Fig. 3 here displays an accuracy plot for a multicycle DNN and it should show that the accuracy steadily improves for each epoch, converging to a consistent value at around 5 epochs.

1 G. F. Kennedy, J. Zhang and A. M. Bond, *Anal. Chem.*, 2019, **91**, 12220, DOI: 10.1021/acs.analchem.9b01891.

Stefano Fornasaro added: Professor Albrecht is right. That procedure would result in a high risk of overfitting.

Frédéric Kanoufi asked: Thanks for this nice journey into artificial intelligence analysis of electrochemical data. You mentioned in your talk, a computer is not watching and analyzing an image (a cyclic voltammogram) as a trained scientist. As your article (DOI: 10.1039/d1fd00050k) suggests that some cases are difficult to classify upon the strategy proposed (using successive cycles), I wonder if a more classical analysis based on variations of CV response at different scan rates would be more helpful?

As a follow up question, increasing the scan rate may indeed increase the noise (and decrease the result of the classification), then could a strategy rely on trying to remove noise (by standard smoothing procedure, *etc.*) from the data before analyzing the data by the neural network?

Finally some mechanistic studies also rely on the evolution of peak potential, width or peak current with scan rate. Instead of analyzing an image could an analysis of the signal by identifying the peak descriptors be possible? Would it be faster to perform than an image analysis, would it be more accurate?

Luke Gundry answered: I believe I answered this question partly in the talk but I will add a few things. For the first part of the question the multiple scan rates or concentrations could be used, as long as it is constant over all simulated data and experimental input. In general these DNNs are very robust and flexible, the difficult part is simply making sure that the process is constant over all the data. As such, any form of data smoothing could be done to the image and I believe a moving average was used in the experimental data we tried. For the final part of the question it would be really hard to predict performance without attempting it as we simply do not know. Applying say an algorithm at simulation that identifies the peak, extracts them and then passes it to the DNN as a normalised 1D array could be possible, but it may bias the outcome due to the data extraction method, which is something you need to be wary of designing these algorithms.

Gabriel Meloni said: At the nanoscale EC mechanisms are usually poorly developed/hard to access. Piggy backing on Frédéric's point, maybe it would be interesting to investigate the scan rate dependency on $E_{1/2}$ for radial diffusion systems (similar to RDE studies for EC) to recover mechanistic information from nanoscale voltammograms. Would that be a possibility?

Luke Gundry responded: It may be possible but it is hard to say, there are two main issues with the suggestion.

One would be simulating all the training data which would be very time consuming due to the radial diffusion and the very high simulation accuracy required due to the very small scale affecting partial differential equation stability. This issue though time consuming can be worked through and with increased use of highly parallel cloud computing methods the time of one simulation would have less effect as you could run say 64 simulations at once.

The second issue is would the information be enough to make the labels in the system distinct enough? In a very recent paper from the Compton group[1] that uses a CE system, specific thermodynamic parameters can be identified. Though different it is a steady state system and they use a single voltammogram, which would suggest that your query could be applied. However, it would require a lot of independent simulations per training point increasing the number of total simulations by a multiple of the number of scan rates required. It would ultimately be possible though time consuming and not feasible at the current point. If there was a way to get more sensitive information from a single experiment, say by using AC voltammetry to get the systems more distinct over a single experiment.

1 H. Chen, E. Kätelhön, H. Le and R. G. Compton, *Anal. Chem.*, 2021, **93**, 13360, DOI: 10.1021/acs.analchem.1c03154.

Tim Albrecht enquired: Your work raises an interesting question about what the network actually learns. For example, increasing the potential window would only add capacitive charging and would not affect the relevant mechanistic information content in the data. Have you investigated this and could you comment on whether the network learns mechanistic information or rather simply what the image looks like? In the latter case, the result should be much more dependent on the exact scaling parameters, whereas in the former case this presumably would be less so.

Luke Gundry replied: Yes, we have tested this effect but its quite hard to pin down due to how the DNN looks at the image, due to the image using a moving kernel of a 3 × 3 array to identify features. If the potential range is changing the image will be compressed down to the size used and then "looked" at by the same 3 × 3 array. This array will then miss a lot of information as it will be different to what it's been trained for. The general question is does the DNN look at what it should be looking at as a chemist, to this extent the feature maps shown in Fig. 5 of the paper (DOI: 10.1039/d1fd00050k) shows it looks at the general image, with the main issue being that it looks at regions with more data points with an emphasis, which contains mechanistic information. This could be fixed quite easily by adding a weighted array that identifies the redox region and changes everything else to zero. Generally, the addition of three cycles adds more mechanistic information which is useful for the case explored in this paper.

Stefano Fornasaro queried: Is the code already available on Github?

Luke Gundry answered: Not as of the moment, DNN architecture can be seen in the supporting information of ref. 1. However, in future the code will be made available on my Github profile (https://github.com/lukegun) once our group have completed some things we are currently working on.

1 G. F. Kennedy, J. Zhang and A. M. Bond, *Anal. Chem.*, 2019, **91**, 12220, DOI: 10.1021/acs.analchem.9b01891.

Dongfei Chen asked: Is it possible to use your method to perform baseline subtractions for square wave voltammograms or differential pulse voltammograms?

Luke Gundry replied: Its difficult to say without knowing more about the system, a Deep Neural Network (DNN) could probably be trained for baseline subtraction. There was a really good paper by Dean *et al.*[1] that used square wave voltammetry and deep neural networks for chemical identification, which is of a similar vein.

I will say that due to the computational time required for DNN training it may be advantageous to try a more direct method first, such as using the first and second derivative of the current for region identification. Feel free to email me further if you want to discuss a more specific application.

1 S. N. Dean, L. C. Shriver-Lake, D. A. Stenger, J. S. Erickson, J. P. Golden and S. A. Trammell, *Sensors*, 2019, **19**, 2392, DOI: 10.3390/s19102392.

Grant Jeffcoat commented: As I understand by reading the paper; the neural networks struggled with voltammetric results that even humans would struggle with. And as you pointed out, the networks are trained on idealized data from models.

Based on this, I was wondering what you envision for immediate applications of this software. Additionally, what kinds of applications do you foresee with future development. You did outline some possible development for the networks. How might these expand or open applications?

Luke Gundry responded: Our group discussed the future application more in a recent *ChemComm* paper.[1] However, the current application of the code presented in the paper would be minimal due to the difficulty in running the code for a general chemist and the prediction made could simply be identified by doing the same experiment at a couple of different analyte concentrations. The usefulness of these study is that inclusion of different channels for a DNN helps its prediction capabilities on tougher systems. Future aspects of this kind of technology would be useful, as testing experimental data is very quick and can be designed to be very general for a wide range of reaction mechanisms. As such, an initial reaction mechanism could be predicted qualitatively in seconds before passing it to more advanced/longer running quantitative algorithms as shown in Fig. 4 here, which is taken from ref. 1.

1 L. Gundry, S.-X. Guo, G. Kennedy, J. Keith, M. Robinson, D. Gavaghan, A. M. Bond and J. Zhang, *Chem. Commun.*, 2021, **57**, 1855, DOI: 10.1039/d0cc07549c.

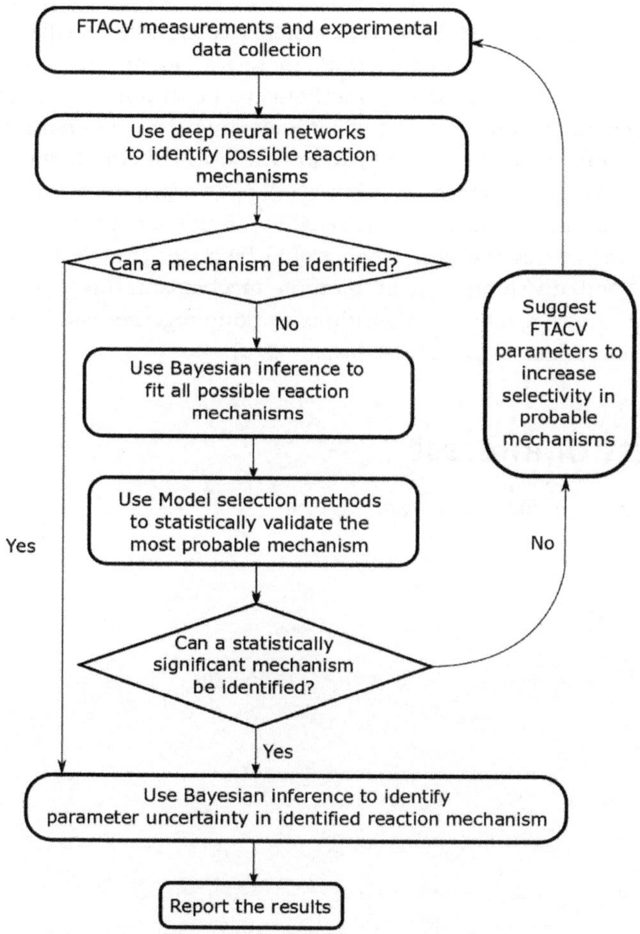

Fig. 4 Proposed artificial intelligence flowchart for analysis of FTAC voltammetric experimental data. Reproduced from Ref. 1 with permission from the Royal Society of Chemistry.

Martin A. Edwards communicated: Thanks for the interesting talk in an exciting area. In your talk, you alluded to one of the next steps for this work being expanding this to a wider range of mechanisms. I am unfamiliar with the networks you use and am curious about how such a system is anticipated to scale with the number of possible outcomes. Would you anticipate a network of similar complexity/depth to have sufficient power to resolve more mechanisms? Or would you have to keep adding layers to achieve comparably good resolution? I presume that this makes it slower/harder to do computationally, but have no idea of the anticipated scaling in the size/complexity of the network and increase in training time with number of outcomes. Do you have an idea how quickly each of these might grow? Does the current approach seem plausible for the number of mechanisms you plan to look at, or do you anticipate having to add something extra to make that step?

Luke Gundry communicated in reply: Originally the DNNs were tested on 7 reaction mechanisms and performed reasonably decently, the largest issue is the generalization when applied to out of sample data/experimental data and there were issues with differentiating between systems which were visually identifiable. In a similar vein to a lot of Savéant's work, there are a lot of voltammograms that can be fit by two different reaction mechanisms depending on the parameters used. Understandably, in this case voltammograms can't differentiate between the two reactions, with the most apparent case of this being E and EE mechanisms where the two electron transfers overlap. To work around this generally smart parameter ranges need to be selected to differentiate identical systems, an automated way to classify these labels would be good. Overall, to get a decent value and good differentiation of reaction mechanisms (as well as identify quantitative values) a range of algorithms working together will be required to correctly differentiate all possible cases.

Conflicts of interest

There are no conflicts to declare.

Faraday Discussions

PAPER

Polyaniline nanowire arrays generated through oriented mesoporous silica films: effect of pore size and spectroelectrochemical response

Wahid Ullah, Grégoire Herzog, Neus Vilà and Alain Walcarius *

Received 15th May 2021, Accepted 4th October 2021
DOI: 10.1039/d1fd00034a

Indium–tin oxide electrodes modified with vertically aligned silica nanochannel membranes have been produced by electrochemically assisted self-assembly of cationic surfactants (cetyl- or octadecyl-trimethylammonium bromide) and concomitant polycondensation of the silica precursors (tetraethoxysilane). They exhibited pore diameters in the 2–3 nm range depending on the surfactant used. After surfactant removal, the bottom of mesopores was derivatized with aminophenyl groups *via* electrografting (*i.e.*, electrochemical reduction of *in situ* generated aminophenyl monodiazonium salt). These species covalently bonded to the ITO substrate were then exploited to grow polyaniline nanofilaments by electropolymerization of aniline through the nanochannels. Under potentiostatic conditions, the length of polyaniline wires is controllable by tuning the electropolymerization time. From cyclic voltammetry characterization performed either before or after dissolution of the silica template, it appeared that both the polyaniline/silica composite and the free polyaniline nanowire arrays were electroactive, yet with much larger peak currents in the latter case as a result of larger effective surface area offered to the electrolyte solution. At identical electropolymerization time, the amount of deposited polyaniline was larger when using the silica membrane with larger pore diameter. All polyaniline deposits exhibited electrochromic properties. However, the spectroelectrochemical data indicated more complete interconversion between the coloured oxidized form and colourless reduced polyaniline for the arrays of nanofilaments in comparison to bulky films. In addition, the template-free nanowire arrays (*i.e.*, after silica dissolution) were characterized by faster electrochromic behaviour than the polyaniline/silica hybrid, confirming the potential interest of such polyaniline nano-brushes for practical applications.

Laboratoire de Chimie Physique et Microbiologie pour les Matériaux et l'Environnement (LCPME), UMR 7564, CNRS – Université de Lorraine, 405 Rue de Vandoeuvre, Villers-lès-Nancy, F-54600, France. E-mail: alain.walcarius@univ-lorraine.fr; Fax: +33 3 83 27 54 44

1. Introduction

Ordered mesoporous materials hold a prominent position in electrochemistry,[1] notably because of their attractive features (very large surface areas offering huge amounts of active sites, open regular structure ensuring fast transport properties, hosting ability for the immobilization of (electro)catalysts or (bio)molecules for instance), which can be exploited primarily in the field of electrochemical sensors and biosensors,[1–4] or for energy conversion and storage.[1,5–7] Electroanalytical applications involve mainly nanomaterials based on either mesoporous carbon[8–11] or mesoporous silica.[12–14] Silica-based materials are insulating but offer the versatility to be easily functionalized with a wide range of organic groups covalently bonded to the internal surfaces of mesopore channels.[15–17] Among the various morphologies available for mesoporous silica and organically modified silicates, thin films appear as very appealing in various fields,[18–20] including electrochemistry.[1] An important parameter that must be taken into account to ensure good performance in many applications is the accurate control of pore orientation, which is ideally obtained with uniaxially oriented one-dimensional (1D) mesopore channels.[21]

Mesoporous silica thin films made of vertically aligned nanochannels with respect to the underlying electrode surface, are indeed attractive for electroanalytical purposes.[14,22–25] They can be produced by electrochemically assisted self-assembly (EASA, combining the sol–gel process with electrochemically driven cooperative self-assembly of surfactant micelles onto an electrode surface),[26,27] a Stöber solution growth mechanism,[28] an oil-induced co-assembly process,[29] or other approaches.[30–32] The EASA method is very fast and versatile, always leading to a high level of hexagonal packing of perpendicularly oriented mesopore channels, typically 2–3 nm in diameter,[27,33] and also enabling the attachment of organo-functional groups onto the silica walls.[34] These oriented nanoporous membranes on electrodes exhibit permselective behaviour based on both size and charge selectivities,[25] which can be otherwise tuned by adjusting the pore diameter,[35] the solution pH or ionic strength,[36–38] or *via* appropriate derivatization of the films.[39–41] When functionalized with redox-active moieties, the vertical nanochannels become an area of unidirectional charge transfer by long-range electron hopping through the insulating silica membrane,[35,42–45] with more effective charge percolation than on other types of mesostructures.[46] Finally, such oriented mesoporous silica films are attractive hosts for the confinement of reagents (nano-objects, nano-sized materials or supramolecular assemblies).[44,47,48] They are also, in principle, ideal hard templates for the growth of nanofilaments with ultra-small diameters, but this is quite challenging, as demonstrated for metal nanowires electrodeposition, for instance.[49] The generation of organic polymers by electropolymerization through vertically ordered mesoporous silica-nanochannel films appeared to be feasible,[50–60] but care has to be taken to restrict the polymer growth to the nanochannels length (*i.e.*, not outside the film) and to ensure the durable binding of polymeric nanofilaments to the underlying electrode surface (if wishing to remove the silica hard template for further use as polymer nanowire arrays).

The first attempts to generate polymeric nanostructures within vertically oriented silica mesochannels as hard templates, have been reported for conducting polymers such as polythiophene[50] or related derivatives,[51,52] polypyrrole,[53] polyaniline,[54] or polyquinone.[55] If the electro-polymerization process was presumably starting on the area of the electrode surface located at the bottom of the mesopore channels, and polymer chains were growing from this area in the direction of the nanochannels, the resulting polymeric nanowires observed by microscopy exhibited diameters much larger than the mesopore diameters,[50,56] suggesting a growth out of the mesoporous membrane. Only with careful control of the electropolymerization conditions (notably the deposition time), was it possible to get polymer chains confined only in the mesopore channels, as reported for polyaniline (PANI).[54,57] However, upon dissolution of the silica template, PANI nanostructures did not stick onto the electrode surface, leaching out themselves in the form of broken pieces of polymer.[54] This therefore did not allow their use as arrays of "free" nanowires attached to the electrode surface (even if such PANI nanofilaments were indeed electroactive once entrapped in the silica membrane[54,57]). A way to improve their adhesion/linkage to the electrode material is by prior deposition of a layer of the polymeric material (*i.e.*, as a bulk coating) before the generation of the mesoporous silica membrane and subsequent electropolymerization, so that the polymeric nanowires bind to the bulk coating of the same composition, even after template removal (as shown for poly(3,4-ethylenedioxythiophene),[58] polyquinone[59] or polypyrrole[56]). Nevertheless, the resulting materials are actually made of two parts (underlying bulky coating and overlying wires) so that the response of the modified electrodes would be not only due to the individual nanofilaments. Note that arrays of single PANI nanowires/nanorods can be otherwise generated, but their lowest diameters are restricted down to ten or some tens of nanometers depending on the template approach used for their synthesis.[61–63] In a preliminary communication,[60] we have proposed a way to circumvent the above limitations, based on the covalent binding of aniline moieties onto the electrode surface prior to electropolymerization of PANI through the vertically aligned silica nanochannels, so that the resulting PANI nanofilaments remain chemically attached to the electrode support after removal of the mesoporous template.

In the present work, we have extended the approach to the generation of PANI nanofilaments covalently attached to indium–tin oxide (ITO) electrodes by using vertically oriented mesoporous silica templates with two distinct pore diameters (in the 2–3 nm range), and investigated their respective electrochemical behaviour and spectroelectrochemical response. The materials preparation involved first the deposition of the mesoporous silica film by EASA on ITO using a surfactant as the structure-directing agent (cetyl- or octadecyltrimethylammonium bromide), then the electrografting of aminophenyl diazonium cations, and finally the electropolymerization of aniline to grow the PANI nanofilaments through the vertically aligned nanochannels. The electropolymerization process was monitored by the recorded *i*–*t* curves during the potentiostatic deposition, while the electrochemical behaviour of the resulting PANI nanostructures was analysed by cyclic voltammetry respectively before and after removal of the silica template. The obtained PANI nanofilaments exhibited attractive electrochromic properties resulting from fast interconversion between oxidized and reduced forms, which was extremely stable for the template-free materials.

2. Experimental section

2.1 Chemicals and reagents

Tetraethoxysilane (TEOS, 98%, Alfa Aesar), ethanol (95–96% Merck), $NaNO_3$ (99%, Fluka), HCl (Riedel de Haan, 1 M solution), cetyltrimethyl ammonium bromide (CTAB, 99%, Acros) and trimethyloctadecyltrimethylammonium bromide (ODAB; 98% Sigma Aldrich) were used for mesoporous silica film synthesis. All reagents for electrografting were purchased from Sigma Aldrich and used as received: *p*-phenylenediamine (98%), $NaNO_2$ (≥97%) and HCl (37%). PANI electropolymerization was performed using aniline (99.5%, Sigma Aldrich), which was distilled under vacuum before use. H_2SO_4 (98%, Sigma) was used as the electrolyte for the electrochemical characterization of PANI. NaOH (VWR International) was used for silica template dissolution. All solutions were prepared with high purity water (18.2 MΩ cm^{-1}) obtained from a Purelab Option-Q from ELGA.

2.2 Preparation of vertically aligned mesoporous silica thin films

The mesoporous silica films with pore diameters of 2.0 and 2.9 nm (Fig. 1) were prepared by electrochemically assisted self-assembly on ITO plates (surface resistivity 8–12 Ω, from Delta Technologies) as described elsewhere.[35] Briefly, typical sols were prepared from mixing in 1 : 1 v/v ratio, an ethanol solution containing 100 mM of the silica precursor, TEOS, and an aqueous medium made of 32 mM of the surfactant (CTAB or ODAB) in the presence of 0.1 M $NaNO_3$ as the electrolyte. After adjusting pH to 3 (by adding the necessary amount of 1 M HCl), the sol solutions were hydrolysed under stirring at room temperature for 2.5 hours, and then used as the electrodeposition medium for the preparation of vertically aligned mesoporous silica films. This was achieved by applying a potential of −1.3 V for 20 s at the ITO working electrode (8 mm in diameter) in contact with the aged sol. After film formation, the electrodes were treated overnight at 130 °C (to ensure good cross-linking of the silica network). Removal of the surfactant species was performed using 0.1 M HCl in ethanol, as previously described.[27]

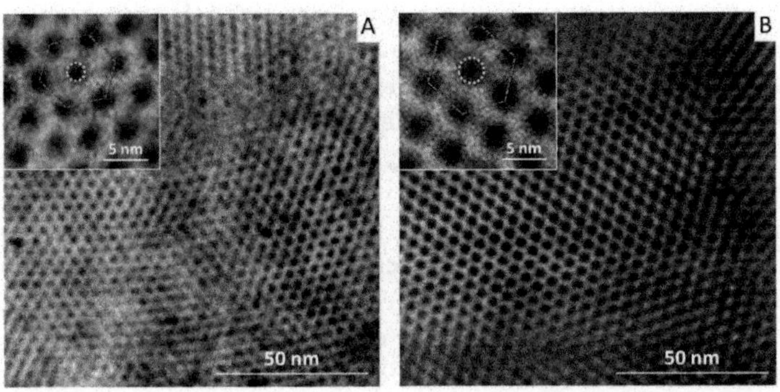

Fig. 1 Transmission electron micrographs (top view) of the ordered and oriented mesoporous silica films prepared by EASA on ITO electrode using CTAB (A) or ODAB (B) as structure-directing agents.

2.3 Generation of PANI nanofilaments

An aqueous solution containing 5 mM *p*-phenylenediamine and 5 mM $NaNO_2$ in 0.5 M HCl was first prepared and allowed to react for 10 min under stirring in order to produce aminophenyl monodiazonium cations. These *in situ* generated species were then electrochemically reduced (typically at −0.2 V for 1 min) using ITO electrodes covered with the vertically aligned mesoporous silica films in order to covalently attach aminophenyl groups on the ITO surfaces located at the bottom of the mesochannels. The grafted electrodes were then rinsed with water and sonicated for 5 min in ethanol to remove the unreacted diazonium species. PANI nanofilaments were generated by exposing the electrode surface (5 mm in diameter) to a solution of 0.1 M aniline (in 0.5 M H_2SO_4) and performing electropolymerization typically under potentiostatic conditions (at +0.85 V for selected periods of time).[60] The same experiments were also carried out with bare ITO electrodes for comparison purposes. In order to get template-free PANI nanofilaments, the mesoporous silica film was dissolved by chemical etching in 1 M NaOH (10 min).

2.4 Apparatus and procedures for materials characterization

All electrochemical experiments (mesoporous silica film synthesis, electrografting, electropolymerization, cyclic voltammetry (CV) characterization and spectroelectrochemical measurements) were performed with a PGSTAT 100 or a μAutolab from Ecochemie (Metrohm, Switzerland) operated by NOVA software (Autolab). Three kinds of three-electrode cells were used: a first homemade one for film synthesis, made of ITO working electrode, a silver rod as the pseudo-reference electrode and stainless steel counter electrode;[35] a second homemade one for electrografting, electropolymerization and cyclic voltammetry, made of modified ITO working electrodes, an Ag|AgCl|1 M KCl reference electrode (purchased from Metrohm, Switzerland) and a stainless steel rod as the counter electrode; and a third one commercially available (the liquid transmission spectroscopy "Omni-Cell" from SPECAC, UK) adapted to host our modified ITO plates as working electrodes, along with Ag|AgCl wire as reference electrode and a platinum grid as the counter electrode. UV-visible spectra were recorded using the Cary 60 UV-Vis spectrometer (Agilent Technologies). Transient profiles of absorbance variations were recorded at a wavelength of 750 nm upon successfully interchanging PANI between its oxidized to reduced forms, by applying alternatively +0.3 V and −0.2 V to the working electrode.

The mesoporous films were observed by transmission electron microscopy (TEM) using a JEOL ARM200F cold FEG TEM/STEM apparatus equipped with GIF Quantum ER (after mechanically removing portions of the film from the ITO surface). The film mesostructure was also analysed by grazing-incidence small-angle X-ray scattering (GISAXS), by collecting data on a "SAXSess mc^2" instrument (Anton Paar) equipped with Cu-Kα source ($\lambda = 0.1542$ nm) operating at 40 kV and 50 mA and a CCD detector (Princeton Instruments).

3. Results and discussion

3.1 Generation of PANI nanofilaments by combining electrografting and electropolymerization

Fig. 2A illustrates schematically how arrays of PANI nanofilaments were generated onto ITO. First, a mesoporous silica film made of vertically aligned hexagonally-

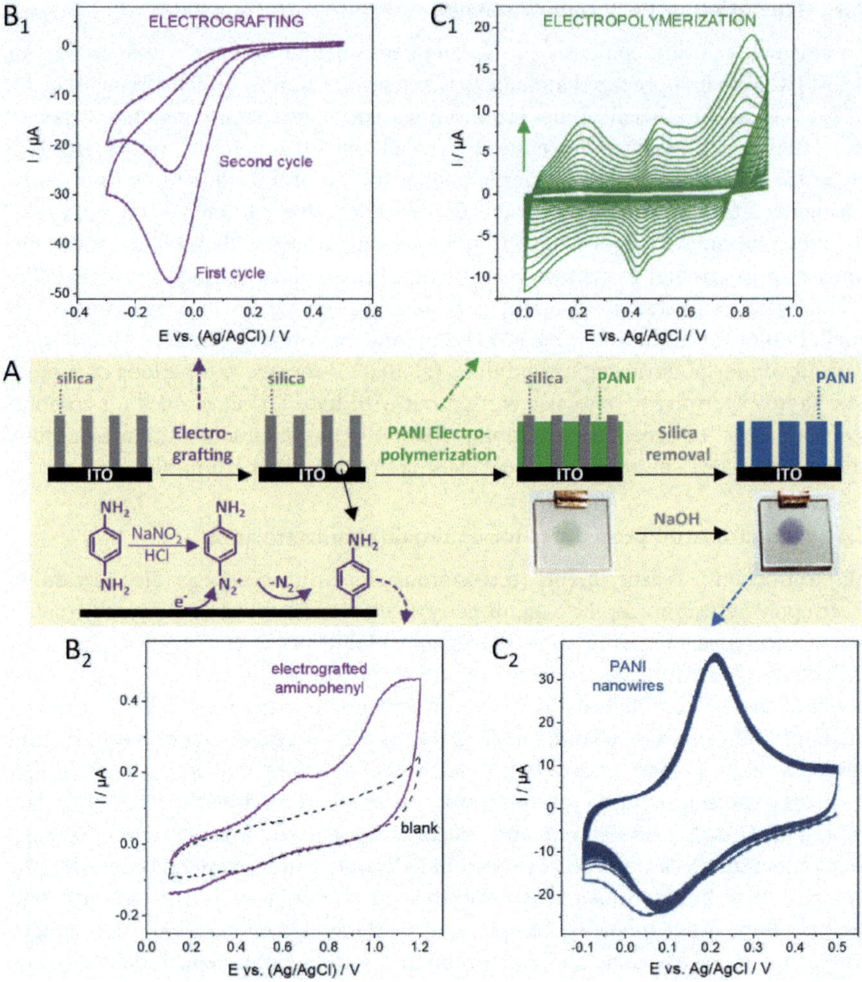

Fig. 2 (A) Illustration of the generation of PANI nanofilaments on an ITO electrode, using a vertically aligned mesoporous silica membrane as the hard template, by combining electrografting of the monodiazonium salt of *p*-phenylenediamine and electro-polymerization of aniline, with subsequent removal of silica to get the template-free nanofilaments (photos of the deposits are also shown). (B) Typical CVs related to the electrografting step: (B_1) two successive CVs recorded at 50 mV s^{-1} in 5 mM *p*-phenylenediamine + 5 mM $NaNO_2$ (in 0.5 M HCl) using the mesoporous silica-modified ITO electrode; (B_2) CVs recorded at 20 mV s^{-1} in 0.5 M H_2SO_4 using the same modified ITO electrode electrografted with aminophenyl groups (blank = CV curve recorded without electrografting). (C) Typical CVs related to the electropolymerization step: (C_1) CV curves (20 cycles) recorded at 50 mV s^{-1} in 0.1 M aniline (in 0.5 M H_2SO_4) using the same electrode as for (B_1); (C_2) 100 consecutive CV scans recorded at 50 mV s^{-1} in 0.5 M H_2SO_4 using an ITO electrode with attached PANI nanofilaments.

packed nanochannels was generated by the nowadays well established EASA method,[27] involving the application of a cathodic potential likely to induce the self-assembly of cationic surfactants onto the electrode surface and concomitant gelification of the sol solution upon electrochemically triggered pH increase, leading to the growth of a silica mesopore channels network oriented perpendicularly to the underlying support. In a second step, aminophenyl groups were electrografted onto the ITO surface areas located at the bottom of the nanochannels, by reduction of the corresponding monodiazonium cations generated *in situ*,[60,64,65] *via* a mechanism given in the left bottom part of Fig. 2A. The electrografting process can be induced, as shown by cyclic voltammetry (Fig. 2B). Fig. 2B$_1$ shows the CV curves obtained for the electrochemical reduction of the monodiazonium cations of *p*-phenylenediamine, with the characteristic peak appearing at −0.08 V (*vs.* Ag/AgCl) corresponding to the formation of the aryl radicals expected to bind to the electrode surface. Its intensity decreased on the second cycle as a result of the partial coverage/passivation of the ITO surface grafted with the aminophenyl groups. The presence of these can be also evidenced by CV (Fig. 2B$_2$), showing their characteristic irreversible oxidation signal around +1.0 V,[60,64] and an additional smaller one at lower potential values originating most probably from free oligomers.[65,66] Electrografting can also be performed potentiostatically, at −0.2 V for 1 min, as previously optimized.[60]

In a third step, PANI electropolymerization was performed in an aniline solution (in H_2SO_4 medium) and CV was used to follow the process (Fig. 2C). If performed in potentiodynamic conditions, the electropolymerization resulted in continuous growth of voltammetric signals in the 0.0–0.9 V potential window (Fig. 2C$_1$), which are characteristics of the formation of polyaniline (*i.e.*, oxidation/reduction peaks for leucoemeraldine/emeraldine around 0.2 V and for emeraldine/pergraniline close to 0.8 V, with some additional signals in-between arising from by-products[66,67]). The growth was rather slow during the first 6–7 CV cycles and then much faster, suggesting two kinetic regimes, but this will be discussed more quantitatively when performing electropolymerization in a potentiostatic mode (see Section 3.2).

The silica template can be removed by treating the electrode in NaOH medium and, thanks to the electrografted aminophenyl moieties acting as a molecular glue between the ITO surface and PANI, the nanofilaments stayed attached to the electrode (see the remaining coloured deposits on the bottom right part of Fig. 2A), the difference in colour originating from the deprotonation of emeraldine salt (green) to form the emeraldine base (blue) at high pH values.[68,69] When performing the same treatment for PANI nanofilaments produced on an ITO surface that was not previously derivatized with aminophenyl groups, they were lost in solution, confirming the importance of the diazonium electrografting to ensure their covalent bonding to the underlying electrode substrate.[60] The durable attachment of PANI nanofilaments to the ITO electrode surface is also supported by the multiple successive CV scans performed in the potential window corresponding to the redox processes of the leucoemeraldine/emeraldine system, showing very reproducible peak currents for 100 consecutive cycles (Fig. 2C$_2$), confirming moreover the reversible transformation of PANI chains between their undoped and doped forms.

The film analysis by GISAXS (Fig. 3) indicated a high level of mesostructural order over wide areas of the silica membranes, with 2D GISAXS patterns exhibiting scattered spots in the equatorial plane, which is indicative of the unique vertical orientation of the nanochannels onto the ITO surface, as is typical for mesoporous silica thin films generated by EASA.[70] These spots arose from scattering of the (10) plane of the mesopore arrays, which appeared as peaks in the 1D

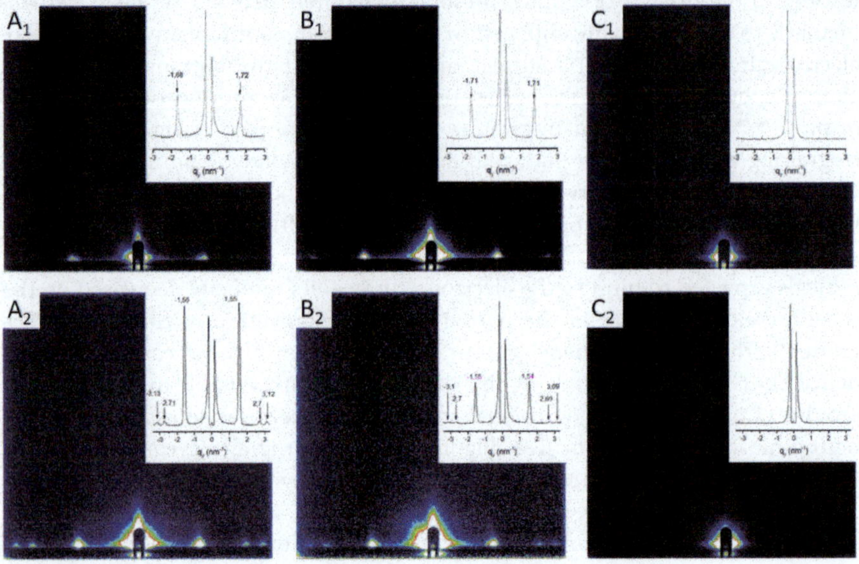

Fig. 3 GISAXS patterns at the incidence angle of 0.25° and corresponding fitting curves (inset) for (A) ITO electrodes covered with the oriented mesoporous silica films and electrografted aminophenyl groups, (B) the same electrodes as in (A) after PANI electropolymerization (at +0.85 V for 140 s), and (C) the same electrodes as in (B) after removal of the silica template (in 1 M NaOH). The subscripts 1 and 2 correspond to the films prepared from CTAB and ODAB surfactants, respectively.

GISAXS traces (see insets in Fig. 3); for ODAB-based films, one can even notice the signals of (11) and (20) planes of the hexagonal mesostructure, confirming the high level of structural order. From values of scattering maxima ($q_{max} = 1.70 \pm 0.02$ nm^{-1} for CTAB films and 1.55 ± 0.01 nm^{-1} for ODAB films), one can calculate *d*-spacings of 3.69 nm and 4.05 nm for the films respectively prepared from CTAB and ODAB surfactants. Considering the hexagonal packing of nanochannels, the pore-to-pore distances were evaluated at 4.26 nm and 4.68 nm, respectively, in agreement with the TEM data (Fig. 1). These observations are consistent with previously reported GISAXS characterization of such oriented films of distinct mesopore diameters.[33]

The film mesostructure was maintained after PANI electropolymerization through both CTAB and ODAB films without any significant change in scattering parameters (compare parts A and B in Fig. 3). Interestingly, the structure of the mesoporous PANI/silica composite films remained intact after a long storage time in air (*i.e.*, same GISAXS patterns as in Fig. 3B obtained after 1 year), suggesting a stabilizing effect of PANI wires (in contrast, the structure of bare mesoporous silica films with empty pores deteriorated within a few weeks). However, after removal of the silica template, if one can be sure that PANI nanofilaments were still attached durably to the ITO surface and are electroactive (Fig. 2C_2), the corresponding GISAXS patterns did not show any noticeable scattering signal (Fig. 3C), suggesting either some flexibility of such wires of ultra-small diameter, or the lack of sensitivity of the GISAXS method to evidence any organization of such small organic filaments.

3.2 Effect of pore size and electrochemical characterization

PANI electropolymerization can be basically performed by cycling potentials in an appropriate window (typically between 0 and 1 V) or by applying a constant anodic potential (typically 0.8–0.9 V), the amount of deposited polymer being related respectively, to the number of CV scans or to the time afforded for the application of the fixed potential. When operating with electrodes covered with a mesoporous silica film, the potentiostatic (or galvanostatic) methods are usually preferred

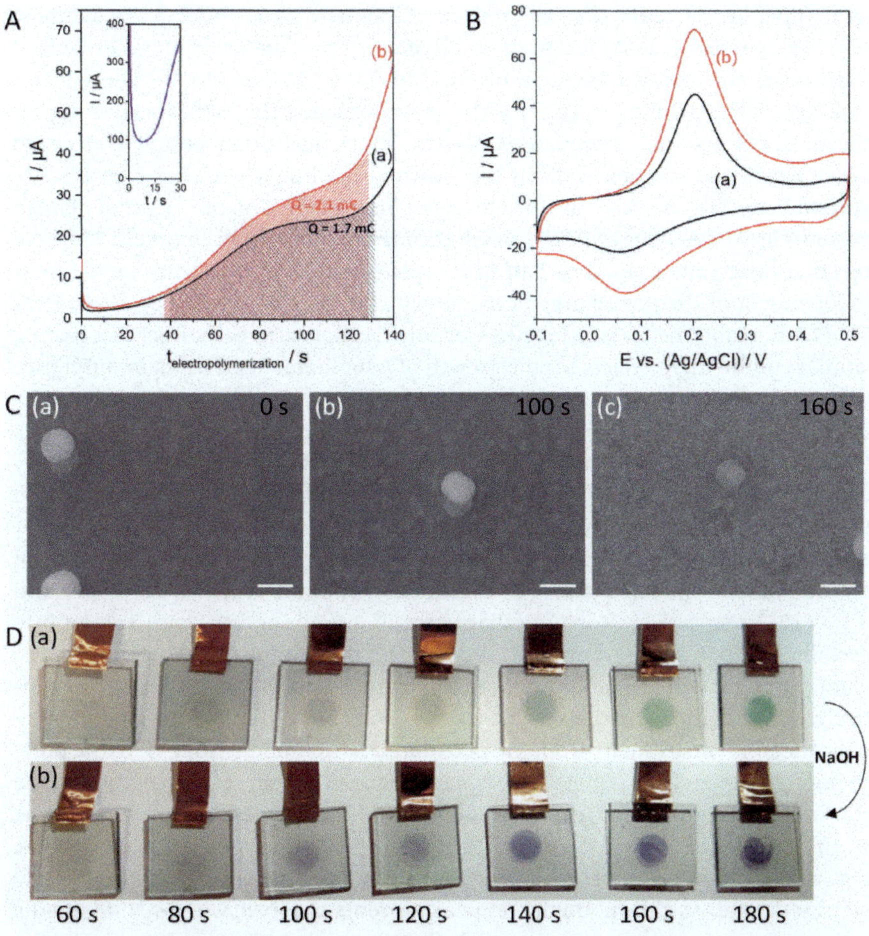

Fig. 4 (A) I–t curves recorded during PANI electropolymerization in 0.1 M aniline (in 0.5 M H_2SO_4) at constant potential (+0.85 V) using ITO electrodes covered with (a) CTAB- and (b) ODAB-based mesoporous silica films, which were previously electrografted with aminophenyl groups (the inset shows the curve obtained in the absence of mesoporous silica film). (B) Corresponding CVs recorded at 50 mV s^{-1} in 0.5 M H_2SO_4 using the same electrodes as in (A) after removal of the silica template by chemical etching (in 1 M NaOH). (C) Typical SEM micrographs of the ITO/mesoporous silica film surface (scale bars = 500 nm), respectively, before (a) or after 100 s (b) or 160 s (c) electropolymerization (CTAB-based films). (D) Photographs of the ITO/mesoporous silica film electrodes obtained after increasing times of PANI electropolymerization: (a) before and (b) after etching of the silica template.

because they enable better control of the electropolymerized material.[54,57,60] Especially, the potentiostatic approach is suited to distinguish the growth rates for PANI generation in the mesopore channels and out of the silica membrane.[57,60] We therefore used this method to investigate the effect of mesopore diameter on the PANI growth process (Fig. 4). Fig. 4A shows the typical *I–t* curves obtained for the electropolymerization of aniline at +0.85 V (as the optimized potential value for PANI growth through hard templates[57,60,71]) using ITO electrodes covered with mesoporous silica films with pore diameters of 2.0 nm (a) and 2.9 nm (b), which were previously grafted with aminophenyl groups. They exhibit the expected characteristic behaviour involving three steps: a first nucleation period (0–40 s) with rather low currents, then an increase of intensity (*ca.* 40–80 s) to reach almost a plateau current (*ca.* 80–120 s) that is related to the rate-determined growth of PANI wires through the mesochannels, and finally (above 120 s) steep current variation corresponding to faster PANI growth outside the membrane on top of the mesoporous silica template. Only the short nucleation period and sharp current increase was observed for the control experiment performed on electrografted bare ITO (*i.e.*, without silica film, see inset in Fig. 4A), proving that plateau currents are indeed due to PANI growth through the mesopore channels. Times to reach current values equal to half of the plateau current, $t_{1/2}$, were very similar independent of the pore diameter ($t_{1/2} = 63$ s for $\varnothing = 2.0$ nm and $t_{1/2} = 60$ s for $\varnothing = 2.9$ nm) and the plateau lengths were also very close to each other, suggesting similar confinement effects for the growth of nanofilaments. The main difference was the intensity of the plateau currents, which were slightly larger for the silica membrane with larger pore diameter, (compare curves (a) and (b) in Fig. 4A). Integrating the *I–t* curves under the plateau gave rise to charges *Q* respectively equal to 1.7 and 2.1 mC. If considering the values of PANI electropolymerization efficiency (65%) and the number of electrons exchanged per aniline monomer during the electropolymerization process (2.7),[72] one can estimate the amounts of aniline units in the PANI grown in the silica membrane at about 2.5×10^{15} and 3.1×10^{15}, respectively, for pore diameters of 2.0 nm and 2.9 nm. Taking into accounts the polyaniline density (1.33 g cm^{-3}),[68] these values would correspond to volumes of generated PANI of about 0.62×10^{-6} and 0.78×10^{-6} cm^3, respectively. When comparing these values to the available volume of mesopore channels (that one can estimate from the pore diameters obtained from TEM, the pore-to-pore distance from GISAXS, the nanochannel length equal to membrane thickness from profilometry measurements, and the exposed electrode area), one realizes that PANI would occupy roughly between $\frac{1}{2}$ and $\frac{1}{3}$ of the pore volume. Fig. 4B displays the CV curves recorded for these two samples after etching of the silica template. As shown, the larger peak currents observed with PANI filaments generated in the larger nanochannels, confirms that the amount of electropolymerized PANI was larger in this case, consistent with the above observations from *I–t* curves (Fig. 4A). Integration of the anodic signals gave values of 0.13 mC for the CTAB-based film (pore $\varnothing = 2.0$ nm) and 0.22 mC for the ODAB-based film (pore $\varnothing = 2.9$ nm). Considering that two aniline units are involved in the one-electron transfer process,[69] these values correspond to respectively, 1.6×10^{15} and 2.7×10^{15} aniline units. If compared to the above amounts of electropolymerized material estimated from chronoamperometric data, it means that about 64% and 84% of the generated PANI were electrochemically accessible after etching of the silica membrane. The lower recovery value obtained for the nanofilaments prepared from the mesoporous film with smaller diameter can be due to some damage/loss of PANI during etching, as otherwise suggested previously for similar materials deposited through very narrow nanochannels.[54] When lengthening too much, the electropolymerization (*e.g.*, for 160 s), PANI grew outside the film and one can observe it by SEM *via* a rather dense granular deposit on top of the smooth mesoporous silica film surface (compare parts (a) and (c) in Fig. 4C), which is not present when stopping the polymer growth on the plateau (*i.e.*, $t_{\text{electropolymerization}} = 100$ s) for which the surface state (see part (b) in Fig. 4C) is almost the same as prior to electropolymerization (a few silica beads are also noticeable, as is common for mesoporous silica films produced by EASA[27]). The increasing amounts of grown PANI can also be evidenced with the naked eye, showing a progressively increasing coloration (Fig. 4D), noticeable before and after removal of the silica template with the characteristic colour change from green to violet/blue (bottom right part of Fig. 2A). Note that the silica etching did

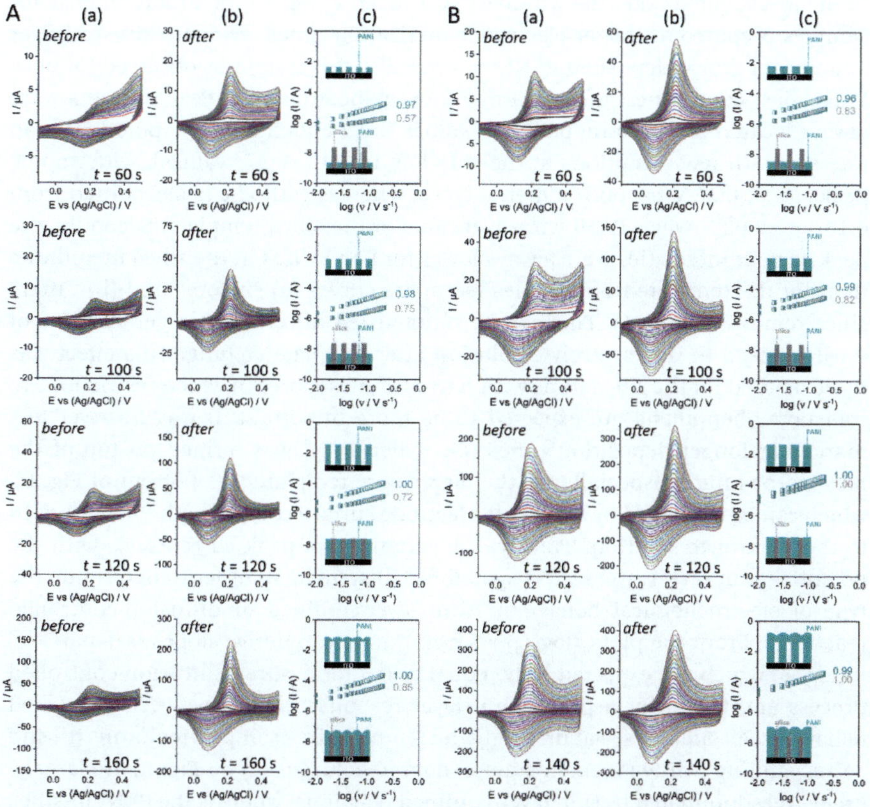

Fig. 5 Cyclic voltammetry characterization of PANI nanowires generated through mesoporous silica films with pore diameters of (A) 2.0 nm and (B) 2.9 nm for various electropolymerization times (60, 100, 120, 140 or 160 s) either (a) 'before' or (b) 'after' etching of the silica membrane; CV curves were recorded at various potential scan rates from 10 to 200 mV s^{-1} (every 10 mV s^{-1}). The variations of anodic peak currents *versus* scan rates (log–log scale) are shown in parts (c) of the figures (▮ – before and ▮ – after template removal, with numbers representing the slopes of the linear fittings), along with schemes illustrating the expected extent of PANI filaments growth in each case.

not affect the mechanical stability or sticking of the PANI deposits, except when applying too long electropolymerization times (*e.g.*, for 180 s) for which the too large amounts of PANI were likely to be damaged during NaOH treatment (see bottom right of Fig. 4D).

CV performed at various potential scan rates (Fig. 5) was then used to characterize the electrochemical behaviour of PANI nanowire arrays obtained using ITO electrodes covered with mesoporous silica films with distinct pore diameters (*i.e.*, 2.0 and 2.9 nm) for different electropolymerization times selected from Fig. 4A, corresponding to PANI filaments of variable length ($60 < t_{\text{electropolymerization}} < 120$ s) and one sample with additional PANI deposits onto the silica membrane (*i.e.*, $t_{\text{electropolymerization}} = 140$–$160$ s). The CV curves were recorded before and after removal of the silica template. The left part (Fig. 5A) corresponds to data obtained for the CTAB-based film (∅ pores = 2.0 nm) while the right part (Fig. 5B) gathers those results originating from the ODAB-based film (∅ pores = 2.9 nm).

General tendencies can be identified from these experiments. First, the CV signals corresponding to the leucoemeraldine/emeraldine redox processes visible in the investigated potential window (−0.1 to 0.5 V) were always more intense for samples prepared at longer electropolymerization times, evidently due to larger amounts of grown/deposited PANI material, and this trend was observed for both PANI/silica composites and silica-free PANI nanostructures. Peak currents were always located at the same potential values independently of the potential scan rate (yet with few exceptions at the very low 10–20 mV s^{-1} values), with anodic peaks at *ca.* 0.22 V for both material types, whereas cathodic peaks shifted from 0.13 V to 0.08 V when PANI wires were free of the silica template. Secondly, the peak current intensities were always lower for PANI wires in the silica membrane than for the template-free samples (compare curves (a) "before" and (b) "after" silica removal in Fig. 5). This can be rationalized in terms of larger surfaces of PANI exposed to the electrolyte solution after template etching. This effect was more marked for the silica films with narrower mesopores in which restrictions to transport phenomena are expected to be more important. It was however less marked at longer deposition times (*i.e.*, when PANI was formed on top of the mesoporous film), especially for the larger pore template (see bottom of Fig. 5), which can be explained by the whole electrode surface covered with PANI exposed to the electrolyte solution. Thirdly, the variations of peak currents, I_p, with the potential scan rate, ν, were investigated for all systems in order to determine the type of electrochemical behaviour (thin layer-confined or diffusion-controlled reactions). From the plots "$\log(I_p)$ *vs.* $\log(\nu)$", one can derive slopes (see parts (c) in Fig. 5) which are expected to be equal to 0.5 for a purely diffusion-controlled process and to 1.0 for a purely thin layer response, and in-between for mixed behaviour. As one can see at first sight, all template-free samples (columns (b) and corresponding 'white-greenish square' data in columns (c) on Fig. 5) are largely/exclusively dominated by thin layer-confined reactions, whereas the PANI-in-silica films (columns (a) and corresponding 'grey-greenish square' data in columns (c) on Fig. 5) are characterized by some diffusion-controlled contributions, especially for the narrow mesopore diameter series and shorter electropolymerization times. To be more quantitative, one can consider that in the case of mixed behaviour the overall current is the sum of contributions from the thin layer (proportional to the scan rate) and from diffusion (proportional to the square root of the scan rate):[73]

$$i = k_1 \nu + k_2 \nu^{1/2} \tag{1}$$

where i is the current measured at a given potential (for example the peak potential), ν the scan rate, and k_1 and k_2 empirical parameters representing both contributions. Eqn (1) can be normalized *versus* the square root of the potential scan rate (eqn (2)) in order to extract the parameters k_1 and k_2 (respective percentages of thin layer and diffusion control) by plotting the variation of $i/\nu^{1/2}$ as a function of $\nu^{1/2}$.

$$\frac{i}{\nu^{1/2}} = k_1 \nu^{1/2} + k_2 \tag{2}$$

Accordingly, one always obtained thin layer contributions in the 99–100% range for all PANI nanofilaments after silica template removal, confirming a real interest of such ultra-small nanostructures for ensuring fast electron transfer processes, contrary to bulkier PANI deposits (even nanostructured ones) for which diffusion-controlled processes have been most often reported (as a result of rather slow counter-ions diffusion through the polymeric material).[71,74–76] For PANI wires in the silica membrane (*i.e.*, before etching), some contributions of diffusion-controlled reactions were evidenced (65%, 25%, 21% and 11%, respectively, for PANI deposits at 60 s, 100 s, 120 s and 160 s in CTAB-based films, and 11% for both PANI deposits at 60 s and 100 s in ODAB-based films (and less than 1% for those PANI deposits at 120 s and 140 s with ODAB films)), confirming the above qualitative observations. Yet the contributions of thin layer behaviour remain in the majority (except for the very short PANI nanofilaments prepared for 60 s in CTAB-based films), suggesting that even in such confined environments the electron transfer processes were quite fast and not significantly restricted by diffusion phenomena, consistent with the above estimation of PANI occupying roughly $\frac{1}{2}$ to $\frac{1}{3}$ of the mesopore section (thus keeping some space for easy mass transport of counter ions). The fast transport processes are also due to the unique vertical orientation of the nanochannels onto the electrode surface.[77,78]

3.3 Spectroelectrochemical characterization and electrochromic properties

The potential-dependent coloration/bleaching of PANI make it an attractive material for electrochromic applications.[79,80] However, the rigidity of PANI chains and the lack of access to aniline units in bulk polymer can hinder the transport of counter ions during doping/dedoping processes, undermining the electrochromic response. Among the strategies developed to overcome such limitations, one can cite the nanostructure of PANI[71,74–76] or the design of PANI-based hybrid nanocomposites.[81–84] We have thus evaluated the possible interest of the PANI nanofilaments developed here as electrochromic materials. Focusing on the leucoemeraldine/emeraldine salt redox system, one can expect its transformation from a nearly transparent material with a slightly yellowish tint in the reduced form, to a greenish colour in the oxidized form. Fig. 6 shows the absorbance spectra recorded for PANI electropolymerized through the mesoporous silica films for two selected polymerization times, leading either to PANI nanofilaments only, or to PANI nanofilaments and PANI deposits grown outside the films (see part A for CTAB-based films and part B for the ODAB-based ones), at applied potentials of −0.2 V and +0.3 V (corresponding respectively to the reduced and

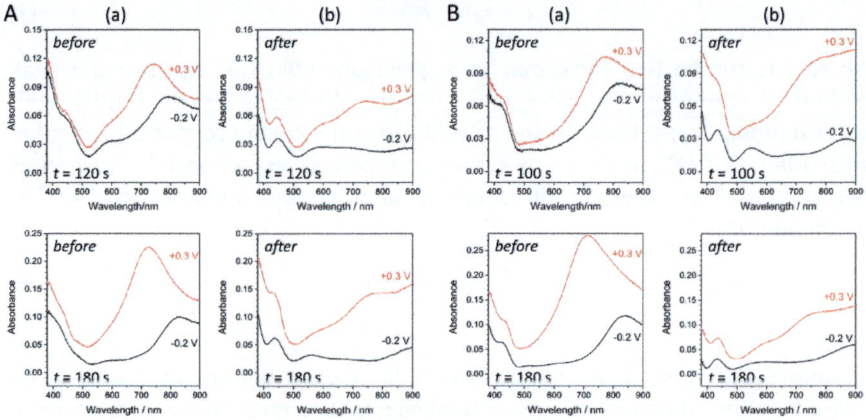

Fig. 6 UV-vis spectra of the oxidized (+0.3 V) or reduced (−0.2 V) forms of PANI nanowires generated through mesoporous silica films with pore diameters of (A) 2.0 nm and (B) 2.9 nm for various electropolymerization times (100, 120 or 180 s), (a) 'before' and (b) 'after' etching of the silica membrane.

oxidized forms of the materials). In addition to the π–π* transition of the aromatic rings (visible at $\lambda = 430$–440 nm independently of the oxidation state of the polymer), the spectra recorded for the oxidized samples revealed a main band in the $\lambda = 720$–750 nm region corresponding to the polaron state of the green emeraldine salt state of PANI (yet with some minor contribution at λ values around 560–580 nm arising from the less conductive emeraldine base).[85] The intensity of this main band was larger for PANI electrogenerated for longer times as a result of larger amounts of polymer on the electrode surface, but the main result was the almost complete vanishing of this band when reducing the PANI deposits at −0.2 V only for the template-free samples (*i.e.*, after removal of the silica membrane, comparing the red and black curves in parts (b) of Fig. 6), suggesting more efficient electrochromic behaviour in this case. The non-negligible absorbance band remaining after reduction of PANI occluded in the silica matrix (see parts (a) of Fig. 6) can be explained by less effective and slower redox processes, in agreement with the above CV data (Fig. 5); this cathodic bias-induced change in the spectra was characterized by a red shift of absorbance towards higher wavelengths, consistent with previous reports.[54,81–83] Similar observations were made for both CTAB- and ODAB-based films. This constitutes a significant improvement with respect to bulk PANI layers, for which the extent of bleaching was strongly limited by the existence of cross-linked and intertwined polymeric chains, expected to induce conductivity loss and incomplete electrochemical reduction, contrary to the present PANI nanofilaments.[57]

Fig. 7A shows typical *in situ* absorbance responses of (a) PANI/ITO and (b and c) PANI nanowires in mesoporous silica films (CTAB-type) on ITO respectively before (b) and after template etching (c), as measured at $\lambda = 750$ nm using a spectroelectrochemical cell containing the ITO working electrode biased successively at −0.2 V (bleaching step) and 0.3 V (coloration step) for 20 s per step (200 cycles). The absorbance change was larger for the bare ITO electrode covered with a bulky PANI layer (in comparison to PANI nanofilaments) but it decreased upon successive measurements, probably due to some mechanical degradation of the

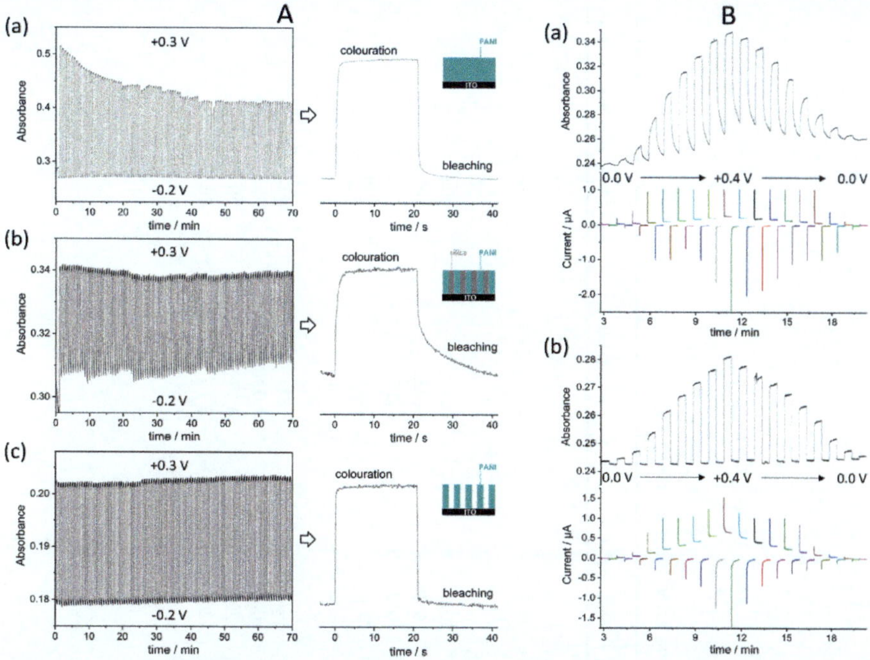

Fig. 7 (A) Cycling stability at 750 nm (200 cycles at switching potentials of +0.3 V and −0.2 V; 20 s each) of PANI generated under 120 s electropolymerization on (a) bare ITO or (b and c) as nanofilaments grown through a mesoporous silica film (pore ∅ = 2.0 nm) on ITO electrode respectively (b) 'before' and (c) 'after' etching of the silica membrane (enlargement of one cycle on the right). (B) Effect of the applied potential on the spectroelectrochemical response of PANI nanowires, measured by switching potentials from −0.2 V to various anodic values (from 0.0 to +0.4 V, by steps of 50 mV), for samples obtained respectively (a) 'before' and (b) 'after' etching of the silica membrane.

material (one can see some losses of PANI from such thick deposits). The stability was much better for PANI nanofilaments in the silica membrane, even if one can notice incomplete bleaching as measurements progressed, whereas perfect cyclability (no noticeable variation in absorbance changes between the 1st and 200th cycles). Enlarging the time window around one cycle enabled us to evidence the dramatic difference in response time depending on whether PANI nanofilaments were occluded in the silica membrane or template-free. For the example depicted in Fig. 7A (parts (b) and (c)), an improvement by *ca.* one order of magnitude for coloration, and almost two orders of magnitude for bleaching, was observed when passing from the PANI/silica nanocomposite to the template-free PANI nanofilaments, and this was the case for all samples investigated here, independently of the diameters of the nanochannel membranes used to prepare the PANI wires (Table 1). This can also be observed when recording the absorbance responses continuously for increased and decreased bias (Fig. 7B), showing an excellent and reversible dynamic for template-free PANI nanofilaments (b), in comparison to the PANI/silica nanocomposite film (a), for which bleaching was incomplete in the time scale of the experiment. Faster response times with template-free PANI nanofilaments is explained by faster redox switching between oxidized and reduced forms of PANI, in agreement with the above CV

Table 1 Electrochromic characteristics of typical PANI materials investigated in this work (same conditions as in Fig. 7A)

Samples		$t_{colouration}{}^{a}$/s	$t_{bleaching}{}^{a}$/s	$\Delta_{absorbance}$	$\%_{response}$ after 200 cycles
PANI on bare ITO					
$t_{electropolymerization} = 100$ s		1.1	1.2	0.25	56
PANI nanofilaments grown through CTAB-type films					
$t_{electropolymerization} = 120$ s	Before template removal	2.1	16.1	0.14	93
	After template removal	0.12	0.10	0.022	100
$t_{electropolymerization} = 180$ s	Before template removal	0.4	0.9	0.15	96
	After template removal	0.5	1.0	0.11	90
PANI nanofilaments grown through ODAB-type films					
$t_{electropolymerization} = 100$ s	Before template removal	1.8	14.4	0.06	84
	After template removal	0.11	0.10	0.038	99
$t_{electropolymerization} = 180$ s	Before template removal	0.6	2.6	0.17	98
	After template removal	0.3	2.0	0.10	86

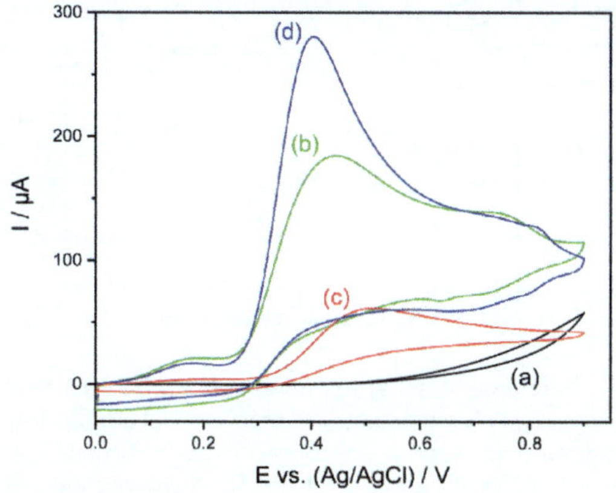

Fig. 8 Cyclic voltammograms recorded in 1 mM ascorbic acid (in 0.1 M H_2SO_4) at a potential scan rate of 50 mV s^{-1}, using (a) a bare ITO electrode, (b) ITO covered with a bulky PANI film or (c and d) ITO modified with PANI nanowires generated through a mesoporous silica film with pore diameters of 2.0 nm for 120 s electropolymerization time, respectively, (c) 'before' and (d) 'after' etching of the silica membrane. (b–d) PANI deposits correspond to an electropolymerization charge of 2.57 mC.

observations. The shortest response times obtained for the most efficient systems (*i.e.*, 0.12 s for coloration and 0.10 s for bleaching with PANI nanofilaments electropolymerized for 120 s using the CTAB-based template; and 0.11 s for coloration and 0.10 s for bleaching with PANI prepared for 100 s from the ODAB-based template) are among the best in comparison to other PANI nanostructures.[71,74–76] Note that lengthening the electropolymerization time for PANI growth (*i.e.*, up to reaching PANI deposits not only in the nanochannels but also on top of the silica membrane) resulted in response times of the same order of magnitude as for a bulk PANI layer on bare ITO and larger variations of absorbance, but it was at the expense of some lack of long-term stability, especially after removal of the silica template (Table 1). A compromise thus has to be found, ensuring an amount of PANI deposits large enough to get significant coloration efficiency with fast response time after template removal, but not too large to avoid problems related to possible degradation over prolonged use which could contribute to the lack of long-term stability.

3.4 Towards other applications?

Just to give an idea on what could be achieved next with such a selection of PANI nanofilaments, and considering the interest of PANI nanowire arrays for applications in the fields of electrocatalysis (*i.e.*, for sensors[86–89]) or energy storage (*e.g.*, as supercapacitors[88–91]), we provide hereafter some preliminary results in support of these possible applications.

Fig. 8 shows the response of various electrodes to 1 mM ascorbic acid. On bare ITO electrode (curve a), no peak can be observed in the investigated potential range. When using PANI modified ITO electrodes, well-defined anodic signals were obtained, thanks to the catalytic properties of PANI for the oxidation of ascorbic acid.[92] However, the efficiency of the electrocatalytic process was strongly dependent on the electrode type. For PANI nanowires in mesoporous silica films

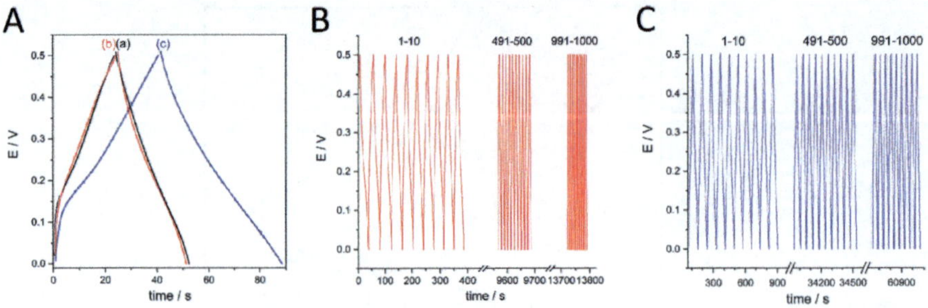

Fig. 9 (A) Galvanostatic charge–discharge curves recorded in 0.5 M H_2SO_4 at a current density of 25 μA cm^{-2} using ITO electrodes covered with (a) a bulky PANI film or (b and c) PANI nanowires generated through a mesoporous silica film with pore diameters of 2.0 nm ((b) 'before' and (c) 'after' etching of the silica membrane). (B and C) Charge–discharge profiles (10 first, 491–500th and 991–1000th cycles) for PANI nanowires prepared as in (A) and recorded in the same conditions, respectively (B) 'before' and (C) 'after' etching of the silica membrane. All PANI deposits correspond to an electropolymerization charge of 2.51 mC.

(CTAB-type) on ITO before etching of the silica membrane (curve c), one observed a peak current of 61 μA located at +0.50 V, whereas the free nanofilaments gave rise to a much better electrocatalytic response with a peak current of 280 μA located at +0.40 V (curve d). This is consistent with the more effective charge transport reported above for template-free PANI as a result of larger exposition to the electrolyte solution. The behaviour of an ITO electrode covered with a bulky PANI layer (prepared by electropolymerization on bare ITO with similar charge as for PANI nanofilaments) resulted in intermediate results (peak of 184 μA at +0.44 V, see curve b), which can be due to less effective doping/dedoping processes in bulky PANI deposits than for PANI nanoarrays.[93] This highlights the attractiveness of 1D PANI nanofilament arrays for possible application in electrochemical sensing.

Fig. 9 illustrates another possible application of PANI nanofilaments, *i.e.*, as pseudocapacitive materials, showing typical triangle-shaped galvanostatic charge–discharge curves (Fig. 9A), with good cycling behaviour (over 1000 cycles) for both PANI nanofilaments in the silica membrane and template-free wires (Fig. 9B and C). The area specific capacitance values evaluated from the discharge curves were 1.30 and 2.35 mF cm^{-2}, respectively, before and after dissolution of the silica membrane, and 1.35 mF cm^{-2} for a bulk PANI film of the same amount of electropolymerized aniline units. For template-free PANI nanofilaments, the electrode surface is covered only with a small amount of the polymer (*i.e.*, 0.55 μg) and, if expressing the specific capacitance per mass of the material, one reached 427 F g^{-1} (at 9 A g^{-1}), which is comparable to PANI–graphene composites.[91] This material retained 64% of its initial specific capacitance after 500 cycles and 47% after 1000 cycles (Fig. 9C), whereas PANI nanowires in the mesoporous film before silica etching (Fig. 9B) were less stable (only 19% and 8% capacitance retention after 500 and 1000 cycles, respectively). These results have been obtained without any optimization and, in view of previous reports on redox-active mesoporous silica coatings prepared by EASA,[45,94] one can expect significant improvement when passing from a flat ITO support to electrode surfaces made of carbon nanomaterials, for instance.

4. Conclusions

In this work, we have shown that one can control the growth of PANI nanofilaments through vertically aligned mesopore channels of silica films of distinct pore diameter (2–3 nm) generated by EASA onto ITO electrodes. Their durable attachment to the electrode surface (even after removal of the silica template) is achieved by electrografting aminophenyl groups at the bottom of the nanochannels prior to PANI electropolymerization. The growth rate and the amount of deposited PANI were larger when using the silica membrane with larger pore diameters, resulting therefore in more intense CV responses. Both PANI nanofilaments in mesoporous silica and after etching of the membrane, were electrochemically accessible, but peak currents were significantly larger after template removal owing to the wider exposure of the polymer to the electrolyte solution. This also had an effect on the electrochromic properties of the materials, showing significantly improved response times for PANI nanofilaments free of the silica membrane, due to easily accessible large surface areas and short diffusion lengths. In the best cases, spectroelectrochemical data corresponding to the variation of optical contrasts recorded at 750 nm for successive interconversion of the emeraldine/leucoemeraldine system, revealed ultrafast response speeds (0.1 s for coloration and bleaching to reach 95% of the maximum absorbance variation) and excellent cycling stability (100% of the optical contrast maintained after 200 cycles between −0.2 and 0.3 V in 0.5 M H_2SO_4). The PANI nanofilaments are also promising for applications in other fields such as electrocatalysis and sensors or energy storage as supercapacitors.

Author contributions

WU: investigation, methodology, formal analysis, visualization, writing (original draft); GH: conceptualization, methodology, validation, supervision, writing (review and editing); NV: conceptualization, methodology, investigation, validation, supervision, writing (review and editing); AW: conceptualization, funding acquisition, validation, supervision, visualization, writing (original draft, review and editing).

Conflicts of interest

There are no conflicts to declare.

Acknowledgements

The authors are grateful to Dr Jaafar Ghanbaja (IJL – CNRS, Université de Lorraine) for TEM micrographs and to Mr Lionel Richaudeau (L2CM – CNRS, Université de Lorraine) for GISAXS measurements. WU acknowledges a PhD grant from the Université de Lorraine. This work was supported partly by the French PIA project "Lorraine Université d'Excellence" (grant no. ANR-15-IDEX-04-LUE).

Notes and references

1 A. Walcarius, *Chem. Soc. Rev.*, 2013, **42**, 4098.
2 A. Walcarius, *Electroanalysis*, 2015, **27**, 1303.
3 M. Etienne, L. Zhang, N. Vila and A. Walcarius, *Electroanalysis*, 2015, **27**, 2028.
4 X. Yang, P. Qiu, J. Yang, Y. Fan, L. Wang, W. Jiang, X. Cheng, Y. Deng and W. Luo, *Small*, 2021, **17**, 1904022.
5 L. Zu, W. Zhang, L. Qu, L. Liu, W. Li, A. Yu and D. Zhao, *Adv. Energy Mater.*, 2020, **10**, 2002152.
6 C. Li, Q. Li, Y. V. Kaneti, D. Hou, Y. Yamauchi and Y. Mai, *Chem. Soc. Rev.*, 2020, **49**, 4681.
7 I. E. Rauda, V. Augustyn, B. Dunn and S. H. Tolbert, *Acc. Chem. Res.*, 2013, **46**, 1113.
8 J. C. Ndamanisha and L.-P. Guo, *Anal. Chim. Acta*, 2012, **747**, 19.
9 A. Walcarius, *Trends Anal. Chem.*, 2012, **38**, 79.
10 X. Bo and M. Zhou, in *Advanced Electrode Materials*, ed. A. Tiwari, F. Kuralay and L. Uzun, Scrivener Publishing, 2017, ch. 6, pp. 213–241.
11 A. Walcarius, *Sensors*, 2017, **17**, 1863.
12 M. Hasanzadeh, N. Shadjou, M. de la Guardia, M. Eskandani and P. Sheikhzadeh, *Trends Anal. Chem.*, 2012, **33**, 117.
13 H. Rao, X. Wang, X. Du and Z. Xue, *Anal. Lett.*, 2013, **46**, 2789.
14 A. Walcarius, *Curr. Opin. Electrochem.*, 2018, **10**, 88.
15 F. Hoffmann, M. Cornelius, J. Morell and M. Froeba, *Angew. Chem., Int. Ed.*, 2006, **45**, 3216.
16 Y. Chen and J. Shi, *Adv. Mater.*, 2016, **28**, 3235.
17 V. B. Cashin, D. S. Eldridge, A. Yu and D. Zhao, *Environ. Sci.: Water Res. Technol.*, 2018, **4**, 110.
18 D. Feng, J. Wei, M. Wang, Q. Yue, Y. Deng, A. M. Asiri and D. Zhao, *Adv. Porous Mater.*, 2013, **1**, 164.
19 G. J. A. A. Soler-Illia, P. C. Angelome, M. C. Fuertes, A. Calvo, A. Wolosiuk, A. Zelcer, M. G. Bellino and E. D. Martinez, *J. Sol-Gel Sci. Technol.*, 2011, **57**, 299.
20 C. Sanchez, C. Boissiere, D. Grosso, C. Laberty and L. Nicole, *Chem. Mater.*, 2008, **20**, 682.
21 K. C.-W. Wu, X. Jiang and Y. Yamauchi, *J. Mater. Chem.*, 2011, **21**, 8934.
22 V. Urbanova and A. Walcarius, *Z. Anorg. Allg. Chem.*, 2014, **640**, 537.
23 C. Mousty and A. Walcarius, *J. Solid State Electrochem.*, 2015, **19**, 1905.
24 F. Yan, X. Lin and B. Su, *Analyst*, 2016, **141**, 3482.
25 P. Zhou, L. Yao, K. Chen and B. Su, *Crit. Rev. Anal. Chem.*, 2020, **50**, 424.
26 A. Walcarius, E. Sibottier, M. Etienne and J. Ghanbaja, *Nat. Mater.*, 2007, **6**, 602.
27 A. Goux, M. Etienne, E. Aubert, C. Lecomte, J. Ghanbaja and A. Walcarius, *Chem. Mater.*, 2009, **21**, 731.
28 Z. Teng, G. Zheng, Y. Dou, W. Li, C.-Y. Mou, X. Zhang, A. M. Asiri and D. Zhao, *Angew. Chem., Int. Ed.*, 2012, **51**, 2173.
29 K.-C. Kao, C.-H. Lin, T.-Y. Chen, Y.-H. Liu and C.-Y. Mou, *J. Am. Chem. Soc.*, 2015, **137**, 3779.

30 M. Dutreilh-Colas, M. Yan, P. Labrot, N. Delorme, A. Gibaud and J.-F. Bardeau, *Surf. Sci.*, 2008, **602**, 829.
31 Y. Yamauchi, M. Sawada, M. Komatsu, A. Sugiyama, T. Osaka, N. Hirota, Y. Sakka and K. Kuroda, *Chem.–Asian J.*, 2007, **2**, 1505.
32 E. K. Richman, T. Brezesinski and S. H. Tolbert, *Nat. Mater.*, 2008, **7**, 712.
33 C. Robertson, R. Beanland, S. A. Boden, A. L. Hector, R. J. Kashtiban, J. Sloan, D. C. Smith and A. Walcarius, *Phys. Chem. Chem. Phys.*, 2015, **17**, 4763.
34 N. Vilà, J. Ghanbaja, E. Aubert and A. Walcarius, *Angew. Chem., Int. Ed.*, 2014, **53**, 2945.
35 N. Vilà, E. André, R. Ciganda, J. Ruiz, D. Astruc and A. Walcarius, *Chem. Mater.*, 2016, **28**, 2511.
36 W. Li, L. Ding, Q. Wang and B. Su, *Analyst*, 2014, **139**, 3926.
37 N. Vilà, P. de Oliveira, A. Walcarius and I. M. Mbomekallé, *Electrochim. Acta*, 2019, **309**, 209.
38 L. Sun, L. Zhou, F. Yan and B. Su, *Langmuir*, 2019, **35**, 14486.
39 M. Saadaoui, I. Fernandez, G. Luna, P. Diez, S. Campuzano, N. Raouafi, A. Sanchez, J. M. Pingarron and R. Villalonga, *Anal. Bioanal. Chem.*, 2016, **408**, 7321.
40 X. Lin, B. Zhang, Q. Yang, F. Yan, X. Hua and B. Su, *Anal. Chem.*, 2016, **88**, 7821.
41 X. Lin, Q. Yang, F. Yan, B. Zhang and B. Su, *ACS Appl. Mater. Interfaces*, 2016, **8**, 33343.
42 N. Vilà and A. Walcarius, *Electrochim. Acta*, 2015, **179**, 304.
43 G. Giordano, N. Vilà, E. Aubert, J. Ghanbaja and A. Walcarius, *Electrochim. Acta*, 2017, **237**, 227.
44 S. Ahoulou, N. Vilà, S. Pillet, D. Schaniel and A. Walcarius, *Chem. Mater.*, 2019, **31**, 5796.
45 J. Wang, N. Vilà and A. Walcarius, *ACS Appl. Mater. Interfaces*, 2020, **12**, 24262.
46 S. Saint-André, F. Albanese, G. J. A. A. Soler-Illia and M. Tagliazucchi, *Phys. Chem. Chem. Phys.*, 2019, **21**, 2743.
47 A. Goux, J. Ghanbaja and A. Walcarius, *J. Mater. Sci.*, 2009, **44**, 6601.
48 L. Lu, L. Zhou, J. Chen, F. Yan, J. Liu, X. Dong, F. Xi and P. Chen, *ACS Nano*, 2018, **12**, 12673.
49 P. N. Bartlett, R. Beanland, J. Burt, M. M. Hasan, A. L. Hector, R. J. Kashtiban, W. Levason, A. W. Lodge, S. Marks, J. Naik, A. Rind, G. Reid, P. W. Richardson, J. Sloan and D. C. Smith, *Nano Lett.*, 2018, **18**, 941.
50 M. A. del Valle, M. Gacitúa, F. R. Díaz, F. Armijo and R. del Río, *Electrochem. Commun.*, 2009, **11**, 2117.
51 M. A. del Valle, A. Ramos, M. Antilen, L. Hernandez, G. Arteaga, F. Diaz and G. Louarn, *Electrochemistry*, 2014, **82**, 146.
52 S. M. Fonseca, T. Moreira, A. J. Parola, C. Pinheiro and C. A. T. Laia, *Sol. Energy Mater. Sol. Cells*, 2017, **159**, 94.
53 H. Kang, H. Lee and J. Kwak, *J. Korean Electrochem. Soc.*, 2011, **14**, 22.
54 L. Ding, W. Li, Q. Wang, Q. Sun, Y. He and B. Su, *Chem.–Eur. J.*, 2014, **20**, 1829.
55 M. A. del Valle, L. A. Hernández, A. M. Ramírez and F. R. Díaz, *Ionics*, 2016, **23**, 191.
56 A. M. R. Ramírez, M. Gacitúa, E. Ortega, F. R. Díaz and M. A. del Valle, *Electrochem. Commun.*, 2019, **102**, 94.
57 A. Gamero-Quijano, C. Karman, N. Vilà, G. Herzog and A. Walcarius, *Langmuir*, 2017, **33**, 4224.

58 M. A. del Valle, L. A. Hernández, F. R. Díaz and A. Ramos, *Int. J. Electrochem. Sci.*, 2015, **10**, 5152.
59 L. A. Hernández, M. A. del Valle, F. R. Díaz, D. J. Fermín and T. A. G. Risbridger, *Electrochim. Acta*, 2015, **166**, 163.
60 W. Ullah, G. Herzog, N. Vilà and A. Walcarius, *Electrochem. Commun.*, 2021, **122**, 106896.
61 N.-R. Chiou, C. Lu, J. Guan, L. J. Lee and A. J. Epstein, *Nat. Nanotechnol.*, 2007, **2**, 354.
62 W. Zhong, J. Deng, Y. Yang and W. Yang, *Macromol. Rapid Commun.*, 2005, **26**, 395.
63 L. Jiang and Z. Cui, *Polym. Bull.*, 2006, **56**, 529.
64 J. Lyskawa and D. Bélanger, *Chem. Mater.*, 2006, **18**, 4755.
65 A. Subrata, A. Veksha, Z. Y. Pong, G. Lisak and R. D. Webster, *ChemElectroChem*, 2020, **7**, 3368.
66 A. Vacca, M. Mascia, S. Rizzardini, S. Palmas and L. Mais, *Electrochim. Acta*, 2014, **126**, 81.
67 K. Darowicki and J. Kawula, *Electrochim. Acta*, 2004, **49**, 4829.
68 J. Stejskal and R. G. Gilbert, *Pure Appl. Chem.*, 2002, **74**, 857.
69 A. Yazdanpanah, A. Ramedani, A. Abrishamkar, P. B. Milan, Z. S. Moghadan, N. P. S. Chauhan, F. Sefat and M. Mozafari, in *Fundamentals and Emerging Applications of Polyaniline*, ed. M. Mozafari and N. P. S. Chauhan, Elsevier, 2019, ch. 6, pp. 105–119.
70 J. Cheng, S. J. Rathi, P. Stradins, G. L. Frey, R. T. Collins and S. K. R. Williams, *RSC Adv.*, 2014, **4**, 7627.
71 D. Ge, L. Yang, Z. Tong, Y. Ding, W. Xin, J. Zhao and Y. Li, *Electrochim. Acta*, 2013, **104**, 191.
72 J. W. Schultze and A. Thyssen, *Synth. Met.*, 1991, **43**, 2825.
73 P. Yu, C. Li and X. Guo, *J. Phys. Chem. C*, 2014, **118**, 10616.
74 S. Zhang, J. Ren, S. Chen, Y. Luo, X. Bai, L. Ye, F. Yang and Y. Cao, *J. Electroanal. Chem.*, 2020, **870**, 114248.
75 M. Jamdegni and A. Kaur, *Thin Solid Films*, 2020, **714**, 138373.
76 X. Huang, Q. Niu, S. Fan and Y. Zhang, *Chem. Eng. J.*, 2021, **417**, 128126.
77 M. Etienne, Y. Guillemin, D. Grosso and A. Walcarius, *Anal. Bioanal. Chem.*, 2013, **405**, 1497.
78 T. Nasir, G. Herzog, L. Liu, M. Hébrant, C. Despas and A. Walcarius, *ACS Sensors*, 2018, **3**, 484.
79 A. Watanabe, K. Mori, Y. Iwasaki, Y. Nakamura and S. Niizuma, *Macromolecules*, 1987, **20**, 1793.
80 S. Zhang, S. Chen, F. Hu, R. Xu, B. Yan, M. Jiang, Y. Gu, F. Yang and Y. Cao, *Sol. Energy Mater. Sol. Cells*, 2019, **200**, 109951.
81 R. C. Silva, M. V. Sarmento, F. A. R. Nogueira, J. Tonholo, R. J. Mortimer, R. Faez and A. S. Ribeiro, *RSC Adv.*, 2014, **4**, 14948.
82 S. Zhang, G. Sun, Y. He, R. Fu, Y. Gu and S. Chen, *ACS Appl. Mater. Interfaces*, 2017, **9**, 16426.
83 S. Mishra, S. Lambora, P. Yogi, P. R. Sagdeo and R. Kumar, *ACS Appl. Nano Mater.*, 2018, **1**, 3715.
84 Y. Wang, S. Xiong, X. Wang, J. Chu, R. Zhang, M. Gong, B. Wu, M. Qu, Z. Li and Z. Chen, *Polym. J.*, 2020, **52**, 783.
85 A. Habib and S. P. Maheshwari, *J. Electrochem. Soc.*, 1989, **136**, 1050.

86 L. Wang, Y. Ye, Y. Shen, F. Wang, X. Lu, Y. Xie, S. Chen, H. Tan, F. Xu and Y. Song, *Sens. Actuators, B*, 2014, **203**, 864.
87 L. Liu, J. Guo and L. Ding, *Electroanalysis*, 2021, **33**, 2048.
88 N. Hui, F. Chai, P. Lin, Z. Song, X. Sun, Y. Li, S. Niu and X. Luo, *Electrochim. Acta*, 2016, **199**, 234.
89 N. Hui, J. Wang, A. Liang and M. Jiang, *Electroanalysis*, 2016, **28**, 279.
90 K. Wang, H. Wu, Y. Meng and Z. Wei, *Small*, 2014, **10**, 14.
91 X. Gao, H. Zhang, H. Yue, F. Yao, X. Zhang, E. Guo, Y. Ma, Z. Wang and Y. Wang, *ChemistrySelect*, 2020, **5**, 11004.
92 C. Sanchis, M. A. Ghanem, H. J. Salavagione, E. Morallón and P. N. Bartlett, *Bioelectrochem*, 2011, **80**, 105.
93 Y. Zhang, W. Chu, Q. Zhou, S. Li, N. Li and J. Zheng, *J. Electroanal. Chem.*, 2016, **775**, 105.
94 J. Wang, N. Vilà and A. Walcarius, *Electrochim. Acta*, 2021, **366**, 137407.

Faraday Discussions

PAPER

In situ surface enhanced Raman spectroscopy study of electrode–polyelectrolyte interfaces

Guangzhe Wang, Yingming Wang, Gongwei Wang, Li Xiao * and Lin Zhuang

Received 13th August 2021, Accepted 5th November 2021
DOI: 10.1039/d1fd00051a

As polyelectrolytes play a more and more important role in electrochemical fields, further understanding of the electrode–polyelectrolyte interface is in high demand. Surface-enhanced Raman spectroscopy (SERS) is utilized widely in electrode–solution interface research due to its ultra-high sensitivity, but is still rarely in the study of the electrode–polyelectrolyte interface due to difficulties in constructing appropriate electrochemical *in situ* devices. Additionally, the reported electrochemical *in situ* Raman works on the electrode–polyelectrolyte interface have a common problem of the coexistence of electrode–solution interfaces and electrode–polyelectrolyte interfaces. Here, we used screen printing electrodes (SPE) with a compact planar three-electrode structure to carry out a new electrochemical *in situ* SERS test method, which was suitable for the study of the electrode–polyelectrolyte interface. Polyelectrolyte membranes can be conveniently and closely coated on the SPE's planar three electrodes to achieve isolated electrode–polyelectrolyte interfaces without electrode–solution interfaces coexisting. Strongly potential-dependent signals were obtained from the Pt–Nafion™ interface directly across the Nafion™ membrane, which verifies that this method is practical for the electrochemical *in situ* SERS study of the electrode–polyelectrolyte interface.

1. Introduction

The electrode–electrolyte interface has always been a focus of electrochemical research, and researchers have been trying to investigate the electrochemical processes occurring in extremely thin electrical double layers (EDLs). As polyelectrolytes play an increasingly important role in various electrochemical fields, such as polymer electrolyte fuel cells (PEFCs), solid polymer electrolyte water electrolyzers and CO_2 reduction electrolyzers, it is necessary to study electrode–polyelectrolyte interfaces to improve their development. In fact, the electrode–

Sauvage Center for Molecular Sciences, College of Chemistry and Molecular Sciences, Hubei Key Lab of Electrochemical Power Sources, Wuhan University, Wuhan, 430072, China. E-mail: chem.lily@whu.edu.cn

polyelectrolyte interface can behave quite differently from the traditional electrode–solution interface, but the essential causes are still not deeply understood. Taking our previous work,[1] for example, the oxygen reduction reaction (ORR) catalytic performance of Pt/C in aqueous KOH solution is much higher than that of $Mn_xCo_yO_z$. However, under alkaline polymer electrolyte fuel cell (APEFC) conditions (60 °C, 100% RH), the cell performance with $Mn_xCo_yO_z$ as the cathode conversely exceeded that with Pt/C as the cathode. This indicates that the activity of the catalyst in a polyelectrolyte system can be tremendously different from, or even opposite to, that in a traditional solution system, which calls for further study on the structure of, and what happens at, the electrode–polyelectrolyte interface.

Generally, SERS[2–5] has been used widely in electrode–solution interface studies but rarely in electrode–polyelectrolyte interface studies. For the latter, researchers usually use commercial spectroelectrochemical cells,[6] in which the electrode–polyelectrolyte interfaces are constructed by immersing electrodes in diluted polyelectrolyte suspensions containing additional electrolytes, such as $HClO_4$ [7] (Fig. 1A). One issue is that this essentially measures a mixture of electrode–solution interfaces with dissolved and discrete polyelectrolyte molecules interacting with the electrode surfaces, rather than the pure electrode–polyelectrolyte interfaces, which is far from the situation in polyelectrolyte-applied devices like polymer electrolyte membrane fuel cells (PEMFCs), where the polyelectrolyte forms a solid membrane phase to contact with the catalysts (Fig. 1B). Although the use of membrane electrode assemblies (MEAs),[8] on the other hand, enables conducting Raman measurements at the working conditions of PEMFCs without additional electrolyte solution, Raman signals from electrode–water interfaces may also be obtained besides those from the electrode–polyelectrolyte interfaces. These mixed interferences are generated by the incomplete encapsulation of catalyst particles by the polymer electrolyte[8] (Fig. 1B), therefore[9] (1) species generated at electrode–polyelectrolyte interfaces can spillover to electrode–water interfaces and (2) some electrochemical processes are likely to occur at catalyst-pure water interfaces due to the enhancement of the ionic conductivity of pure water around nanoscale or microscale catalyst particles closely contacting with each other, where the enhancement can be considered to be led by the EDLs of the

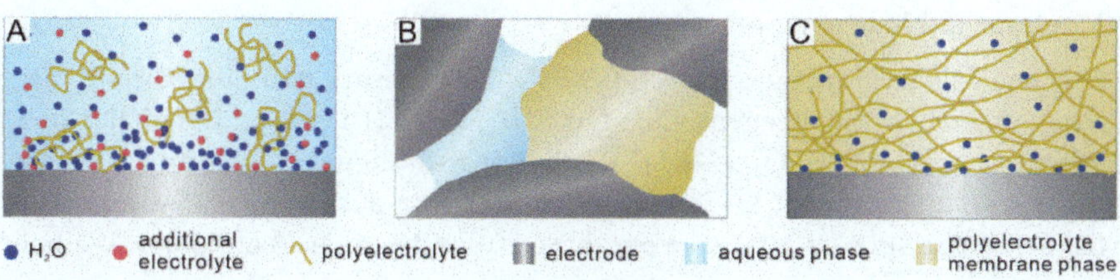

Fig. 1 Schematics of the electrode–electrolyte interfaces in different test methods. (A) An electrode is immersed in a diluted polyelectrolyte suspension with an additional electrolyte. (B) A simplified microscopic structure of a MEA. The solid polyelectrolyte and liquid water are dispersed between the electrode (catalyst) particles. The liquid water can be imported by the capillary condensation of introduced highly humidified gases besides directly introduced water. (C) An electrode is completely in close contact with a solid polyelectrolyte membrane to construct an isolated electrode–polyelectrolyte interface.

overlapped catalyst particles. Given the high sensitivity of SERS, these possible mixed interferences cannot be ignored.

What exactly happens at the isolated electrode–polyelectrolyte interface (Fig. 1C)? It is bound to be thoroughly studied by researchers trying to investigate in detail its specific EDL structure, which is different from that of the electrode–solution interface. However, as far as we know, no electrochemical *in situ* Raman device or method has been invented for such a isolated interface. We conclude that there are two main barriers to designing such a method. On the one hand, the bulk Raman signal of a solid polyelectrolyte with all kinds of organic functional groups is much more complicated than that of an aqueous solution. On the other hand, operational difficulty exists in that the surfaces of a working electrode (usually a disk electrode), a counter electrode (usually Pt wire) and a reference electrode are supposed to be in close contact with a thin and fragile sheet of polyelectrolyte membrane simultaneously. A very thin membrane is required to decrease the bulk signal, but will meanwhile make the operation process more difficult.

To overcome the barriers, herein we combine a screen printing electrode (SPE) with the SERS technique.[10] A SPE is selected to replace the traditional electrodes *e.g.* the disk electrode and Pt wire, in order to establish a compact planar three-electrode structure, for which it is commonly used in the quick and convenient electrochemical analysis of solution systems. On a SPE, a polyelectrolyte membrane can be simply cast by drop coating so that the entire three electrodes are naturally in close contact with the membrane to construct the electrode–polyelectrolyte interfaces (Fig. 1C). Its thickness can be flexibly and conveniently adjusted by changing the dosage and concentration of the drop coating polyelectrolyte solution. Besides decreasing the thickness of the membrane, enhancing the interface signal with the SERS technique is another approach to overcome the bulk interference. The planar system can serve as a simplified model to study all kinds of electrode–polyelectrolyte interfaces, with its simple structure and convenience for loading various electrochemical catalysts used in fuel cells. Here, we choose to investigate Pt–H_2SO_4(aq) and Pt–Nafion™ interfaces with Au NPs as the SERS-active nanoparticles to assess the feasibility of this method.

2. Experimental section

Chemicals and instruments

The chemicals used include $HAuCl_4$ (99.99%, Alfar Aesar), trisodium citrate (GR, Sinopharm Chemical Reagent Co., Ltd.), $H_2PtCl_6 \cdot 6H_2O$ (AR, Shanghai No. 1 Reagent Factory), $Na_2HPO_4 \cdot 12H_2O$ (AR, Sinopharm Chemical Reagent Co., Ltd.), Nafion™ solution (D1010, 10 wt%, DuPont) and ultra-pure water (UP water, 18.25 $M\Omega\ cm^{-1}$). The SPEs (Au101) were purchased from the Botan Technology (Weihai, China) Co., Ltd.

For the *in situ* SERS experiments, the electrochemical tests were carried out on a computer-controlled CHI600A electrochemical workstation (CH Instruments), and the Raman spectra were obtained with a XPLORA PLUS confocal microprobe Raman system (HORIBA), which utilizes a backscattering configuration to collect Raman signals through a microscope with a 50× objective vertically and excitation wavelength of 638 nm provided by an air-cooled He–Ne laser. Other electrochemical tests were performed on a computer-controlled CHI634a electrochemical workstation (CH Instruments). The reversible hydrogen electrodes (RHE) as the RE were all homemade.

Preparation of the Au NPs

We utilized Frens' sodium citrate reduction method[11] to synthesize 55 nm Au NPs: 100 mL of 0.01 wt% $HAuCl_4$ solution was added to a 150 mL round bottom flask and heated to boiling and refluxing, then 1.20 mL of 1 wt% trisodium citrate solution was quickly added to it, followed by 30 min of boiling and refluxing. Finally, it was cooled with a water bath to room temperature. The Au NP solution was centrifuged at 2000*g* and rinsed with UP water, then centrifuged again at 2000*g* to be concentrated. The concentrated solution was used to prepare the SERS-active substrates.

Preparation of the SPE for testing

The pristine SPE consisted of a working electrode (WE) and a counter electrode (CE) of Au and a reference electrode (RE) of Ag/AgCl (Fig. 2). Firstly, we electroplated Pt on the Au WE and CE in an electroplating solution of 4 mM H_2PtCl_6 and 0.7 M Na_2HPO_4 by chronoamperometry. For the Au WE, a cathodic current density of 0.4 mA cm^{-2} for 300 s and a cathodic current density of 0.29 mA cm^{-2} for 180 s were successively applied, and for the Au CE a cathodic current density of 0.4 mA cm^{-2} for 1200 s was applied. Thus, we obtained Pt-plated-on-Au SPEs (denoted as Au@Pt-SPE) (Fig. 2). Note that the deposited Pt layer is thick enough to shield the Au substrate from the Raman incident light, so that no Raman signal from the Au substrate should be obtained. As the enhancement efficiency of the Au NPs decreases exponentially with the distance from their surface,[12] that they should be located right at the Pt surface we want to analyze. Thus, 2 μL of the Au NPs solution was dropped on the WE of the Au@Pt-SPE and it was vacuum dried to get a SERS-active Au@Pt-SPE. Based on that, 50 μL of 10 wt% Nafion™ solution was dropped and spread on all three electrodes of the SERS-active Au@Pt-SPE and it was vacuum dried to get a SERS-active Nafion™-membrane-coated Au@Pt-SPE (denoted as Au@Pt@Nafion™-SPE) (Fig. 2). Note that the Nafion™ membrane was ~30 μm thick (as determined by SEM).

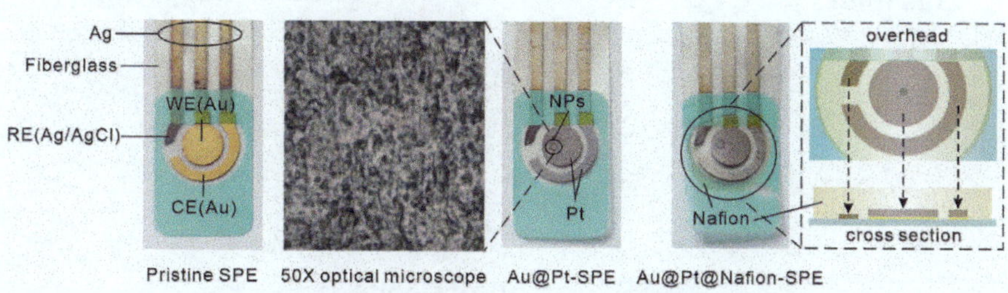

Fig. 2 Images of the SPEs at different preparation steps with the materials labeled and the 50× optical microscopy image of the WE surface with Au NPs on it. The schematic overhead and cross section diagrams of the Au@Pt@Nafion™-SPE are also included, showing that the three electrodes are completely covered with the membrane. The green area is inert insulation paint.

Electrochemical *in situ* SERS tests

The device for the electrochemical *in situ* SERS tests is shown in Fig. 3. The above prepared SPE, with a WE on which the Au NPs had been already introduced, was fixed at the inner base of the chamber after being rinsed with UP water and dried with Ar stream. A round quartz wafer was fixed on the inner top of the chamber. Before the test, 10 μL of 0.5 M H_2SO_4 solution was dropped on the three electrodes of the Au@Pt-SPE, while 10 μL of UP water was dropped on the Nafion™ membrane of the Au@Pt@Nafion™-SPE. The drops were then covered with slices of a square quartz wafer. In addition, several drops of UP water were set around the SPE to maintain a high humidity in the chamber, in order to avoid the drying of the electrolyte solution or UP water under the square quartz wafer. All of the tests in this chamber were performed in Ar purging.

3. Results and discussion

The electrochemical characteristics of the SPEs

We adopted Pt, one of the most widely used and efficient electrochemical catalysts, as the model material to verify the feasibility of this method, because of the characteristic and distinguishable electrochemical adsorption/desorption of hydrogen/oxygen on the Pt surface. As shown in Fig. 4A, cyclic voltammetry (CV) curves were measured for the Au@Pt-SPE in 0.5 M H_2SO_4 and the Au@Pt@Nafion™-SPE in UP water to demonstrate the electrochemical behaviors of Pt in the electrolytes with different structures but similar acidity. In this part, the three electrodes of the SPEs were immersed in the solution or UP water while CV scanning.

The Au@Pt-SPE shows a typical and characteristic hydrogen region (0.05–0.35 V) of Pt, as well as the EDL region (0.35–0.7 V) and the oxygen region (0.7–1.3 V)[13–15] in 0.5 M H_2SO_4 solution (Fig. 4A). The pair of redox peaks at 0.3 V in the

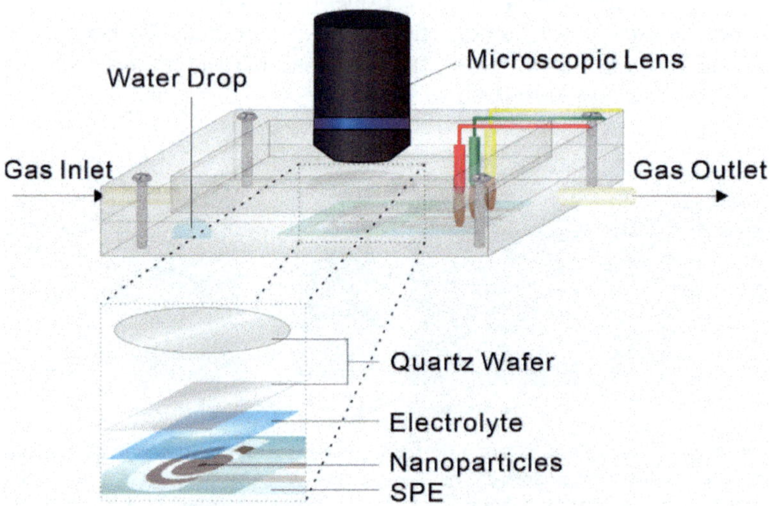

Fig. 3 The schematic diagram of the electrochemical *in situ* Raman testing device. The quartz wafers were 0.1 mm thick.

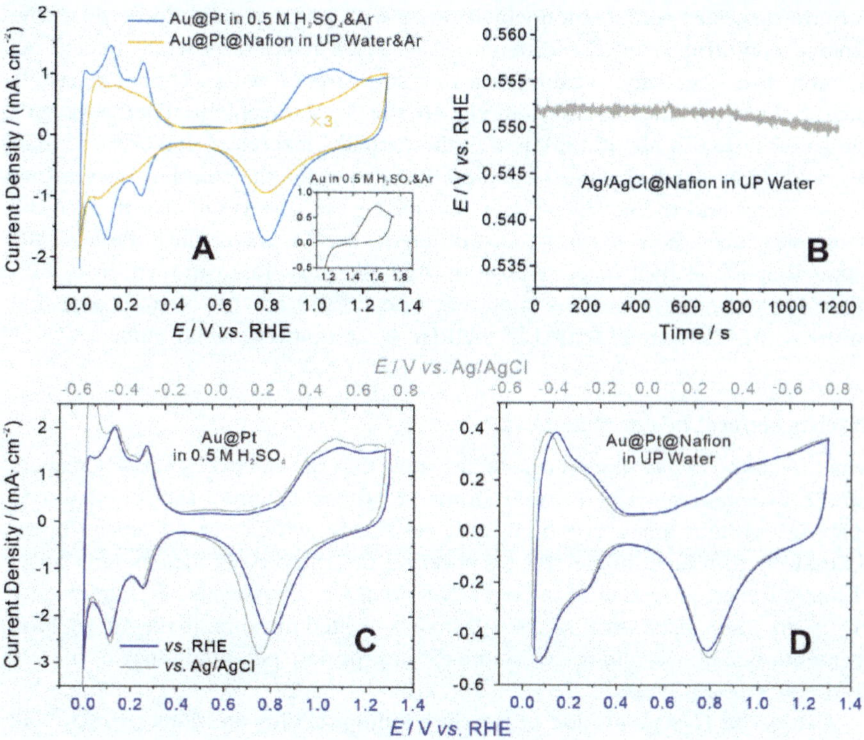

Fig. 4 (A) The CV curves of the Au@Pt-SPE in different electrolytes. The current density of the curve for the Nafion™ system is magnified by 3 times. The WE and CE of the SPEs were activated in advance by CV scanning for 50 cycles at 50 mV s^{-1} in the corresponding systems. The inset shows the CV curve of the pristine Au SPE at high potentials. (B) The variation of the electrode potential of the Ag/AgCl electrode of the Au@Pt@Nafion™-SPE in UP water *vs.* RHE with time. (C and D) The CV curves of the SPEs in different systems using a RHE and the Ag/AgCl electrode as the RE, respectively. The scanning rate of the CV curves was 50 mV s^{-1}, and all of the above tests were performed under an Ar atmosphere.

hydrogen region originates from the adsorption and desorption of strongly adsorbed H on Pt(100), while the pair of redox peaks at 0.1 V originates from the adsorption and desorption of weakly adsorbed H on Pt(110). In the oxygen region, two overlapping oxidation peaks for O adsorption appear at 1.0 V and 1.1 V. The reduction peak at 0.8 V is for O desorption. For the Nafion™ system, the three regions above still clearly exist, although the peaks are far vaguer than those in the solutions but are still distinguishable. This may result from the adsorption of organic functional groups on Pt. Some infrared spectroscopy and electrochemical studies have shown that the $-SO_3H$ group in Nafion™ can strongly adsorb on Pt, thus occupying some of the adsorption sites and reducing its H and O adsorption and desorption activity.[7,16–18] In addition, the Au substrate starts to be oxidized at 1.4 V and is reduced at 1.3 V, as can be seen from the inset.

For the electrochemical *in situ* SERS test, the choice of the RE is important for the SPE. The RE has to be stable and easy to prepare for the planar structure. In this work, a Ag/AgCl electrode was used for the above reasons. However, the electrode potential of the Ag/AgCl electrode can differ after preparation, due to its uncertain surface composition which varies with its preparation procedure, the storage conditions and the different electrolytes it contacts while in use. As a result, it is necessary to calibrate it to a standard RE with a certain electrode potential before using. As shown in Fig. 4B, the electrode potential of the Ag/AgCl electrode varied by about 2 mV in 20 min. Actually, it rarely varied by more than 10 mV within 1–2 h during our *in situ* SERS tests for the solution system (not shown here) and the polyelectrolyte system. Fig. 4C and D indicate that, in the same electrolyte, even if the RE was different, the CV curves have the basically same shape (the relative peak position and the peak intensity) within certain electrode potential ranges, meaning that we can calibrate the electrode potential of the Ag/AgCl electrode to that of the RHE by comparing the CV curves.

Electrochemical *in situ* SERS tests

The CVs of the *in situ* Raman test device show that the electrochemical behaviors of the two systems (Fig. 5) were similar to those obtained for the solutions (Fig. 4A) without any overly high ionic resistance, which can be caused by an excessively thin layer of electrolytes covering the three electrodes as ions travel through the cross section of the layer. Note that CV scanning is necessary before the *in situ* SERS tests, because the CV curve is needed to calibrate the Ag/AgCl RE, as mentioned above, and to determine the appropriate potential range of the WE in the *in situ* SERS tests.

As a 0.5 M H_2SO_4 solution possesses a similar acidity and ions (H^+/SO_4^{2-} in H_2SO_4 and $H^+/-SO_3^-$ in Nafion™) to those of the Nafion™ membrane, the *in situ* SERS spectra for the Au@Pt-SPE in 0.5 M H_2SO_4 were firstly obtained in order to compare the difference between the Pt–H_2SO_4(aq) interface and the Pt–Nafion™ interface. It can be seen in Fig. 6 that several Raman peaks obviously depend on the scanning potential, indicating that they were generated from the Pt–H_2SO_4(aq) interface, where the potential matters. The broad peak at 580 cm^{-1} is attributed to the Pt–O stretching mode,[19,20] which starts to appear at 1.2 V during

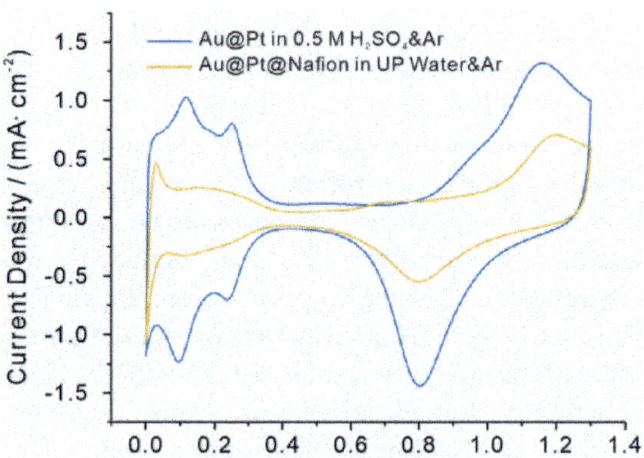

Fig. 5 The CV curves obtained for the *in situ* Raman test device at a scanning rate of 50 mV s^{-1} with the electrodes of the SPEs not immersed in the solution or UP water but covered with a thin layer of the solution or UP water. The potential was recorded with a Ag/AgCl RE during the tests and afterwards calibrated to RHE.

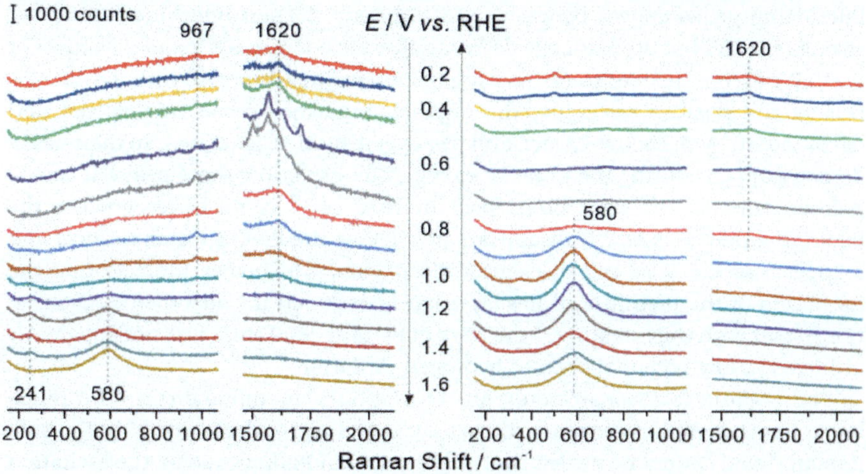

Fig. 6 The potential-dependent *in situ* SERS spectra for the Au@Pt-SPE in 0.5 M H_2SO_4. The initial potential was 0.2 V and the potential step was 0.1 V. The potential was recorded with the Ag/AgCl RE during the test and afterwards calibrated to RHE.

the positive sweep. Actually, the O adsorption starts at 0.8 V, as shown in the CV (Fig. 5), but the correlated potential is delayed by about 0.3 V before this Raman peak appears. This is because the intermediates generated on the Pt surface had too short lives and were at too low concentrations to feed back sufficiently strong signals at the point of their generation without accumulation.[21] The intensity of this peak gradually increases from 1.2 V to 1.5 V, indicating the increase in the coverage of the adsorbed O. When sweeping negatively from 1.6 V, the adsorbed O coverage should not have further increased, but actually its peak obviously increases from 1.6 V to 1.1 V, which might result from the formation of an oxide layer on the Au NP surface at high potentials, increasing the distance between the detected species, reducing the "hot spot" effect, and thus weakening the enhancement efficiency.[22] In addition, the varying potential itself[23] and the relatively low reflectivity of the oxidized Pt surface[22] can also affect the enhancement efficiency and the signal intensity. During the negative sweep, the oxide layer was reduced and the enhancement efficiency was restored, so the intensity of the peak at 580 cm^{-1} increases. When negatively sweeping to 0.8 V, that is, the peak potential of O desorption, the Raman peak almost completely disappears. However, it is obvious that not all the adsorbed O had been reduced at this potential, again indicating the importance of the accumulation of active intermediates before they can be detected. The weak peak at 241 cm^{-1} is assigned to the O–O of the peroxygen species adsorbed on the surface of Pt,[20] which appears during the positive sweep at 1.0 V, increases to its strongest at 1.3 V, then decreases and disappears at 1.6 V. The decrease of this peak is basically accompanied by the increase of the Pt–O peak, which might reflect the process of O–O bond cleavage and conversion to adsorbed O. However, the O–O peak does not reappear during the negative sweep, indicating that the conversion from O–O to O could be irreversible. The weak peak at 967 cm^{-1} originates from the stretching mode of SO_4^{2-}.[24] When positively sweeping to 0.5 V, 0.6 V and 0.7 V, dense and fluctuating peaks appear between 1135–1750 cm^{-1}. Their intensities are rather strong at 0.7 V, but dramatically decrease afterwards from 0.8 V and disappear at 1.0 V. We assign them to some trace organic impurities imported with the Au NPs. During the positive sweeping, the organic impurities could be oxidized and decomposed by the active oxygen and peroxygen species generated, so their peaks do not reappear during the negative sweep. After excluding these peaks, it can be pointed out that the weak broad peak at 1620 cm^{-1} could be assigned to the bending mode of H_2O.[23] It disappears at 1.2 V as the Pt–O appears to form and reappears at 0.8 V as the Pt–O peak disappears, which may indicate that the formation of the platinum oxides layer partly excluded the adsorbed H_2O on Pt. Consistent with the literature,[12] the peak of Pt–H at 2000 cm^{-1} was not apparently observed at low potentials before hydrogen evolution.

The organic functional groups in the Nafion™ membrane can have much larger Raman scattering cross sections than H_2O does, so that, inevitably, the bulk Raman signal cannot be neglected. For the confocal Raman system, the excitation ability of the laser at its focus is the strongest, while that at the non-focus points is far weaker. When the laser is focused precisely on the Pt–Nafion™ interface, the molecules at the interface would receive the strongest excitation, and the bulk molecules would be excited far more weakly. Additionally, with SERS-active nanoparticles, the interface signal would be further enhanced, leaving the bulk signal unchanged. Therefore, theoretically, the Pt–Nafion™ interface signal can be directly observed and distinguished even across the Nafion™ membrane, which is proven, as shown in Fig. 7. When the laser is focused on the top side of the Nafion™ membrane, we mainly obtain the bulk Nafion™ signal, while the interface signal is mainly obtained as the laser is focused on the Pt–Nafion™ interface. The latter shows a strong band between around 1000–1700 cm^{-1}, which is assigned to the impurities from the Au NPs and does not appear in the former. In addition, the relative intensity of the peak at 729 cm^{-1}, assigned to the bending mode of the Nafion™ $-SO_3^-$,[7] is slightly higher for the latter than for the former. This indicates that, with the laser focused on the Pt–Nafion™ interface and with

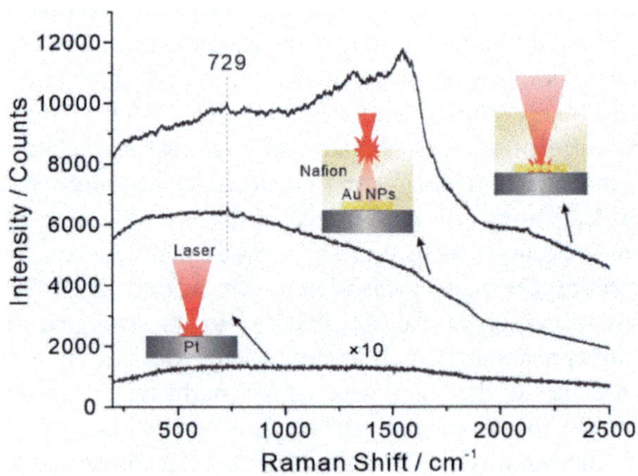

Fig. 7 The Raman spectra with the incident laser focused at different planes. The intensity is magnified by 10 times for the case when the laser is focused at the Pt surface of the Au@Pt-SPE.

the enhancement by the Au NPs, the interface signal can be obtained and distinguished.

This is also actually proven, as shown in Fig. 8. The peak of the Pt–O stretching mode at ~580 cm^{-1} in the H_2SO_4 solution is blueshifted to 618 cm^{-1} in the Nafion™ membrane. This might be explained by the smaller water content of the Nafion™ membrane than that of the H_2SO_4 solution, which leads to a weaker hydrogen bond interaction between Pt–O and H_2O that makes the Pt–O peak redshift. This peak appears in the positive sweep at 0.9 V, and then weakens up to 1.3 V due to the partial oxidation of the Au NP surface (and even the potential change and the decrease in the surface reflectivity); in the negative sweep, it gradually increases and reaches a stable point at 1.0 V, then starts to weaken from 0.7 V, and finally disappears at 0.5 V, which is 0.3 V lower than that for the H_2SO_4 system. This might be related to the strong adsorption of the Nafion™ $-SO_3^-$ on Pt.[7] The strongly adsorbed $-SO_3^-$ increases the negative charge density on the surface of Pt, which hinders the process of the adsorbed O obtaining electrons as the adsorbed O could undergo a partially negatively charged transition state during desorption, making the reduction of the adsorbed O fully completed at a lower potential. Actually, SO_4^{2-} could also adsorb on Pt to cause such an effect. But the aqueous SO_4^{2-} would be excluded and freely migrate away from the Pt surface when the potential turns negative enough, thus eliminating the effect. In contrary, the $-SO_3^-$ is fixed on the polymer chains of Nafion™, and their migration is greatly limited in the solid Nafion™ membrane. As a result, even if the potential turns negative, the adsorbed $-SO_3^-$ is unable to travel far from the Pt surface. The two weak peaks at 729 cm^{-1} and 1059 cm^{-1} originate from the bending mode and symmetric stretching mode of the $-SO_3^-$,[7] respectively, and

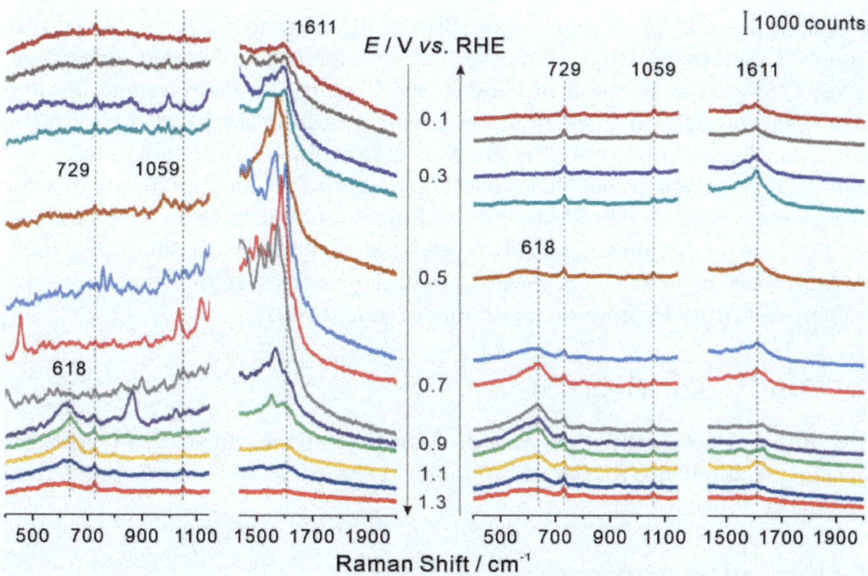

Fig. 8 The potential-dependent *in situ* SERS spectra for the Au@Pt@Nafion™-SPE in UP water. The initial potential was 0.1 V and the potential step was 0.1 V. The potential was recorded with the Ag/AgCl RE during the test and afterwards calibrated to RHE.

their intensities are constant, indicating that the concentration of the $-SO_3^-$ at the interface hardly varied. Despite the platinum oxide layer already appearing to form at 0.9 V, the H_2O bending mode signal at 1611 cm^{-1} does not disappear until the potential rises to 1.2 V, and, when scanning back, the H_2O peak reappears at 1.2 V prior to the disappearance of the Pt–O peak at 0.5 V, which is quite different from the situation for the H_2SO_4 system. This might be caused by the co-adsorption of H_2O with the adsorbed $-SO_3^-$. The strongly adsorbed $-SO_3^-$ constantly stays at the Pt–Nafion™ interface so that the signal of the co-adsorbed H_2O is maintained even with the presence of the platinum oxide layer. In addition, the relative intensity of the H_2O peak to the Pt–O peak is obviously higher than that for the H_2SO_4 system, also demonstrating the co-adsorption mentioned above. Although Nafion™ still has many characteristic peaks in the range of 1000–1600 cm^{-1},[7] due to the possible presence of the substrate impurities they could not be determined.

As discussed above, the electrochemical characteristics and behaviors of the molecules at the Pt–Nafion™ interface are significantly different from those at the Pt–H_2SO_4(aq) interface. Therefore, it is necessary to test the isolated electrode–polyelectrolyte interface without the coexistence of any electrode–solution interfaces, only by which can we acquire fundamental and distortion-free information on the structure of the electrode–polyelectrolyte interface.

4. Conclusions

We established a practical and convenient electrochemical *in situ* SERS test method suitable for the electrode–polyelectrolyte interface without the coexistence of the electrode–solution interface by combining planar SPE with the SERS technique. A Nafion™ membrane was simply cast on the SPE by drop coating to construct the isolated Pt–Nafion™ interface, from which a distinguishable *in situ* SERS signal was successfully obtained directly across the Nafion™ membrane, overcoming the bulk interference. Prominent differences were discovered and compared between the Pt–Nafion™ interface and the Pt–H_2SO_4(aq) interface to elucidate the necessity of testing the isolated electrode–polyelectrolyte interface by our method so as to acquire distortion-free information on the structure of it. This model method provides an opportunity to load diverse electrochemical catalysts and cast different polyelectrolyte membranes to study the structure of various isolated electrode–polyelectrolyte interfaces, thus revealing the mechanism of electrochemical processes at the interfaces and helping to further improve the performance of fuel cells.

Conflicts of interest

The authors declare that they have no known competing financial interests or personal relationships that could have appeared to influence the work reported in this paper.

Acknowledgements

(1) National Natural Science Foundation of China (21991154, 22122204, 21872108); (2) National Key Research and Development Program

(2018YFB1502302); (3) Wuhan University Innovation Team (2042017kf0232, 2042020kf1073, 2042019kf0270).

References

1 Y. Wang, Y. Yang, S. Jia, X. Wang, K. Lyu, Y. Peng, H. Zheng, X. Wei, H. Ren, L. Xiao, J. Wang, D. A. Muller, H. D. Abruña, B. J. Hwang, J. Lu and L. Zhuang, *Nat. Commun.*, 2019, **10**, 6–13.
2 M. Fleischmann, P. J. Hendra and A. J. McQuillan, *Chem. Phys. Lett.*, 1974, **26**, 163–166.
3 D. L. Jeanmaire and R. P. Van Duyne, *J. Electroanal. Chem.*, 1977, **84**, 1–20.
4 M. G. Albrecht and J. A. Creighton, *J. Am. Chem. Soc.*, 1977, **99**, 5215–5217.
5 R. P. Van Duyne, in *Chemical and Biochemical Applications of Lasers*, Elsevier, 1979, pp. 101–185.
6 J. F. Li, X. D. Tian, S. B. Li, J. R. Anema, Z. L. Yang, Y. Ding, Y. F. Wu, Y. M. Zeng, Q. Z. Chen, B. Ren, Z. L. Wang and Z. Q. Tian, *Nat. Protoc.*, 2013, **8**, 52–65.
7 J. Zeng, D. I. Jean, C. Ji and S. Zou, *Langmuir*, 2012, **28**, 957–964.
8 I. Kendrick, J. Fore, J. Doan, N. Loupe, A. Vong, N. Dimakis, M. Diem and E. S. Smotkin, *J. Electrochem. Soc.*, 2016, **163**, H3152–H3159.
9 Q. Wang, PhD thesis, Wuhan University, 2010.
10 Z. Q. Tian, B. Ren, J. F. Li and Z. L. Yang, *Chem. Commun.*, 2007, 3514–3534.
11 G. Frens, *Nat. Phys. Sci.*, 1973, **241**, 20.
12 J. F. Li, Z. L. Yang, B. Ren, G. K. Liu, P. P. Fang, Y. X. Jiang, D. Y. Wu and Z. Q. Tian, *Langmuir*, 2006, **22**, 10372–10379.
13 Y. Furuya, T. Mashio, A. Ohma, M. Tian, F. Kaveh, D. Beauchemin and G. Jerkiewicz, *ACS Catal.*, 2015, **5**, 2605–2614.
14 L. García-Cruz, V. Montiel and J. Solla-Gullón, *Phys. Sci. Rev.*, 2019, **4**, 1–34.
15 C. A. Campos-Roldán and N. Alonso-Vante, *J. Solid State Electrochem.*, 2021, **25**, 187–194.
16 A. M. Gómez-Marín, A. Berná and J. M. Feliu, *J. Phys. Chem. C*, 2010, **114**, 20130–20140.
17 R. Subbaraman, D. Strmcnik, A. P. Paulikas, V. R. Stamenkovic and N. M. Markovic, *ChemPhysChem*, 2010, **11**, 2825–2833.
18 H. Yano, E. Higuchi, H. Uchida and M. Watanabe, *J. Phys. Chem. B*, 2006, **110**, 16544–16549.
19 R. Muralidharan, M. McIntosh and X. Li, *Phys. Chem. Chem. Phys.*, 2013, **15**, 9716–9725.
20 Y. F. Huang, P. J. Kooyman and M. T. M. Koper, *Nat. Commun.*, 2016, **7**, 1–7.
21 J. C. Dong, X. G. Zhang, V. Briega-Martos, X. Jin, J. Yang, S. Chen, Z. L. Yang, D. Y. Wu, J. M. Feliu, C. T. Williams, Z. Q. Tian and J. F. Li, *Nat. Energy*, 2019, **4**, 60–67.
22 U. Zhumaev, A. V. Rudnev, J. F. Li, A. Kuzume, T. H. Vu and T. Wandlowski, *Electrochim. Acta*, 2013, **112**, 853–863.
23 D. Y. Wu, J. F. Li, B. Ren and Z. Q. Tian, *Chem. Soc. Rev.*, 2008, **37**, 1025–1041.
24 A. Mineshige, N. Ohata, T. Tanaka, M. Kobune and T. Yazawa, *Electrochemistry*, 2007, **75**, 179–181.

Faraday Discussions

ROYAL SOCIETY OF CHEMISTRY

DISCUSSIONS

State of the art energy conversion at the nanointerface: general discussion

Tim Albrecht, Paul W. Bohn, Mark A. Buckingham, Xiangkun Elvis Cao, Dongfei Chen, Qianjin Chen, Manuel Corva, Martin A. Edwards, Ali Reza Kamali, Frédéric Kanoufi, Steffi Krause, Steven Linfield, Xu Liu, Hui Ma, Bing-Wei Mao, Popular Pandey, Kristina Tschulik, Swathi Naidu Vakamulla Raghu, Alain Walcarius, Li Xiao and Yi-Lun Ying

DOI: 10.1039/d2fd90002e

Steven Linfield opened a general discussion of the paper by Alain Walcarius: Pore diameter can be a limiting factor during metal deposition in nanopores. Did you encounter any problems filling the nanopores with the aniline solution? What percentage of pores did you observe the electropolymerisation occurring in? If not all pores were filled, what effect do you expect the density of nanofilaments in the template-free sample to have on its electrochromic properties?

Alain Walcarius responded: We feel that most (if not all) mesopores should be filled with PANI, but before answering the question in more detail, just a few words on metal nanowire electrodeposition. We also tried to grow metal nanowires in very small diameter nanochannels but if metal cations can be effectively reduced at the mesopore bottom (on the underlying electrode surface) they have the tendency to form nanoparticles rather than mechanically stable wires (perhaps for crystallographic reasons?); actually we sent some of our mesoporous films to Phil Bartlett, and his group attempted to grow metal nanowires by their powerful method of electrodeposition from a supercritical fluid, but they concluded that sub 7 nm nanowires formed in mesoporous films are not stable.[1]

For PANI, no such limitation was observed and we are confident that most (if not all) mesopores should be filled with PANI, on the basis of rough calculations on the amount of PANI filaments (from integration of CV curves) compared to the volume of the nanochannels (as estimated from pore density, length and diameter). Actually, there is no limitation due to water ingress in the mesoporous membrane. A non-functionalized mesoporous material prepared from cationic surfactants (as we used here for film preparation) is highly hydrophilic and its pores are easily filled with water (thus explaining their rather poor chemical stability). Actually, just being in air, at 70% moisture, the mesopores are completely filled (this can be easily checked by water adsorption experiments, as described in ref. 2). So there is no problem at all to get water entering such mesoporous films, explaining the possible electropolymerization of PANI and easy growth through the silica membrane because linear oligomers are compatible (in size) with the long mesochannels.

1 P. Bartlett, *et al.*, *Nano Lett.*, 2018, **18**, 941–947.
2 M. Etienne, Y. Guillemin, D. Grosso and A. Walcarius, *Anal. Bioanal. Chem.*, 2013, **405**, 1497–1512.

Ali Reza Kamali commented: Very nice work! Have you noticed any damage to the ITO substrate caused by the SiO_2 removal process? Thank you.

Alain Walcarius replied: Thank you. No, the treatment in NaOH (to dissolve the silica membrane) did not result in any noticeable damage to the ITO surface (it would have been the case if stronger etching agents such as HF had been used for example). We have just to avoid applying too strong cathodic potentials that can indeed contribute to some degradation of the electrode response.

Mark A. Buckingham asked: I was wondering if you observe localised electrolyte concentration effects within the material channels? Does this lead to restricted mass transport of electrolyte into/out of the channels at all?

Alain Walcarius answered: Mass transport processes in such oriented membranes are rather fast thanks to the vertical alignment of the nanochannels. They have been largely investigated (without PANI inside). For instance, using redox probes with different charge ($[Ru(NH_3)_6]^{3+}$, $Fc(MeOH)_2$, $[Fe(CN)_6]^{3-}$), distinct permselective properties were evidenced, with almost no interaction with the neutral probe, some accumulation of the positively charged one and significant rejection of the anionic one (as a result of attractive and repulsive electrostatic interactions, respectively, because the internal silica surface is negatively charged due to the presence of silanolate groups). Permselectivity can be somehow tuned by adjusting the ionic strength of the medium (affecting the thickness of the electrical double layer relative to the channel diameter). When filled with PANI filaments, resistance to mass transport in the membrane is expected to occur during the doping/dedoping processes, explaining the lower current values observed for PANI entrapped in the nanochannels in comparison to the free PANI nanofilaments (*i.e.*, after dissolution of the silica membrane). As the dopant species are hydrogen sulfonate anions (we are working in sulfuric acid medium), one can expect significant resistance to mass transport due to electrostatic repulsions by the silica walls, and this should be even more restrictive than for the pristine mesoporous silica film due to the presence of nanofilaments in the small volume of mesopores. So yes, the localised environment within the channels is expected to affect the mass transport of electrolytes (for the PANI/silica composite, not for free PANI nanofilaments).

Qianjin Chen enquired: Are these PANI nanofilaments still mechanically robust after the removal of the silica membrane? I assume it is such "free" PANI nanofilaments structure that provides sufficient spaces for fast mass transfer during electrochromic investigation? Will they collapse after cycles of experiment?

Alain Walcarius replied: Thank you. Yes, the PANI nanofilaments stayed durably attached to the ITO support, resulting in stable colouration/bleaching in electrochromic experiments, but most probably they did not stay as vertically oriented brushes due to their particular aspect ratio (long filaments (100 nm range) with diameter at the (sub)nanometer scale); I would expect them to be more in the form of a spaghetti-like deposit, while keeping a morphology open enough to ensure fast mass transport processes. As successive CV experiments did not result in any noticeable current decrease, no significant collapsing would happen. The only cases where some lack of mechanical robustness was observed are those corresponding to thicker deposits (*i.e.*, where 3D growth of PANI occurred on the top of the mesoporous silica membrane) for which some damage can be encountered when dissolving the silica material.

Steffi Krause asked: Were the bulk films made in the same way as the nanofilaments with a graft of aminophenyl underneath and was the length of the filaments comparable to the thickness of the bulk film? I am trying to understand why the film seemed to degrade, but the nanofilaments did not.

Alain Walcarius answered: We compared the PANI nanofilaments with two kinds of bulk PANI deposits: one prepared at the same electropolymerization charge (thus expected to lead to the same amount of aniline moieties formed on the electrode surface) and one prepared for the same electropolymerization time (thus expected to be much thicker than the nanofilaments layer), both synthesized with a graft of aminophenyl underneath. In both cases, similar observations were made. Actually, for "bulk" PANI deposits, we think that the variations in the electrochromic response are not really due to the film degradation but most probably originate from approaching a steady-state (as doping/dedoping processes might induce structural changes in bulk PANI, they are expected to stabilize after multiple sweeps): after an initial decrease in the electrochromic response, the subsequent signals of bulk films were rather stable (see right part of Fig. 7A part (a), DOI: 10.1039/d1fd00034a).

Martin A. Edwards queried: Do you find a thickness where the electrochemical behavior of the films degrades – perhaps though the resistance on the nanofilaments, or resistance/diffusional transport in the interstitial space, *etc.*?

Alain Walcarius replied: There are actually two limiting cases, with an optimal situation in-between. When PANI nanofilaments are not very long, they are located at the mesopore bottom and their electrochemical transformation within the silica membrane is somehow restricted by diffusional transport (during doping/dedoping processes); these limitations do not exist anymore when the silica membrane is dissolved (this is demonstrated by the transition from diffusion-limited processes to thin-layer behavior when passing from the PANI/silica composite to the free PANI wires). When PANI deposits are too large (*i.e.*, when they are also formed outside the membrane), they may suffer from mechanical degradation upon dissolution of the silica membrane but they are still electrochemically reversible. The optimal situation is when PANI nanofilaments are filling the whole channels (*i.e.*, with a length corresponding to the silica membrane thickness) for which no significant restriction to either charge transfer or mass transport has been observed.

Frédéric Kanoufi commented: This work is a very elegant controlled synthesis of nanostructured materials. One of the interesting aspects is the change in transport regime, and the possibility to remove the mass-transfer limitation during the doping/dedoping of polyaniline. I wonder if it is not related to the formation of a much less dense polymer structure (increase of free volume surrounding each filaments) when removing the template. I then wonder if you cannot observe the dynamics of the polymer brush restructuring: for example could it be probed by successive cycles during the electrochemical characterization or during the GISAXS experiment, or any other *in situ* measurement?

Alain Walcarius responded: Thank you. Indeed, removal of the silica membrane will for sure lead to a less dense polymer structure (also less dense than a bulk PANI film formed directly on a bare electrode) and this is the reason for fewer restrictions to mass transport during doping/dedoping processes and thus more intense voltammetric responses. Thank you for the idea to observe the dynamics of the polymer brushes. So far, we performed successive cyclic voltammetry measurements and no noticeable change in the response was observed after 100 consecutive scans. Unfortunately, GISAXS did not give any exploitable response: the individual nanofilaments are probably more in the form of a spaghetti-like deposit than vertically oriented brushes, due to their particular aspect ratio (long filaments (100 nm range) with diameter at the (sub)nanometer scale). One would probably need another *in situ* method…

Paul W. Bohn remarked: These are very elegant experiments. One might imagine that the characteristics of the electropolymerization reaction would depend on the local environment. Do you have any evidence of radial variations in structure or composition across individual polyaniline nanowires?

Alain Walcarius replied: Thank you very much. Indeed, electropolymerization reactions in such confined spaces would certainly be influenced by the local environment and we would have been more than happy to provide a molecular view about that but we don't have a definite answer so far. What we can currently state is that the 1D growth of PANI nanofilaments through the nanochannels is kinetically slower than the 3D growth process, primarily because of some resistance to mass transport through the membrane but perhaps also due to reactivity changes resulting in slower polymerization. Dealing with radial variations in structure or composition, previous EDS analyses performed from TEM cross-section views indicated an increase in the nitrogen content (*i.e.*, amount of PANI) from the electrode surface (bottom of the mesopores) to the top of the membrane, suggesting the possible branching of PANI filaments along the channels.[1] This is also consistent with a rough estimation of the number of individual PANI nanofilaments of about 2 per mesopore channel.

Additionally, in collaboration with a Polish group (K. Łępicka and P. S. Sharma), we are currently investigating the electropolymerization of Ni–salen derivatives into oriented mesoporous silica films, showing restrictions at the molecular level for monomers that are all likely to penetrate the nanochannels but react differently depending on their substituents (submitted).

1 A. Gamero-Quijano, C. Karman, N. Vilà, G. Herzog and A. Walcarius, *Langmuir*, 2017, **33**, 4224–4234.

Manuel Corva said: Thank you for presenting this very interesting work. Have you noticed and countered any inhomogeneity in the ITO substrate affecting your silica and/or PANI films order/properties?

Alain Walcarius responded: Thank you. No, we did not notice any inhomogeneity in the ITO substrate, just some small variations in the plateau currents observed on *i*–*t* curves (during potentiostatic electropolymerization), which can be easily managed by controlling the effective charge transferred during PANI deposition.

Dongfei Chen asked: Could you please share with us some of your thoughts about its potential utility? Also, is it possible to use the interface to sense analytes from complex samples given that the membrane tends to present a filtering function?

Alain Walcarius replied: Yes, there are many examples of applications of such oriented mesoporous silica membranes, thanks to their molecular sieving properties (size and charge selectivity, permselective behaviour, protection against biofouling, *etc.*), possibility of functionalization with redox-active moieties (long-range charge transfer, energy storage, molecular electronics), and use as mould for the growth of nanowires (electrocatalysis, electrochromism). More information on these applications can be found in some reviews.[1–3] Two examples of using the interface to sense analytes from complex samples (by exploiting the membrane filtering function) are, for instance: the detection of propranolol in the presence of haemoglobin[4] or the determination of chloramphenicol in whole blood.[5] For the particular case of PANI nanofilaments generated in mesoporous silica thin films, the expected applications are in the fields of electrochromism, electrocatalysis and electrochemical energy storage.

1 A. Walcarius, *Curr. Opin. Electrochem.*, 2018, **10**, 88–97.
2 P. Zhou, L. Yao, K. Chen and B. Su, *Crit. Rev. Anal. Chem.*, 2020, **50**, 424–444.
3 A. Walcarius, *Acc. Chem. Res.*, 2021, **54**, 3563–3575.
4 M. Bueno Serrano, C. Despas, G. Herzog and A. Walcarius, *Electrochem. Commun.*, 2015, **52**, 34–36.
5 Q. Sun, F. Yan, L. Yao and B. Su, *Anal. Chem.*, 2016, **88**, 8364–8368.

Steven Linfield commented: You mentioned a stabilising effect of PANI filaments on the structure of mesoporous silica and how this lead the PANI/silica composite films in air to remain intact over a year (in contrast to the rapid deterioration of empty mesoporous silica). What is the reason for this stabilisation effect?

Alain Walcarius answered: We observed two kinds of "stabilization". First, the structure of the PANI/silica composite was maintained for a year (information from GISAXS measurements) and, secondly, the voltammetric response of PANI inside the membrane and its electrochromic response were also stable for a long time. The reasons for stabilization are not fully understood but this might be a combination of mechanical effects (presence of nanofilaments in the nanochannels, preventing their collapsing) and a barrier to aqueous dissolution processes enabled by the presence of PANI (similar stabilization was observed for thiol-functionalized mesoporous silica anchoring mercury(II)).

Frédéric Kanoufi noted: From the diazonium electrografting curve it seems that you block only half the area of the pores (decrease of current by factor of two). It then suggests that the polymer grafting density is indeed quite low which likely favors its spreading on the surface (chain entanglements). Is this sufficient to produce a sufficiently stable grafted material?

Alain Walcarius responded: Yes, electrografting by electro-reduction of a diazonium salt on ITO electrode is much less effective than on carbon substrates for instance, and this is most probably lucky for us because on the one hand it did not block the electrode surface and on the other hand this contributed to limit the number of anchor points and therefore the number of nanofilaments per pore (keeping some space available for diffusion). Nevertheless, in spite of "incomplete" coverage, the grafting density was apparently enough to keep stable, covalently bonded, PANI filaments on the electrode surface, even after dissolution of the silica membrane.

Tim Albrecht opened a general discussion of the paper by Li Xiao: In the conclusion of your paper, you mention that you are looking forward to new applications. Could you elaborate on how one might use the new capabilities you are describing and what you have learned more generally from your results so far?

Li Xiao answered: The significance of our work is to establish a relatively simple method to study the information of isolated electrode–polyelectrolyte interface without introducing any interference from solution electrolyte. Based on this method, we can change the types of electrode materials and polymer electrolytes to study the characteristics of different interfaces, and then explain some unknown phenomena.

For example, in electrocatalysis research, we often encounter a catalyst that has good performance in aqueous solution, but poor performance in fuel cells. Our research may help us understand the behavior differences of this catalyst at the two electrode–electrolyte interfaces, then help us understand the intrinsic causes of its performance differences in different electrolytic cells and devices, and provide reasonable ideas for designing catalysts with better performance.

Popular Pandey asked: What are the advantages of studying electrode–polyelectrolyte interfaces over electrode–solution interfaces? How did you remove the bulk electrode–solution interface contributions in the SERS signals from those of the electrode–polyelectrolyte interface?

Li Xiao replied: It is not an advantage, but necessary to study the electrode–polyelectrolyte interface. Nowadays, a lot of modern technologies use polyelectrolyte instead of traditional electrolyte solutions, such as fuel cells, solid polymer electrolyte water electrolysis and CO_2 membrane electrolyzers, *etc.* Many experimental results show that the same catalyst can perform completely differently in electrolyte–solution and electrolyte–polyelectrolyte interfaces, such as our previous works published in ref. 1 and 2. Studying this novel interface can help us to better understand the design principles of the catalysts. To answer your second question, we didn't use electrolyte solution in our SERS study. Instead, we only used ultra-pure water and Nafion. So there was no bulk electrode–solution signal to be excluded. Another thing is that only the surface signal was enhanced by Au NPs, therefore the bulk signal from Nafion can also be excluded. We include a discussion about this in our paper.

1 Z. Yin, H. Peng, X. Wei, H. Zhou, J. Gong, M. Huai, L. Xiao, G. Wang, J. Lu and L. Zhuang, *Energy Environ. Sci.*, 2019, **12**, 2455.
2 Y. Wang, Y. Yang, S. Jia, X. Wang, K. Lyu, Y. Peng, H. Zheng, X. Wei, H. Ren, L. Xiao, J. Wang, D. A. Muller, H. D. Abruña, B. J. Hwang, J. Lu and L. Zhuang, *Nat. Commun.*, 2019, **10**, 1506.

Ali Reza Kamali commented: This is a very interesting study. Did you measure the temperature of the device during the Raman spectroscopy measurements? Have you considered the possible influence of heating caused by the laser irradiation?

Li Xiao answered: Thanks for the question. We didn't particularly control the temperature during the test. All the tests were carried out under room temperature. Since there was a quantity of water that able to absorb heat on the SPE during the Raman measurements, we don't think the laser could have lead to a prominent change of temperature on the testing surface.

Hui Ma enquired: Could you give more explanation about the potential-dependent Raman shift? Why the Raman peak of water under high potential is lower than that under low potential?

Li Xiao replied: The potential dependence of Raman peaks can reflect the characteristics of adsorbed species on the electrode surface. We can combine the Raman-potential signal with CV profiles for analysis. In the double-layer region before the oxygen region of CV, water molecules are mainly adsorbed on the electrode surface. When the potential becomes gradually positive, oxygen-containing adsorption species are gradually adsorbed on the Pt surface, which will form competitive adsorption with water molecules, and finally push out water molecules. Therefore, in the high potential region, the signal of water molecules is weaker than that in the low potential region, but at the same time, the signal of oxygen-containing adsorbed species is enhanced.

Frédéric Kanoufi followed up on the question by Hui Ma: Do you mean that you are desorbing/decomposing the citrate capping agent of the Au NPs used for Raman enhancement? Do the Raman bands fit with some bands related to citrate bond vibrations? What other organic contamination could it be? Once this contamination is decomposed upon oxidation, could you form a new organic shell at the Au NP surface, for example have you inspected further cycles and could you detect a new Au–organic (with Nafion?) interaction?

Li Xiao answered: It is difficult for us to assign the contamination to the citrate capping agent. We found the Raman peaks of the contamination varied irregularly during the experiment. At certain potentials, some peaks suddenly appear and seem to fit with the tri-sodium citrate peaks. But once we change the scanning potential, those peaks can suddenly disappear without any gradual change in their intensity or wave-number. We also found that multiple CV cycles can help us clean the electrode surface, and no new organic shell was found in our experiment.

Swathi Naidu Vakamulla Raghu enquired: Does the area of measurement due to the laser offer any local differences/inhomogeneity that is significantly different to the rest of the interface?

Li Xiao replied: We have measured Raman signals from different areas of the interface. The results show that overall the Raman peaks of the species are independent of the area. The peak intensity can vary probably due to the inhomogeneous morphology of the deposited Pt. The diameter of the laser spot on the interface was about 1 μm, which should be large enough to obtain averaged signals over the detected area.

Kristina Tschulik commented: You have attributed some of the spectroscopic features you detect to impurities. Which impurities do you consider to be most likely in your experiments?

Li Xiao responded: We believe that the most likely impurities could be the small organic molecules introduced in the synthesis of Au NPs. However, we did not make a detailed analysis in this regard, because impurities are not our main concern. It is more important for us to use methods like multi-cycle CV to remove the interference of these impurities.

Bing-Wei Mao said: It is very important to study electrode–polyelectrolyte interfaces. But such kinds of interfaces are solid–solid interfaces and their characterizations are very challenging. You have shown that a confocal SERS method can be nicely applied to study Pt–Nafion interfaces. Since the laser incidence is from the Nafion side, it is expected Nafion would absorb light to certain extent and also produce some bulk signals. Would you please comment on whether an ATR (attenuated total reflection) mode could be applicable and more suitable to study electrode–polyelectrolyte interfaces?

Li Xiao replied: We believe that the influence of bulk Nafion signals is limited. In our paper, we have compared the Raman signals obtained by focusing the laser at different positions. The results show that when the laser focus was at the electrode–polyelectrolyte interface, the interface signal would not be significantly disturbed by the bulk signal of Nafion membrane.

We agree that ATR-IR can be a suitable method for this interface study. Its advantage is that the optical path does not pass through the bulk Nafion phase and will not introduce the interference of the body phase signal. Its disadvantage is that the ATR-IR method needs to use silicon column and a complex electroplating process, hence the experimental cost and time consumption can increase a lot compared to our method.

Yi-Lun Ying asked: Would the pollution affect the hydrophobicity of the electrode under potential? After cleaning at high voltage, is it possible that the water peak is expected to appear?

Li Xiao replied: Actually for the electrode–Nafion interface, we can constantly find water peaks in the entire potential scanning range. According to the Raman peaks of Pt–O in the electrode–Nafion interface, it was clear that Pt–O bond is much stronger than that in the electrode–H_2SO_4 interface, meaning the hydrogen bond in Pt–O–H_2O was weakened. This result indicates that there was less water in the (hydrophobic) electrode–Nafion interface. Therefore the hydrophobicity of the electrode was not affected by the pollution.

Tim Albrecht commented: In your talk, you mentioned that some of the features in your spectra could be due to impurities or contamination. Do you think you could have done more to purify or clean the sample, when taking the spectra?

Li Xiao responded: Yes. Multi-cycle cyclic voltammetry scanning is able to remove some pollutant interference. However, because the Raman instrument needed to be repaired, we failed to complete the SERS experiment to remove the pollution before my talk.

Xu Liu communicated: You mentioned in the paper that "it is necessary to test the isolated electrode–polyelectrolyte interface without the coexistence of any electrode–solution interfaces". Could you provide insights into potential pathways that people can pursue to achieve these tests?

Also, you mentioned that "Nafion™ still has many characteristic peaks in the range of 1000–1600 cm^{-1}, due to the possible presence of the substrate impurities they could not be determined". What could the substrate impurities be?

Li Xiao communicated in reply: At present, traditional electrochemistry can be used to study the isolated electrode–polyelectrolyte interface. Electrochemistry can also be combined with Raman or attenuated total reflection infrared (ATR-IR) methods. Electrochemistry can further combine with spectroscopic methods such as electrochemical impedance spectroscopy (EIS), *etc.* The combination of these methods is the potential pathway, but to realize it we need to overcome different types of technical obstacles, such as the design of the electrolytic cell, *etc.*

We consider "the substrate impurities" the same as the contaminants in the H_2SO_4 system, which is introduced by the Au nanoparticles, but we are not sure what they could be. In the synthesis of the Au nanoparticles, trisodium citrate is used as reductant and it can be oxidized into other organics. We haven't found a literature on what organics can be produced in this reaction. We compared the obtained spectra with that of trisodium citrate but find weak relationship between them. And besides the possible peaks of trisodium citrate, there are many other peaks with changeable and irregular Raman shift and intensity. Thus, it's hard to point out what the contaminants could be.

Xiangkun Elvis Cao communicated: Would you be able to include a high-level summary of the key pros and cons comparing surface-enhanced Raman spectroscopy (SERS) and other methods in probing the electrode–polyelectrolyte interface? Also, what would the potential shortcomings be of combining a printing electrode (SPE) with the SERS technique, as described in the current study?

Li Xiao communicated in reply: Other potential methods for studying the electrode–polyelectrolyte interface include the attenuated total reflection infrared (ATR-IR) method. Its advantage is that the optical path does not pass through the bulk Nafion phase and will not introduce the interference of the body phase signal. Its disadvantage is that the ATR-IR method needs to use silicon columns and a complex electroplating process, hence the experimental cost and time consumption can increase a lot compared to our method.

The combination of SPE and SERS technology has many advantages. We have optimized a lot of experimental details, including improving the simplicity of the method, the stability of the reference electrode, and the rationality of the laser focusing position. At present, this method has low cost, and the electrodes can be purchased in batches without the interference of solution electrolytes. As for disadvantages, Au NPs may introduce some impurity signals at the interference.

Conflicts of interest

There are no conflicts to declare.

Faraday Discussions

PAPER

Hybrid scanning electrochemical cell microscopy-interference reflection microscopy (SECCM-IRM): tracking phase formation on surfaces in small volumes

Dimitrios Valavanis, †[a] Paolo Ciocci,†[b] Gabriel N. Meloni, †[a] Peter Morris,[a] Jean-François Lemineur,[b] Ian J. McPherson, [a] Frédéric Kanoufi *[b] and Patrick R. Unwin *[a]

Received 12th September 2021, Accepted 5th October 2021
DOI: 10.1039/d1fd00063b

We describe the combination of scanning electrochemical cell microscopy (SECCM) and interference reflection microscopy (IRM) to produce a compelling technique for the study of interfacial processes and to track the SECCM meniscus status in real-time. SECCM allows reactions to be confined to well defined nm-to-μm-sized regions of a surface, and for experiments to be repeated quickly and easily at multiple locations. IRM is a highly surface-sensitive technique which reveals processes happening (very) close to a substrate with temporal and spatial resolution commensurate with typical electrochemical techniques. By using thin transparent conductive layers on glass as substrates, IRM can be coupled to SECCM, to allow real-time *in situ* optical monitoring of the SECCM meniscus and of processes that occur within it at the electrode/electrolyte interface. We first use the technique to assess the stability of the SECCM meniscus during voltammetry at an indium tin oxide (ITO) electrode at close to neutral pH, demonstrating that the meniscus contact area is rather stable over a large potential window and reproducible, varying by only *ca.* 5% over different SECCM approaches. At high cathodic potentials, subtle electrowetting is easily detected and quantified. We also look inside the meniscus to reveal surface changes at extreme cathodic potentials, assigned to the possible formation of indium nanoparticles. Finally, we examine the effect of meniscus size and driving potential on $CaCO_3$ precipitation at the ITO electrode as a result of electrochemically-generated pH swings. We are able to track the number, spatial distribution and morphology of material with high spatiotemporal resolution and rationalise some of the observed deposition patterns with finite element method modelling of reactive-transport. Growth of solid phases on surfaces from solution is an important pathway to functional materials and SECCM-IRM provides a means for *in situ* or *in operando* visualisation and tracking of these processes with

[a]*Department of Chemistry, University of Warwick, Coventry CV4 7AL, UK. E-mail: p.r.unwin@warwick.ac.uk*
[b]*Université de Paris, ITODYS, CNRS, F-75006 Paris, France. E-mail: frederic.kanoufi@u-paris.fr*
† These authors contributed equally.

improved fidelity. We anticipate that this technique will be particularly powerful for the study of phase formation processes, especially as the high throughput nature of SECCM-IRM (where each spot is a separate experiment) will allow for the creation of large datasets, exploring a wide experimental parameter landscape.

Introduction

The ability to visualise and study phase formation and change is extremely important in materials science, to understand processes such as crystallisation, precipitation, and deposition. Practical applications include the development of new pharmaceuticals,[1] crop protection agents,[2] new energy storage materials,[3] among others. Many techniques are available to track phase formation, including optical,[4] fluorescence,[5] and scanning electron and transmission electron microscopies (SEM/TEM),[6] as well as Raman and infrared spectroscopy,[7,8] or scanning probe techniques like scanning tunnelling microscopy or atomic force microscopy (STM/AFM).[9,10] Application of microscopy with electrochemical control is particularly attractive as it provides a means to trigger, monitor and track (electro)chemical processes, ultimately at the single entity level.[11,12] More generally, the use of co-located microscopy techniques that probe and correlate structure and activity in the same space have the power to reveal surface processes in quantitative microscopic detail,[13] particularly when applied *in operando*.[14–17]

Electrochemical techniques can be used to change solution composition in a controlled manner, by either consuming/producing reagents at the surface of an electrode immersed in a solution,[18–20] or by precisely mixing reagents using migration and electroosmotic flow in a nanopipette.[21–24] A particular advantage of electrochemistry is that the driving force can usually be controlled *via* the applied potential and mass transport can be varied over a wide range, and modelled with a high degree of accuracy,[25–27] allowing spatiotemporal changes in solution composition to be predicted quantitatively.

Recently a hybrid approach, using interference reflection microscopy (IRM) coupled to electrochemistry, was used to visualise and study the growth of ensembles of nanoparticles at optically transparent electrodes, *in situ*, at different overpotentials.[28–30] As a surface sensitive technique, IRM is particularly suited to the investigation of phenomena at the electrode/electrolyte interface.[31–34] At optically transparent electrodes, such as indium tin oxide (ITO) immersed in solution, local phase formation and/or change at the electrode surface modify the refractive index, giving rise to a difference in contrast in the observed reflected light image. For example, this phase change can arise from the growth of nanoparticles or crystals at the electrode surface providing real-time *in situ* observation of the phase formation event.[31] The sensitivity of IRM, or its analogue known as iSCAT, named for interferometric scattering microscope,[35] can be pushed towards single molecule imaging, or in the context of electrochemistry to the imaging of the restructuring of the electrochemical double layer of nanoparticles.[34] Although IRM is usually performed in large electrolyte volume, IRM can be coupled with spatially resolved electrochemical techniques to provide local electrochemical control and monitoring of phase formation events.

Scanning electrochemical cell microscopy (SECCM) provides a powerful platform to create very well defined and consistent microscale and nanoscale-sized electrochemical cells on various substrates.[36,37] Since its inception,[38,39] SECCM, and the scanning micropipette contact method (SMCM) as an earlier variant was named, has been widely employed to investigate electrochemical processes at the nanoscale, revealing heterogeneous properties of many features of materials including 2D materials,[36,40–43] nanotubes,[44,45] grain boundaries and crystallographic facets in electrodes and electrocatalysts,[46–48] single particles,[49–53] among a rapidly expanding range of applications.[37] For phase formation and phase change processes, the ability to confine electrochemical reactions to small volumes allows the investigation of a few or even single events.[54,55] In this context, the use of SECCM in hopping mode,[56] where the SECCM meniscus is landed at a series of spots on a surface, is particularly powerful, as it is possible to build up large datasets with the same, or different, experimental conditions applied to each spot.

In this work, we combine SECCM with the surface sensitivity of IRM to investigate phase formation in tiny droplet cells at an ITO electrode surface. First, we employ this new hybrid technique to visualise and monitor the evolution of the SECCM meniscus dimension during electrochemical actuation. Subtle variations in the meniscus surface area are detected when faradaic current flows at the ITO/electrolyte interface. Meanwhile IRM can track, *in situ* with high temporal resolution (ms), changes to the electrode within the meniscus footprint, with the indication of indium (In) nanoparticle (NP) formation at potential regions associated with the solvent breakdown at the ITO electrode. Second, as proof of concept of the use of SECCM-IRM to track phase formation, we demonstrate the nucleation and growth of calcium carbonate ($CaCO_3$), driven by a pH change caused primarily by the oxygen reduction reaction (ORR) at the ITO surface. We observe a dependence of the number of crystallisation sites on the electrochemical cell geometry and ORR driving potential. Supported by finite element method (FEM) simulations, we further demonstrate that the pattern of crystals formed within the SECCM meniscus is related to spatiotemporal variations in supersaturation within the meniscus, governed by the interplay between oxygen transport across the meniscus/air interface, and its reduction at ITO, mass transport from the SECCM pipette, coupled to carbonate buffer speciation.

The combination of SECCM and IRM opens up many new opportunities for the investigation of electro(chemical) dynamics at the nanoscale. The successful development and application of SECCM-IRM in this work serves as a foundation for future studies.

Experimental

Reagents and materials

Potassium nitrate (KNO_3, Sigma), hydrochloric acid (HCl, 37%, Honeywell Fluka), anhydrous calcium chloride ($CaCl_2$, 96%, Acros Organics), and sodium bicarbonate ($NaHCO_3$, 99+%, Acros Organics), were used as supplied by the manufacturer. Solutions were prepared with ultrapure deionised water (PURELAB Chorus, ELGA, UK).

ITO coated coverslips (20 × 20 mm, 70 nm coating thickness, SPI Supplies, USA) served as the working electrode (WE)-IRM substrate. They were first cleaned by sonication in isopropanol and subsequently in deionised water. Electrical connection to the ITO was made by connecting a copper wire to the coverslip with silver conductive paint (RS Components, UK). To secure the wire, two-part epoxy adhesive was used (Araldite, UK). An Ag/AgCl quasi-reference counter electrode (QRCE) was prepared by anodizing 0.125 mm diameter annealed silver wire (99.99%, Goodfellow, UK) in a saturated KCl solution for 10 min. All potentials reported herein are *versus* this QRCE, which is stable over long experimental run times.[57] All SECCM experiments were performed with a single barrel pipette, loaded with electrolyte solution, as the mobile probe.[37] Three different sizes of pipettes (end aperture) were used in the experiments, all pulled from borosilicate capillary tubes: (i) 200 nm diameter pipettes from outer diameter (O.D.) 1.2 mm, inner diameter (I.D.) 0.69 mm tubes; (ii) 5 μm diameter pipettes from O.D. 1.0 mm, I.D. 0.5 mm tubes, both pulled using a laser pipette puller (P-2000, Sutter Instrument, USA). (iii) Pipettes with a diameter in the range of 20–30 μm were pulled from O.D. 1.0 mm, I.D. 0.5 mm tubes, using a vertical resistive heating pipette puller (PC-10, NARISHIGE, Japan). Micropipette diameters were measured using an optical microscope before the experiments, while nanopipettes were routinely visualised using SEM and TEM.[58]

SECCM-IRM instrumentation

An inverted fluorescence microscope (DMI4000B, Leica, Germany) was utilised as the base for the SECCM-IRM workstation. For the IRM measurements, 2 light sources were used to back-illuminate the ITO electrodes: either the blue channel LED (intensity peak at wavelength of 470 nm, full-width at half-maximum (FWHM) bandwidth of 19 nm) from a multi-LED light source (Niji, Bluebox Optics, UK); or a mercury metal halide white light source (EL6000, Leica) in conjunction with a bandpass filter (ET480/20x, intensity peak at 480 nm, FWHM bandwidth of 20 nm, Chroma, USA). Images were captured using a 20× objective (N Plan, numerical aperture (NA) 0.35, Leica) or a 63× oil immersion objective (HCX Plan Apochromatic, NA 1.4, Leica). An additional 1.6× tube lens was used, to achieve up to 100× effective magnification. IRM images were recorded with a CMOS camera (C11440-42U30, Hamamatsu Photonics, Japan) at acquisition rates ranging from 30 to 300 frames per second (FPS). The microscope, camera and light sources were controlled using MicroManager (release 1.4.22, open source).[59]

A two-axis piezoelectric positioning stage (Nano-Bio300, with a Nano-Drive controller, Mad City Labs, USA) was mounted atop the microscope and used to move the ITO substrate (WE) in the *XY* plane. A single axis piezoelectric actuator (P-753.3, with an E-665 controller, Physik Instrumente, Germany) was used for moving the SECCM probe along the *Z* axis, perpendicular to the WE plane. For coarse SECCM probe positioning, *XYZ* micropositioners (M-461-XYZ-M, Newport, USA) were used in conjunction with a *Z* axis stepper motor (8303 Picomotor Actuator, Newport). All high resolution SECCM approaches were performed using the *Z* axis piezoelectric actuator.

Electrochemical measurements were performed using custom built current followers. Data acquisition and SECCM probe positioning control were performed by an FPGA card (USB-7856R OEM, National Instruments, USA), in turn governed by the Warwick Electrochemical Scanning Probe Microscopy (WEC-SPM) platform,[60] controlled by LabVIEW software (release 2019, National Instruments). Image recording was triggered from the WEC-SPM platform, by outputting a TTL signal to the camera input trigger port. The camera output trigger's pulse train was also fed to the FPGA card, and recorded concurrently with the electrochemical measurements, allowing synchronous analysis and presentation of SECCM and IRM data. The entire equipment (microscope, stage, positioners, and camera) was enclosed in a Faraday cage, lined with thermal isolation panels, and supported on an active optical table (CleanBench, TMC, USA).

Electrochemical data processing and plotting was carried out using the MATLAB (release R2019b, The MathWorks, Inc., USA) and OriginPro (release 2021, OriginLab, USA) software packages. Image analysis and treatment was performed with custom Python routines based on Skimage and Trackpy packages (release 2021, open source).

Raman spectroscopy

Raman spectra were recorded using a Raman microscope (Horiba LabRam HR Evolution) fitted with a charged couple device detector and a 660 nm OPSS laser. A 50× (0.5 NA) objective and 1800 lines mm^{-1} grating was used, with 10 × 1 s acquisitions averaged per spectrum.

Reactive-transport modelling

The time-dependent speciation inside the meniscus and pipette were simulated by numerically solving the diffusion equation for the system using the FEM in COMSOL Multiphysics (version 5.6, COMSOL Inc., USA). A typical axisymmetric geometry consisting of the end of the pipette and the SECCM meniscus was based on optical images of the pipette and the wetted area observed by IRM. We applied the reactive-transport model we previously reported for time-dependent $CaCO_3(aq)$ speciation.[21] Simulation details are further described in the FEM results section below.

Results and discussion

SECCM-IRM principles

The SECCM-IRM setup is presented schematically in Fig. 1. It consists of an inverted optical microscope which focuses the light source, through an objective lens, onto the back of an ITO-coated glass coverslip. The top conductive side of the coverslip is used as the WE. The SECCM probe, loaded with electrolyte solution (50 mM KNO_3, for the study of ITO; and 5 mM $CaCl_2$ and 5 mM $NaHCO_3$ (pH 2.7 HCl) for the crystallisation experiments) and containing a QRCE, was coarsely brought closer to the ITO substrate using the micromanipulators and the stepper motor. The SECCM QRCE potential was controlled during the experiments, while the ITO substrate (WE) was held at a virtual ground. The resulting current was measured at the WE with a custom-built current follower. For ease of interpretation, all potentials reported are for the WE electrode, which is the negative value of the SECCM QRCE potential. Final tip approach was performed by applying a small bias between the QRCE and WE (+0.2 V for ITO studies and −0.1 V for the crystallisation experiments). The probe movement stopped automatically as soon the current was above a threshold, set at the noise level (ranging between 100 fA and 30 pA, depending on the sensitivity of the current follower that was appropriate for each probe size). The event indicated contact between the

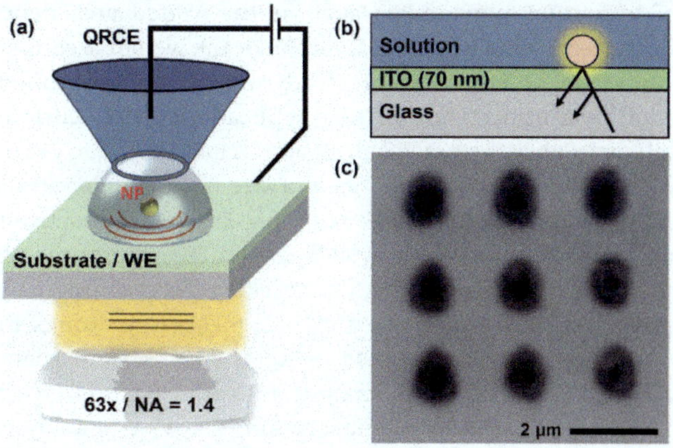

Fig. 1 Schematic (not to scale) of (a) the SECCM-IRM setup showing the SECCM probe, ITO WE and optical microscope. (b) Principle of operation of IRM to produce high contrast images of events and features at the WE interface. (c) Typical IRM image, recorded *in situ*, of an SECCM array section, with replicates of individual experiments. The darker electrolyte residue, left after probe retraction, is contrasted against the brighter ITO substrate background.

electrolyte meniscus formed at the end of the probe and the WE, resulting in a small electrochemical cell, with dimensions related to the tip-end diameter used. This event could also be detected in the IRM images (*vide infra*).

The optical detection principle has been detailed in previous works.[28,31,32] It is based on the reflection of the light wave at the ITO–electrolyte interface and its possible interference with light wave scattered by objects present on the ITO surface, which disturb the local optical conditions and are highlighted within the background, with the camera acting as an interferometric detector. Optical features can show either a negative contrast, the feature appearing darker than the background, or a positive contrast, where the feature appears brighter than the background. The sign of the contrast depends on the dielectric properties of the object relative to its environment and its size.[31,61]

Once the meniscus had landed on the WE (no contact from the pipette itself), a short pause allowed monitoring of the meniscus stability and provided a background for later image treatment (*vide infra*). The potential was then swept or stepped to drive ORR (and water reduction) at the WE, with synchronous IRM visualisation of the meniscus footprint region. Once the potential sweep or step was finished, the potential was switched back to the approach value. The probe was then retracted, to detach the meniscus from the surface, and the WE shifted laterally and the approach recommenced. In this way, the experiment could be repeated on a fresh ITO region, allowing replicates of the experiment (Fig. 1). IRM images were acquired synchronously throughout the entire experiment.

Real-time imaging of the SECCM meniscus

The use of IRM opens up the possibility of real-time, *in situ*, tracking of the SECCM meniscus, allowing monitoring of its size and shape during experiments and the (electro)chemical phenomena occurring within. This opens the possibility of important phenomena such as the wetted area dependence on potential (electrowetting) to be studied. Although the wetted area may have little impact on the SECCM currents for steady-state transport-limited measurements,[62] it is of great importance for new applications where current density is essential, such as in energy storage/conversion materials.[63]

Previously, the meniscus contact area and size has been deduced or inferred from a variety of means, including: (i) scanning sharp electroactive features on surfaces, such as individual single walled nanotubes, the border between an electrode and insulator, or step edges, and tracking the resulting current–lateral distance profiles with nm-resolution;[36,44,64–66] (ii) measuring droplet footprints or line tracks left on a surface, with SEM or AFM, after SECCM scanning under certain conditions;[36,63,67] (iii) AFM analysis of surface patterns after SECCM molecular-grafting, polymerisation or adsorption;[68–72] (iv) AFM analysis of SECCM etch features;[73,74] (v) *in situ* visualisation of the meniscus and nanopipette with laser scanning fluorescence microscopy;[62] (vi) analysis of dual-channel ion conductance tip approach curves.[36,62,75] Nonetheless, as we show herein, IRM provides unprecedented views of the SECCM meniscus dynamics under operation.

To demonstrate IRM SECCM-meniscus imaging in real-time, we repeatedly approached a 50 mM KNO_3-loaded 200 nm diameter nanopipette to an ITO substrate and recorded cyclic voltammograms between +0.2 V and −1.6 V at 1.7 V s^{-1} while IRM images were synchronously recorded at a rate of 325 FPS. This procedure was repeated at 17 different locations over the ITO substrate. Fig. 2a shows IRM images of the meniscus (dark contrasted area), which has a diameter of *ca.* 400 nm, at selected potentials. The voltammograms (Fig. 2b) show no faradaic contribution before solvent breakdown at the extreme cathodic potential. The evolution of the dimension of the meniscus footprint was monitored during each voltammogram. The area of the meniscus was evaluated either from the surface area (in pixels) of the dark region in the centre of the images, or from the super-localization of its edges along two orthogonal directions.

The meniscus was rather reproducible among the 17 different experiments with an overall ±5% variation in area, measured during the short pause after meniscus landing. Fig. 2c gives the evolution of the ensemble-averaged surface area of 17 menisci during the potential cycle. The meniscus was rather stable during the forward potential scan from 0 V to −1.47 V, but its surface area increased when faradaic current flowed, in the solvent breakdown region. On the reverse scan, the meniscus did not retract to its original dimension. Note that the sudden increase in the meniscus size to a stable value at the start of the scan is associated with meniscus landing that we will analyse in detail elsewhere.

The inset of Fig. 2c shows the evolution (in a narrower potential region from −1 V to −1.6 V) of the width of an individual meniscus estimated from the FWHM of the intensity profile. If the surface area provides an estimate with a quantum pixel resolution, the edge localization procedure allows sub-pixel resolution, here the individual meniscus diameter increases by 0.31 pixel, 20 nm, above the usual 5 nm resolution of the procedure.[76,77] The observed electrowetting is not attributed to double layer charging effects since it rather occurs in the solvent electrochemical decomposition. However, the observed phenomena are in general agreement with the Lippmann equation predicting a decrease of the droplet contact angle (and therefore increase in its footprint) with applied voltage.[78–80]

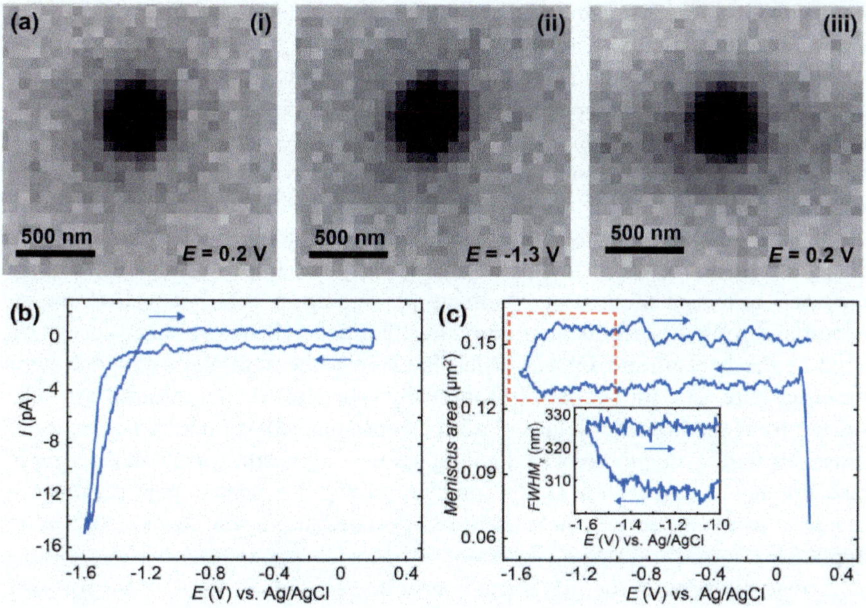

Fig. 2 (a) IRM images of an SECCM meniscus (200 nm diameter tip) at selected potentials during cyclic voltammetry (scan rate 1.7 V s^{-1}) at ITO with 50 mM KNO_3 solution. (b) Representative cyclic voltammogram recorded at one of the 17 spots of the experiment array. (c) Mean SECCM meniscus area (averaged over the 17 spots) as a function of the electrode potential as calculated from the corresponding IRM images. Inset in (c) shows the evolution of the FWHM of the meniscus, in the selected potential window, evaluated from its edge-tracking along the horizontal axis, during the cyclic voltammogram corresponding to (b). Arrows denote potential scan direction. IRM imaging recorded at 325 FPS (*ca.* 5 mV potential resolution). Light source: mercury metal halide white light, filtered around the wavelength of 480 nm.

SECCM-IRM offers a non-invasive dynamic observation of the meniscus with unprecedented >300 FPS image acquisition rate of the whole meniscus and, through superlocalization strategy, with sub-10 nm relative size variation. Other high spatial resolution methods used to infer the SECCM meniscus size, such as SEM or AFM are performed *ex situ* on the meniscus footprint.

NP growth within a nanometre sized meniscus

The faradaic current flowing at the most cathodic potentials can be attributed to solvent breakdown, and there may also be the reduction of ITO producing surface confined In NPs. This process has been evidenced by *ex situ* SEM images of the electroreduced NPs for a variety of electrolyte composition and pH.[81] The dynamics of the NP formation and its competition with H_2 nanobubble electrogeneration during the hydrogen evolution reaction (HER) was more recently monitored *in situ* by IRM.[32]

From the previously reported optical and SEM images, the density of In NPs is of the order of 10^8 cm^{-2}.[31] Thus, in principle, the confined region on the ITO WE accessed by SECCM facilitates the study of a single NP formation and growth event. For example, within a meniscus of diameter, d < 500 nm, such as the one monitored in Fig. 2a, the probability of finding one nucleation site is below 1 for an average nucleation site density lower than $4/\pi d^2 = 5 \times 10^8$ cm^{-2}.

Although previous studies of nucleation/growth of NPs with optical microscopy monitoring could optically isolate a single event, the electrochemical information (current) associated to such event was recorded for the entire electrode, averaged over the ensemble of NPs.[28–30,82] In contrast, with SECCM-IRM, the current is confined to the small SECCM meniscus, allowing the synchronous electrochemical monitoring and optical imaging of a single event.[83] Image background subtraction is then a helpful technique to improve analyses of noisy data sets and it is often used in image processing in IRM.[28] The IRM image captured during the last instant before the potential cycle was switched on can be used as the background image, which is taken to be equivalent of the system baseline state. The individual pixel intensity values of the background are subtracted from the intensity values of all the subsequent IRM images. For periodic noise, created by fluctuations in the light source output, time domain filtering is also applied: images with sharp intensity changes repeated periodically are excluded from the image stack; and pixel-by-pixel sliding-window smoothing is applied.

A treated version of the IRM snapshots at 0.2, −1.3 and −1.5 V for the forward scan and 0.2 V for the reverse scan can be seen in Fig. 3a, with an indicative delimitation of the meniscus border (red dashed circle). These images can be compared to the raw images for 0.2, −1.3 (forward) and 0.2 V (reverse) in Fig. 2a. The image is featureless during the forward scan until the potential reaches

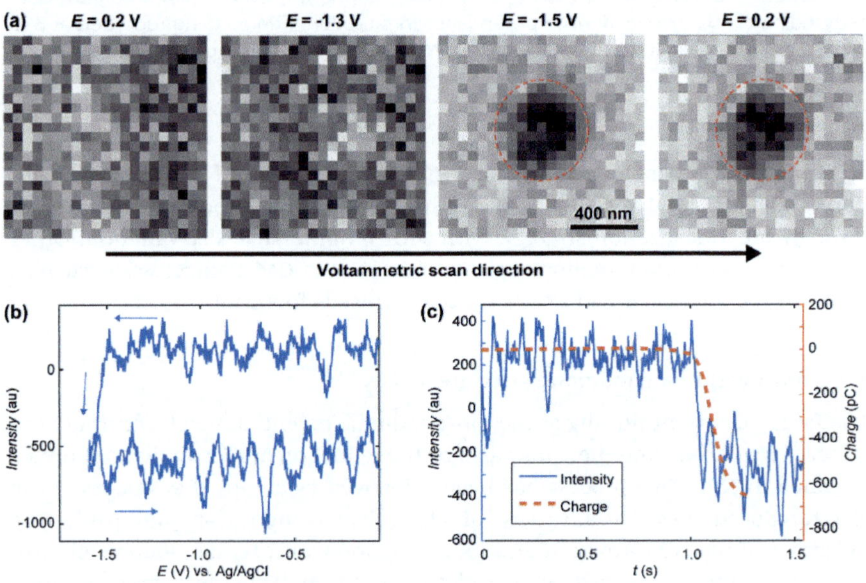

Fig. 3 (a) Treated IRM images recorded during the cyclic voltammetry in Fig. 2a, showing the presence of a feature inside the SECCM meniscus after −1.5 V was reached on the forward scan. (b) Opto-voltammogram showing the intensity in the corresponding ROI (red circle in part a) over the applied potential range. (c) Charge transient (calculated from integrating the current in Fig. 2a) compared to the optical intensity transient ($t = 0$ at the onset of the cyclic voltammetry).

−1.5 V, or until a significant cathodic current is flowing (Fig. 2b). Once cathodic current flows, an optical feature is clearly visible inside the meniscus perimeter, which was not visible in the untreated data (Fig. 2a(ii)), and this feature remains during the whole reverse scan (at 0.2 V in Fig. 3a). By inspecting the intensity (Fig. 3b) of the region of interest (ROI) within the meniscus (red dashed circle in Fig. 3a) throughout the entire image series, a decrease in the intensity was initiated at −1.5 V and is maintained over the cycle. Assigning $t = 0$ to the onset of the potential cycle, the intensity of the ROI is also plotted *vs.* time in Fig. 3c. These data show that the intensity decreases over time after the feature appears on the image, indicating that the feature is growing in size with time, alongside the electrode plane.[31,32,61] Furthermore, by integrating the current over the cyclic voltammogram, one obtains the charge flowing through the ITO. The variation of the ROI intensity profile is remarkably similar to the morphology of the charge–time transient in Fig. 3c.

The feature may have two possible chemical origins. It could be related to HER and nanobubble formation or the reduction of ITO into In NP. The formation of a nanobubble at such a small volume would likely block the pipette opening which would be detected by a sudden drop in electrochemical current.[84] Also, once HER had stopped (in the reverse scan) the nanobubble would dissolve back into solution quickly, which would be discerned by the IRM intensity.[85,86] The current magnitude increases along the forward potential cycle and a crossing of the forward trace is detected on the reverse. Such crossing is often observed during the electrodeposition of metals and is usually attributed to the formation of a material with increased electroactive area (and electroactivity). From the electrowetting analysis, in Fig. 2c, one could evaluate the variation in current density during voltammetry, knowing the dynamic meniscus surface area, obtained from Fig. 2c. When this is done, the same voltammetric shape presenting the crossing branches is observed, indicating that the crossing is not related to the change in meniscus surface area. It is then due to the formation of a material with higher electrochemical activity than the ITO, as would most likely be the metal In formed during ITO reduction. The ability to visualise, track and record the current flow for a single sub-diffraction limited event corresponding to ITO reduction into In NP with high temporal resolution of better than 5 ms, showcases the capabilities of SECCM-IRM for single entity opto-electrochemical dynamic measurements.

Moreover, from observations of both the meniscus surface area and surface degradation, we conclude that ITO is optically stable for IRM during electrochemical experiments ($E > -1.5$ V in KNO_3 at neutral pH), enabling us to monitor the electrodeposition of various nanomaterials in a rather large potential window. However, any interference or influence from possible ITO instability has to be considered for each particular set of conditions, as illustrated in the next section.

Controlled $CaCO_3$ crystallisation in a microscale meniscus

We now turn to the electrochemically-induced deposition of insulating materials on ITO. For this we take $CaCO_3$ as our model system and use the slightly larger volumes provided by the meniscus of a few μm (1–8 μm) diameter SECCM probe. The use of a larger meniscus allows us to probe dynamics within a population of nano-entities within the electrolyte.

$CaCO_3$ is found throughout the natural world where it exists in a variety of structural forms (polymorphs) and morphologies. Many biological systems utilise $CaCO_3$ in their structural, defensive and optical structures, and have the ability to control and even switch between different polymorphs.[87] The product of crystallisation is also heavily dependent on the reaction volume, with small (micron-size) volumes stabilising early-stage, less stable forms (*e.g.* amorphous calcium carbonate, ACC).[16] The early stages of $CaCO_3$ precipitation remain hotly debated,[88–90] although it is known to proceed *via* rapid formation of ACC at moderate supersaturations.

To control the precipitation of $CaCO_3$ we modify the local supersaturation, S:

$$S = \sqrt{\frac{a_{\mathrm{Ca}^{2+}} a_{\mathrm{CO_3}^{2-}}}{K_{\mathrm{SP}}}} \tag{1}$$

where a_i is the activity of species i, and K_{SP} is the solubility product of calcite, by changing the pH.[18–20] Starting with an acidified (*ca.* pH 2.7, therefore undersaturated, Fig. 4a) mixture of $CaCl_2$ and $NaHCO_3$, a cathodic current at the ITO consumes H^+ and produces OH^- by reduction of O_2 (ORR), increasing the pH, converting HCO_3^- to CO_3^{2-} and causing the initially undersaturated solution to become supersaturated (Fig. 4b), although far from stoichiometrically (*i.e.* $[Ca^{2+}] \gg [CO_3^{2-}]$). At this low starting pH most of the solution carbonate is in the form of CO_2 (Fig. 4a) and the solution buffer capacity and pH is expected to drift over time towards less acidic (due to CO_2 loss to the atmosphere). Polarizing the ITO at different potentials allows us to tune the rate at which the supersaturation is changed and therefore to probe the heterogeneous nucleation/precipitation mechanism *in situ*. Previous studies have demonstrated that different electrochemical activity emerges from different compositional surface heterogeneities on ITO.[32] SECCM-IRM allows simultaneous events to be detected at the nanoscale within the meniscus. Indeed, the formation of crystals at certain locations on ITO may serve as a useful marker of local electrochemical activity.

Electrochemically-induced $CaCO_3$ precipitation was investigated across a series of applied potentials at the ITO electrode starting from the same initial solution. In each case a small potential (−0.1 V) was used to approach the SECCM probe to the surface. This potential was held for 1 s to allow the SECCM meniscus to stabilise before being switched to the potential of interest for 10 s, and then back to −0.1 V while the probe retracted, ensuring a constant reaction time in each case. To maintain the pipette solution conditions over the course of several depositions, on alternate approaches of the probe to the surface, the potential was not switched, so as to allow the solution to re-equilibrate and any precipitate within the tip to be removed by deposition *via* meniscus contact with the ITO surface. This procedure was repeated by means of successive approaches at different locations of the ITO surface to form an array of points, typically with 5–8 repeats at each potential.

Whole, *ca.* 8 μm wide, meniscus IRM images, after 10 s of polarisation at three different potentials, are shown in Fig. 5, along with the current–time (I–t) traces recorded (Fig. 5a). The dark contrasted area delineates the meniscus region. At less negative potential (Fig. 5b(i)), the current was negligible and only a few bright spots are observed by IRM. At −0.7 V (Fig. 5b(ii)), however, a more significant current was recorded, and multiple spots, now dark, are observed in the images. This trend continues at −0.8 V (Fig. 5b(iii)), but at −0.85 V (not shown), although

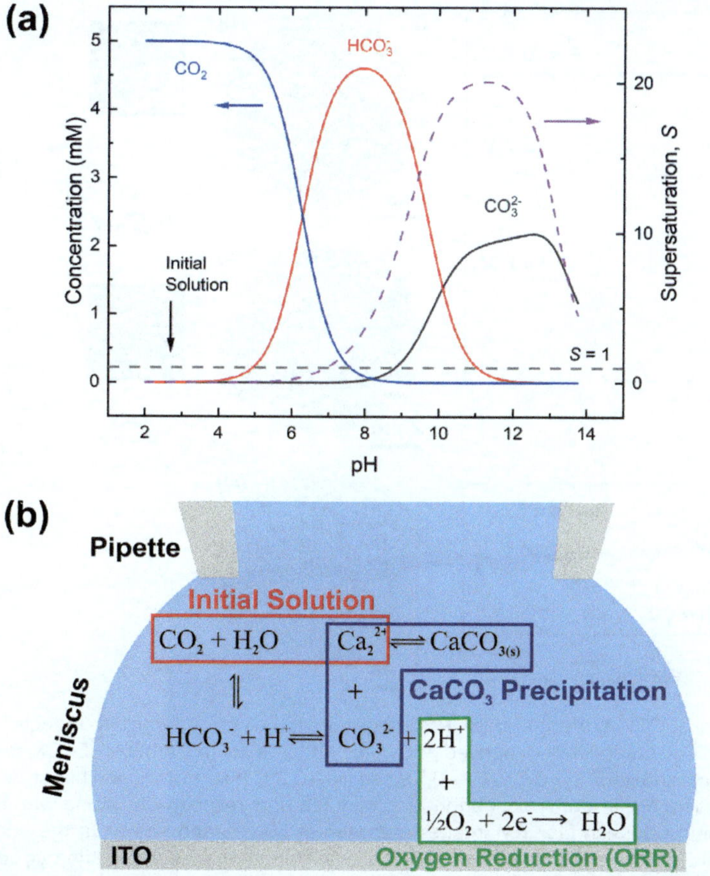

Fig. 4 The principle of electrochemically-induced $CaCO_3$ precipitation. (a) Speciation diagram (left axis) and supersaturation (right axis) for the $CaCO_3$(aq) system as a function of pH. Only principal species are shown, with the remaining mass balance made up of ion pairs (not shown). (b) Equilibria relating O_2 reduction with $CaCO_3$ precipitation.

the current magnitude was still increasing, no clear spots inside the meniscus were seen. Feature detection was hampered by a general darkening of the spot area at this potential and below. The I–t traces in all cases where $E < -0.5$ V show non-monotonic behaviour. An initial spike to negative currents is followed by a rapid decay, and the onset of a sustained increase towards more negative currents. This is unexpected for an ORR pulse at ITO and could suggest that either the crystallisation events or a separate process at the ITO, possibly the formation of In NPs (*vide infra*), are increasing the surface activity.

The size of the optical features seen in the IRM images can be estimated by their FWHM intensity, which is calculated by fitting a Gaussian function to the intensity profiles. The smallest FWHM associated with the dark features in Fig. 5b(ii) and (iii) are 185 ± 25 nm (from 30 spots – with an example shown in Fig. 5b(ii)). Optical features characterized by this low limit value of FWHM can be considered as diffraction-limited, suggesting that nanometric particles (<200 nm) have been generated. The bright features observed at −0.5 V (Fig. 5) increase in

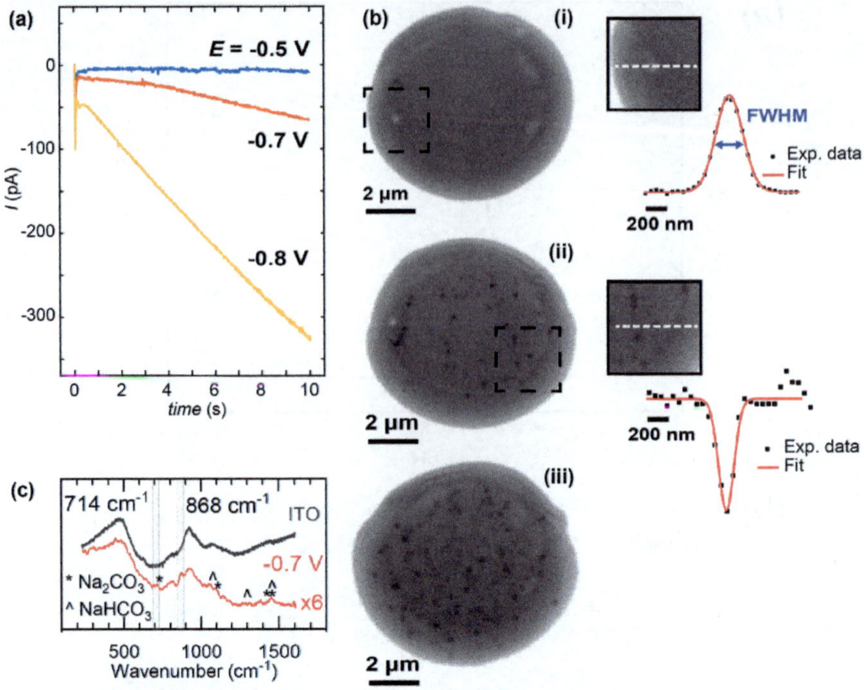

Fig. 5 (a) SECCM current and (b) IRM images for the electrochemically-induced precipitation of $CaCO_3$ at different applied potentials with a *ca.* 8 μm diameter SECCM meniscus. Applied potentials, *E*, (i) −0.5 V, (ii) −0.7 V and (iii) −0.8 V. Insets of b(i) and b(ii) show optical intensity line profiles of typical bright and dark features respectively, along with the corresponding Gaussian fits. IRM images recorded at 50 FPS and shown at the end of the potential pulse (10 s). Light source: blue LED (wavelength of 470 nm). (c) Raman spectrum from a dried meniscus deposited at −0.7 V (with the spectrum from bare ITO subtracted). The bare ITO spectrum is also shown.

intensity with time, in contrast to a decrease observed for the lower potentials. Such behaviour is attributed to the difference in size of the particles formed, as was demonstrated in earlier work for gas nanobubbles: bubbles larger than *ca.* 300 nm appeared lighter than the background, those smaller appeared darker.[32] Despite a difference in the shape and refractive index of the $CaCO_3$ deposits, we might expect a similar behaviour. Indeed, the bright features also show two-fold larger FWHM (365 nm – see example in Fig. 5b(i)), indicating a particle > 300 nm in size at $E = -0.5$ V, while the FWHM remains diffraction-limited during the whole course of experiments at the other potentials.

For $E < -0.85$ V, in addition to the formation of NPs, a homogeneous decrease of the optical intensity was observed over the entire meniscus area (no change detected for the outer ITO/air interface region). This can be explained if a homogeneous decrease of the reflective properties of the ITO/electrolyte interface is considered, and most likely results from the deposition of a homogeneous layer of $CaCO_3$, or a highly dense array of nanoobjects separated by <500 nm such that they cannot be resolved by the optical visualisation, onto the ITO surface. These large negative driving potentials will lead to more rapid formation of the supersaturation condition,[19] and faster nucleation generally results in more nuclei and hence particles. Dense layers of ACC are also known to form at interfaces at appropriately high supersaturations.[91] At such negative potentials and low pH, the reductive degradation of ITO may also occur.[32] The formation of $CaCO_3$ crystals is supported by ITO background-subtracted Raman micro-spectroscopy of a dried meniscus deposited at −0.7 V, which reveals spectra with peaks consistent with the presence of $CaCO_3$ particles (Fig. 5c). A broad peak at 868 cm^{-1} is consistent with either the ν_3 mode of amorphous $CaCO_3$ (863 cm^{-1}) or calcite (873 cm^{-1}), while a small feature at 714 cm^{-1} is also consistent with calcite.[92]

Controlled $CaCO_3$ crystallisation at a larger meniscus

Using pipette probes of tens of μm in diameter allows investigation of even larger areas and the production of a population of hundreds or thousands of electrochemically-induced crystallisation events. At this length scale surface heterogeneities, concentration gradients over the meniscus volume and localized fluxes may contribute to the SECCM-IRM observations. Fig. 6 shows example IRM images taken after 8 s for two different applied potentials at the ITO electrode.

When the applied potential is −0.8 V (Fig. 6a(ii)), a number of features, are observed. Most features show bright contrast and we observe a few aggregates, where neighbouring features coalesce. Experiments at −1.0 V (Fig. 6b) show no homogeneous drop in optical intensity over the meniscus: a much larger amount of nanometric features are observed at the beginning of the potential pulse (Fig. 6b(iii)) which quickly grow to form a variety of μm-sized crystals with different size and shape, formed from the aggregation of nanometric deposits. For both −0.8 V (Fig. 6a(ii)) and −1.0 V (Fig. 6b(ii)) dark contrasted features are also detected. This behaviour is similar to that in the small meniscus and, as suggested by the background experiments and cyclic voltammograms in acidic

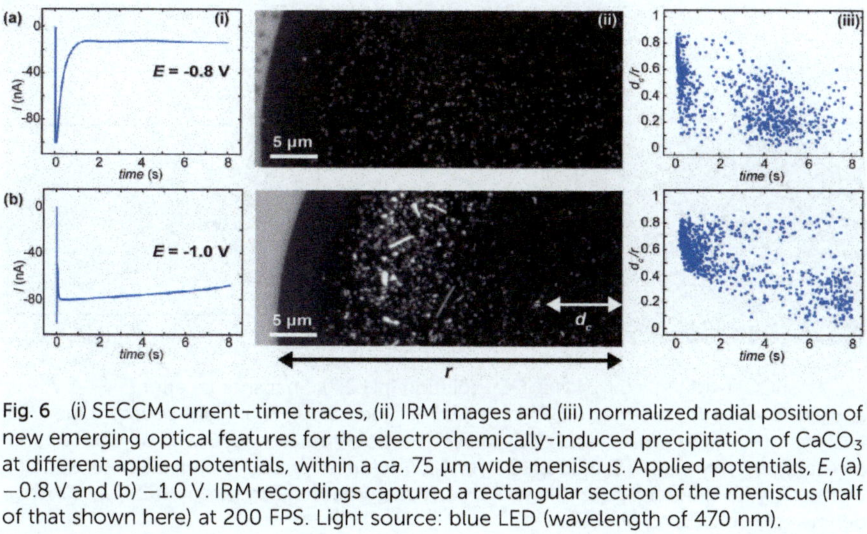

Fig. 6 (i) SECCM current–time traces, (ii) IRM images and (iii) normalized radial position of new emerging optical features for the electrochemically-induced precipitation of $CaCO_3$ at different applied potentials, within a *ca.* 75 μm wide meniscus. Applied potentials, *E*, (a) −0.8 V and (b) −1.0 V. IRM recordings captured a rectangular section of the meniscus (half of that shown here) at 200 FPS. Light source: blue LED (wavelength of 470 nm).

media (*vide infra*), the features could be associated with indium oxide (In_2O_3) reduction into metallic In.

Further spatiotemporal analysis of the crystallisation events show that the position of emerging new optical features in the meniscus, plotted as the radial position (d_c) divided by the meniscus radius (r), as a function of time, has a non-stochastic trend with the onset time of crystallisation for −0.8 V (Fig. 6a(iii)) and −1.0 V (Fig. 6b(iii)). In the first 0.5 s after the potential was applied, a large percentage of the events initiate in a circular crown between $0.4 < d_c/r < 0.9$ for $E = -0.8$ V and between $0.5 < d_c/r < 0.8$ for $E = -1.0$ V. Other crystals start to be formed in areas closer to the meniscus centre only at longer times.

When the experiment was repeated without $CaCl_2$ present in the solution (Fig. 7a) a few spots were observed in the IRM images for $E \leq -1.0$ V. For more negative potential, and for the considered acidic pH, it is known that ITO is unstable (with respect to reduction of In to metallic NPs), which has been observed, along with H_2 nanobubbles, by IRM previously.[32] The features observed here (Fig. 7a) are different in shape and appear in much smaller numbers than those observed when Ca^{2+} is present. To assess the effect of possible ITO degradation to the overall intensity change seen in IRM, cyclic voltammetry in acid media (HCl solution, pH 2.7) was carried out for aerated and deaerated (with nitrogen) solutions and over 2 different potential windows. Fig. 7b shows the voltammograms recorded from 0.8 to −0.8 V. From this experiment it is clear the SECCM current is predominantly carried by ORR as a clear cathodic wave at around −0.4 V, present in aerated solution, is absent when the solution is deaerated. No other voltammetric feature alludes to ITO degradation, although the relatively large capacitive current might obscure some features. In contrast, for a larger potential window (Fig. 7c), from 1.2 V to −1.2 V, a clear anodic peak at around −0.55 V is observed in the reverse scan, for both the aerated and deaerated solutions. This suggests that under the pH conditions for the $CaCO_3$ studies ITO reduction (leading to the re-oxidation peak at −0.55 V) only starts to become significant at potentials more negative than −0.8 V and that at less cathodic potentials, many (if not all) of the diffraction-limited optical spots can be attributed to $CaCO_3$ particles.

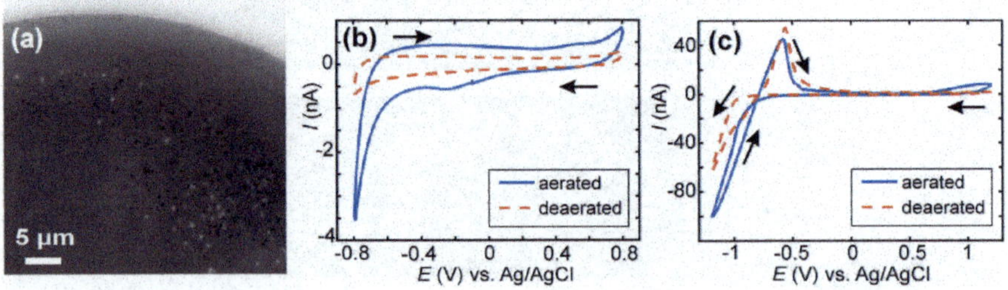

Fig. 7 SECCM-IRM in a 5 mM $NaHCO_3$ solution (pH 2.7). Snapshot (a) after $E = -1$ V was applied for 10 s for a series of IRM images recorded at 10 FPS. Light source: blue LED (wavelength of 470 nm). (b and c) Cyclic voltammograms recorded with an aerated and a deaerated 5 mM $CaCl_2$, 5 mM $NaHCO_3$ (pH 2.7) solution for a *ca.* 80 µm diameter SECCM meniscus with different potential windows. Voltammetric scan rate = 0.2 V s^{-1}. Arrow denotes potential scan direction.

FEM model of $CaCO_3$ precipitation in the larger meniscus

To understand the differences observed when applying −0.8 V and −1.0 V to the larger menisci, reactive transport modelling of the pipette and meniscus was carried out. The model used an axisymmetric geometry based on microscopic characterisation of the pipette and the wetted area observed with IRM. Two approach distances (meniscus heights – d_{app}) corresponding to the outer radius of the micropipette (31 μm), and half of this value, were modelled. This is because the meniscus thickness impacts O_2 transport at the air–water interface (and hence the current distribution).[75] We did not measure the meniscus thickness in this work, but for this size pipette it should be achievable in the future through the use of laser scanning confocal microscopy and an appropriate fluorophore.[62] Indeed, the use of a pH-sensitive fluorophore would also confirm the pH distribution in these types of experiments.[93]

To achieve a realistic estimate of the supersaturation we consider the 12 most significant species in the solution (the ions Ca^{2+}, Cl^-, Na^+, HCO_3^-, CO_3^{2-}, H^+ and OH^-, the ion pairs $NaCO_3^-$, $CaCO_3^0$, $NaHCO_3$ and $Ca(HCO_3)^+$ and the gas phase species O_2 and CO_2) based on prior modelling with the PHREEQC speciation code.[94] We solved the diffusion eqn (2) for the flux of the ith species, J_i, at concentration, c_i, and time, t. Equilibria between the species was maintained by the continuity equation (eqn (3)), where R_i is the reaction flux necessary to maintain equilibrium (although deviations of up to 400% in equilibrium constants at long times were observed as there was a limitation on the rate of HCO_3^- dissociation that could be applied for reasonable computational efficiency, which would lead to an underestimate of supersaturation). Equilibrium constants were based on the local species activity a_i (eqn (4)), calculated from the activity coefficient γ_i based on the local ionic strength, I (eqn (5)) using the Davies equation (eqn (6)) with $A = 1.82 \times 10^6\,(\varepsilon T)^{-3/2}$, where ε is the solution permittivity. For charged species $b = 0.3$, for uncharged species $b = 0.1$. Note that direct capture of CO_2 by OH^- to form HCO_3^- (as opposed to equilibration with H_2O to form HCO_3^- and H^+ based on the local pH) was not considered, based on modelling of similar systems where the rate was negligible on these timescales due to the low OH^- concentrations.[46]

$$j_i = -D_i \nabla c_i \tag{2}$$

$$\frac{\partial c_i}{\partial t} + \nabla j_i = R_i \tag{3}$$

$$a_i = \gamma_i c_i \tag{4}$$

$$I = \frac{1}{2}\sum z_i^2 c_i \tag{5}$$

$$\log_{10}\ \gamma_i = -A z_i^2 \left(\frac{\sqrt{I}}{1+\sqrt{I}} - bI \right) \tag{6}$$

The equilibrium concentrations given by the PHREEQC code were applied as the boundary condition at a plane 3500 μm inside the micropipette. Given the apparent importance of spatial variation across the wetted area, we imposed atmospheric equilibrium concentrations of O_2 and CO_2 at the air/meniscus interface (273 μM and 13.6 μM, respectively). A consequence of this approximation is that the total concentration of carbon species in the meniscus is depleted relative to that in the micropipette. We note that the alternative limiting treatment of CO_2 (no exchange with the atmosphere) would result in greater buffering and therefore longer timescales for the pH swing.

Using the simplifying assumptions that all current arises from the ORR, and that for a given potential, ORR is first order in O_2, we imposed an inwards flux of O_2 and an outwards flux of OH^- at the ITO surface:

$$J_{OH^-,ITO} = -4J_{O_2,ITO} = k_{ORR}[O_2] \tag{7}$$

where we chose k such that the simulated current over the ITO area, A, ($i = F\int nJ_{OH^-}\,dA$) matched the experimental current. For all other boundaries and species not mentioned, a no flux boundary condition was used.

The expected evolution of pH and supersaturation, for fluxes corresponding to the current observed at −0.8 V and −1.0 V, are shown in Fig. 8a. At $d_{app} = r$, both pH and S change much earlier at −1.0 V than at −0.8 V, with the −1.0 V case reaching saturation ($S = 1$) in less than 0.5 s, while −0.8 V takes closer to 4 s. These timescales correspond well with appearance times of new features presented in Fig. 6a(iii) and b(iii), with a clustering of new features around 1 s in the

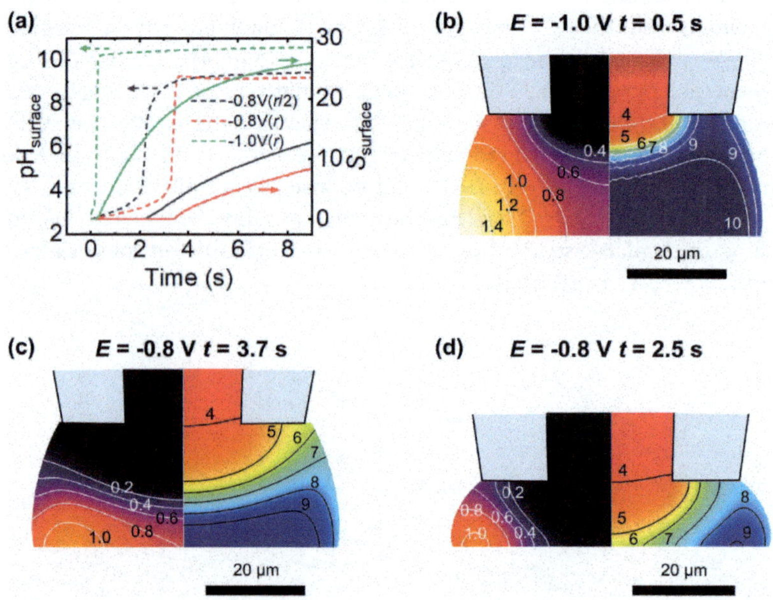

Fig. 8 Reactive transport modelling of $CaCl_2$ + $NaHCO_3$ (5 mM, pH 2.7) droplets at the surface of ITO during oxygen reduction. (a) The time-dependence of the average surface pH (left axis, dashed lines) and average surface supersaturation (right axis, solid lines) for a 62 μm diameter pipette for 2 different applied potentials (ORR rates; see text) and approach heights ($d_{app} = r$ or $r/2$). (b) Profile of the supersaturation, S (left half, and contour lines) and pH (right half) at the first time point where supersaturation with respect to calcite was achieved with an ORR rate corresponding to −1.0 V for $d_{app} = r$. (c) same as (b) but for an ORR rate corresponding to −0.8 V. (d) Same as for (c) but with $d_{app} = r/2$.

−1.0 V case, and a similar clustering at 4 s in the −0.8 V case. The effect of a closer approach ($d_{app} = r/2$) was examined and found to shift the observed changes to earlier times, (*ca.* 2 s at −0.8 V, Fig. 8a, black line, and *ca.* 0.4 s at −1.0 V, data not shown for clarity) and corresponding lower values of S. Electroneutrality was also considered and found to be violated as OH^- flux into the meniscus is not counter balanced by inward diffusion of cations; however, including electroneutrality (*via* the coupled Poisson–Nernst–Planck equations, data not shown) led only to minimal changes in timescale, with a slight increase in time taken to reach supersaturation.

The spatial distribution of pH and S at the first timepoint where saturation is observed for both −1.0 V and −0.8 V cases, at $d_{app} = r$, are shown in Fig. 8b and c, respectively. Both parameters have a strong spatial dependence, with change starting at the meniscus/air interface where the O_2 flux is greatest. Combined with the buffering action at the centre of the droplet due to diffusion from inside the micropipette, this explains the experimentally observed preference for nucleation to occur around the edges of the meniscus rather than at its centre. The pH buffering by atmospheric CO_2 may also contribute to the absence of particles at the very edge in the −1.0 V case (Fig. 6b(ii)). At longer times, at −1.0 V the pH of the meniscus rapidly equilibrates and the spatial variation in S and pH is lost, however at −0.8 V slight variation in S and pH across the surface is observed to persist. Closer approaches (to $d_{app} = r/2$, Fig. 8d) show a larger degree of spatial variation in pH and S. This simplified model is able to rationalise the timescale over which new features are observed and, to some extent, their spatial variation, emphasising the importance of FEM modelling in interpreting SECCM deposition experiments.

Conclusion

We have demonstrated the combination of SECCM and IRM to produce a new instrumental method for the investigation of electrochemically-driven phase formation phenomena with high spatial and temporal resolution. The already proven capabilities of SECCM to precisely and repeatedly position a small scale meniscus from a pipette tip on a substrate has been married to the IRM surface-sensitive imaging technique, enabling high spatiotemporal imaging of the SECCM footprint and processes occurring within it.

Making use of an optically-transparent ITO electrode, we have been able to image and track the stability of the SECCM meniscus size during voltammetry, resolving an important question in SECCM, which has hitherto mainly been addressed with *ex situ* analysis of droplet footprints or the analysis of SECCM scanning droplet images at a fixed potential. The ability to view the meniscus *in situ* provides a more holistic view of meniscus contact. Moreover, careful analysis of the IRM images unambiguously showed the growth of diffraction-limited features within the meniscus, attributed to the reduction of ITO into In NP at the most extreme cathodic potentials.

To demonstrate new possibilities for SECCM-IRM, we studied the precipitation of $CaCO_3$ by electrochemically changing the solution pH to create locally supersaturated solution. IRM *in situ* imaging provided evidence of the number and spatial distribution of crystallisation events over the SECCM meniscus in real-time, revealing striking patterns for different driving potentials and tip sizes.

Smaller reaction volumes showed increased density of fairly evenly distributed nanoscale $CaCO_3$ crystals (confirmed by Raman spectroscopy), with increased applied cathodic potential, which created high supersaturation. For larger reaction volumes (SECCM tips), the location of the crystallisation events become dependent on the driving potential hinting at the dynamic interplay between the pipette mass transport, ORR rates at the ITO surface, and CO_2 and O_2 transport across the air/meniscus interface. FEM simulations have highlighted the origins of the crystallisation patterns and we were able to predict the onset time and, to an extent, the location of the crystallisation events.

More generally, this work establishes a foundation for the use of SECCM-IRM in nanoscale (electro)chemistry. In future work we shall explore the landing and pull-off of the SECCM meniscus on (electrode) surfaces, during translation of the probe to and from the surface, also expanding the work on ITO to other transparent electrode materials. Particularly, we will investigate electrode materials resistant to the electrochemically generated corroding environment, *e.g.* in this work the $CaCO_3$ electrocrystallisation and its visualisation are complicated by the electrochemical etching of ITO under some conditions. Transparent thin Au coated glass coverslips could be used, as previously demonstrated for the optical monitoring of the electrochemistry of single NPs.[28,29,31,95] Coverslips coated with graphene-based materials are also possible electrode candidates, as shown in the IRM imaging of the electrochemistry of graphene oxide.[33,95] Finally, reflectivity microscopy can also be operated at reflective and non-transparent electrodes[96] and the combination with a modified SECCM setup (*e.g.*, with an angled nanopipette, as used in some scanning ion conductance microscopy studies)[97] could be explored. Among other applications, we shall also investigate direct electrodeposition (*e.g.* of metals), where we shall be able to tie together electrochemical signatures to an optical view of the entire meniscus. SECCM-IRM should reveal considerable new detail on the mechanisms of nanoscale electrodeposition.

Conflicts of interest

There are no conflicts to declare.

Acknowledgements

We acknowledge use of the Spectroscopy Research Technology Platform, University of Warwick. D. V., P. C., F. K. and P. R. U. acknowledge support from the European Union's Horizon 2020 Research and Innovation Programme under the Marie Skłodowska-Curie MSCA-ITN Single-Entity Nanoelectrochemistry, SENTINEL [812398]. P. M., I. J. M. and P. R. U. acknowledge an EPSRC Programme Grant (grant EP/R018820/1) which funds the Crystallisation in the Real World consortium. P. R. U. thanks the Royal Society for a Wolfson Research Merit Award.

References

1 L. S. Taylor, D. E. Braun and J. W. Steed, Crystals and Crystallization in Drug Delivery Design, *Cryst. Growth Des.*, 2021, **21**(3), 1375–1377, DOI: 10.1021/acs.cgd.0c01592.

2 M. A. McDonald, H. Salami, P. R. Harris, C. E. Lagerman, X. Yang, A. S. Bommarius, M. A. Grover and R. W. Rousseau, Reactive Crystallization: A Review, *React. Chem. Eng.*, 2021, **6**(3), 364–400, DOI: 10.1039/d0re00272k.
3 J. Zhou and B. Wang, Emerging Crystalline Porous Materials as a Multifunctional Platform for Electrochemical Energy Storage, *Chem. Soc. Rev.*, 2017, **46**(22), 6927–6945, DOI: 10.1039/c7cs00283a.
4 E. R. Ravenhill, M. Adobes-Vidal and P. R. Unwin, Calcium Carbonate Crystallisation at Charged Graphite Surfaces, *Chem. Commun.*, 2017, **53**(93), 12552–12555, DOI: 10.1039/c7cc07441g.
5 C. Zheng, Q. Zang, H. Nie, W. Huang, Z. Zhao, A. Qin, R. Hu and B. Z. Tang, Fluorescence Visualization of Crystal Formation and Transformation Processes of Organic Luminogens with Crystallization-Induced Emission Characteristics, *Mater. Chem. Front.*, 2018, **2**(1), 180–188, DOI: 10.1039/c7qm00435d.
6 H. E. M. Hussein, R. J. Maurer, H. Amari, J. J. P. Peters, L. Meng, R. Beanland, M. E. Newton and J. V. Macpherson, Tracking Metal Electrodeposition Dynamics from Nucleation and Growth of a Single Atom to a Crystalline Nanoparticle, *ACS Nano*, 2018, **12**(7), 7388–7396, DOI: 10.1021/acsnano.8b04089.
7 Y. Hu, J. K. Liang, A. S. Myerson and L. S. Taylor, Crystallization Monitoring by Raman Spectroscopy: Simultaneous Measurement of Desupersaturation Profile and Polymorphic Form in Flufenamic Acid Systems, *Ind. Eng. Chem. Res.*, 2005, **44**(5), 1233–1240, DOI: 10.1021/ie049745u.
8 V. Liotta and V. Sabesan, Monitoring and Feedback Control of Supersaturation Using ATR-FTIR to Produce an Active Pharmaceutical Ingredient of a Desired Crystal Size, *Org. Process Res. Dev.*, 2004, **8**(3), 488–494, DOI: 10.1021/op049959n.
9 R. M. Penner, Mesoscopic Metal Particles and Wires by Electrodeposition, *J. Phys. Chem. B*, 2002, **106**(13), 3339–3353, DOI: 10.1021/jp013219o.
10 J. V. Zoval, R. M. Stiger, P. R. Biernacki and R. M. Penner, Electrochemical Deposition of Silver Nanocrystallites on the Atomically Smooth Graphite Basal Plane, *J. Phys. Chem.*, 1996, **100**(2), 837–844, DOI: 10.1021/jp952291h.
11 Y. Wang, X. Shan and N. Tao, Emerging Tools for Studying Single Entity Electrochemistry, *Faraday Discuss.*, 2016, **193**, 9–39, DOI: 10.1039/c6fd00180g.
12 L. A. Baker, Perspective and Prospectus on Single-Entity Electrochemistry, *J. Am. Chem. Soc.*, 2018, **140**(46), 15549–15559, DOI: 10.1021/jacs.8b09747.
13 C. L. Bentley, M. Kang and P. R. Unwin, Nanoscale Surface Structure–Activity in Electrochemistry and Electrocatalysis, *J. Am. Chem. Soc.*, 2019, **141**(6), 2179–2193, DOI: 10.1021/jacs.8b09828.
14 J. T. Mefford, A. R. Akbashev, M. Kang, C. L. Bentley, W. E. Gent, H. D. Deng, D. H. Alsem, Y.-S. Yu, N. J. Salmon, D. A. Shapiro, P. R. Unwin and W. C. Chueh, Correlative Operando Microscopy of Oxygen Evolution Electrocatalysts, *Nature*, 2021, **593**(7857), 67–73, DOI: 10.1038/s41586-021-03454-x.
15 Y. Yang, Y. Xiong, R. Zeng, X. Lu, M. Krumov, X. Huang, W. Xu, H. Wang, F. J. DiSalvo, J. D. Brock, D. A. Muller and H. D. Abruña, Operando Methods in Electrocatalysis, *ACS Catal.*, 2021, **11**(3), 1136–1178, DOI: 10.1021/acscatal.0c04789.

16 F. C. Meldrum and C. O'Shaughnessy, Crystallization in Confinement, *Adv. Mater.*, 2020, **32**(31), 2001068, DOI: 10.1002/adma.202001068.
17 S. Santesson, E. S. Cedergren-Zeppezauer, T. Johansson, T. Laurell, J. Nilsson and S. Nilsson, Screening of Nucleation Conditions Using Levitated Drops for Protein Crystallization, *Anal. Chem.*, 2003, **75**(7), 1733–1740, DOI: 10.1021/ac020496y.
18 C. Gabrielli, G. Maurin, G. Poindessous and R. Rosset, Nucleation and Growth of Calcium Carbonate by an Electrochemical Scaling Process, *J. Cryst. Growth*, 1999, **200**(1–2), 236–250, DOI: 10.1016/S0022-0248(98)01261-5.
19 B. Blount, K. Kilner, H. Hu, D. Gohmann, E. Gordon, Y. Wang and H. Ren, Electrochemically Induced Nucleation of a Nanoscopic Ionic Solid, *J. Phys. Chem. C*, 2020, **124**(31), 17413–17417, DOI: 10.1021/acs.jpcc.0c05009.
20 O. Oloye and A. P. O'Mullane, Electrochemical Capture and Storage of CO2 as Calcium Carbonate, *ChemSusChem*, 2021, **14**(7), 1767–1775, DOI: 10.1002/cssc.202100134.
21 P. D. Morris, I. J. McPherson, G. N. Meloni and P. R. Unwin, Nanoscale Kinetics of Amorphous Calcium Carbonate Precipitation in H_2O and D_2O, *Phys. Chem. Chem. Phys.*, 2020, **22**(38), 22107–22115, DOI: 10.1039/d0cp03032e.
22 P. D. Morris, I. J. McPherson, M. A. Edwards, R. J. Kashtiban, R. I. Walton and P. R. Unwin, Electric Field-Controlled Synthesis and Characterisation of Single Metal–Organic-Framework (MOF) Nanoparticles, *Angew. Chem., Int. Ed.*, 2020, **59**(44), 19696–19701, DOI: 10.1002/anie.202007146.
23 B. Babakinejad, P. Jönsson, A. López Córdoba, P. Actis, P. Novak, Y. Takahashi, A. Shevchuk, U. Anand, P. Anand, A. Drews, A. Ferrer-Montiel, D. Klenerman and Y. E. Korchev, Local Delivery of Molecules from a Nanopipette for Quantitative Receptor Mapping on Live Cells, *Anal. Chem.*, 2013, **85**(19), 9333–9342, DOI: 10.1021/ac4021769.
24 B. Chen, D. Perry, J. Teahan, I. J. McPherson, J. Edmondson, M. Kang, D. Valavanis, B. G. Frenguelli and P. R. Unwin, Artificial Synapse: Spatiotemporal Heterogeneities in Dopamine Electrochemistry at a Carbon Fiber Ultramicroelectrode, *ACS Meas. Sci. Au*, 2021, **1**(1), 6–10, DOI: 10.1021/acsmeasuresciau.1c00006.
25 L. I. Stephens and J. Mauzeroll, Demystifying Mathematical Modeling of Electrochemical Systems, *J. Chem. Educ.*, 2019, **96**(10), 2217–2224, DOI: 10.1021/acs.jchemed.9b00542.
26 E. Kätelhön and R. G. Compton, Testing and Validating Electroanalytical Simulations, *Analyst*, 2015, **140**(8), 2592–2598, DOI: 10.1039/c4an02276a.
27 A. Molina, J. Gonzalez, E. O. Barnes and R. G. Compton, Simple Analytical Equations for the Current–Potential Curves at Microelectrodes: A Universal Approach, *J. Phys. Chem. C*, 2014, **118**(1), 346–356, DOI: 10.1021/jp409167m.
28 J.-F. Lemineur, J.-M. Noël, C. Combellas and F. Kanoufi, Optical Monitoring of the Electrochemical Nucleation and Growth of Silver Nanoparticles on Electrode: From Single to Ensemble Nanoparticles Inspection, *J. Electroanal. Chem.*, 2020, **872**, 114043, DOI: 10.1016/j.jelechem.2020.114043.
29 J.-F. Lemineur, J.-M. Noël, D. Ausserré, C. Combellas and F. Kanoufi, Combining Electrodeposition and Optical Microscopy for Probing Size-Dependent Single-Nanoparticle Electrochemistry, *Angew. Chem., Int. Ed.*, 2018, **57**(37), 11998–12002, DOI: 10.1002/anie.201807003.

30 L. Godeffroy, P. Ciocci, A. Nsabimana, M. Miranda Vieira, J. M. Noël, C. Combellas, J. F. Lemineur and F. Kanoufi, Deciphering Competitive Routes for Nickel-Based Nanoparticle Electrodeposition by an Operando Optical Monitoring, *Angew. Chem.,Int. Ed.*, 2021, **60**(31), 16980–16983, DOI: 10.1002/anie.202106420.

31 J. F. Lemineur, J. M. Noël, A. Courty, D. Ausserré, C. Combellas and F. Kanoufi, In Situ Optical Monitoring of the Electrochemical Conversion of Dielectric Nanoparticles: From Multistep Charge Injection to Nanoparticle Motion, *J. Am. Chem. Soc.*, 2020, **142**(17), 7937–7946, DOI: 10.1021/jacs.0c02071.

32 P. Ciocci, J.-F. Lemineur, J.-M. Noël, C. Combellas and F. Kanoufi, Differentiating Electrochemically Active Regions of Indium Tin Oxide Electrodes for Hydrogen Evolution and Reductive Decomposition Reactions. An in Situ Optical Microscopy Approach, *Electrochim. Acta*, 2021, **386**, 138498, DOI: 10.1016/j.electacta.2021.138498.

33 W. Li, M. Wojcik and K. Xu, Optical Microscopy Unveils Rapid, Reversible Electrochemical Oxidation and Reduction of Graphene, *Nano Lett.*, 2019, **19**(2), 983–989, DOI: 10.1021/acs.nanolett.8b04216.

34 K. Namink, X. Meng, M. T. M. Koper, P. Kukura and S. Faez, Electric-Double-Layer-Modulation Microscopy, *Phys. Rev. Appl.*, 2020, **13**(4), 044065, DOI: 10.1103/physrevapplied.13.044065.

35 R. W. Taylor and V. Sandoghdar, Interferometric Scattering Microscopy: Seeing Single Nanoparticles and Molecules via Rayleigh Scattering, *Nano Lett.*, 2019, **19**(8), 4827–4835, DOI: 10.1021/acs.nanolett.9b01822.

36 N. Ebejer, A. G. Güell, S. C. S. Lai, K. McKelvey, M. E. Snowden and P. R. Unwin, Scanning Electrochemical Cell Microscopy: A Versatile Technique for Nanoscale Electrochemistry and Functional Imaging, *Annu. Rev. Anal. Chem.*, 2013, **6**(1), 329–351, DOI: 10.1146/annurev-anchem-062012-092650.

37 O. J. Wahab, M. Kang and P. R. Unwin, Scanning Electrochemical Cell Microscopy: A Natural Technique for Single Entity Electrochemistry, *Curr. Opin. Electrochem.*, 2020, **22**, 120–128, DOI: 10.1016/j.coelec.2020.04.018.

38 C. G. Williams, M. A. Edwards, A. L. Colley, J. V. Macpherson and P. R. Unwin, Scanning Micropipet Contact Method for High-Resolution Imaging of Electrode Surface Redox Activity, *Anal. Chem.*, 2009, **81**(7), 2486–2495, DOI: 10.1021/ac802114r.

39 N. Ebejer, M. Schnippering, A. W. Colburn, M. A. Edwards and P. R. Unwin, Localized High Resolution Electrochemistry and Multifunctional Imaging: Scanning Electrochemical Cell Microscopy, *Anal. Chem.*, 2010, **82**(22), 9141–9145, DOI: 10.1021/ac102191u.

40 A. G. Güell, A. S. Cuharuc, Y. R. Kim, G. Zhang, S. Y. Tan, N. Ebejer and P. R. Unwin, Redox-Dependent Spatially Resolved Electrochemistry at Graphene and Graphite Step Edges, *ACS Nano*, 2015, **9**(4), 3558–3571, DOI: 10.1021/acsnano.5b00550.

41 J. W. Hill and C. M. Hill, Directly Visualizing Carrier Transport and Recombination at Individual Defects within 2D Semiconductors, *Chem. Sci.*, 2021, **12**(14), 5102–5112, DOI: 10.1039/d0sc07033e.

42 M. Brunet Cabré, A. E. Paiva, M. Velický, P. E. Colavita and K. McKelvey, Electrochemical Kinetics as a Function of Transition Metal Dichalcogenide Thickness, *Electrochim. Acta*, 2021, **393**, 139027, DOI: 10.1016/j.electacta.2021.139027.

43 Y. Takahashi, Y. Kobayashi, Z. Wang, Y. Ito, M. Ota, H. Ida, A. Kumatani, K. Miyazawa, T. Fujita, H. Shiku, Y. E. Korchev, Y. Miyata, T. Fukuma, M. Chen and T. Matsue, High-Resolution Electrochemical Mapping of the Hydrogen Evolution Reaction on Transition-Metal Dichalcogenide Nanosheets, *Angew. Chem.*, 2020, **132**(9), 3629–3636, DOI: 10.1002/ange.201912863.

44 A. G. Güell, K. E. Meadows, P. V. Dudin, N. Ebejer, J. V. Macpherson and P. R. Unwin, Mapping Nanoscale Electrochemistry of Individual Single-Walled Carbon Nanotubes, *Nano Lett.*, 2014, **14**(1), 220–224, DOI: 10.1021/nl403752e.

45 J. C. Byers, A. G. Güell and P. R. Unwin, Nanoscale Electrocatalysis: Visualizing Oxygen Reduction at Pristine, Kinked, and Oxidized Sites on Individual Carbon Nanotubes, *J. Am. Chem. Soc.*, 2014, **136**(32), 11252–11255, DOI: 10.1021/ja505708y.

46 R. G. Mariano, M. Kang, O. J. Wahab, I. J. McPherson, J. A. Rabinowitz, P. R. Unwin and M. W. Kanan, Microstructural Origin of Locally Enhanced CO2 Electroreduction Activity on Gold, *Nat. Mater.*, 2021, **20**(7), 1000–1006, DOI: 10.1038/s41563-021-00958-9.

47 B. D. B. Aaronson, C.-H. Chen, H. Li, M. T. M. Koper, S. C. S. Lai and P. R. Unwin, Pseudo-Single-Crystal Electrochemistry on Polycrystalline Electrodes: Visualizing Activity at Grains and Grain Boundaries on Platinum for the Fe^{2+}/Fe^{3+} Redox Reaction, *J. Am. Chem. Soc.*, 2013, **135**(10), 3873–3880, DOI: 10.1021/ja310632k.

48 Y. Wang, E. Gordon and H. Ren, Mapping the Potential of Zero Charge and Electrocatalytic Activity of Metal–Electrolyte Interface via a Grain-by-Grain Approach, *Anal. Chem.*, 2020, **92**, 2859–2865, DOI: 10.1021/acs.analchem.9b05502.

49 B. Tao, L. C. Yule, E. Daviddi, C. L. Bentley and P. R. Unwin, Correlative Electrochemical Microscopy of Li-Ion (De)Intercalation at a Series of Individual $LiMn_2O_4$ Particles, *Angew. Chem.*, 2019, **131**(14), 4654–4659, DOI: 10.1002/ange.201814505.

50 S. C. S. Lai, P. V. Dudin, J. V. Macpherson and P. R. Unwin, Visualizing Zeptomole (Electro)Catalysis at Single Nanoparticles within an Ensemble, *J. Am. Chem. Soc.*, 2011, **133**(28), 10744–10747, DOI: 10.1021/ja203955b.

51 M. Choi, N. P. Siepser, S. Jeong, Y. Wang, G. Jagdale, X. Ye and L. A. Baker, Probing Single-Particle Electrocatalytic Activity at Facet-Controlled Gold Nanocrystals, *Nano Lett.*, 2020, **20**(2), 1233–1239, DOI: 10.1021/acs.nanolett.9b04640.

52 C. L. Bentley, Scanning Electrochemical Cell Microscopy for the Study of (Nano)Particle Electrochemistry: From the Sub-particle to Ensemble Level, *Electrochem. Sci. Adv.*, 2021, 1–18, DOI: 10.1002/elsa.202100081.

53 P. Saha, M. M. Rahman and C. M. Hill, Borohydride Oxidation Electrocatalysis at Individual, Shape-controlled Au Nanoparticles, *Electrochem. Sci. Adv.*, 2021, 1–10, DOI: 10.1002/elsa.202100120.

54 J. Ustarroz, M. Kang, E. Bullions and P. R. Unwin, Impact and Oxidation of Single Silver Nanoparticles at Electrode Surfaces: One Shot versus Multiple Events, *Chem. Sci.*, 2017, **8**(3), 1841–1853, DOI: 10.1039/c6sc04483b.

55 J. C. Byers, B. Paulose Nadappuram, D. Perry, K. McKelvey, A. W. Colburn and P. R. Unwin, Single Molecule Electrochemical Detection in Aqueous Solutions

and Ionic Liquids, *Anal. Chem.*, 2015, **87**(20), 10450–10456, DOI: 10.1021/acs.analchem.5b02569.

56 C.-H. Chen, L. Jacobse, K. McKelvey, S. C. S. S. Lai, M. T. M. M. Koper and P. R. Unwin, Voltammetric Scanning Electrochemical Cell Microscopy: Dynamic Imaging of Hydrazine Electro-Oxidation on Platinum Electrodes, *Anal. Chem.*, 2015, **87**(11), 5782–5789, DOI: 10.1021/acs.analchem.5b00988.

57 C. L. Bentley, D. Perry and P. R. Unwin, Stability and Placement of Ag/AgCl Quasi-Reference Counter Electrodes in Confined Electrochemical Cells, *Anal. Chem.*, 2018, **90**(12), 7700–7707, DOI: 10.1021/acs.analchem.8b01588.

58 D. Perry, D. Momotenko, R. A. Lazenby, M. Kang and P. R. Unwin, Characterization of Nanopipettes, *Anal. Chem.*, 2016, **88**(10), 5523–5530, DOI: 10.1021/acs.analchem.6b01095.

59 A. D. Edelstein, M. A. Tsuchida, N. Amodaj, H. Pinkard, R. D. Vale and N. Stuurman, Advanced Methods of Microscope Control Using Manager Software, *J. Biol. Methods*, 2014, **1**(2), e10, DOI: 10.14440/jbm.2014.36.

60 Warwick Electrochemisitry and Interfaces Group, *WEC-SPM*, https://warwick.ac.uk/fac/sci/chemistry/research/unwin/electrochemistry/wec-spm/, accessed Apr 21, 2021.

61 J.-F. Lemineur, P. Ciocci, J.-M. Noël, H. Ge, C. Combellas and F. Kanoufi, Imaging and Quantifying the Formation of Single Nanobubbles at Single Platinum Nanoparticles during the Hydrogen Evolution Reaction, *ACS Nano*, 2021, **15**(2), 2643–2653, DOI: 10.1021/acsnano.0c07674.

62 M. E. Snowden, A. G. Güell, S. C. S. Lai, K. McKelvey, N. Ebejer, M. A. O'Connell, A. W. Colburn and P. R. Unwin, Scanning Electrochemical Cell Microscopy: Theory and Experiment for Quantitative High Resolution Spatially-Resolved Voltammetry and Simultaneous Ion-Conductance Measurements, *Anal. Chem.*, 2012, **84**(5), 2483–2491, DOI: 10.1021/ac203195h.

63 C. L. Bentley, M. Kang, F. M. Maddar, F. Li, M. Walker, J. Zhang and P. R. Unwin, Electrochemical Maps and Movies of the Hydrogen Evolution Reaction on Natural Crystals of Molybdenite (MoS_2): Basal vs. Edge Plane Activity, *Chem. Sci.*, 2017, **8**(9), 6583–6593, DOI: 10.1039/c7sc02545a.

64 A. G. Güell, N. Ebejer, M. E. Snowden, K. McKelvey, J. V. Macpherson and P. R. Unwin, Quantitative Nanoscale Visualization of Heterogeneous Electron Transfer Rates in 2D Carbon Nanotube Networks, *Proc. Natl. Acad. Sci. U. S. A.*, 2012, **109**(29), 11487–11492, DOI: 10.1073/pnas.1203671109.

65 E. Daviddi, Z. Chen, B. Beam Massani, J. Lee, C. L. Bentley, P. R. Unwin and E. L. Ratcliff, Nanoscale Visualization and Multiscale Electrochemical Analysis of Conductive Polymer Electrodes, *ACS Nano*, 2019, **13**(11), 13271–13284.

66 S. C. S. Lai, A. N. Patel, K. McKelvey and P. R. Unwin, Definitive Evidence for Fast Electron Transfer at Pristine Basal Plane Graphite from High-Resolution Electrochemical Imaging, *Angew. Chem., Int. Ed.*, 2012, **51**(22), 5405–5408, DOI: 10.1002/anie.201200564.

67 L. C. Yule, V. Shkirskiy, J. Aarons, G. West, C. L. Bentley, B. A. Shollock and P. R. Unwin, Nanoscale Active Sites for the Hydrogen Evolution Reaction on Low Carbon Steel, *J. Phys. Chem. C*, 2019, **123**(39), 24146–24155, DOI: 10.1021/acs.jpcc.9b07216.

68 P. M. Kirkman, A. G. Güell, A. S. Cuharuc and P. R. Unwin, Spatial and Temporal Control of the Diazonium Modification of sp^2 Carbon Surfaces, *J. Am. Chem. Soc.*, 2014, **136**(1), 36–39, DOI: 10.1021/ja410467e.
69 G. Zhang, P. M. Kirkman, A. N. Patel, A. S. Cuharuc, K. McKelvey and P. R. Unwin, Molecular Functionalization of Graphite Surfaces: Basal Plane versus Step Edge Electrochemical Activity, *J. Am. Chem. Soc.*, 2014, **136**(32), 11444–11451, DOI: 10.1021/ja505266d.
70 A. N. Patel, K. McKelvey and P. R. Unwin, Nanoscale Electrochemical Patterning Reveals the Active Sites for Catechol Oxidation at Graphite Surfaces, *J. Am. Chem. Soc.*, 2012, **134**(50), 20246–20249, DOI: 10.1021/ja3095894.
71 E. E. Oseland, Z. J. Ayres, A. Basile, D. M. Haddleton, P. Wilson and P. R. Unwin, Surface Patterning of Polyacrylamide Gel Using Scanning Electrochemical Cell Microscopy (SECCM), *Chem. Commun.*, 2016, **52**(64), 9929–9932, DOI: 10.1039/c6cc05153g.
72 K. McKelvey, M. A. O'Connell and P. R. Unwin, Meniscus Confined Fabrication of Multidimensional Conducting Polymer Nanostructures with Scanning Electrochemical Cell Microscopy (SECCM), *Chem. Commun.*, 2013, **49**(29), 2986, DOI: 10.1039/c3cc00104k.
73 A. S. Parker, R. Al Botros, S. L. Kinnear, M. E. Snowden, K. McKelvey, A. T. Ashcroft, M. Carvell, A. Joiner, M. Peruffo, C. Philpotts and P. R. Unwin, Combinatorial Localized Dissolution Analysis: Application to Acid-Induced Dissolution of Dental Enamel and the Effect of Surface Treatments, *J. Colloid Interface Sci.*, 2016, **476**, 94–102, DOI: 10.1016/j.jcis.2016.05.018.
74 S. L. Kinnear, K. McKelvey, M. E. Snowden, M. Peruffo, A. W. Colburn and P. R. Unwin, Dual-Barrel Conductance Micropipet as a New Approach to the Study of Ionic Crystal Dissolution Kinetics, *Langmuir*, 2013, **29**(50), 15565–15572, DOI: 10.1021/la403630u.
75 C.-H. Chen, K. E. Meadows, A. Cuharuc, S. C. S. Lai and P. R. Unwin, High Resolution Mapping of Oxygen Reduction Reaction Kinetics at Polycrystalline Platinum Electrodes, *Phys. Chem. Chem. Phys.*, 2014, **16**(34), 18545, DOI: 10.1039/c4cp01511h.
76 H. Wang, X. Shan, H.-Y. Chen and N. Tao, Pauli Repulsion-Induced Expansion and Electromechanical Properties of Graphene, *Nano Lett.*, 2017, **17**(1), 236–241, DOI: 10.1021/acs.nanolett.6b03955.
77 V. Brasiliense, J. Clausmeyer, P. Berto, G. Tessier, C. Combellas, W. Schuhmann and F. Kanoufi, Monitoring Cobalt-Oxide Single Particle Electrochemistry with Subdiffraction Accuracy, *Anal. Chem.*, 2018, **90**(12), 7341–7348, DOI: 10.1021/acs.analchem.8b00649.
78 G. Lippmann, Relations Entre Les Phénomènes Électriques et Capillaires, *Ann. Chim. Phys.*, 1875, **5**, 494–549.
79 A. Quinn, R. Sedev and J. Ralston, Influence of the Electrical Double Layer in Electrowetting, *J. Phys. Chem. B*, 2003, **107**(5), 1163–1169, DOI: 10.1021/jp0216326.
80 A. Quinn, R. Sedev and J. Ralston, Contact Angle Saturation in Electrowetting, *J. Phys. Chem. B*, 2005, **109**(13), 6268–6275, DOI: 10.1021/jp040478f.

81 L. Liu, S. Yellinek, I. Valdinger, A. Donval and D. Mandler, Important Implications of the Electrochemical Reduction of ITO, *Electrochim. Acta*, 2015, **176**, 1374–1381, DOI: 10.1016/j.electacta.2015.07.129.

82 C. M. Hill and S. Pan, A Dark-Field Scattering Spectroelectrochemical Technique for Tracking the Electrodeposition of Single Silver Nanoparticles, *J. Am. Chem. Soc.*, 2013, **135**(46), 17250–17253, DOI: 10.1021/ja4075387.

83 P. Saha, J. W. Hill, J. D. Walmsley and C. M. Hill, Probing Electrocatalysis at Individual Au Nanorods via Correlated Optical and Electrochemical Measurements, *Anal. Chem.*, 2018, **90**(21), 12832–12839, DOI: 10.1021/acs.analchem.8b03360.

84 Y. Liu, C. Jin, Y. Liu, K. H. Ruiz, H. Ren, Y. Fan, H. S. White and Q. Chen, Visualization and Quantification of Electrochemical H_2 Bubble Nucleation at Pt, Au, and MoS_2 Substrates, *ACS Sens.*, 2021, **6**(2), 355–363, DOI: 10.1021/acssensors.0c00913.

85 Q. Chen, L. Luo, H. Faraji, S. W. Feldberg and H. S. White, Electrochemical Measurements of Single H_2 Nanobubble Nucleation and Stability at Pt Nanoelectrodes, *J. Phys. Chem. Lett.*, 2014, **5**(20), 3539–3544, DOI: 10.1021/jz501898r.

86 Y. Qiu, H. Ren, M. A. Edwards, R. Gao, K. Barman and H. S. White, Electrochemical Generation of Individual Nanobubbles Comprising H_2, D_2 , and HD, *Langmuir*, 2020, **36**(22), 6073–6078, DOI: 10.1021/acs.langmuir.0c00232.

87 J. W. Morse, R. S. Arvidson and A. Lüttge, Calcium Carbonate Formation and Dissolution, *Chem. Rev.*, 2007, **107**(2), 342–381, DOI: 10.1021/cr050358j.

88 D. Gebauer, A. Völkel and H. Cölfen, Stable Prenucleation Calcium Carbonate Clusters, *Science*, 2008, **322**(5909), 1819–1822, DOI: 10.1126/science.1164271.

89 P. J. M. Smeets, A. R. Finney, W. J. E. M. Habraken, F. Nudelman, H. Friedrich, J. Laven, J. J. De Yoreo, P. M. Rodger and N. A. J. M. Sommerdijk, A Classical View on Nonclassical Nucleation, *Proc. Natl. Acad. Sci. U. S. A.*, 2017, **114**(38), E7882–E7890, DOI: 10.1073/pnas.1700342114.

90 F. Sebastiani, S. L. P. Wolf, B. Born, T. Q. Luong, H. Cölfen, D. Gebauer and M. Havenith, Water Dynamics from THz Spectroscopy Reveal the Locus of a Liquid-Liquid Binodal Limit in Aqueous $CaCO_3$ Solutions, *Angew. Chem., Int. Ed.*, 2017, **56**(2), 490–495, DOI: 10.1002/anie.201610554.

91 F. H. Y. Lui, Y. Wang, Y. Yao, R. J. Mobbs, R. E. Pogson, P. Koshy, F. P. Lucien, D. Zhou and C. C. Sorrell, Quiescent Mineralisation for Free-Standing Mineral Microfilms with a Hybrid Structure, *J. Colloid Interface Sci.*, 2021, **604**, 327–339, DOI: 10.1016/j.jcis.2021.06.171.

92 M. Farhadi Khouzani, D. M. Chevrier, P. Güttlein, K. Hauser, P. Zhang, N. Hedin and D. Gebauer, Disordered Amorphous Calcium Carbonate from Direct Precipitation, *CrystEngComm*, 2015, **17**(26), 4842–4849, DOI: 10.1039/c5ce00720h.

93 N. C. Rudd, S. Cannan, E. Bitziou, I. Ciani, A. L. Whitworth and P. R. Unwin, Fluorescence Confocal Laser Scanning Microscopy as a Probe of pH Gradients in Electrode Reactions and Surface Activity, *Anal. Chem.*, 2005, **77**(19), 6205–6217, DOI: 10.1021/ac050800y.

94 D. L. Parkhurst and C. A. J. Appelo, Description of Input and Examples for PHREEQC Version 3—A Computer Program for Speciation, Batch-Reaction, One-Dimensional Transport, and Inverse Geochemical Calculations, in *U.S.*

Geological Survey Techniques and Methods, Book 6, U.S. Geological Survey Techniques and Methods, 2013, ch. A43, p. 497.

95 S. Campidelli, R. Abou Khachfe, K. Jaouen, J. Monteiller, C. Amra, M. Zerrad, R. Cornut, V. Derycke and D. Ausserré, Backside Absorbing Layer Microscopy: Watching Graphene Chemistry, *Sci. Adv.*, 2017, **3**(5), e1601724, DOI: 10.1126/sciadv.1601724.

96 S. Chakri, A. N. Patel, I. Frateur, F. Kanoufi, E. M. M. Sutter, T. T. M. Tran, B. Tribollet and V. Vivier, Imaging of a Thin Oxide Film Formation from the Combination of Surface Reflectivity and Electrochemical Methods, *Anal. Chem.*, 2017, **89**(10), 5303–5310, DOI: 10.1021/acs.analchem.6b04921.

97 A. Shevchuk, S. Tokar, S. Gopal, J. L. Sanchez-Alonso, A. I. Tarasov, A. C. Vélez-Ortega, C. Chiappini, P. Rorsman, M. M. Stevens, J. Gorelik, G. I. Frolenkov, D. Klenerman and Y. E. Korchev, Angular Approach Scanning Ion Conductance Microscopy, *Biophys. J.*, 2016, **110**(10), 2252–2265, DOI: 10.1016/j.bpj.2016.04.017.

Faraday Discussions

ROYAL SOCIETY OF CHEMISTRY

PAPER

iR drop in scanning electrochemical cell microscopy†

Brandon Blount,‡ Gabriel Juarez,‡ Yufei Wang and Hang Ren *

Received 31st July 2021, Accepted 27th August 2021
DOI: 10.1039/d1fd00046b

Nanoscale electrochemical mapping techniques, *e.g.*, scanning electrochemical cell microscopy (SECCM), have been increasingly used to study the local electrochemistry in electrocatalysis. Its capability for local electrochemistry mapping helps to reveal the heterogeneity in the electrode kinetics and mechanisms, which are otherwise averaged out in ensemble measurements. Accurate determination of the electrode kinetics requires the careful assessment of the ohmic potential drop in the solution, *i.e.*, the *iR* drop. Herein, the *iR* drop in SECCM experiments is assessed. We showed that the *iR* drop in single-barrel SECCM can be estimated using the solution conductivity and pipette geometry, or the mass transfer limiting current without the assumption of pipette geometry. For dual-barrel SECCM, we developed a method of measuring the solution resistance directly, which can be used to compensate for the *iR* drop and the potential shift in the experiments. These methods offer a convenient way to estimate and compensate for the *iR* drop in SECCM, allowing the more accurate measurement of local electrode kinetics for the determination of local mechanisms in electrocatalysis.

1. Introduction

The kinetic measurement of electrode processes plays an important role in the development of new electrocatalysts.[1,2] Electrocatalytic mechanisms can often be inferred from the current–overpotential relationship, *e.g.*, *via* Tafel analysis.[3] One good practice to ensure the accurate measurement of electrode kinetics is to assess and compensate for the ohmic drop between the working electrode and the reference electrode, *i.e.*, the *iR* drop.[4] The importance of the *iR* drop in electrocatalysis has been recently re-emphasized, and failure to assess and compensate for the *iR* drop can lead to inaccurate kinetic measurements and the wrong assignment of mechanisms.[5]

Department of Chemistry, The University of Texas at Austin, 105 E 24th St, Austin, TX 78712, USA. E-mail: hren@utexas.edu

† Electronic supplementary information (ESI) available: Experimental methods, pipette characterization, derivation of solution resistance, additional results in single-barrel SECCM, finite element simulation, measuring solution resistance in dual-barrel SECCM and effect of noise. See DOI: 10.1039/d1fd00046b

‡ BB and GJ contributed equally to the work.

Another challenge in revealing the true mechanism from the conventional ensemble measurement of electrode kinetics is that practical electrode materials often show large degree of heterogeneity.[6–8] Consequently, the measured kinetics is an average over the whole electrode surface and the detailed distribution of local activity and the corresponding mechanisms, including those at hot spots, are inevitably lost during the averaging.[9] Therefore, assigning one mechanism to the whole electrode based on the average kinetic data measured from bulk electrochemical techniques can often be oversimplifying, if not erroneous. To circumvent the issue of heterogeneity, localized electrochemical mapping techniques, including scanning electrochemical cell microscopy (SECCM), have emerged as powerful techniques in assessing local electrochemistry.[10,11] In SECCM, the electrochemical measurement can be confined to a nanoscopic droplet and the scanning of the droplet on the electrode surface can generate high-resolution maps of local activity. When combined with colocalized imaging techniques, the structure and activity can be correlated in a site-by-site manner.[8,12–14] Despite the great success in mapping the local activity for a range of electrocatalytic reactions and energy storage at the single-particle level,[8,15–17] the relevance of the *iR* drop in SECCM measurements is often not discussed in detail. While this is warranted when a high concentration of the supporting electrolyte is used, the significance of the *iR* drop in the SECCM setup has rarely been systematically assessed,[18] nor has its compensation method been discussed.[19–21] The existence of a significant *iR* drop can lead to errors in the determination of the local electrocatalytic activity, undermining the great potential of scanning electrochemical probe microscopy in understanding the local electrocatalytic mechanism.

In this paper, we assess the effect of the *iR* drop in single-barrel and dual-barrel SECCM voltammetric experiments. In each kind of experiment, we first present the theoretical treatment of the *iR* drop and then validate the theory using finite element simulation and experiments. The theory developed here allows the facile and semi-quantitative estimation of the significance of the *iR* drop in SECCM measurements. Lastly, we present an experimental method of measuring and compensating for the *iR* drop in dual-barrel SECCM experiments.

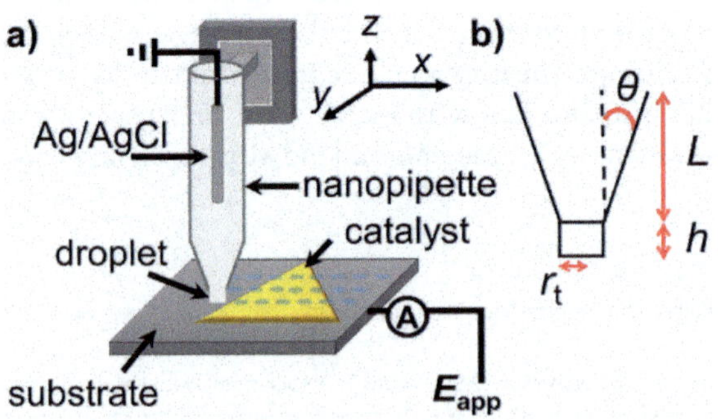

Fig. 1 (a) Schematic of a single-barrel SECCM setup. (b) Ideal cross-sectional geometry of a single-barrel pipette with a cylindrical droplet.

2. Results and discussion

2.1 *iR* drop in single-barrel SECCM

A schematic for a typical single-barrel SECCM setup is shown in Fig. 1a. An electrolyte-filled nanopipette with a nanodroplet at its opening is used as the probe. The active working electrode is defined by the contact between the droplet and the substrate electrode, and a quasi-reference counter electrode (QRCE), often in the form of a wire, is placed inside the pipette barrel. The *iR* drop in SECCM experiments is the ohmic potential drop in the solution between the working electrode and the QRCE. In this section, we first derive the analytical expression of the solution resistance in a single-barrel pipette based on an ideal geometry of a truncated cone (Fig. 1b). Such an analytical expression helps to semi-quantitatively estimate whether the *iR* drop is important in the experiment. We then show a more general way to estimate the *iR* drop based on the limiting current without the assumption of the pipette geometry. Finally, both methods are validated using finite element simulation and applied to experiments.

2.1.1 Analytical expression of solution resistance in SECCM. The analytical expression of the solution resistance (R_s) in the single-barrel SECCM is obtained by approximating the geometry of the SECCM probe by a truncated conical nanopipette with a cylindrical droplet, as shown in Fig. 1b. Accordingly, the R_s is composed of the droplet resistance (R_d) and the pipette resistance (R_p) in series, which can be expressed as:

$$R_s = R_d + R_p = \int_{-h}^{0} \frac{dz}{\kappa A} + \int_{0}^{L} \frac{dz}{\kappa A} \tag{1}$$

In eqn (1), κ is the conductivity of the solution, A is the cross-section area at the height z and h and L are the heights of the droplet and the pipette, respectively, as shown in Fig. 1b.

Eqn (1) can be simplified by assuming a constant κ (details are provided in Section 3 of the ESI†), which yields:

$$R_s = R_d + R_p = \frac{h}{\kappa\pi r_t^2} + \frac{L}{\kappa\pi r_t(r_t + L\tan(\theta))} \tag{2}$$

The variables in eqn (2) are consistent with those labeled in Fig. 1b, where r_t is the pipette tip radius, θ is the half-cone angle and L is the length of the pipette. When $L\tan(\theta) \gg r_t$, eqn (2) can be further simplified as:

$$R_s = \frac{h}{\kappa\pi r_t^2} + \frac{1}{\kappa\pi r_t \tan(\theta)} \tag{3}$$

To validate the expression of the R_s in eqn (2), we perform finite element simulations of single-barrel SECCM experiments. The Nernst–Planck equation is solved using COMSOL Multiphysics. Electroneutrality is assumed since the thickness of the double layer (*i.e.*, the Debye length) when the electrolyte is >1 mM is negligible when compared with the opening of the pipette in the simulation. The details of the model are provided in Section 3 of the ESI .† As shown in Fig. 2, the analytical expression of the R_s (eqn (2)) matches very well with the simulation results when the half-cone angle (θ) of the nanopipette is smaller than 20°, which

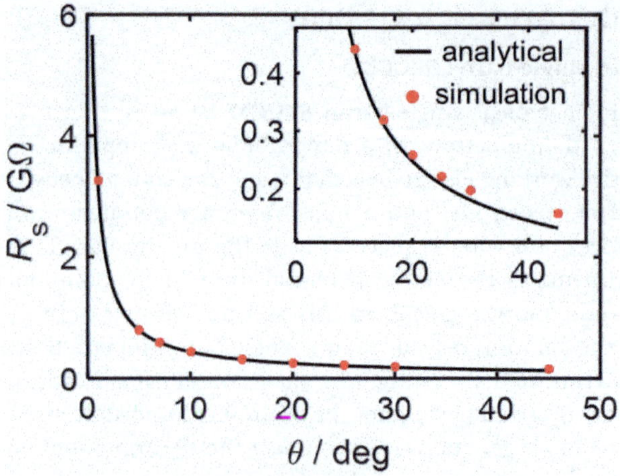

Fig. 2 Comparison of the analytical expression (black line, eqn (2)) and finite element simulation (red dot) of the R_s in single-barrel SECCM at various half-cone angles (θ). Inset: zoomed-in portion of the figure between 0 and 0.5 GΩ. The geometry is shown in Fig. 1b with $h = 350$ nm, $r_t = 350$ nm and $L = 1$ μm. The solution is 1 mM KCl.

is most common in SECCM experiments. The differences between the analytical solution and the finite element results at $\theta > 20°$ are explained by the radial diffusion that is not accounted for in the derivation of eqn (2).[22]

Eqn (2) or (3) provide semi-quantitative ways to quickly determine whether the *iR* drop in an SECCM experiment is negligible. In practical SECCM experiments, the pipettes are often sufficiently long that eqn (2) and (3) provide the same result. Under this scenario, the R_s can be readily estimated with the knowledge of θ and the solution conductivity (κ), which in turn can be calculated based on the solution composition. The *iR* drop is obtained by multiplying the R_s by the measured current.

Despite the simplicity, the assumptions in the analytical expression should be noted. First, eqn (2) is derived based on an ideal geometry, as drawn in Fig. 1b. Although it captures the essence of most single-barrel SECCM experiments, it can be an oversimplification. For more accurate results, one can apply finite element simulation using the exact pipette geometry obtained from TEM, as described by Unwin and coworkers.[23] Secondly, a constant conductivity everywhere is assumed when deriving eqn (2). However, a change of the local solution composition near the electrode surface is generally expected when faradaic current flows, which alters the local conductivity. Indeed, the R_s is never truly constant when faradaic reactions occur, leading to over- or under-compensation depending on the exact faradaic reaction (see Section 2.1.3).

2.1.2 Assessing the *iR* drop from the limiting current. An alternative method to assess the *iR* drop in single-barrel SECCM without the assumption of the pipette geometry is to use the mass transfer limiting current (i_{lim}). The principle is based on the similarity in the mathematical treatment of steady-state mass transport and current conduction: both reduce to Laplace equations, *i.e.*, $\nabla^2 C = 0$ and $\nabla^2 \Phi = 0$, respectively. The i_{lim} for a redox-active species j in SECCM can be expressed as:

$$i_{\text{lim}} = GnFD_jC_j^* \tag{4}$$

In this equation, G is a geometric factor accounting for the shape of the droplet and the nanopipette, n is the number of electrons transferred per one molecule of j reacted, F is Faraday's constant and D_j and C_j^* are the diffusion coefficient and the bulk concentration of the redox species j, respectively.

Similarly, the steady-state ionic current can be expressed as:

$$i = G\kappa\Phi_0 \tag{5}$$

where Φ_0 is the potential difference between the electrode surface and the ground, and κ is the solution conductivity.

By rearranging eqn (5), the solution resistance (R_s) is obtained:

$$R_s = \frac{\Phi_0}{i} = \frac{1}{G\kappa} \tag{6}$$

Comparing eqn (4) and (6), the solution resistance (R_s) is related to the limiting current (i_{lim}) *via*:

$$R_s = \frac{1}{G\kappa} = \frac{nFDC^*}{i_{\text{lim}}\kappa} \tag{7}$$

Eqn (7) is the basis for calculating the R_s from the limiting current without the need to know the exact geometry of the droplet or the nanopipette. Because of this, it should apply to other electrochemical systems as well. Since the largest steady-state current for a redox process is measured at i_{lim}, the largest iR drop is $i_{\text{lim}}R_s$, which can be readily obtained by rearranging eqn (8):

$$i_{\text{lim}}R_s = \frac{nFDC^*}{\kappa} \tag{8}$$

Examples of the largest iR drops calculated for various electrolyte compositions using eqn (8) are shown in Table 1.

The solution conductivity (κ) can be further calculated using

$$\kappa = F\sum|z_i|u_iC_i = \frac{F^2}{RT}\sum z_i^2D_iC_i \tag{9}$$

Table 1 Calculated maximum *iR* drops in different solutions

Redox species	Electrolyte	κ (S cm^{-1})	$i_{\text{lim}}R_s$ (mV)
100 mM $HClO_4$	1 mM KCl	4.26×10^{-2}	21.1
100 mM $HClO_4$	1 mM KCl + 100 mM KNO_3	5.73×10^{-2}	15.7
5 mM $Ru(NH_3)_6Cl_3$	1 mM KNO_3	2.85×10^{-3}	1.5
5 mM $Ru(NH_3)_6Cl_3$	10 mM KNO_3	4.18×10^{-3}	1.0
5 mM ferrocene	1.5 mM $TBAPF_6$	6.88×10^{-5}	21.0
5 mM ferrocene	50 mM $TBAPF_6$	2.29×10^{-3}	0.6
Reduction of H_2O	100 mM KCl	4.26×10^{-2}	2.7×10^{4} [a]

[a] Assuming 55.5 M of H_2O and the limiting current for water reduction (hydrogen evolution reaction) can be achieved.

In eqn (9), z_i is the charge, u_i the mobility, C_i the concentration and D_i the diffusion coefficient of the ionic species i.

By substituting eqn (9) into eqn (8) and using $T = 293$ K, one obtains:

$$i_{\lim} R_s = 0.0252 \frac{n D_{\text{redox}} C^*_{\text{redox}}}{\sum z_i^2 D_i C_i} \tag{10}$$

where D_{redox} and C^*_{redox} are the diffusion coefficient and the bulk concentration of the redox species, respectively. By assuming $n = 1$ and a 1 : 1 electrolyte, eqn (10) is further simplified as:

$$i_{\lim} R_s = 0.0252 \frac{D_{\text{redox}} C^*_{\text{redox}}}{\sum D_i C_i} \tag{11}$$

Eqn (11) suggests that the maximum *iR* drop is 25 mV times the ratio between the product of *DC* for the redox species over the sum of the product of *DC* for all of the ions when $n = 1$ and a 1 : 1 electrolyte is used.

2.1.3 Validation using simulation and experiments. We use finite element simulation to validate the two analytical methods of the *iR* drop estimation/correction in single-barrel SECCM (eqn (2) and (7)). Three example systems are used, including the oxidation of ferrocene (Fc), the reduction of proton and the reduction of $Ru(NH_3)_6^{3+}$. Briefly, both the ideal *iR*-free and the uncorrected raw voltammograms can be simulated numerically. Eqn (2) or (7) can then be used to correct the electrode potential in the uncorrected voltammogram using:

$$E_{\text{corr}} = E_{\text{app}} - iR_s \tag{12}$$

where E_{corr} is the *iR*-corrected potential, E_{app} is the applied uncorrected potential, i is the voltammetric current and R_s is the solution resistance calculated using eqn (2) or (7). Using Fc oxidation as an example, the corrected voltammograms using either eqn (2) or (7) agree relatively well with the ideal *iR*-free one, as shown in Fig. 3a. We do notice an overcompensation when the voltammetric current is high. To reveal any small errors, we also directly compare the R_s obtained using eqn (2) and (7) with that from the simulation. As shown in Fig. 3b, both eqn (2) and (7) predict very similar R_s values that are close to those of the simulation between 0 and 0.3 V, where the faradaic current is low. At a potential >0.3 V, where Fc oxidation results in significant faradaic current, the simulated R_s decreases, while eqn (2) or (7) predict a constant R_s. A decrease in the R_s during Fc oxidation is expected because the neutral Fc is converted to an ionic species, ferrocenium (Fc^+), and its accumulation at the electrode surface increases the local conductivity. This change of the local conductivity during the voltammetry is captured in the simulation, as shown in Fig. 3c and d. The fact that eqn (2) and (7) do not capture any change in the R_s is not surprising because both equations assume a constant κ during their derivation for simplicity. In principle, a location dependent κ (as shown in Fig. 3c and d) should be used when deriving these analytical expressions, however, the tedious mathematical treatment will result in complicated expression of R_s, which defeat the purpose of being a simple method to estimate the *iR* drop. The validation of the *iR* drop compensation using the reduction of H^+ and $Ru(NH_3)_6^{3+}$ as examples are shown in Section 5 of the ESI.† In both cases, the analytical expression provides a relatively accurate estimation of

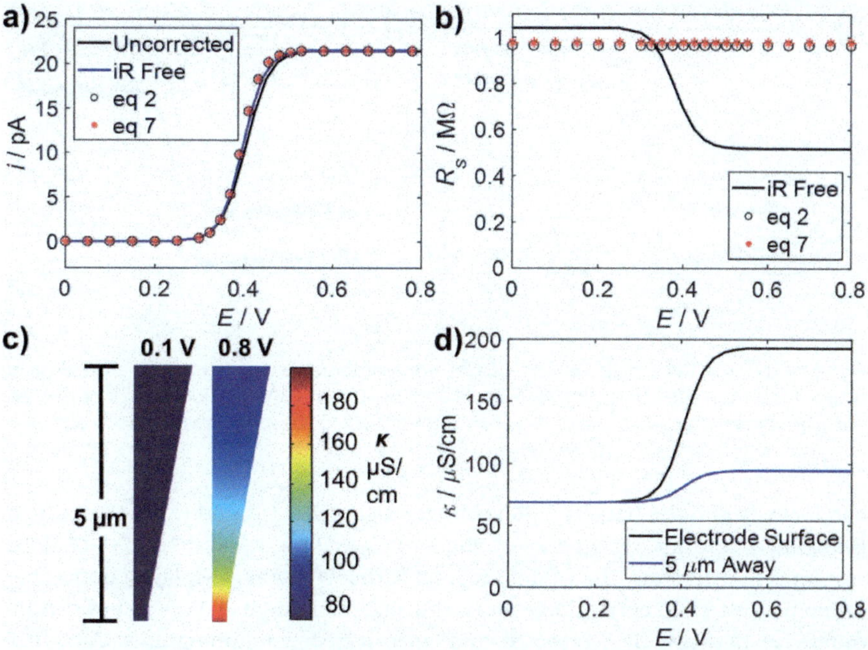

Fig. 3 Simulated result of 5 mM Fc in 1.5 mM $TBAPF_6$. (a) Steady-state voltammograms with uncorrected (black) and *iR*-free (blue) results and corrections by eqn (2) (black circles) and eqn (7) (red stars). (b) Solution resistance during the voltammetry. (c) Solution conductivity profiles at the electrode surface up to a distance of 5 μm along the pipette. (d) Solution conductivity plots as a function of the applied potential at the electrode surface and at 5 μm along the axisymmetric line.

the R_s, although it does not capture the change of the R_s during the voltammetry. An increase in the R_s is expected during proton reduction and $Ru(NH_3)_6{}^{3+}$ reduction because the net charge on the redox species decreases after the reduction, which is captured in the finite element simulation, as shown in Fig. S5 and S6.† For $Ru(NH_3)_6{}^{3+}$ and proton reduction, a small discrepancy in the R_s is observed between eqn (2) and (7), which is likely due to the ignorance of the migration current when deriving eqn (7), as discussed in Section 5a of the ESI.† Nonetheless, the analytical expressions are very convenient to use and the errors are generally small. We will show in the next section that the subtle change in the solution resistance can be captured experimentally using a dual-barrel SECCM setup.

We further apply the analytical *iR* estimation/correction methods to experiments. Voltammograms were obtained for the SECCM of Fc oxidation at two concentrations of the supporting electrolyte (tetrabutylammonium hexafluorophosphate, $TBAPF_6$), as shown in Fig. 4. Qualitatively, the voltammogram with 1.5 mM $TBAPF_6$ shows a larger *iR* drop than the one with 50 mM $TBAPF_6$. To quantify the effect of the *iR* drop on the voltammogram, we use the difference of the potentials at 75% and 25% of the limiting current (*i.e.*, $\Delta E = E_{3/4} - E_{1/4}$) to measure the steepness of the rise of the voltammetric wave. A ΔE of 62 mV is observed for the solution with 1.5 mM $TBAPF_6$ and 54 mV for that with 50 mM $TBAPF_6$ (Fig. 4). The solution resistance is calculated based on the solution

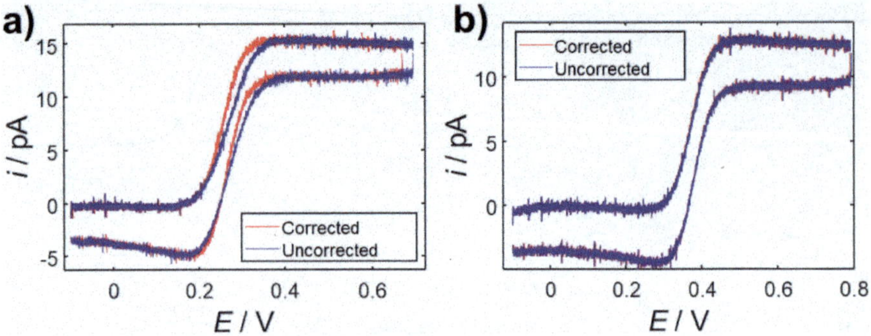

Fig. 4 Experimental single-barrel SECCM voltammograms for 5 mM Fc oxidation in propylene carbonate containing (a) 1.5 mM and (b) 50 mM $TBAPF_6$. Blue: raw uncorrected voltammogram. Red: iR-corrected voltammogram using the calculated R_s from eqn (2).

conductivity and the nanopipette geometry. Using $\theta = 10°$ and $r_t = 315$ nm, eqn (2) predicts solution resistances of 960 and 29 MΩ for 1.5 and 50 mM $TBAPF_6$, respectively. Therefore, the voltammogram with low $TBAPF_6$ displays a noticeable difference after iR correction, while the one with high $TBAPF_6$ shows little difference between the corrected and uncorrected voltammograms. After the correction, the voltammograms at the two different concentrations of $TBAPF_6$ are very similar, with a ΔE of 47 mV and 50 mV for 1.5 mM and 50 mM $TBAPF_6$, respectively. The compensation using eqn (7) yields very similar results compared to those from eqn (2) and therefore is not shown here. Another experimental example of iR compensation in the proton reduction reaction is shown in Section 4 of the ESI.†

2.2 *iR* drop in dual-barrel SECCM

2.2.1 General model. Dual-barrel nanopipettes (also known as theta pipettes) have also been used in SECCM for electrochemical mapping, especially when the droplet needs to be scanned over a surface without breaking contact.[10] A schematic of a dual-barrel setup is shown in Fig. 5.

The major difference between the dual- and single-barrel SECCM is that two QRCEs are used in the dual-barrel setup, with each residing in one of the barrels. One way of electrical connection is shown in Fig. 5a: potentials of V_1 and V_2 are applied to the working electrode and one QRCE, respectively, with the other QRCE as the ground (the right barrel in Fig. 5a). The current flowing through the working electrode is recorded as i_1 by a current follower and the current flowing through the QRCE on the left is recorded as i_2. The potential bias V_2 and the current i_2 offer a handle for assessing the solution resistance. Note that i_2 is considered to mostly reflect the ionic current flowing between the pipette barrels and has been used as a feedback mechanism to maintain a consistent droplet contact at the electrode surface during SECCM.[18] On the other hand, V_2 also causes a shift of the working electrode potential away from V_1,[10] but can be corrected as discussed below.

An equivalent circuit diagram for a dual-barrel SECCM experiment is shown in Fig. 5b. Each barrel has a pipette solution resistance (R_{pip} or R'_{pip}). A charge transfer resistance (R_{CT}), governed by the kinetics of the faradaic reaction, is in

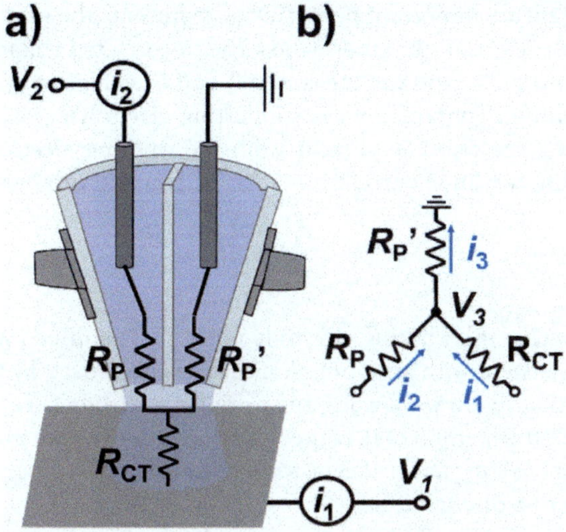

Fig. 5 (a) Schematic of the dual-barrel SECCM setup. (b) An equivalent circuit model.

series with the pipette resistance. The circuit does not include capacitors (*e.g.*, a double-layer capacitor) because their contribution is negligible at the steady state. From the circuit in Fig. 5b, it is observed that:

$$i_1 + i_2 = i_3 \tag{13}$$

$$V_1 - V_3 = i_1 R_{CT} \tag{14}$$

$$V_2 - V_3 = i_2 R_p \tag{15}$$

$$V_3 = i_3 R_p' \tag{16}$$

Assuming a symmetrical pipette geometry and therefore $R_p = R_p'$, we obtain:

$$R_p = \frac{V_2}{i_1 + 2i_2} \tag{17}$$

$$R_{CT} = \frac{i_1(V_1 - V_2) + i_2(2V_1 - V_2)}{{i_1}^2 + 2i_1 i_2} \tag{18}$$

The potential on the working electrode/electrolyte interface with the *iR* drop correction is then expressed as:

$$V_{WE} = i_1 R_{CT} = V_1 - \frac{V_2(i_1 + i_2)}{(i_1 + 2i_2)} \tag{19}$$

And the *iR* drop between the working electrode and the QRCE at the ground is:

$$V_{ir} = \frac{V_2(i_1 + i_2)}{(i_1 + 2i_2)} \tag{20}$$

Eqn (19) and (20) are important because V_{WE} represents the electrode potential with the iR drop correction. In other words, the iR corrected voltammogram can be calculated from V_1, V_2 and the measured i_1 and i_2 using eqn (19).

When the solution contains excess supporting electrolyte, *i.e.*, $R_p \ll R_{CT}$, or equivalently when the faradaic current from the working electrode (i_1) is very small compared to the ionic current, *i.e.*, $i_1 \ll i_2$, eqn (19) can be simplified as:

$$V_{WE} = V_1 - \frac{1}{2}V_2 \tag{21}$$

This suggests that the working electrode potential is shifted away from V_1 by $\frac{1}{2}V_2$, which is consistent with an expression that has been used by Unwin and co-authors.[18] Note that in Unwin's work, the working electrode is connected as the ground. However, the method of iR compensation we derive here works regardless of the connection, as long as one keeps track of the direction of current flow, and therefore will not be discussed here.

2.2.2 Validation using finite element simulation. The expression for the iR compensation in dual-barrel SECCM is validated by finite element simulation. We solve the Nernst–Planck equation in 3-D using COMSOL Multiphysics. The model is described in Section 5 of the ESI.† The simulated iR-free and iR-uncorrected voltammograms for the oxidation of 20 mM $Ru(NH_3)_6{}^{3+}$ in 1 mM KNO_3 at a potential bias (V_2) of 0.05 V are shown in Fig. 6. The uncorrected voltammogram is iR compensated using eqn (19) and the compensated voltammogram is then compared with the simulated iR-free voltammogram. As shown in Fig. 6, eqn (19) sufficiently compensates for the iR drop in the voltammogram. The iR compensation is not affected by the potential bias, as shown by the simulation and validation in Fig. S8.† Additional cross-sectional potential profiles and voltammogram simulations at various supporting electrolyte concentrations can be found in Section 5b of the ESI.†

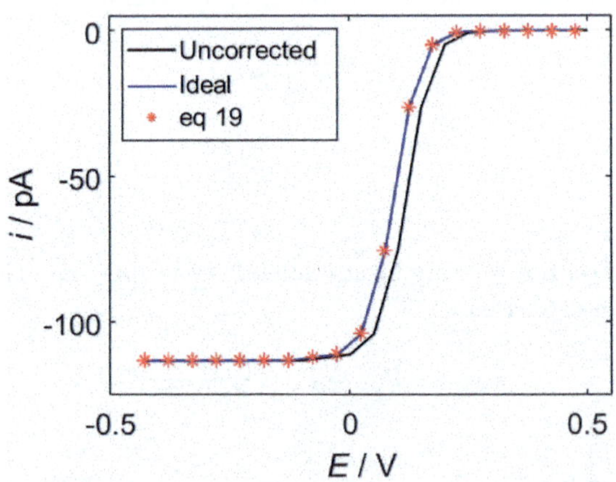

Fig. 6 Simulated steady-state uncorrected (black) and ideal iR-free (blue) voltammograms of 20 mM $Ru(NH_3)_6{}^{3+}$ with 1 mM KNO_3 in a dual-barrel pipette. $V_2 = 0.05$ V is applied between the two QRCEs. Eqn (19) is applied to correct the raw voltammogram for the iR drop and the corrected voltammogram is shown by the red stars.

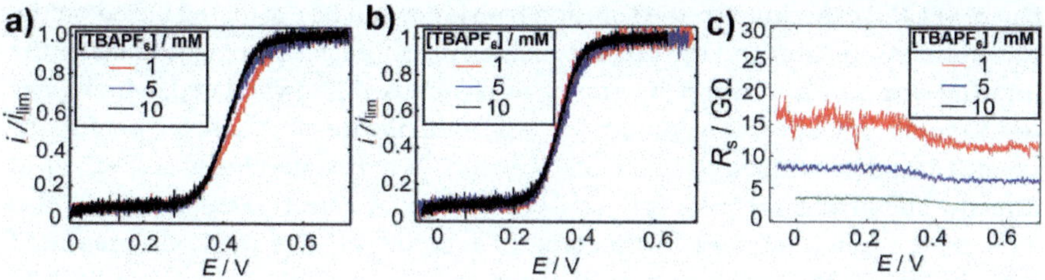

Fig. 7 Normalized dual-barrel SECCM voltammograms of 20 mM Fc in propylene carbonate at various concentrations of $TBAPF_6$ (a) before and (b) after *iR* correction using eqn (19). (c) Calculated solution resistance (R_s) from eqn (17) as a function of the electrode potential. The substrate is Pt with $V_2 = 0.1$ V.

2.2.3 Application of *iR* compensation in dual-barrel SECCM experiments. We now apply the *iR* compensation method to dual-barrel SECCM experiments in three model systems: the oxidation of Fc, the reduction of $Ru(NH_3)_6^{3+}$ and the reduction of H^+. Fig. 7 shows the oxidation of 20 mM Fc with different concentrations of $TBAPF_6$. At 1 mM $TBAPF_6$, a significant *iR* drop is observed (Fig. 7a). This is expected for such a low concentration of the supporting electrolyte and the *iR* drop 126 mV is calculated using eqn (9) when it is a single-barrel system. When the solution is more conductive at 5 mM and 10 mM $TBAPF_6$, the *iR* drop is indeed much smaller, as shown by a sharper increase of the voltammetric current (Fig. 7a). The *iR* drop can be mostly compensated for using eqn (19) at all of the different concentrations of $TBAPF_6$, as shown by the convergence of the corrected voltammograms at different concentrations of $TBAPF_6$, as shown in Fig. 7b.

The derived *iR* compensation method for the dual barrel pipette also allows us to measure the R_s during the voltammetry using eqn (17). As shown in Fig. 7c, the measured R_s decreases as the concentration of the electrolyte increases at all potentials. The same trend is also observed for the reduction of $Ru(NH_3)_6^{3+}$ and a proton, as shown in Fig. S5 and S6, respectively.† For the Fc system at a given electrolyte concentration, a decrease in the R_s is observed when the electrode potential is swept more positively to induce the oxidation of Fc, as shown in Fig. 7c. As discussed previously in Section 2.1.3, this is expected as the oxidation of Fc generates the ionic species Fc^+ and therefore decreases the solution resistance. Our method also successfully captures the change of the R_s during voltammetry for other systems, including the reduction of $Ru(NH_3)_6^{3+}$ and a proton. In both systems, increases in the R_s are observed as expected, which are shown in Section 6 of the ESI.†

Note that to obtain a good *iR* compensation, i_2 needs to be measured accurately, especially when the solution is highly resistant and i_2 is low. Under this scenario, the noise of i_2, as well as any offset and leakage current from the amplifier itself, can cause issues in *iR* compensation using eqn (19). We refer the readers to Section 7 in the ESI† for the discussion of noise during *iR* compensation.

3. Conclusions

We have shown several methods of *iR* drop compensation in SECCM experiments, which are validated using finite element simulation and experiments. In single-barrel SECCM experiments, the solution resistance can be estimated based on the solution composition and the pipette geometry. A more general method using the limiting current for *iR* compensation is also proposed and validated, which does not assume any specific geometry of the nanopipette and can be potentially applied to other nanoelectrochemical systems. We also developed a method of directly measuring the solution resistance in dual-barrel SECCM, which also corrects the potential shift in this type of experiment. The methods presented here can provide a fast estimation of the significance of the *iR* drop in a SECCM experiment, as well as a method to compensate for the *iR* drop. We envision that this will allow more accurate local electrode kinetics measurements by SECCM, contributing to the revelation of local structure-dependent electrocatalytic mechanisms.

Appendix 1: common symbols used in this manuscript

We list below some of the common symbols used throughout the manuscript.

Symbol	Meaning
V_1	Potential applied at the electrode surface
V_2	Potential applied between the two QRCEs in the dual-barrel SECCM
i_1	Current flowing through the substrate electrode
i_2	Current flowing from one pipette to the other under the bias V_2
R_s	Solution resistance
R_p	Solution resistance in one barrel of the dual-barrel pipette
R_{CT}	Charge transfer resistance
i_{lim}	Mass transfer limiting current
V_{WE}	The actual potential at the working electrode without the *iR* drop
V_{iR}	The ohmic potential drop in the solution (*i.e.*, the *iR* drop)
G	Geometric factor for the single-barrel pipette
κ	Conductivity of the solution
E	Electrode potential
θ	Half-cone angle
r_t	Tip radius of the pipette

Author contributions

BB and GJ performed the key experiments and data analysis. GJ performed the simulation. YW characterized the nanopipette. HR conceived of the project and drafted the manuscript.

Conflicts of interest

There are no conflicts to declare.

Acknowledgements

The research was sponsored by the Defense Advanced Research Project Agency (DARPA) and the Army Research Office and was accomplished under grant number W911NF-20-1-0304. The views and conclusions contained in this document are those of the authors and should not be interpreted as representing the official policies, either expressed or implied, of the Defense Advanced Research Project Agency (DARPA) and the Army Research Office or the U.S. Government. The U.S. Government is authorized to reproduce and distribute reprints for Government purposes notwithstanding any copyright notation herein. Acknowledgment is made to the Donors of the American Chemical Society Petroleum Research Fund for the partial support of this research under award number 61155-DNI5. We thank Dr Kim McKelvey for the help with the WEC-SPM software.

References

1 Z. Yang, J. Zhang, M. C. W. Kintner-Meyer, X. Lu, D. Choi, J. P. Lemmon and J. Liu, *Chem. Rev.*, 2011, **111**, 3577–3613.
2 S. P. S. Badwal, S. S. Giddey, C. Munnings, A. I. Bhatt and A. F. Hollenkamp, *Front. Chem.*, 2014, **2**, 1–28.
3 T. Shinagawa, A. T. Garcia-Esparza and K. Takanabe, *Sci. Rep.*, 2015, **5**, 13801.
4 A. J. Bard and L. R. Faulkner, *Electrochemical Methods: Fundamentals and Applications*, Wiley, New York, 1980.
5 G. Marcandalli, A. Goyal and M. T. M. Koper, *ACS Catal.*, 2021, **11**, 4936–4945.
6 C. L. Bentley, M. Kang and P. R. Unwin, *J. Am. Chem. Soc.*, 2019, **141**, 2179–2193.
7 M. J. S. Farias, W. Cheuquepan, G. A. Camara and J. M. Feliu, *ACS Catal.*, 2016, **6**, 2997–3007.
8 C. L. Bentley, M. Kang, F. M. Maddar, F. Li, M. Walker, J. Zhang and P. R. Unwin, *Chem. Sci.*, 2017, **8**, 6583–6593.
9 Y.-L. Ying, J. Wang, A. R. Leach, Y. Jiang, R. Gao, C. Xu, M. Edwards, A. D. Pendergast, H. Ren, K. T. Weatherly Connor, W. Wang, P. Actis, L. Mao, H. S. White and Y.-T. Long, *Sci. China: Chem.*, 2020, **63**(5), 589–618.
10 N. Ebejer, M. Schnippering, A. W. Colburn, M. A. Edwards and P. R. Unwin, *Anal. Chem.*, 2010, **82**, 9141–9145.
11 L. C. Yule, C. L. Bentley, G. West, B. A. Shollock and P. R. Unwin, *Electrochim. Acta*, 2019, **298**, 80–88.
12 E. Daviddi, K. L. Gonos, A. W. Colburn, C. L. Bentley and P. R. Unwin, *Anal. Chem.*, 2019, **91**, 9229–9237.
13 R. G. Mariano, M. Kang, O. J. Wahab, I. J. McPherson, J. A. Rabinowitz, P. R. Unwin and M. W. Kanan, *Nat. Mater.*, 2021, **20**, 1000–1006.
14 J. T. Mefford, A. R. Akbashev, M. Kang, C. L. Bentley, W. E. Gent, H. D. Deng, D. H. Alsem, Y.-S. Yu, N. J. Salmon, D. A. Shapiro, P. R. Unwin and W. C. Chueh, *Nature*, 2021, **593**, 67–73.
15 A. Kumatani, C. Miura, H. Kuramochi, T. Ohto, M. Wakisaka, Y. Nagata, H. Ida, Y. Takahashi, K. Hu, S. Jeong, J.-i. Fujita, T. Matsue and Y. Ito, *Adv. Sci.*, 2019, **6**, 1900119.
16 Y. Takahashi, A. Kumatani, H. Munakata, H. Inomata, K. Ito, K. Ino, H. Shiku, P. R. Unwin, Y. E. Korchev, K. Kanamura and T. Matsue, *Nat. Commun.*, 2014, **5**, 5450.
17 R. G. Mariano, K. McKelvey, H. S. White and M. W. Kanan, *Science*, 2017, **358**, 1187–1192.

18 M. E. Snowden, A. G. Güell, S. C. S. Lai, K. McKelvey, N. Ebejer, M. A. O'Connell, A. W. Colburn and P. R. Unwin, *Anal. Chem.*, 2012, **84**, 2483–2491.

19 B. Tao, L. C. Yule, E. Daviddi, C. L. Bentley and P. R. Unwin, *Angew. Chem., Int. Ed.*, 2019, **58**, 4606–4611.

20 B. D. B. Aaronson, S. C. S. Lai and P. R. Unwin, *Langmuir*, 2014, **30**, 1915–1919.

21 S.-y. Tan, P. R. Unwin, J. V. Macpherson, J. Zhang and A. M. Bond, *Anal. Chem.*, 2017, **89**, 2830–2837.

22 B. Zhang, Y. Zhang and H. S. White, *Anal. Chem.*, 2006, **78**, 477–483.

23 D. Perry, D. Momotenko, R. A. Lazenby, M. Kang and P. R. Unwin, *Anal. Chem.*, 2016, **88**, 5523–5530.

Faraday Discussions

ROYAL SOCIETY OF CHEMISTRY

PAPER

Electrochemically probing exciton transport in monolayers of two-dimensional semiconductors†

Chloe L. Tolbert and Caleb M. Hill *

Received 13th August 2021, Accepted 29th September 2021
DOI: 10.1039/d1fd00052g

Two-dimensional semiconductors (2DSCs) are attractive for a variety of optoelectronic and catalytic applications due to their ability to be fabricated as wide-area, monolayer-thick films and their unique optical and electronic properties which emerge at this scale. One important class of 2DSCs are the transition metal dichalcogenides (TMDs), which are of particular interest as absorbing layers in ultrathin optoelectronic devices. While TMDs are known to exhibit excellent photovoltaic properties at the bulk level, it is not yet clear how carriers are transported in these materials at thicknesses approaching the monolayer limit, where distinct changes in band structure and the nature of photogenerated carriers occur. Here, it is demonstrated that electrochemical microscopy techniques can be employed as powerful tools for visualizing these processes in 2DSCs, even within individual monolayers. Carrier generation-tip collection scanning electrochemical cell microscopy (CG-TC SECCM), which utilizes spatially-offset optical and pipet-based electrochemical probes to locally generate and detect photogenerated carriers, was applied to visualize carrier generation and transport within well-defined n-WSe_2 samples prepared *via* mechanical exfoliation. Data from these experiments directly reveal how carrier transport varies within complex 2DSC structures as layer thicknesses approach the monolayer limit. These results not only provide valuable new insights into carrier transport within monolayer TMD materials, but also demonstrate electrochemical imaging to be a powerful, yet underutilized approach for visualizing solid-state processes in semiconducting materials.

Introduction

Two-dimensional semiconductors (2DSCs) such as the transition metal dichalcogenides (TMDs) are attractive materials for a variety of optoelectronic applications due to their ability to be prepared as atomically-thin layers and the

Department of Chemistry, University of Wyoming, 1000 E. University Ave., Laramie, WY 82071, USA. E-mail: caleb.hill@uwyo.edu

† Electronic supplementary information (ESI) available: Additional data, details on finite element simulations, and SECCM movies for figures 2, 3, and 4. See DOI: 10.1039/D1FD00052G

interesting properties they exhibit at this scale.[1,2] At monolayer thicknesses, the band structure of many 2DSCs changes significantly, adopting direct bandgaps which yield strong photoluminescence (PL)[3–5] as well as exhibiting multiple local minima, or "valleys", which could potentially be utilized for information processing.[6,7] This, along with the large absorption coefficients exhibited by these materials, make them especially attractive as absorbing layers in photovoltaic cells and other optoelectronic devices.[8–14] In order to design efficient 2DSC-based devices for these applications, it is critically important to understand how photogenerated carriers are transported within 2DSCs and how transport can be controlled in a rational manner.

At the monolayer limit, photogenerated carriers in TMDs become excitonic in nature due to strong confinement in the *xy*-plane and weak dielectric screening, resulting in strong exciton binding energies on the order of several hundred meV.[15–20] This presents unique challenges in optoelectronic applications, as excitons are neutral carriers which cannot be directly influenced by applied fields. Excitons thus move primarily *via* diffusion, so understanding this process, and how different structural motifs/defects within TMDs affect it, is of critical importance. Previous studies carried out to visualize exciton transport within TMDs have employed different optical microscopy techniques (PL or transient absorption) which leverage the favourable optical properties of monolayer samples.[21–25] While this approach has been successful in visualizing carrier transport within pristine monolayer systems, it cannot be applied in more complex structures which lack a clear optical signature for the carriers of interest.

Recently, our laboratory has been exploring how pipet-based electrochemical techniques, such as scanning electrochemical cell microscopy (SECCM),[26–37] can be employed as powerful tools for visualizing carrier transport in 2DSCs. In this approach, carrier generation-tip collection (CG-TC) SECCM, an electrolyte-filled pipet is utilized to detect photogenerated carriers as they drive reactions at a nm-scale electrochemical interface. CG-TC SECCM allows carrier transport to be directly visualized by measuring these photocurrents as a function of probe position, which was recently demonstrated in studies of bulk n-WSe_2.[38] Here, studies are described applying CG-TC SECCM to visualize carrier transport in WSe_2 structures approaching the monolayer limit, ultimately revealing surprisingly long exciton transport distances in WSe_2 monolayers.

Experimental methods

Sample preparation

Bulk n-WSe_2 crystals with dopant densities of $\sim 10^{17}$ cm^{-3} were donated by Prof. Bruce Parkinson. The crystals were mechanically exfoliated immediately before experiments using Scotch tape and transferred to clean indium tin oxide (ITO)-coated cover glass slides (22 × 22 mm, #1.5, 30–60 Ω sq.$^{-1}$, SPI) using a PDMS-assisted dry transfer approach to help preserve the integrity of the nanoflakes (Gel-Pak, Gel Film, Pf-40/17-X4). The samples were left for a minimum of 24 hours to allow for complete van der Waals adherence between the nanoflakes and the ITO. Samples were stored in enclosed plastic containers when not in use, and all data was acquired within one week of initial preparation.

Sample characterization

Atomic force microscopy (AFM) imaging was carried out using a Cypher ES AFM in tapping mode with standard probes (Nanosensors, PPP-NCHR-20, n-Si, 0.01–0.02 Ω cm). We note that AFM imaging showed no significant evidence of organic contaminants on the surface of the samples discussed in this report. PL imaging was carried out using a custom PL lifetime microscope. Samples were mounted on a 2-axis piezo stage (PI P546.2C7) on an inverted optical microscope and illuminated with a pulsed laser diode (485 nm, FWHM = 110 ps, Picoquant) through a 100×, NA = 0.9 objective. Emitted PL from the sample was collected through the same objective, filtered to remove excitation light, and directed into a non-polarizing 90 : 10 beamsplitter. One beam from the beamsplitter was directed onto the slit of a spectrometer-CCD combo (Andor SR-303i-B, Newton EMCCD) and the other was focused onto an avalanche photodiode (MPD) connected to a TCSPC system (Picoquant Picoharp 300). All PL and SECCM data were acquired under low excitation power (<35 μW) to avoid effects due to exciton–exciton annihilation (see Fig. S1 in ESI†).

Probe fabrication and characterization

SECCM probes were fabricated from quartz capillaries (1.2 mm outer diameter, 0.9 mm inner diameter, Sutter) using a Sutter P-2000 pipet puller. A two-line program was employed to yield *ca.* 500 nm diameter probes (heat = 750, fil = 4, vel = 30, delay = 150, pull = 80/heat = 685, fil = 3, vel = 30, delay = 135, pull = 150). These probes were filled with an electrolyte consisting of 0.01 M I_2 (Mallinckrodt, U.S.P Grade) and 0.10 M NaI (Sigma Aldrich, ≥ 99%). The geometry of the fabricated probes evaluated using SEM imaging (FEI Quanta FEG 450, 2 keV accelerating voltage).

CG-TC SECCM studies

A home-built setup was used to conduct all SECCM studies. ITO slides containing the WSe_2 samples were attached to a ceramic holder, mounted onto an inverted optical microscope, and illuminated locally with a 633 nm laser (ThorLabs) focused through a 40×, 0.5 NA objective. An optical chopper (Thorlabs) set at 20 Hz was used to distinguish photocurrents during the experiments depicted in Fig. 3 and 4. Probes were completed by inserting a Ag/AgI wire into the electrolyte solution, and all potentials were referenced *vs.* this electrode. Completed probes were then mounted to a 3-axis piezo system (PI NanoCube). The SECCM probes were translated towards the sample while applying a cathodic bias to the ITO substrate, and probe–sample contact was detected as a sudden spike in current. A triangular potential waveform was then applied to the sample and the resulting current was recorded before retracting the probe and repeating this process at another location on the sample. Electrochemical measurements were made using a patch-clamp style amplifier (Dagan CHEM-CLAMP) and all instrumentation was controlled through custom LabView software.

Results and discussion

Carrier generation-tip collection SECCM

The basic principle of CG-TC SECCM is illustrated in Fig. 1. A semiconducting sample is immobilized on a conducting substrate and illuminated locally with

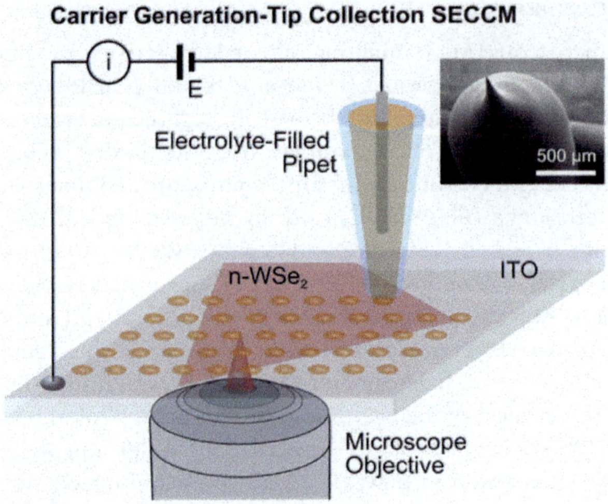

Fig. 1 Carrier generation-tip collection scanning electrochemical cell microscopy. An electrolyte-filled pipet is brought into contact with a semiconducting sample, establishing a local photoelectrochemical interface in the vicinity of a focused excitation source. By examining the resulting photocurrents as a function of probe location, carrier transport within a material can be directly visualized. The inset shows an SEM image taken of a representative SECCM probe.

a focused laser beam. If the energy of incident photons is greater than the bandgap of the semiconducting material, the excitation of electrons from the valence band to the conduction band will create mobile carriers. These photogenerated carriers will then begin to move outward from the excitation source *via* diffusion and migration (field driven transport). In CG-TC SECCM, the presence of carriers is interrogated using an SECCM probe to define a photoelectrochemical interface at a defined distance from the excitation source. Photogenerated carriers which reach this interface can drive electrochemical reactions, giving rise to a measurable current. By carrying out voltammetric measurements with the probe positioned across an array of points spanning the excitation source, carrier transport within the material can be directly visualized through electrochemical "movies" which show how photocurrents vary as a function of spatial position and applied potential. More details on the fundamentals of this technique can be found in a recent publication from our lab.[38]

Transport within bulk WSe_2

CG-TC SECCM was first applied to probe carrier transport within exfoliated layers of n-type WSe_2 (n-WSe_2) with thicknesses greater than 10 nm. At these thicknesses, 2DSCs such as WSe_2 behave electronically as "bulk" materials. Probe–sample contact thus generates a space-charge layer in the 2DSC which extends outwardly from the solid–electrolyte interface. For an n-type semiconductor under depletion conditions, the thickness of this layer can be estimated using the following equation:

$$d_{sc} = \sqrt{\frac{\varepsilon\varepsilon_0(E - E_{fb})}{2qn_D}}$$

Here, ε is the dielectric constant of the material, ε_0 is the vacuum permittivity, E is the applied potential, E_{fb} is the flatband potential, q is the elementary unit of charge, and n_D is the dopant concentration in the semiconductor (*ca.* 10^{17} cm^3). Taking ε to have in-plane and out-of-plane values of 16 and 8,[39,40] respectively, the corresponding space charge layer thicknesses are ~110 nm and 80 nm. In the experiments presented here, the WSe_2 layers are well below this thickness, resulting in roughly linear potential profiles along the *z*-axis. These fields are highly confined, however, to within a few hundred nm from the pipet in the *xy*-plane. Given the μm-scale distances involved in these studies, the exact shape of this space charge region has a negligible impact on these experiments.

Example data from CG-TC SECCM studies of a 25 nm thick, defect-free n-WSe_2 layer is given in Fig. 2. While the layer itself exhibited lateral dimensions on the order of several hundred μm, measurements were obtained within a small, 30 μm × 30 μm square region free from any visible defects. The probe reaction employed was iodide oxidation/reduction:

$$3I^- + 2h^+ \rightleftharpoons I_3^-$$

Free carriers created within the excitation region diffused outward within the n-WSe_2 sheet. Carriers reaching the space charge region surrounding the pipet are separated by the large fields present and holes are driven to the electrochemical interface. At the interface, these holes oxidize I^- to I_2/I_3^-, resulting in a measurable photocurrent.

Photocurrent maps obtained in the vicinity of the excitation are provided in Fig. 2b. At cathodic potentials (0.45 V), photocurrents are highly localized,

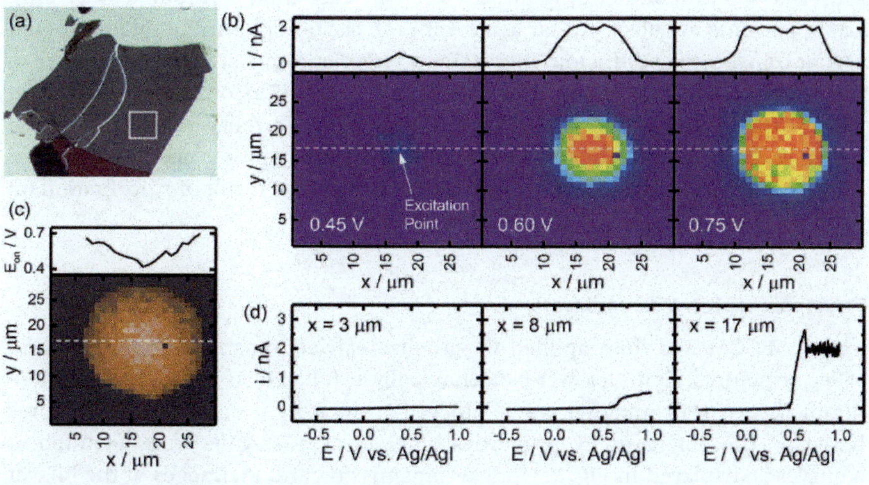

Fig. 2 CG-TC SECCM imaging of transport with bulk n-WSe_2. (a) Optical transmission image of an exfoliated n-WSe_2 flake with a thickness of 25 nm. The region within this flake investigated *via* SECCM is indicated. (b) Photocurrent images at different potentials obtained in the vicinity of a focused 633 nm laser (1.12 μW). (c) Image of onset potentials (E_{on}), calculated as the potential necessary to reach a current of 100 pA. (d) Voltammograms obtained at different points along the dashed line in (b). The sweep rate was 0.20 V s^{-1}. All SECCM data was obtained using a ~600 nm diameter probe filled with 0.1 M NaI, 0.01 M I_2.

exhibiting a peaked profile that extends only a few μm from the excitation centroid. At more anodic potentials, the photocurrent profile significantly widens, eventually adopting a "top-hat" profile with a width of *ca.* 15 μm. A photocurrent movie is provided in the ESI† which provides a continuous visualization of this process (similar movies are also provided for Fig. 3 and 4).

As described in our previous report, the shape of this profile can be attributed to limitations on the mass transfer of redox species to the active interface in the SECCM geometry.[38] This mass-transfer limited current is:

$$i_{\mathrm{mt}} \approx \frac{2\pi n q_{\mathrm{e}} D_{\mathrm{r}} C_{\mathrm{r}}^{*} r_{\mathrm{p}}\left(1 - \cos\theta_{\mathrm{p}}\right)}{\tan\theta_{\mathrm{p}}}$$

Here, n is the number of transferred electrons, q_e is the electronic charge, D_r is the diffusion coefficient of the redox active species, C_r^* is its bulk concentration, r_p is the pipet radius, and θ_p is the half-angle of the pipet. Using $n = 2/3$, $D = 2 \times 10^{-5}$ $cm^2\ s^{-1}$,[41] $C^* = 100$ mM, $r_p = 250$ nm, and $\theta_p = 10°$ yields a value of ~2.4 nA, in good agreement with the plateaus in the observed profiles.

The widths of these profiles can be used to estimate the diffusion lengths of carriers in the material *via* comparison to finite element simulations which model carrier generation, recombination, and transport. The 15 μm widths observed here correspond to lateral carrier diffusion lengths of ~2.1 μm (see ESI† for details).

Fig. 2c presents these results in terms of onset potentials, defined as the potential necessary to reach an anodic photocurrent of 100 pA. Approaching the excitation centroid, the potential required to reach this threshold lowers at a rate of *ca.* 30 mV μm^{-1}. This reflects the larger potentials needed to concentrate holes at the electrochemical interface and drive the oxidation of iodide to an appreciable degree. Example voltammograms provided in Fig. 2d show the evolution of the photocurrent response starting at a distance far removed from the excitation source ($x = 3$ μm) and ending directly over the excitation centroid ($x = 17$ μm). Lower currents are observed far away from the excitation, attributable to limited carrier transport *via* diffusion. Increasingly larger currents are observed closer to the excitation source, eventually becoming limited by diffusion of redox species in the SECCM probe as discussed above. In all cases, a fairly simple sigmoidal response is obtained, though oscillatory behaviour can be observed at high current densities within the plateau region, which is the result of I_2 formation on the WSe_2 surface.[42]

Transport within WSe_2 bilayers

CG-TC SECCM was then applied to visualize carrier transport within few-layer WSe_2 structures. Samples were mechanically exfoliated as before, and photoluminescence (PL) imaging was employed to identify mono/bilayer structures based on the emergent PL properties of WSe_2 at this scale. Data acquired at a bilayer is displayed in Fig. 3. Optical transmission and PL images of the bilayer are provided in Fig. 3a. The dimensions of the bilayer region were roughly 10 × 90 μm, and the sample displayed significant PL across this entire region. The PL spectra (Fig. 3c) shows a broad (FWHM ≈ 100 meV) peak centered around 1.6 eV, attributable to an overlap of exciton emission and indirect band gap emission.[43,44]

Photocurrent images obtained in a continuous region of the bilayer are displayed in Fig. 3b. In the bilayer sample, the observed photocurrent profiles are

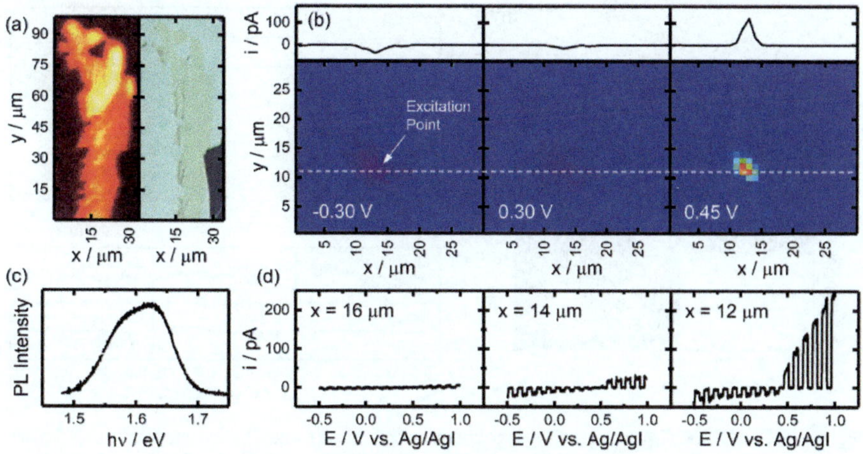

Fig. 3 CG-TC SECCM imaging of transport within a WSe_2 bilayer. (a) Optical transmission and photoluminescence (PL) images of an exfoliated WSe_2 bilayer. (b) Photocurrent images at different potentials obtained in the vicinity of a chopped (20 Hz), focused 633 nm laser (0.5 µW). (c) PL spectrum acquired from the layer. (d) Voltammograms obtained at different points along the white dashed line in (b). The sweep rate was 0.20 V s^{-1}. All SECCM data was obtained using a ~600 nm diameter probe filled with 0.1 M NaI, 0.01 M I_2.

much narrower than those observed in bulk samples, essentially overlapping with the laser profile (see Fig. S2 in the ESI†). This suggests the lateral diffusion of photogenerated excitons is profoundly limited compared to the bulk example. This is likely attributable to an increase in exciton recombination arising from a higher density of defects in the bilayer sample. The PL image provided in Fig. 3a supports this assertion, as significant intensity fluctuations are seen across the bilayer structure, reflecting the influence of nonradiative exciton recombination at defects.

Example voltammograms are provided in Fig. 3d. The bilayer exhibits a photoconductivity-type response, generating photocurrents at both cathodic and anodic potentials. The polarity of the photocurrents changes at a potential of ~0.5 V *vs.* Ag/AgI, which corresponds to the flat-band potential within the WSe_2 bilayer. The observed photocurrents, which reach maximum values of *ca.* 200 pA, are also much lower than one would expect. At 633 nm, a bilayer can be estimated to absorb roughly 12% of incident radiation,[44] yet these photocurrents reflect an incident photon to electron conversion efficiency of only 0.07%. This low efficiency, as well as the significant potential dependence in the observed photocurrents, may be attributable to an additional transport mechanism which drives carriers away from the probe–TMD interface, arising from the formation of interlayer excitons (see below).

Transport of excitons generated within monolayered WSe_2

Results from experiments of a more complex structure incorporating a WSe_2 monolayer are depicted in Fig. 4. An optical reflection image of the exfoliated flake is given in Fig. 4a, which allows six distinct regions of varying thickness to be

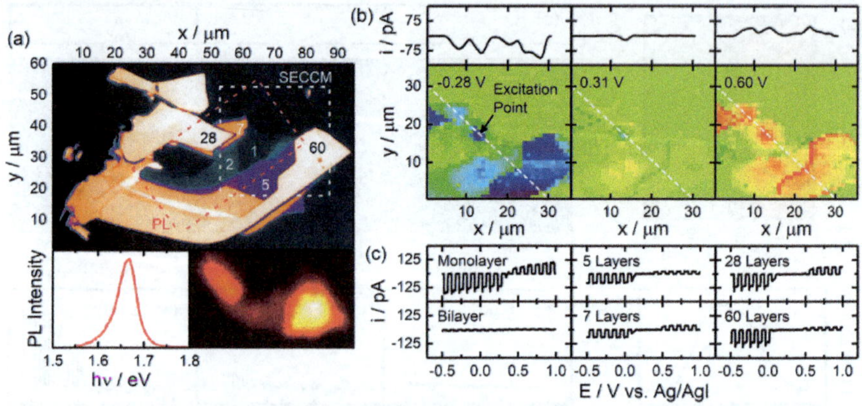

Fig. 4 CG-TC SECCM imaging of transport within a complex WSe_2 structure. (a) False color optical reflection and photoluminescence (PL) images of an exfoliated WSe_2 structure. The number of WSe_2 layers in different regions established *via* AFM measurements are indicated. A PL spectrum acquired from the monolayer is given below. (b) Photocurrent images at different potentials obtained in the vicinity of a chopped (20 Hz), focused 633 nm laser (0.4 µW). (c) Voltammograms obtained at different points along the white dashed line in (b). The sweep rate was 2.0 V s^{-1}. All SECCM data was obtained using a ~600 nm diameter probe filled with 0.1 M NaI, 0.01 M I_2.

easily distinguished. The thickness of each region was confirmed *via* AFM, and layer numbers are indicated in the provided image. A trapezoidal monolayer region was identified and confirmed *via* PL spectroscopy, exhibiting the sharp, intense emission typical of monolayered WSe_2.[44]

CG-TC SECCM images obtained for this sample are given in Fig. 4b. The sample was illuminated within the monolayer region, and SECCM measurements were obtained within a 35 µm × 35 µm region surrounding the excitation centroid, encompassing the area indicated in Fig. 4a. Significant cathodic (blue) and anodic (red) photocurrents were observed within the monolayer region surrounding the excitation centroid. However, in stark contrast to the bilayer sample, photocurrents can be observed throughout the entire sample area imaged *via* SECCM. Photocurrents observed in thicker regions of the sample are comparable in magnitude to those in the immediate vicinity of the excitation, while thinner regions display diminished photocurrents. Some spatial heterogeneity within these regions could be observed, likely due to the proximity of step-edge defects between adjacent regions. Due to the large carrier diffusion lengths observed here, it is unlikely that variations in sample–substrate contact could explain these effects. At anodic potentials, enhanced currents can be visualized at step-edge defects between adjacent regions, attributable to enhanced iodide oxidation kinetics. A notable feature in these images is the almost complete absence of observable photocurrents in the bilayer region (left-middle).

Example voltammograms obtained within each of the distinct thickness regimes are given in Fig. 4c. The general behaviour is similar to that of the bilayer, exhibiting photocurrents at both cathodic and anodic potentials. However, a significant splitting between the cathodic and anodic branches is observed which increases with increasing layer thickness. These trends are summarized in

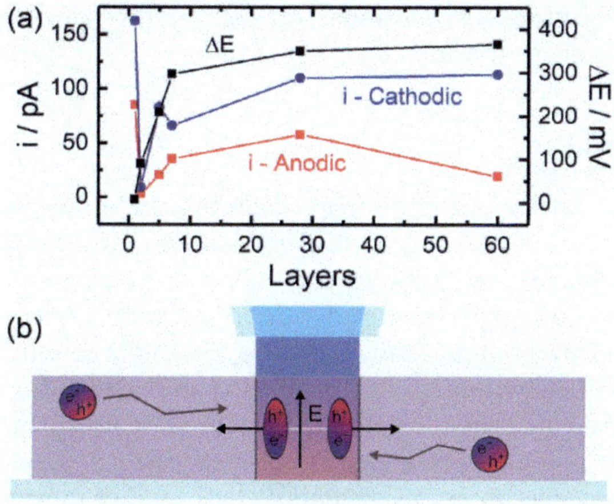

Fig. 5 (a) Relationships between layer thickness, photocurrents, and potential splitting between cathodic and anodic photocurrent branches (ΔE). (b) Illustration of the proposed mechanism responsible for negligible photocurrents in WSe_2 bilayers observed *via* SECCM.

Fig. 5a. This splitting can be attributed to the establishment of a field necessary to drive the transport of carriers vertically within the WSe_2 structure, requiring greater potentials within thicker regions.

Anomalous transport within monolayered WSe_2

Two features of the data presented in Fig. 4 warrant further discussion. First, the CG-TC experiments reveal that photogenerated excitons in monolayered WSe_2 can effectively drive photoelectrochemical reactions across distances in excess of 20 μm, much farther than the diffusion lengths estimated in previous studies on samples supported on inert substrates using optical techniques (<1 μm).[21,22] The ITO-supported WSe_2 monolayer studied here displayed PL lifetimes of *ca.* 1 ns (see Fig. S3 in the ESI†), which are roughly an order of magnitude greater than commonly reported values.[43,45] This discrepancy suggests long-lived charge transfer states exist which increase the effective lifetime and/or diffusion of photogenerated excitons. While more studies are needed to elucidate the origin of these effects, the data presented here nevertheless shows studies on idealized, encapsulated model systems may not meaningfully reflect how these materials will perform in practical devices.

The second notable feature is the significantly lower photocurrents observed at WSe_2 bilayers in the CG-TC SECCM experiments. This may be attributable to the field-driven formation of interlayer excitons in a bilayer structure, an effect which has been demonstrated to occur previously in homobilayers.[45] This idea is depicted schematically in Fig. 5b. In a bilayer structure, photogenerated excitons which diffuse to the pipet interface may be converted into interlayer excitons due to the large (>10^6 V cm^{-1}) fields present. These interlayer species would then be repelled from one another due to electrostatic interactions, which has been shown to enhance transport in bilayer systems as well as heterostructures.[6,24,45] In the present experiments, similar effects would cause excitons within bilayers to be repelled from the probe–sample interface, significantly lowering the currents observable within these regions.

Conclusions

In this report, carrier generation-tip collection scanning electrochemical microscopy (CG-TC SECCM) was applied to visualize carrier transport in WSe_2 structures with thicknesses spanning the bulk (>10 nm) and monolayer levels. Using CG-TC SECCM, photogenerated carriers in bulk n-WSe_2 were found to exhibit diffusion lengths of *ca.* 2 µm. In the bilayer sample investigated, very short diffusion lengths were observed, attributable to a high density of defects present in the bilayer structure. The most striking results were obtained for the investigated monolayer sample, where photogenerated carriers were found to drive reactions over distances in excess of 20 µm. These studies provide powerful new insight into the transport of photogenerated carriers in complex 2DSC structures, and further demonstrate the power of electrochemical approaches for studying transport in these systems.

Author contributions

CLT and CMH designed the experiments. CLT acquired all the data. CLT and CMH analysed the data and wrote the manuscript.

Conflicts of interest

The authors have no conflicts to declare.

Acknowledgements

The authors acknowledge generous support for this work from the National Science Foundation (CHE-2045593), the University of Wyoming, and the Wyoming NASA Space Grant Consortium (NASA Grant #NNX15AI08H). The authors would also like to thank Dr Bruce Parkinson for his generous donation of WSe_2 crystals.

References

1 K. S. Novoselov, D. Jiang, F. Schedin, T. J. Booth, V. V. Khotkevich, S. V. Morozov and A. K. Geim, *Proc. Natl. Acad. Sci. U. S. A.*, 2005, **102**, 10451–10453.

2 Z. Zeng, Z. Yin, X. Huang, H. Li, Q. He, G. Lu, F. Boey and H. Zhang, *Angew. Chem.*, 2011, **123**, 11289–11293.

3 K. F. Mak, C. Lee, J. Hone, J. Shan and T. F. Heinz, *Phys. Rev. Lett.*, 2010, **105**, 2–5.

4 W. Zhao, Z. Ghorannevis, L. Chu, M. Toh, C. Kloc, P.-H. Tan and G. Eda, *ACS Nano*, 2013, **7**, 791–797.

5 A. Splendiani, L. Sun, Y. Zhang, T. Li, J. Kim, C. Y. Chim, G. Galli and F. Wang, *Nano Lett.*, 2010, **10**, 1271–1275.

6 D. Unuchek, A. Ciarrocchi, A. Avsar, Z. Sun, K. Watanabe, T. Taniguchi and A. Kis, *Nat. Nanotechnol.*, 2019, **14**, 1104–1109.
7 K. Behnia, *Nat. Nanotechnol.*, 2012, **7**, 488–489.
8 D. Jariwala, A. R. Davoyan, J. Wong and H. A. Atwater, *ACS Photonics*, 2017, **4**, 2962–2970.
9 K. Sivula and R. van de Krol, *Nat. Rev. Mater.*, 2016, **1**, 15010.
10 T. Mueller and E. Malic, *npj 2D Mater. Appl.*, 2018, **2**, 1–12.
11 D. Akinwande, N. Petrone and J. Hone, *Nat. Commun.*, 2014, **5**, 5678.
12 D. Jariwala, V. K. Sangwan, L. J. Lauhon, T. J. Marks and M. C. Hersam, *ACS Nano*, 2014, **8**, 1102–1120.
13 H. Hosono, *Thin Solid Films*, 2007, **515**, 6000–6014.
14 L. Wang, L. Huang, W. C. Tan, X. Feng, L. Chen, X. Huang and K.-W. Ang, *Small Methods*, 2018, **2**, 1700294.
15 G. Wang, A. Chernikov, M. M. Glazov, T. F. Heinz, X. Marie, T. Amand and B. Urbaszek, *Rev. Mod. Phys.*, 2018, **90**, 21001.
16 A. Steinhoff, M. Rösner, F. Jahnke, T. O. Wehling and C. Gies, *Nano Lett.*, 2014, **14**, 3743–3748.
17 M. Koperski, M. R. Molas, A. Arora, K. Nogajewski, A. O. Slobodeniuk, C. Faugeras and M. Potemski, *Nanophotonics*, 2017, **6**, 1289–1308.
18 K. He, N. Kumar, L. Zhao, Z. Wang, K. F. Mak, H. Zhao and J. Shan, *Phys. Rev. Lett.*, 2014, **113**, 1–5.
19 T. Cheiwchanchamnangij and W. R. L. Lambrecht, *Phys. Rev. B: Condens. Matter Mater. Phys.*, 2012, **85**, 205302.
20 A. Ramasubramaniam, *Phys. Rev. B: Condens. Matter Mater. Phys.*, 2012, **86**, 115409.
21 D. F. Cordovilla Leon, Z. Li, S. W. Jang, C.-H. Cheng and P. B. Deotare, *Appl. Phys. Lett.*, 2018, **113**, 252101.
22 F. Cadiz, C. Robert, E. Courtade, M. Manca, L. Martinelli, T. Taniguchi, K. Watanabe, T. Amand, A. C. H. Rowe, D. Paget, B. Urbaszek and X. Marie, *Appl. Phys. Lett.*, 2018, **112**, 152106.
23 L. Yuan, T. Wang, T. Zhu, M. Zhou and L. Huang, *J. Phys. Chem. Lett.*, 2017, **8**, 3371–3379.
24 L. Yuan, B. Zheng, J. Kunstmann, T. Brumme, A. B. Kuc, C. Ma, S. Deng, D. Blach, A. Pan and L. Huang, *Nat. Mater.*, 2020, **19**, 617–623.
25 Q. Cui, F. Ceballos, N. Kumar and H. Zhao, *ACS Nano*, 2014, **8**, 2970–2976.
26 J. T. Mefford, A. R. Akbashev, M. Kang, C. L. Bentley, W. E. Gent, H. D. Deng, D. H. Alsem, Y.-S. Yu, N. J. Salmon, D. A. Shapiro, P. R. Unwin and W. C. Chueh, *Nature*, 2021, **593**, 67–73.
27 R. G. Mariano, M. Kang, O. J. Wahab, I. J. McPherson, J. A. Rabinowitz, P. R. Unwin and M. W. Kanan, *Nat. Mater.*, 2021, **20**, 1000–1006.
28 J. W. Hill, Z. Fu, J. Tian and C. M. Hill, *J. Phys. Chem. C*, 2020, **124**, 17141–17149.
29 L. E. Strange, J. Yadav, S. Garg, P. S. Shinde, J. W. Hill, C. M. Hill, P. Kung and S. Pan, *J. Phys. Chem. Lett.*, 2020, **11**, 3488–3494.
30 O. J. Wahab, M. Kang and P. R. Unwin, *Curr. Opin. Electrochem.*, 2020, **22**, 120–128.
31 E. Daviddi, Z. Chen, B. Beam Massani, J. Lee, C. L. Bentley, P. R. Unwin and E. L. Ratcliff, *ACS Nano*, 2019, **13**, 13271–13284.

32 C. L. Bentley, J. Edmondson, G. N. Meloni, D. Perry, V. Shkirskiy and P. R. Unwin, *Anal. Chem.*, 2019, **91**, 84–108.
33 C. L. Bentley, M. Kang and P. R. Unwin, *J. Am. Chem. Soc.*, 2017, **139**, 16813–16821.
34 B. Tao, P. R. Unwin and C. L. Bentley, *J. Phys. Chem. C*, 2020, **124**, 789–798.
35 M. Kang, D. Perry, C. L. Bentley, G. West, A. Page and P. R. Unwin, *ACS Nano*, 2017, **11**, 9525–9535.
36 M. E. Snowden, A. G. Güell, S. C. S. Lai, K. McKelvey, N. Ebejer, M. A. O'Connell, A. W. Colburn and P. R. Unwin, *Anal. Chem.*, 2012, **84**, 2483–2491.
37 J. W. Hill and C. M. Hill, *Nano Lett.*, 2019, **19**, 5710–5716.
38 J. W. Hill and C. M. Hill, *Chem. Sci.*, 2021, **12**, 5102–5112.
39 H. J. Lewerenz, S. D. Ferris, C. J. Doherty and H. J. Leamy, *J. Electrochem. Soc.*, 1982, **129**, 418.
40 A. Laturia, M. L. Van de Put and W. G. Vandenberghe, *npj 2D Mater. Appl.*, 2018, **2**, 6.
41 *CRC Handbook of Chemistry and Physics*, ed. D. R. Lide, CRC Press, 89th edn, 2008.
42 H. Tributsch, *J. Electrochem. Soc.*, 1978, **125**, 1086–1093.
43 T. Yan, X. Qiao, X. Liu, P. Tan and X. Zhang, *Appl. Phys. Lett.*, 2014, **105**, 101901.
44 W. Zhao, Z. Ghorannevis, L. Chu, M. Toh, C. Kloc, P.-H. Tan and G. Eda, *ACS Nano*, 2013, **7**, 791–797.
45 Z. Wang, Y.-H. Chiu, K. Honz, K. F. Mak and J. Shan, *Nano Lett.*, 2018, **18**, 137–143.

Faraday Discussions

PAPER

Electrochemical characterization of individual oil micro-droplets by high-frequency nanocapacitor array imaging†

Christophe Renault,[ab] Cecilia Laborde,[a] Andrea Cossettini,[cd] Luca Selmi,[e] Frans Widdershoven[f] and Serge G. Lemay*[a]

Received 19th July 2021, Accepted 19th August 2021

DOI: 10.1039/d1fd00044f

CMOS-based nanocapacitor arrays allow local probing of the impedance of an electrolyte in real time and with sub-micron spatial resolution. Here we report on the physico-chemical characterization of individual microdroplets of oil in a continuous water phase using this new tool. We monitor the sedimentation and wetting dynamics of individual droplets, estimate their volume and infer their composition based on their dielectric constant. From measurements before and after wetting of the surface, we also attempt to estimate the contact angle of individual micron-sized droplets. These measurements illustrate the capabilities and versatility of nanocapacitor array technology.

Introduction

Electrochemical impedance spectroscopy (EIS) is a powerful and flexible technique for probing electrochemical processes at the surface and in the vicinity of an electrode. The nature of the information that is probed depends on the frequency f at which a measurement is performed. This is characterized by the frequency $f_1 = 1/2\pi R_e C_e$, where C_e and R_e are the electrical double layer (EDL) capacitance and the solution access resistance of the electrode, respectively. For $f < f_1$, the measured response is dominated by C_e and the technique is primarily sensitive to the structure of the EDL. For $f > f_1$, the response is instead dominated by R_e, and in this case the composition of the solution away from the electrode is

[a]*MESA+ Institute for Nanotechnology, University of Twente, P.O. Box 217, 7500 AE Enschede, The Netherlands. E-mail: s.g.lemay@utwente.nl*

[b]*Laboratoire de Physique de la Matière Condensée, Ecole Polytechnique, CNRS, IP Paris, Route de Saclay, Palaiseau, 91128, France*

[c]*Integrated Systems Laboratory, ETH Zürich, Gloriastrasse 35, 8092 Zürich, Switzerland*

[d]*DPIA, Università degli Studi di Udine, via delle Scienze 206, 33100 Udine, Italy*

[e]*DIEF, University of Modena and Reggio Emilia, Via Vivarelli 10, Modena, 41125, Italy*

[f]*NXP Semiconductors, Technology & Operations/CTO office, High Tech Campus 46, Eindhoven 5656 AE, The Netherlands*

† Electronic supplementary information (ESI) available. See DOI: 10.1039/D1FD00044F

probed. Since the electric field falls off slowly with distance r from the electrode, the interaction with the sample is long-range and extends many times the radius of a disk electrode, r_e, into the solution. EIS measurements at a nanoelectrode thus provide a very local probe of either the properties of a solid–liquid interface or of an electrolyte in the vicinity of the electrode, depending on the choice of frequency. In practice, however, these measurements are technically very challenging to implement for a nanoelectrode for two main reasons:

(1) The desired signal is easily obscured by parasitic current pathways. The capacitance of a nanoelectrode in aqueous media is of the order $C_e \sim 10^{-15}$ F. In conventional measurements based on standard EIS instrumentation, this is dwarfed by the capacitance of the interconnects to the electrode. As a rough illustration, the capacitance of a coaxial cable is $\sim 10^{-12}$ F cm^{-1}, so even 1 cm of cable causes a background current that is 1000 times larger than the target signal. A nanoelectrode thus contributes only a very minor fraction of the total current flowing through the system, making it difficult to distinguish its contribution from the large background current.

(2) Reaching the regime $f > f_1$ becomes increasingly difficult with decreasing electrode size. The parallel-plate component of the electrode's EDL capacitance scales with electrode radius as $C_e \propto r_e^2$ while the spreading access resistance scales as $R_e \propto r_e^{-1}$, hence $f_1 \propto r_e^{-1}$. Shrinking the electrode size from $r_e = 10$ µm to 10 nm thus requires accessing 1000 times higher frequencies in order to probe the solution beyond the EDL.

Two main approaches have been used to overcome these challenges. Frequency mixing methods employ an active material (a field-effect transistor) as the electrode. The nonlinear response of the current through the transistor to the EDL potential allows applying high-frequency voltages yet reading out the resulting response at much lower frequencies *via* heterodyne methods.[1] In a more direct approach, we have previously reported a platform in which nanoelectrodes and their readout circuitry are closely integrated on a single silicon chip. This allows reducing the length of the interconnects between the electrodes and the relevant readout circuit elements to ~1 µm, in turn reducing the stray capacitance to a very small value of the order $\sim 10^{-15}$ F.[2–6]

Our platform is based on complementary metal–oxide-semiconductor (CMOS) technology and offers important additional capabilities. Using a switched capacitor architecture allowed us the implementation of an array of 65 536 individually addressable 180 nm diameter nanoelectrodes. These can be actuated at frequencies of up to 50 MHz, which is sufficient to reach $f > f_1$ even at high (*e.g.* physiological) salt conditions. When interfaced to external pulse generators, the platform also demonstrated operating frequencies of up to 500 MHz.[7] Simultaneous measurements over the array, function in a manner analogous to a digital camera, where each nanoelectrode plays the role of a single pixel. This permits a form of EIS imaging where objects ranging from synthetic microparticles to living cells, can be imaged in real time and at sub-µm spatial resolution. Here we further demonstrate the capabilities of this platform using oil micro-droplets as a model system.

Emulsions are ubiquitous in daily life. They are present in food,[8] cosmetics[9] and drugs.[10] They also find important roles in petroleum recovery,[11] nonaqueous phase liquid transport[12] and material synthesis.[13] The study of the composition, structure, dynamics and more importantly the interaction of oil (or water) droplets with a given environment is thus of critical importance. Different techniques are usually necessary to fully characterize an emulsion. Droplet size can be obtained by dynamic light scattering (DLS), provided that the sample is reasonably mono-disperse, or by more time-intensive and costly techniques such as cryo-electron microscopy.[14] Optical microscopy, Coulter counters, ultrasound, NMR and sedimentation are also commonly used depending on the size and transparency of the sample.[14] The interaction between an emulsion and a surface is typically characterized by the contact angle measured with a single large droplet deposited on a surface of interest. Despite all of these techniques, however, the study of emulsions still suffers several limitations. First, only the combination of these relatively expensive methods (*k*$5–100/instrument + highly qualified operator) allows complete characterization of an emulsion. Second, most of these measurements are performed in a laboratory and cannot be easily implemented in demanding environments (high temperature, humidity, vibrations or small spaces). Third, none of the aforementioned techniques can probe emulsions with both high temporal and spatial resolution. Alternative methods of characterization are thus desirable.

Pushing in this direction, Bard and coworkers proposed a purely electrochemical method to characterize individual sub-micron droplets of ionic liquid, nitrobenzene, benzene, cyclohexane and tripropylamine in water.[15–18] Analysis of the electrochemical response produced by the impact of droplets at the surface of an ultra-microelectrode (UME) was employed to estimate the size and redox content of the droplets as well as the area of contact between a droplet and a surface.[17,18] This method has the advantage of working in nearly any liquid medium (even with very low conductivity) for both insulating and redox-active droplets, and allows the study of highly disperse populations at the single-particle level. However, the low currents (pA to nA) and the mode of detection based on faradaic reactions requires a highly controlled environment and the presence of a redox reporter either in the droplet or in the continuous phase.

Impedance measurements provide an elegant and reagent-free alternative to the faradaic approach. Variations of the capacitance of the electric double layer (EDL) are commonly used to monitor surface functionalization (*e.g.* self-assembled monolayers),[19] or for detection of small analytes such as proteins.[20] However, this kind of measurement is not selective toward the adsorbate and thus extremely sensitive to nonspecific adsorption. Moreover the capacitance is affected by small changes of the electrolyte concentration in the sample. Finally, probing the ionic composition of the few nanometers thick EDL is relatively inefficient for the detection of objects much larger than the EDL, such as microspheres, a common label for sensors. One way to simultaneously tackle these limitations is to measure impedance at high frequencies. Particularly interesting are frequencies so high that the ions do not have time to respond significantly to the alternating electric field. In this case the electrical current is dominated by the displacement current caused by the polarization of the solvent, as characterized by its permittivity ε. This occurs when the frequency satisfies $f > f_2 = \sigma/2\pi\varepsilon$, where σ is the conductivity of the electrolyte (in practice $f_2 > f_1$). An application of high-frequency measurements is the detection of water content in oil.[21,22] Because the measurement is less affected by the presence of ions in solution, simple analytical models can be used to determine the average content of water in oil. However, no information at a single water droplet level is obtained in this case.

Here we employ polydisperse oil-in-water emulsions made from different oils, namely 1,2-dichlorobenzene (*o*DCB), nitrobenzene and silicone oil. We demonstrate the capabilities of the nanocapacitor array platform by simultaneously determining the size and shape of individual oil micro-droplets as well as their dielectric constant at the micron scale. Simultaneously, we evidence various dynamical behavior of individual droplets including sedimentation, convection, wetting and merging. Finally, we explore a way to estimate *in situ*, the contact angle of individual micro-droplets of only a few pL while typical contact angle measurements require μL's.

Materials and methods

Chemicals and materials

Ortho-dichlorobenzene (>99%), nitrobenzene (>99%), isopropanol (99.9%), toluene (anhydrous, 99.8%) and chloroform (anhydrous, >99%) were purchased from Sigma-Aldrich. Silicone oil (BC 5 cS) was purchased from Basildon Chemicals, and tetrabutylammonium toluene-4-sulfonate (>99%) was obtained from Fluka. All chemicals were used without further purification. DI-water (18.2 MΩ cm, 2–4 ppb total organic content) was produced with a Milli-Q Advantage A10 (Millipore) purification system. The capacitance imaging was performed with CMOS-based microchips and readout board described previously.[2,5]

Emulsion preparation

Emulsions were prepared by adding 10 μL of the desired oil in 990 μL of DI-water and sonicating for one minute in a sonication bath (Branson 1510DTH). The size of droplets that sedimented on a glass slide were measured with an optical microscope (50× objective). The stability of the emulsion was monitored by visual inspection of the turbidity of the solution over a period of 1 h, and by optical microscopy. For all the emulsions the size of the droplet was stable at least up to 15 min. The emulsions were prepared immediately before the measurement, and the chip was incubated within less than a minute after sonication.

Characterization

The scanning electron microscope (SEM) image of the nanoelectrode array was obtained with a LEO Gemini 1550 FEG-SEM equipped with a field emission gun. The electron acceleration voltage and the probing current were 1.5 kV and 50 pA, respectively. The lateral resolution was approximately 2 nm. The static contact angle between a 1 μL *o*DCB and the surface of a CMOS chip was measured with an OCA-15 goniometer (DataPhysics Instruments) in a chamber filled with DI-water at 25 °C. The surface of the chip was cleaned immediately before the measurement by rinsing with acetone and isopropanol and drying in an air stream. Three independent chips were used and the measurement was repeated three times on each chip. The contact angle was determined using the software Scan20. A baseline correction and 5 points ellipsoid fitting procedure was used.

Data processing and simulations

Data analysis was performed with OriginPro 9.1 and in-house codes running on Matlab 2012. Numerical simulations were performed with ENBIOS, a custom 3D control-volume finite element solver of the Poisson–Boltzmann (PB) and Poisson–Nernst–Planck (PNP) equations.[23–25] ENBIOS solves self-consistently the PB–PNP equations in DC and small-signal AC regimes, also accounting for the salinity, temperature and frequency dependence of the electrolyte. Nanoelectrodes were modelled as disks on the simulation boundary; Dirichlet boundary conditions set the electrodes' voltage. We considered simulation domains of 25 × 35 nanoelectrodes, with an intra-electrodes pitch of 600 nm and 890 nm along the x- and y-directions, respectively. Oil droplets were modelled as spheres at a vertical distance $dz = 10$ nm from the array surface. We mimicked the operating principle

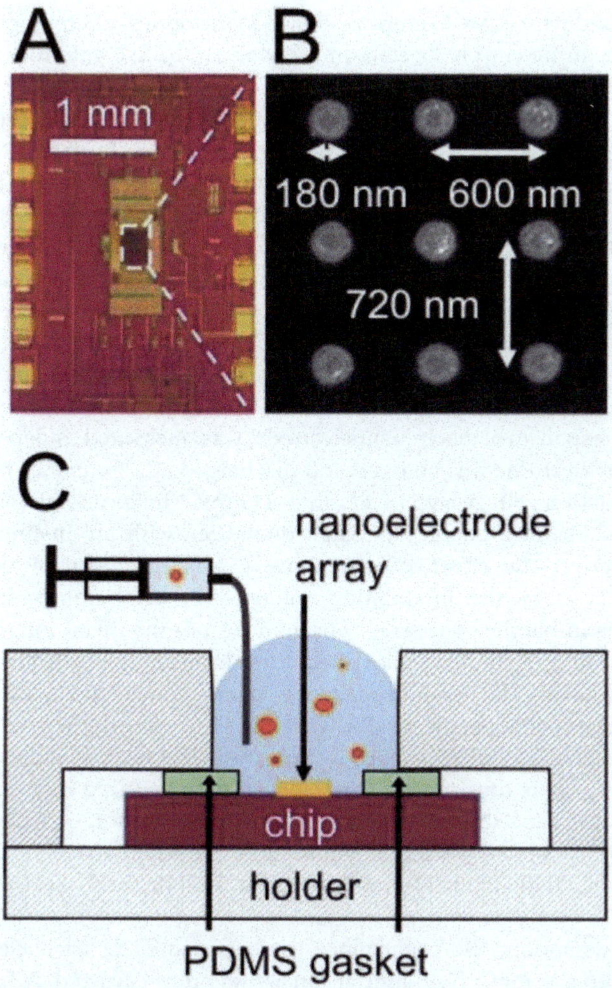

Fig. 1 (A) Optical micrograph of the CMOS chip. (B) SEM image of a portion of the nanoelectrode array. (C) Cross-sectional schematic of the chip holder showing how the array of nanoelectrode is incubated in an emulsion solution.

of the biosensor platform by applying a small-signal AC voltage (V_{AC}) on a row of nanoelectrodes, while keeping all others grounded; simulations were repeated multiple times, each time activating a different row of the simulated array. From each simulation we recorded the AC current (I_{AC}) flowing through the active electrodes. Results were then provided as an effective capacitance $C_{eff} = |I_{AC}/\omega V_{AC}|$ measured at each nanoelectrode, where ω is the angular excitation frequency. Simulating the whole array allowed us to obtain effective capacitance maps, which could be directly compared to experiments. Two procedures were attempted to estimate the size of the droplets based on the simulations (second moment method; projected area method), as reported in the ESI.†

Results and discussion

High-frequency capacitive imaging of oil micro-droplets

The electrochemical measurement was performed with complementary-metal–oxide-semiconductor (CMOS) chips as shown in the optical micrograph of Fig. 1A. At the center of the chip is located an array of 256 by 256 gold disk electrodes, 180 nm in diameter. The array has a pitch of 600 nm by 890 nm or 600 nm by 720 nm. A portion of the array, imaged by scanning electron microscopy (SEM), is shown in Fig. 1B. Here nine gold nanoelectrodes appear as bright disks, while the Si_3N_4 between the nanoelectrodes appears dark. For our experiments the chip was inserted in a polyether ether ketone (PEEK) holder (CSP/μBGA Test & Burn socket, Aries Electronics) with an opening at the center to expose the nanoelectrode array to the liquid. Fig. 1C shows a cross-sectional schematic of the holder (striped pattern) and chip (purple rectangle). The nanoelectrode array (thick orange line) is exposed to an emulsion solution (blue with red dots). A polydimethylsiloxane (PDMS) gasket (green rectangles) ensured the seal between the holder and the chip.

The AC response of each nanoelectrode was measured independently as follows. Each electrode was charged and discharged at a frequency $f = 50$ MHz, with a modulation voltage step of $\Delta E_{mod} = 245$ mV. The discharging current was integrated to obtain a charge and then digitalized with an on-chip analog-to-digital converter. The effective capacitance C was calculated by dividing the charge per cycle by the modulation voltage. The total impedance has two contributions in parallel: the stray capacitance from the metal interconnects in the CMOS chip and the impedance resulting from the nanoelectrode/solution interface. Although the stray capacitance is maintained at an extremely low level (typically 400 aF/nanoelectrode), we removed its contribution by subtracting a reference signal recorded in pure isopropanol. The relative capacitance scale used in the article is thus the effective capacitance measured with an emulsion, C_{ems}, minus the capacitance measured in pure isopropanol, C_{IPA} ($\Delta C = C_{ems} - C_{IPA}$). This procedure allows not only to remove the stray capacitance but also to calibrate each individual electrode with a well-defined electrode/solution interface.

For representation, the capacitance value measured at each electrode was converted into a colored pixel and an image was reconstructed. A full frame (2 electrodes, area 228 μm × 154 μm (ref. 16)) was acquired every 0.2 s. Fig. 2A shows a typical capacitance image taken with a chip incubated for 1 min in an *o*DCB in water emulsion. An example of raw data (before C_{IPA} subtraction) is also provided

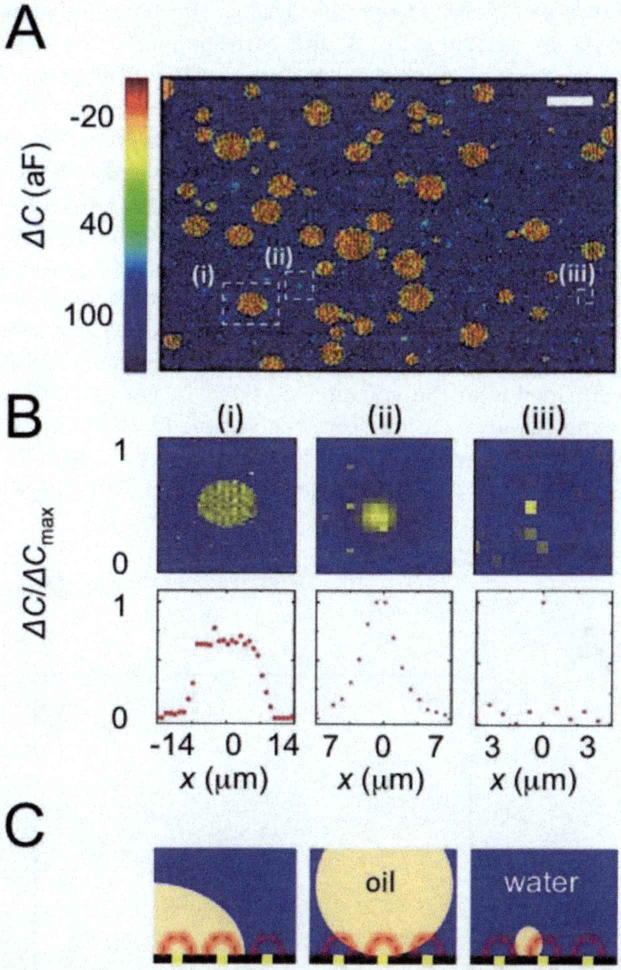

Fig. 2 (A) Capacitance image obtained with a chip incubated for 1 min in an *o*DCB emulsion. Each pixel in this image represents one separately addressable electrode. The scale bar represents 20 µm. (B) Capacitance image (first row) and line profile (second row) of the individual *o*DCB droplets shown in the dashed white rectangles in (A). The capacitance image is plotted with a color scale that maximizes the contrast between *o*DCB and water. The line profiles have been normalized to the maximum capacitance in the whole array while the zero is set to the minimum value of the array. (C) Cross-sectional schematic depicting the systems producing the capacitance responses shown in (B). The red dashed line schematically represents the volume of solution probed by each nanoelectrode (yellow squares). The schematic is not to scale.

as Fig. S1 in the ESI.† This emulsion is used as a benchmark because of its relative stability (>15 min), its simple chemical composition (neither surfactant nor ions added) and its large range of droplet sizes. The diameter of the *o*DCB droplets, measured by optical microscopy (Fig. S2 in the ESI†), is between 1 and 40 µm. Here the polydispersity of the emulsion and the rapid sedimentation of the large droplets (density of *o*DCB 1.3 g mL^{-1})[26] forbids characterization by DLS. The presence of *o*DCB droplets sedimented at the surface of the array is revealed by large capacitance variations, as seen in Fig. 2A. The blue background has an average capacitance of 120 ± 29 aF and corresponds to the water/electrode interface. The presence of *o*DCB droplets at the surface of the array is revealed by a decrease of the capacitance (green to red color Fig. 2A). Three distinct capacitance fingerprints are produced by the *o*DCB droplets. For each type of fingerprint, a typical capacitance image and the associated capacitance profile are provided in the top and bottom row of Fig. 2B, respectively. Cross-sectional schematics depicting the corresponding systems are shown in Fig. 2C.

The capacitance image, ΔC, and the corresponding line profile in Fig. 2B, column (i), shows a large spot with a sharp transition between water and *o*DCB as well as an homogeneous relative capacitance value (-10 ± 26 aF) inside the *o*DCB droplet. The typical diameter of such droplets varies between 3 and 20 μm. This fingerprint corresponds to the situation depicted in Fig. 2C column (i), with a droplet wetting the array. In this case the volume of solution probed by the nanoelectrodes (volume represented by the dashed red hemispheres in Fig. 2C) contains either pure *o*DCB or water and hence a two-level capacitance response is observed.

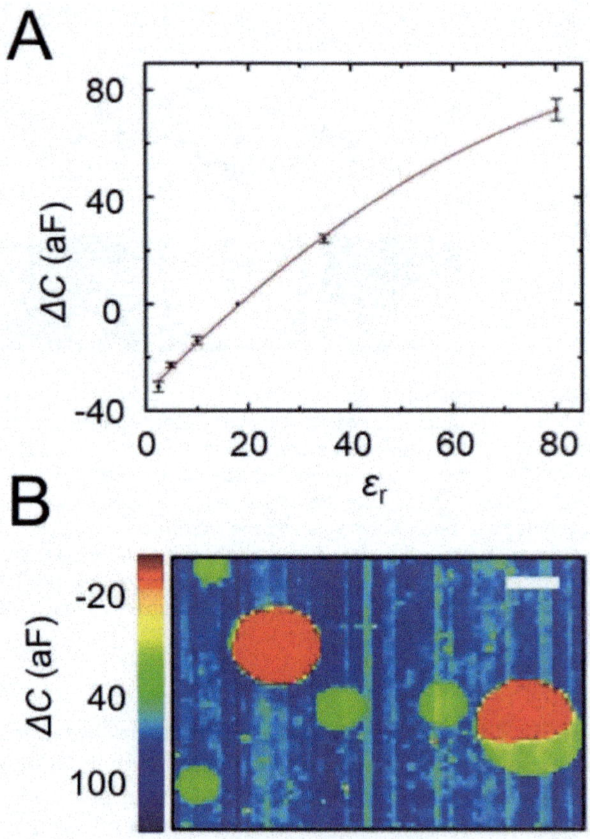

Fig. 3 (A) Capacitance as a function of the relative dielectric constant of the medium. The red line is a guide to the eye. (B) Capacitance image of a chip incubated successively in a silicone oil and nitrobenzene emulsion. The scale bar represents 20 μm.

On the other hand, droplets that are not wetting the array produce a bell-shaped capacitance fingerprint (Fig. 2B, column (ii)). This profile, observed in a previous study of rigid polystyrene spheres, is caused by the spherical shape of the droplet in suspension (scheme in Fig. 2C column (ii)). It can be intuitively understood by the increasing content of water in the probing volume of a nanoelectrode, when the electrode is further from the center of the droplet. The FWHM of these Gaussian-like profiles is typically 5–10 μm. The actual diameter of a suspended droplet can be estimated by 3D finite-element simulations. Further details about the ENBIOS simulations and the determination of the droplet diameter are provided in the ESI (Fig. S3 and S4†). The estimated radius of the fully resolved, non-wetting *o*DCB droplets (see below and the ESI†) is between 2 μm and 10 μm. Another important value is the capacitance at the center of the droplet, about 65 ± 37 aF. This value, intermediate between the capacitance of pure water (120 aF) and pure *o*DCB (−10 aF), confirms that the droplet remains suspended and that a small amount of water lies between the nanoelectrode array and the bottom of the suspended droplet.

Finally, individual pixels (or less than 3 adjacent pixels) display capacitance changes ranging from −35 to 90 aF, as shown in Fig. 2B column (iii). These spikes are attributed to droplets of *o*DCB smaller than the pitch of the array and hence not spatially resolved (schematic Fig. 2C). In this case the droplets are too small to contribute significantly to the signal at remote electrodes.

Mapping the local relative permittivity

The large contrast of capacitance between *o*DCB and water observed in Fig. 2A clearly indicates that the magnitude of the capacitance change strongly depends on the composition of the solution near each electrode. The influence of the permittivity on the capacitance response of our CMOS chip was investigated by preparing emulsions with different oils having a relative permittivity between 2.4 (toluene) and 34.8 (nitrobenzene). A list of the different oils and their permittivity is provided in Table 1. Fig. 3A shows the capacitance measured at the center of large oil droplets and plotted as a function of the relative permittivity of the oil. The error bars represent the standard deviation between three independent measurements on different chips. The measured signal increases monotonically with the permittivity of the solution. A perfectly linear relationship between measured capacitance and permittivity is not observed, however. This could be caused by nonlinearities in the chip's readout circuitry[5] that become particularly significant here because of the significant variations in current levels encountered

Table 1 Parameters associated with different emulsions[27]

Solvent	ε_r (at 25 °C)
Water	80.1
Nitrobenzene	34.8
Isopropanol	17.9
o-Dichlorobenzene	9.9
Chloroform	4.8
Toluene	2.4

for the different oils. Nonetheless, the experimental curve of Fig. 3A can be used to calibrate the response of the chips and hence determine the relative permittivity of individual droplets a few microns in diameter. Interestingly, the coefficient of variation between chips (standard deviation/average value) is ~12%, indicating that a reliable estimate of the dielectric constant can be obtained.

To illustrate this point, a chip was successively incubated with an emulsion of silicone oil ($\varepsilon_r = 2.5$) and an emulsion of nitrobenzene ($\varepsilon_r = 34.8$). The capacitance image of the chip, shown in Fig. 3B, clearly evidences two distinct populations of droplets corresponding to the two different solvents. Silicone oil droplets display an electrode capacitance of -21 ± 2 aF (red spots) while nitrobenzene droplets have a capacitance of 24 ± 3 aF relative to isopropanol. Based on the calibration curve (Fig. 3A) the corresponding value of ε_r for the silicone oil and nitrobenzene droplets are 6 ± 1 and 34 ± 3, respectively. The permittivity of nitrobenzene is in excellent agreement with the calibration value ($\varepsilon_r = 34.8$), while for the silicone oil a larger discrepancy is found (expected value $\varepsilon_r = 2.5$). The origin of this discrepancy in the mixture is not known at this time. Interestingly, the capacitance image of Fig. 2B also shows a large droplet composed of two phases with distinct values of permittivity (bottom right), one corresponding to silicone oil and the other to nitrobenzene. The boundary between these two phases can be clearly observed with an accuracy of about one micron, which corresponds to the pitch of the array.

Dynamics of individual droplets

We can take advantage of the temporal resolution of the capacitance imaging (0.2 s) to observe the electrostatic interactions between the surface of the chip and the *o*DCB droplets. This is illustrated in Fig. 4. Since the Si_3N_4 surface (~95% of the total surface of the array) is negatively charged at pH 5.5–7 (isoelectric point 8)[28] and the droplets do not instantly wet the surface, we conclude that the droplets are also negatively charged. This is in agreement with a value of zeta-potential of −53 mV measured on chlorobenzene in water droplets stabilized with a nonionic surfactant.[29]

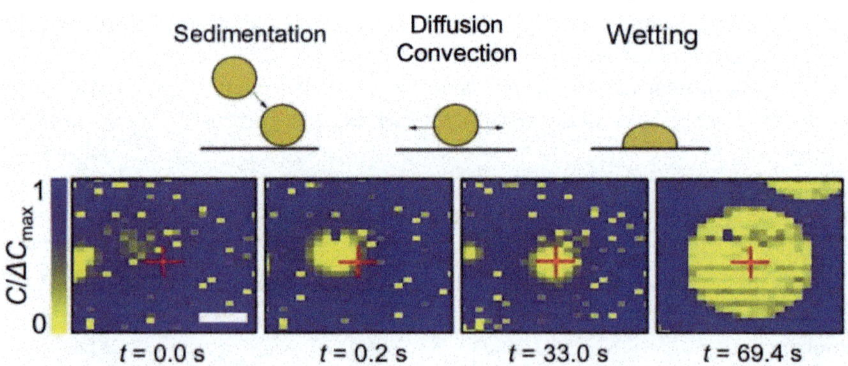

Fig. 4 Illustration of the wetting dynamics. A droplet gradually approaches the surface of the array (first frame from the left), hovers and diffuses for some time above the surface (second and third frames) and eventually wets the array (right-most frame). The scale bar represents 3.3 μm.

While the rapid dynamics of the wetting event itself do not allow its detailed analysis, it is possible to monitor the droplets' movement before wetting. We used a particle tracking code (see Movie S2 in the ESI†) to automatically find the position of the droplets in each frame and reconstitute their trajectories over an entire movie. Fig. 4 shows typical frames in such a trajectory. The capacitance images display small droplets (less than 3 pixels) and larger ones comprised between 3 and 20 μm. The small droplets were not moving, while the movement of the larger droplets was sometimes caused by convection as evidenced by their quasi-linear trajectories and the linear dependence of the displacement with time. Convection was caused by the injection of the emulsion solution with a syringe and the subsequent elastic relaxation, inducing flow in the incubation chamber. Note also that some large droplets seemed not to move at all, or stopped

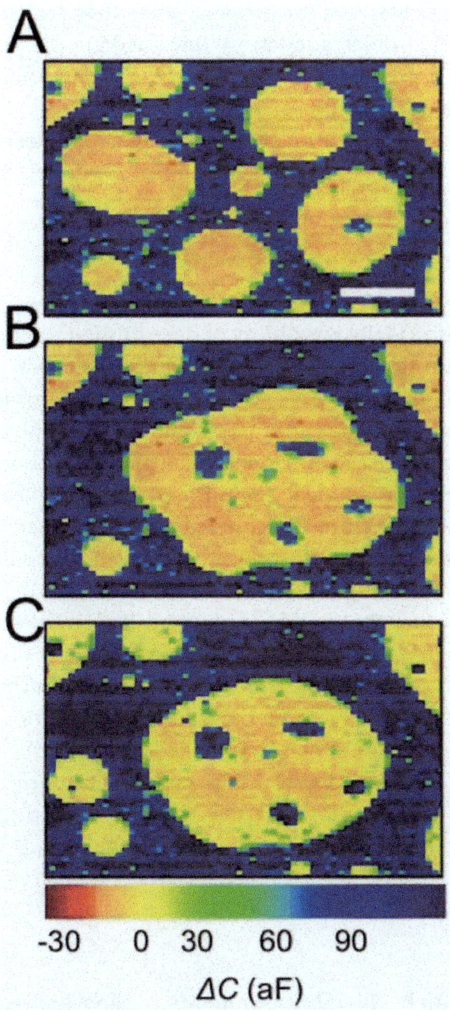

Fig. 5 Capacitance images of *o*DCB droplets taken at 0 s (A), 0.2 s (B) and 5.1 s (C). The scale bar is the same for all images and represents 10 μm. The *o*DCB contains 1 mM of tetrabutylammonium toluene-4-sulfonate.

moving after a period of time. These droplets could be either pinned by a local defect on the array or move extremely slowly solely by diffusion.

The capacitance movies also reveal interactions between droplets. Frames taken with an *o*DCB emulsion containing 1 mM of tetrabutylammonium toluene-4-sulfonate are shown in Fig. 5. These consecutive images show five *o*DCB droplets that merge together to form one large droplet. Interestingly, the first frame reveals that some *o*DCB droplets have high-capacitance inclusions inside the *o*DCB. The value of the capacitance, *ca.* 100 aF, corresponds to the permittivity of water. We thus conclude that some *o*DCB droplets contain smaller water droplets. Examination of Fig. 5B provides an important clue about the formation of these inclusions. The large droplet formed after merging contains four large inclusions and a smaller one near its center. We observed in several events like the one shown in Fig. 5, that the number of inclusions roughly matches the number of merging droplets. We thus conclude that these inclusions are formed by trapping some water between the droplets when they are merging. As shown in Fig. 5B the resulting droplet adopts an intermediate shape between a hemispherical shape that minimizes its surface area and the original position of the parent-droplets. The examination of several frames reveals that several seconds are necessary for such a large droplet to finally reach a quasi-spherical shape as shown in Fig. 5C. Such slow dynamics are typical for capillary relaxation.[30]

Electrochemical contact angle measurements

An interesting consequence of being able to spatially resolve an individual droplet before and after wetting is the possibility of determining its contact angle θ. Some advantages of measuring contact angles at the μm scale, which is difficult using optical methods, include the possibility to reveal local (μm) inhomogeneities of the surface, simultaneous measurements on many droplets, and completely excluding gravitational effects (droplet deformation under its own weight). Spatially resolved electrochemical techniques such as SECM could also be employed in theory, to measure the contact angle. However, this would require the use of sub-micron sized tips (fragile and prone to fouling), an extremely skilled operator and several hours per droplet.

To attempt to measure the contact angle, we followed the strategy presented in Fig. 6. The micrograph on the left shows a microdroplet of *o*DCB hovering at the surface of the array. At this stage the droplet did not wet the array and its shape was still spherical (sphere of radius r_0) as depicted in the schematic on the left. Once the droplet wetted the array the situation depicted on the right was obtained. The droplet still tends to keep a spherical curvature to minimize its surface but the radius r_{wet} is increased while its height h is reduced so that the total volume of the droplet is kept constant. The relation between θ, h and, r_{wet} is given by (ESI†)

$$\theta = 2 \quad \arctan\left(\frac{h}{r_{\mathrm{wet}}}\right) \tag{1}$$

The value of r_{wet} can be directly measured from capacitance images. The radius of the droplet before spreading, r_0, is deduced from numerical simulations of the capacitance image (ESI†). The value of h can then be calculated from r_0 and r_{wet} using eqn (5) in the ESI.†

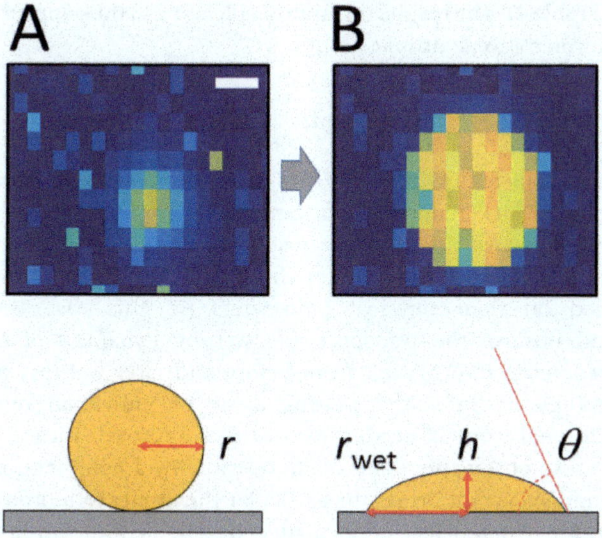

Fig. 6 Capacitance image of an oDCB droplet before (A) and after (B) wetting the array. The schematics show the geometrical parameters that are obtained from the analysis of the CCD view. Scale bar represents 2.4 μm.

This procedure is illustrated for several droplets in Table 2. Unfortunately, a significant systematic uncertainty is introduced in practice by the determination of the radius before wetting, r_0. Two independent numerical approaches were employed based on either the second moment of the capacitive signal or the projected area of the particle in the image (ESI†). Table 2 indicates that the first method systematically yields droplet radii that are smaller than those of the second method. The origin of this discrepancy is not known at this point. It is possible that some physical properties of the system are not captured by the simulations. For example Maxwell–Wagner–Sillars polarization, which enhances the dielectric response of multiphasic systems at 50 MHz, could play a role.[31]

For comparison we also measured the contact angle of large *o*DCB droplets with a goniometer (Fig. S6†). The average contact angle measured on three different chips and droplets (three per chip) is 115 ± 1°. The contact angle measured with the chip on 6 droplets is estimated at 122 ± 9° using the projected

Table 2 Estimated contact angle of individual oDCB microdroplets from capacitance imaging

		Second moment method			Projected area method		
#	r_{wet} (μm)	r_{col} (μm)	h (μm)	θ	r_{col} (μm)	h (μm)	θ
1	2.1	3.6	6.6	145	2.2	3.4	117
2	2.4	4.2	7.7	145	3.1	5.3	131
3	3.0	4.8	8.7	142	3.8	6.4	130
4	4.0	6.9	12.6	145	4.8	8.0	127
5	5.8	7.7	13.2	133	5.7	8.5	112
6	5.9	8.1	14.1	134	5.9	8.9	113

area method (Table 2). The results on large droplets (1 μL) thus appear compatible with those on microscopic droplets (<pL).

Conclusions and outlook

Using oil micro-emulsions as prototypical systems, we have demonstrated several capabilities of the CMOS nanocapacitor array platform. These include the ability to image droplets of a few micrometers in size; detect individual sub-micrometer-sized droplets; obtain information on the composition of individual micro-droplets based on their physical parameters (in this case the dielectric constant); and observe the transport and wetting dynamics of droplets. By combining geometric information from before and after wetting, it is also, in principle, possible to infer the wetting angle of individual microdroplets, although additional work is needed to turn this approach into a quantitative assay. All of these observations are made possible by a combination of several strengths of nanocapacitor arrays, in particular the ability to reproducibly fabricate large arrays of nanoelectrodes, a high degree of integration that allows independently addressing these electrodes, and high-frequency operation. If mass-produced, these can serve as low-cost consumables for the characterization of a variety of hard-, soft- and biological-matter systems.

Conflicts of interest

F. P. W. declares the following financial interest: he is the (co-)inventor of multiple patents on which the CMOS nanocapacitor platform is based and is employed by NXP, where it was developed.

Acknowledgements

The authors thank H. A. Verhoeven for providing the holder and PDMS gaskets, M. Cirelli for the contact angle measurements and M. A. Smithers for the SEM measurements. We further gratefully acknowledge financial support from the European Research Council (ERC) under project number 278801.

References

1 G. S. Kulkarni, W. Zang and Z. Zhong, *Acc. Chem. Res.*, 2016, **49**, 2578–2586.
2 F. Widdershoven, D. Van Steenwinckel, J. Uberfeld, T. Merelle, H. Suy, F. Jedema, R. Hoofman, C. Tak, A. Sedzin, B. Cobelens, E. Sterckx, R. van der Werf, K. Verheyden, M. Kengen, F. Swartjes, F. Frederix and Ieee, in *2010 International Electron Devices Meeting – Technical Digest*, 2010.
3 C. Laborde, F. Pittino, H. A. Verhoeven, S. G. Lemay, L. Selmi, M. A. Jongsma and F. P. Widdershoven, *Nat. Nanotechnol.*, 2015, **10**, 791–795.
4 S. G. Lemay, C. Laborde, C. Renault, A. Cossettini, L. Selmi and F. P. Widdershoven, *Acc. Chem. Res.*, 2016, **49**, 2355–2362.
5 F. Widdershoven, A. Cossettini, C. Laborde, A. Bandiziol, P. P. v. Swinderen, S. G. Lemay and L. Selmi, *IEEE Trans. Biomed. Circuits Syst.*, 2018, **12**, 1369–1382.

6 A. Cossettini, C. Laborde, D. Brandalise, F. Widdershoven, S. G. Lemay and L. Selmi, *IEEE Sens. J.*, 2021, **21**, 4696–4704.
7 A. Cossettini, D. Brandalise, P. Palestri, A. Bertacchini, M. Ramponi, F. Widdershoven, L. Benini and L. Selmi, in *2020 Ieee Sensors*, 2020, pp. 1–4.
8 R. S. H. Lam and M. T. Nickerson, *Food Chem.*, 2013, **141**, 975–984.
9 J. Knowlton and S. Pearce, in *Handbook of Cosmetic Science & Technology*, Elsevier, 1993, pp. 95–118, DOI: 10.1016/B978-1-85617-197-7.50013-7.
10 C. W. Pouton, *Adv. Drug Delivery Rev.*, 1997, **25**, 47–58.
11 P. K. Kilpatrick, *Energy Fuels*, 2012, **26**, 4017–4026.
12 Y. Ouyang, R. S. Mansell and R. D. Rhue, *Crit. Rev. Environ. Sci. Technol.*, 1995, **25**, 269–290.
13 A. Imhof and D. J. Pine, *Nature*, 1997, **389**, 948–951.
14 D. J. McClements, *Food emulsions: principles, practices, and techniques*, CRC Press, 2015.
15 J. E. Dick, C. Renault, B.-K. Kim and A. J. Bard, *Angew. Chem., Int. Ed.*, 2014, **53**, 11859–11862.
16 B.-K. Kim, A. Boika, J. Kim, J. E. Dick and A. J. Bard, *J. Am. Chem. Soc.*, 2014, **136**, 4849–4852.
17 B.-K. Kim, J. Kim and A. J. Bard, *J. Am. Chem. Soc.*, 2015, **137**, 2343–2349.
18 Y. Li, H. Deng, J. E. Dick and A. J. Bard, *Anal. Chem.*, 2015, **87**, 11013–11021.
19 T. Kakiuchi, M. Iida, S.-i. Imabayashi and K. Niki, *Langmuir*, 2000, **16**, 5397–5401.
20 B. A. Ivarsson, P.-O. Hegg, K. I. Lundström and U. Jönsson, *Colloids Surf.*, 1985, **13**, 169–192.
21 P. Patil, M. Gorek, M. Folberth, A. Hartmann, S. Forgang, C. Fulda and K. Reinicke, *SPE Drill. Completion*, 2010, **25**, 380–390.
22 W. Q. Yang, A. L. Stott, M. S. Beck and C. G. Xie, *Rev. Sci. Instrum.*, 1995, **66**, 4326–4332.
23 F. Pittino and L. Selmi, *Comput. Methods Appl. Mech. Eng.*, 2014, **278**, 902–923.
24 P. Scarbolo, F. Pittino, M. Dalla Longa, A. Cossettini and L. Selmi, *ENBIOS-1D Lab*, DOI: 10.4231/D3Z02Z963.
25 A. Hoxha, P. Scarbolo, A. Cossettini and L. Selmi, *ENBIOS-2D Lab*, DOI: 10.4231/D3V11VM7D.
26 W. M. Haynes, *CRC handbook of chemistry and physics*, CRC Press, 2014.
27 *Engineering_ToolBox, Dielectric constants of liquids*, https://www.engineeringtoolbox.com/liquid-dielectric-constants-d_1263.html.
28 P. Greil, R. Nitzsche, H. Friedrich and W. Hermel, *J. Eur. Ceram. Soc.*, 1991, **7**, 353–359.
29 P. H. Elworthy, A. T. Florence and J. A. Rogers, *J. Colloid Interface Sci.*, 1971, **35**, 23–33.
30 V. V. Khatavkar, P. D. Anderson and H. E. H. Meijer, *J. Fluid Mech.*, 2007, **572**, 367–387.
31 C. Grosse and A. V. Delgado, *Curr. Opin. Colloid Interface Sci.*, 2010, **15**, 145–159.

Faraday Discussions

PAPER

Formation sequence of solid electrolyte interphases and impacts on lithium deposition and dissolution on copper: an *in situ* atomic force microscopic study†

Wei-Wei Wang,[a] Yu Gu,[a] Hao Yan,[a] Kai-Xuan Li,[a] Zhao-Bin Chen,[a] Qi-Hui Wu,[b] Christine Kranz,[c] Jia-Wei Yan *[a] and Bing-Wei Mao *[a]

Received 12th July 2021, Accepted 6th October 2021
DOI: 10.1039/d1fd00043h

Copper is the most widely used substrate for Li deposition and dissolution in lithium metal anodes, which is complicated by the formation of solid electrolyte interphases (SEIs), whose physical and chemical properties can affect Li deposition and dissolution significantly. However, initial Li nucleation and growth on bare Cu creates Li nuclei that only partially cover the Cu surface so that SEI formation could proceed not only on Li nuclei but also on the bare region of the Cu surface with different kinetics, which may affect the follow-up processes distinctively. In this paper, we employ *in situ* atomic force microscopy (AFM), together with X-ray photoelectron spectroscopy (XPS), to investigate how SEIs formed on a Cu surface, without Li participation, and on the surface of growing Li nuclei, with Li participation, affect the components and structures of the SEIs, and how the formation sequence of the two kinds of SEIs, along with Li deposition, affect subsequent dissolution and re-deposition processes in a pyrrolidinium-based ionic liquid electrolyte containing a small amount of water. Nanoscale *in situ* AFM observations show that sphere-like Li deposits may have differently conditioned SEI-shells, depending on whether Li nucleation is preceded by the formation of the SEI on Cu. Models of integrated-SEI shells and segmented-SEI shells are proposed to describe SEI shells formed on Li nuclei and SEI shells sequentially formed on Cu and then on Li nuclei, respectively. "Top-dissolution" is observed for both types of shelled Li deposits, but the integrated-SEI shells only show wrinkles, which can be recovered upon Li re-deposition, while the segmented-SEI shells are apparently top-opened due to mechanical stresses introduced at the junctions of the top regions and become "dead" SEIs, which forces subsequent Li nucleation and growth in the interstice

[a]*State Key Laboratory of Physical Chemistry of Solid Surfaces, Collaborative Innovation Center of Chemistry for Energy Materials (iChEM), Department of Chemistry, College of Chemistry and Chemical Engineering, Xiamen University, Xiamen, 361005, China. E-mail: bwmao@xmu.edu.cn; jwyan@xmu.edu.cn*
[b]*College of Mechanical and Energy Engineering, Jimei University, Xiamen, 361021, China*
[c]*Institute of Analytical and Bioanalytical Chemistry, Ulm University, Ulm, 89081, Germany*
† Electronic supplementary information (ESI) available. See DOI: 10.1039/d1fd00043h

of the dead SEIs. Our work provides insights into the impact mechanism of SEIs on the initial stage Li deposition and dissolution on foreign substrates, revealing that SEIs could be more influential on Li dissolution and that the spatial integration of SEI shells on Li deposits is important to improving the reversibility of deposition and dissolution cycling.

Introduction

Lithium metal anodes are most promising candidate of anodes for next generation high energy density secondary batteries because of their lowest electrode potential and extra high capacity.[1–3] As a metal anode, the basic processes are metal deposition and dissolution, and the cycling stability is crucial for maintaining a safe and long-lived battery. However, Li deposition is prone to dendrite growth and solid electrolyte interphases (SEIs)[4,5] are almost inevitably formed due to electrolyte decomposition at Li metal anodes. The composition, structure and resultant physicochemical properties of an SEI significantly influence the interfacial thermodynamics and kinetics of Li deposition and dissolution.[6–11] Ideally, SEI is electron insulator and Li ion conductor, and is regarded as a passivation layer that prevents further decomposition of electrolytes and meanwhile facilitates Li ion transport for deposition and dissolution beneath the SEI.[5] In practice, the dendritic growth nature of Li and imperfect SEIs, such as those with a mosaic-like structure,[4] often promote each other to produce a non-uniform morphology of Li deposits upon deposition–dissolution cycling, which degrades the performance and limits the practical application of Li metal anodes.[12–15]

Tremendous efforts have been devoted in past decades to gain a deeper understanding of Li metal anode processes towards solving key issues for their practical applications.[16,17] Base- or top-growth/dissolution models have been proposed under the consideration of SEI influences,[18–20] which can be used to explain the often observed mossy-like and whisker-like morphologies of Li deposits. Optimizing electrolyte composition, including types of solvent and Li salt, as well as additives, has also been recognized as most effective way of tuning the structure and property of the SEI for improving the cycling stability of Li deposition–dissolution.[21,22] Meanwhile, advanced electrolyte systems with concepts such as high-concentration, localized high-concentration and fluorinated electrolytes showing improved Coulombic efficiency have been recently developed, which has also aroused the investigation of the working mechanism of advanced electrolytes in terms of the role of the SEI in improving the uniformity of Li deposits.[23–26]

As a matter of fact, electrolyte decomposition to form SEIs can proceed chemically and/or electrochemically depending on the electrode potential and on whether there is Li metal for participation.[27–29] Therefore, the exact manner of forming a SEI could be equally important in tuning the component and structure of the SEI. Compared with traditional soaking[30] and chemical aging[22] pretreatments often employed for Li metal anodes, the active control of the current and potential or their combination could serve as more flexible ways of modifying the morphology, component, structure and property of SEIs. Implantable SEIs[31] and ultra-smooth SEIs,[32] which were constructed utilizing current and potential control, respectively, are two successful examples to mention. Nevertheless, these two studies all work with Li metal foils, meaning that SEI formation involves electrolyte reduction with the participation of Li metal. In fact, the situation could be totally different when working with substrates such as copper that initially contain no active anode material of Li, known as an "anode-free" type of Li anode, which is regarded as an ultimate configuration in rechargeable Li metal batteries in the field.[33–37] Additional complexities and challenges are introduced with such a configuration, as each time during charging Li deposition has to start with nucleation and growth on a foreign substrate, which also promotes the initial formation of two different types of SEIs, *i.e.* with Li participation on the Li nuclei surface and without Li participation on the substrate surface. The different formation kinetics and formation sequences of the two different types of SEIs, along with Li deposition, are expected to affect subsequent Li deposition and dissolution processes uniquely and are worthy, yet lack of investigations.

To understand Li metal anode processes, direct *in situ* monitoring of the morphology evolution during SEI formation and Li deposition and dissolution is indispensable.[38,39] Atomic force microscopy (AFM) is powerful in not only nanoscale imaging but also probing the properties of the sample surface.[40–42] Since the early representative AFM work of Li deposition on Cu by Aurbach and coworkers,[43] a number of *in situ* AFM studies of the interfaces/interphases of batteries have been reported in the past two decades.[44–49] In particular, *in situ* AFM studies on the direct observation of morphology evolution during Li[50–53] and Na[54] deposition have been reported in recent years. In a very recent study by the group of Wen and Wan,[53] SEI-shell evolution during Li deposition/stripping on Cu was studied by *in situ* AFM, providing evidence and insights on the degradation of Li anodes. Nevertheless, *in situ* AFM studies on Li metal anodes are still rather limited; in particular, there is a lack of in-depth investigation into the dissolution processes in the field of Li anode research, which would otherwise be equally important to further advance Li metal anodes.[55]

In this paper, we employ *in situ* AFM to investigate how the formation kinetics and formation sequence of SEIs impact the subsequent Li deposition and dissolution processes. We adopted pyrrolidinium-based ionic liquids containing Li salt and a small amount of water[56,57] as the electrolyte for the investigation of Li deposition and dissolution on Cu. As shown in Fig. 1a, the main SEI formation initiates at *ca.* 1.75 V, with a significant cathodic peak at *ca.* 1.33 V, which extends towards a more negative potential. Following the Li underpotential deposition (UPD), which peaked at 0.5 V, Li overpotential deposition (OPD) takes place at potentials below 0 V *vs.* Li/Li^+. The current loop upon the reversal of potential after Li OPD is a typical signal of nucleation and the growth of metal deposits on a foreign substrate. The transient *i–t* recorded after applying a cathodic potential step down to where Li OPD can occur further confirms a progressive nucleation of initial stage Li deposition[58,59] (Fig. S1 and 1b†). As a result, such a nucleation process causes two different conditions for SEI formation: the SEI formation on Li nuclei (SEI-Li) and the SEI formation on the regions not yet taken for Li nucleation (SEI-Cu), whose formation kinetics and, thus, compositions and structures are expected to differ from each other. More remarkably, the subsequent Li nucleation would have to proceed underneath the SEI-Cu, but the growth of nuclei would eventually rupture the SEI-Cu so that additional fresh SEI-Li could be further formed on the side surface of Li nuclei, allowing the two different types of SEI to be sequentially formed on a single Li nucleus. Fig. 1c illustrates the initial stage of Li nucleation and growth on Cu associated with the formation of SEI-Li

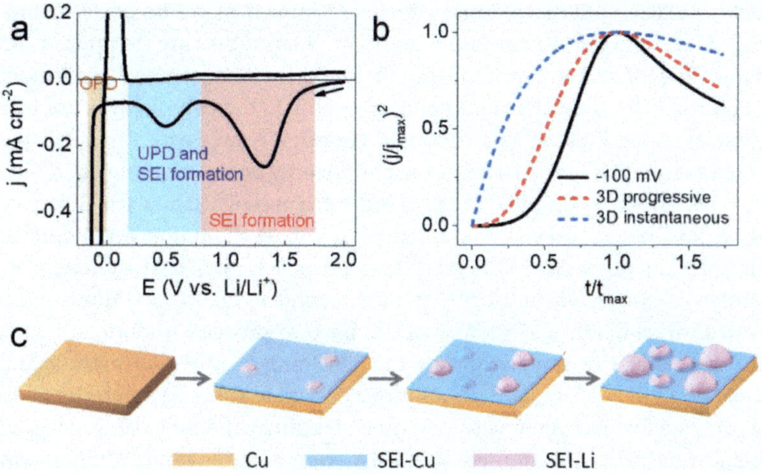

Fig. 1 (a) Cyclic voltammogram showing the formation of SEI and Li deposition and dissolution on a Cu electrode. (b) Dimensionless plot of the transient current after potential stepping from 2 V to −100 mV after removing the transient current associated with SEI formation by fitting, together with the plots of the mathematical expressions for instantaneous and progressive nucleation. (c) Schematic diagram showing Li nucleation and growth and SEI formation on Cu and Li nuclei. The electrolyte: 0.5 M LiTFSI/Py_{14}TFSI-Py_{14}FSI (1 : 1, v/v) electrolyte containing 1300 ppm H_2O. Sweep rate: 10 mV s^{-1}.

and SEI-Cu and their development in the later stage of deposition. We mention that the Li layer formed by UPD serves as a wetting layer, which reduces the overpotential for Li OPD, but cannot prevent Li OPD from nucleation and growth. Hence, SEI-Cu is indistinctly used to represent the SEI initially formed on surfaces other than the Li nuclei.

Within the framework of this paper, we performed X-ray photoelectron spectroscopy (XPS) analysis on the components and structures of two different types of SEIs formed under potential control with and without the participation of Li metal by OPD, and then employ *in situ* AFM to monitor the morphology evolution of Li deposits during Li deposition, as well as the dissolution. Our results show that integrated-SEI shells or segmented-SEI shells could be developed on sphere-like Li deposits depending on whether Li nucleation is preceded by the formation of SEI on Cu. The segmented-SEI shells become apparently top-opened due to stresses introduced between segments and form "dead" SEIs hindering further Li deposition, while the integrated-SEI shells only show wrinkles which can be recovered more or less upon Li redeposition. These observations reveal that SEIs could be more influential to Li dissolution, and spatial integration of SEI shell on Li deposit is important to maintain reversible deposition and dissolution cycles.

Results and discussion

Three kinds of SEIs and XPS profile analysis

Pyrrolidinium-based ionic liquids, such as Py_{14}TFSI and Py_{14}FSI, have been frequently used as electrolytes in the studies of Li ion and Li metal batteries;[60,61] SEIs formed in such a type of electrolyte are largely composed of the reduction products of TFSI and FSI anions, which are known as anion-derived and are composed of inorganic-rich components.[26] Such SEIs are supposed to be mechanically stiff and are more favorable to aid the formation of shell-like SEIs on Li deposits. Upon potential stepping to enable Li deposition, SEIs are formed on Li nuclei *via* both electrochemical and chemical routes with the participation of Li; meanwhile, SEI may also be formed on Cu only by electrochemical reduction without the participation of OPD Li. In order to investigate how the SEIs formed on the regions not yet taken for Li nucleation impact the follow-up Li deposition and dissolution processes, SEIs were also purposely prepared by linear sweep voltammetry (LSV) down to 0.1 V (Fig. S2†). In this case, all the follow-up initial stages of Li nucleation and growth would have to proceed underneath the SEI, which is beneficial to understanding the mechanism of SEI formation and its influence on Li deposition and dissolution. To differentiate between the two SEIs on Cu, the one formed along with potential stepping is named SEI-Cu-step, while the one formed by LSV is named SEI-Cu-lsv. Hence, three kinds of SEIs, namely SEI-Li, SEI-Cu-step and SEI-Cu-lsv, are employed in the follow-up measurements.

The components and structures of the three kinds of SEIs formed under different conditions were first inspected by XPS measurements. XPS depth

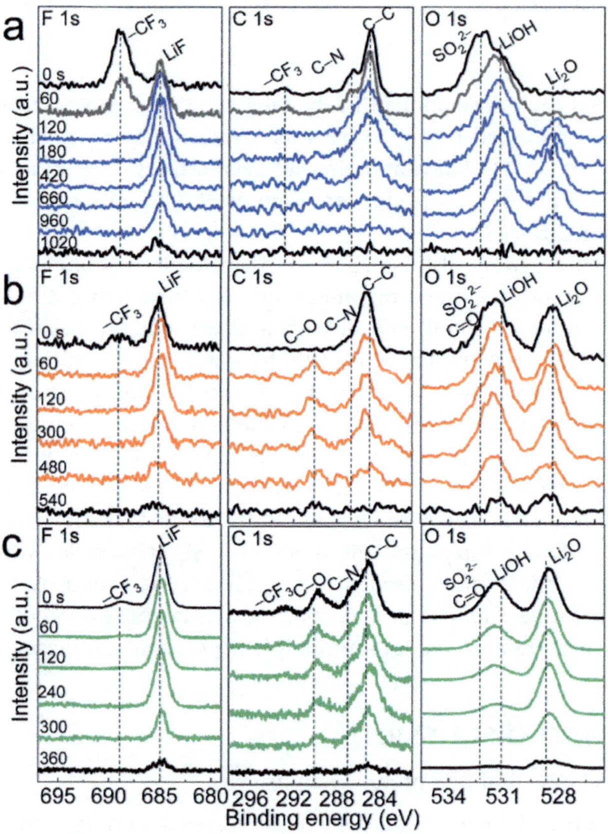

Fig. 2 XPS depth profiles of F 1s, C 1s and O 1s elements recorded after different times of argon ion sputtering for (a) SEI-Cu-lsv, (b) SEI-Cu-step and (c) SEI-Li.

profiles of the binding energies for several elements of the three SEIs are provided in Fig. 2 and S3.† Several pieces of information can be drawn from these as follows. First, all three SEIs are enriched with inorganic compounds, such as LiF, Li_2O and Li_2S_n, which are hybridized with small amounts of organic species of $Li_2S_xO_y$ and $LiCF_3$, as well as C–C and N–S moieties, with varying amounts in different locations across the thickness. The inorganic species are mostly anion-derived,[26] but Li_2O and LiOH may arise additionally from the reduction of the small amount of water. Second and more specifically, SEI-Cu-lsv bears a I–I′-O structure, *i.e.* it is composed of a thick inorganic rich inner layer dominated by Li_2O and LiOH (~70 nm), and another inorganic rich middle layer with increased LiF (~10 nm), and finally a thin organic rich outer layer (~5 nm), dominated by $LiCF_3$ and C–C compounds or moieties. Although SEI-Cu-step shares a similar mechanism of electrolyte reduction without Li participation, direct potential stepping promotes the formation of low-oxidation state products, so that an I–O hybrid SEI structure is more likely with a reduced thickness (~45 nm). For SEI-Li, the participation of OPD Li further accelerates the formation of LiF and Li_2O throughout the thickness (~30 nm), which is hybridized by small amount of organic rich moieties though. Overall, the thickness of the three types of SEIs follows the sequence of SEI-Cu-lsv (~85 nm) > SEI-Cu-step (~45 nm) > SEI-Li (~30 nm). Meanwhile, due to the enrichment of inorganic components and different spatial arrangements along the thickness, the three types of SEIs are expected to have different stiff mechanical properties, with SEI-Li being stiffer than the two SEIs on Cu.

In situ AFM monitoring of the effect of SEI-Cu-lsv

In the following, we employed *in situ* AFM to investigate Li deposition, dissolution and re-deposition under the influences of the three different kinds of SEIs. The initial stage nucleation and growth of Li on the Cu surface with preformed SEI-Cu-lsv was first monitored by *in situ* AFM. The pristine Cu(100) surface at 2 V in the IL electrolyte is smooth (Fig. S4†), and the SEI-Cu-lsv is compact and composed of granules with sizes ranging from 50 to 250 nm, as shown in Fig. 3a. The initial Li nucleation and growth is observed after 0.011 mA h cm^{-2} capacity, equivalent of 0.04 C cm^{-2} charge density, is spent. Zooming into a small area of a nucleus, marked by the red circle at 0.018 mA h cm^{-2} in Fig. 3c, discloses the granular top surface of the SEI (Fig. 3d), which is consistent with the morphology of the SEI-Cu-lsv (Fig. 3a). This confirms that the initial stage of Li deposition proceeds underneath the SEI-Cu-lsv. Li nuclei grow stably in a three-dimensional manner towards the formation of sphere-like deposits. Eventually, the red circled nucleus can reach a typical size of about 1.5 μm after 0.082 mA h cm^{-2} (Fig. 3e and S5†). It can be expected that, in the later stage of deposition, the growth of the nucleus is accompanied by the formation of a new SEI on the fresh Li surface. In other words, the nucleus bears an SEI layer shell which has been sequentially formed on Cu and then on Li. We term such an SEI shell the "segmented-SEI" shell, presumably with SEI-Cu-lsv at the top in blue and SEI-Li on the side surface in pink, as shown in Fig. 3h.

More obvious influences of the SEI are seen in the subsequent dissolution process, as shown in Fig. 4. Starting from the Li deposit formed with a total capacity of 0.06 mA h cm^{-2} (Fig. 4a), an anodic current density of 0.2 mA cm^{-2}

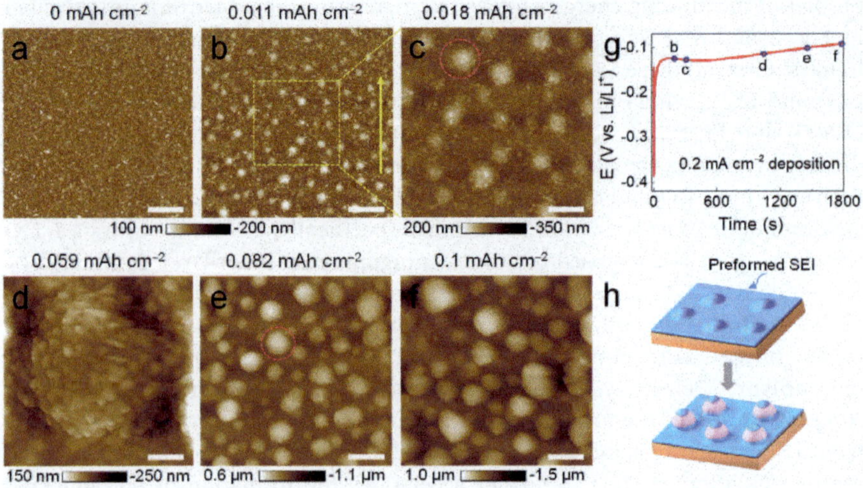

Fig. 3 (a–f) *In situ* AFM monitoring of Li deposition processes on the Cu(100) surface after SEI-Cu-lsv formation. The location of image (f) is somewhat shifted. (g) Potential profile during Li deposition at 0.2 mA cm^{-2} for 1800 s. (h) Schematic diagram of the growth of Li nuclei under the preformed SEI-Cu-lsv. Scale bars are 2 μm in (a, b, e and f), 1 μm in (c) and 300 nm in (d). Note 1 mA h cm^{-2} is equivalent of 3.6 C cm^{-2}.

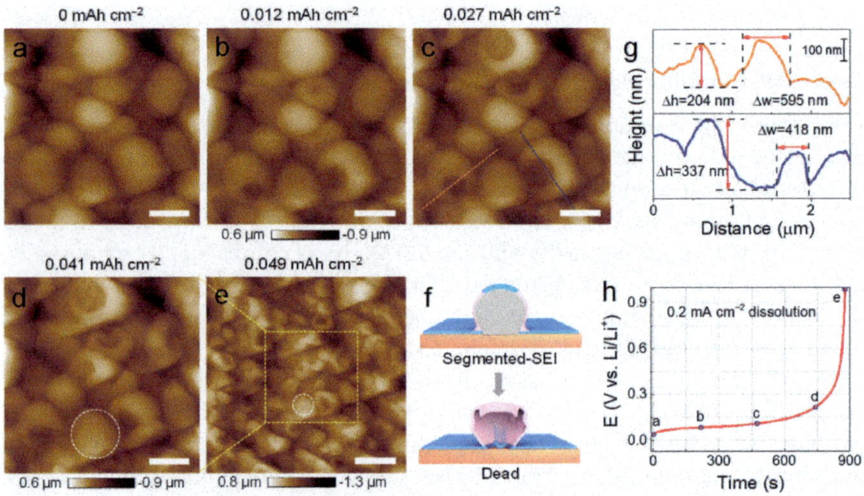

Fig. 4 (a–e) *In situ* AFM monitoring of Li dissolution after deposition for 0.06 mA h cm^{-2} at 0.2 mA cm^{-2} in the presence of SEI-Cu-lsv. (f) Schematic diagram of the Li dissolution process. (g) Cross-sectional analysis of open-shelled SEIs with cut-off heads. (h) Potential profile during Li dissolution at 0.2 mA cm^{-2}. Scale bars are 1 μm in (a–d) and 2 μm in (e).

was applied to enable the dissolution of the Li deposits. At the stage of dissolution by ~24% (Fig. 4b), some of the sphere-like Li deposits are dissolved, leaving behind SEI shells whose tops are apparently lost. As dissolution proceeds, more SEI shells open their tops, while the as-formed open-shells become deformed until finally collapsing upon complete dissolution (Fig. 4d and e). There are also deposits showing almost no changes during dissolution, *e.g.* the white circled ones in Fig. 4d and e, indicating they are likely "dead Li". The observed dissolution behavior with the opening of the shell tops suggests a top-dissolution mechanism for the Li deposit on Cu with the preformed SEI-Cu-lsv. The thickness and depth of the open-shells are roughly measured, as shown in Fig. 4g, but the shells (*ca.* 400–600 nm) are much thicker than SEI-Cu-lsv (*ca.* 85 nm, measured by XPS), which cannot be possible, suggesting that the shells likely curl inward after top opening, and the "lost" tops could be accumulated inside the shell. This phenomenon confirms the assumption of the segmented-SEI shells with SEI-Cu-lsv at the top and SEI-Li on the side surface, Fig. 4f.

To further investigate whether such open-shelled SEIs could be recovered and how they would influence the subsequent processes, we performed another experiment on the surface after experiencing the first cycle of deposition and complete dissolution. As shown in Fig. 5a, prolonged deposition (0.2 mA h cm^{-2}) left the surface with crowded open-shelled SEIs after dissolution. On the surface, upon re-deposition almost all the open-shells remain unrecovered, as if they became "dead" for accommodating Li deposition within the shell. Instead, Li re-deposition occurs preferentially only in the interstices of the crowded open-shells, *e.g.* those marked by yellow circles in Fig. 5b. One of the possible reasons for the open-shelled SEI to become "dead" could be that the broken shell top remnants fall to the bottom of the shell inside, which results in resistance for Li deposition from the inside shell bottom; dynamic evolution of the SEI at the root, causing closure of the shell bottom, could be another reason responsible for the inability of Li to deposit inside the shell. Also refer to Fig. 4f and *vide infra*.

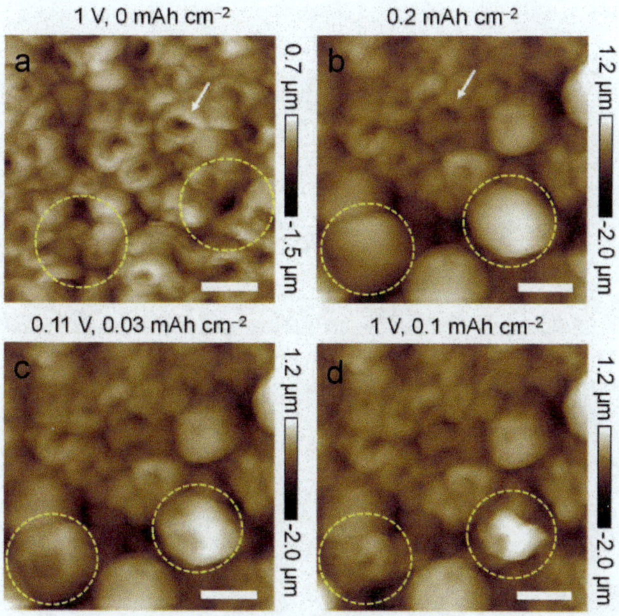

Fig. 5 *In situ* AFM images showing the second cycle of Li deposition–dissolution. (a) After Li deposition for 3600 s at 0.2 mA cm^{-2} then dissolution to 1 V. (b) Second Li deposition for 3600 s at 0.2 mA cm^{-2} after (a). (c, d) Li dissolution for 0.03 and 0.1 mA h cm^{-2}, respectively, after (b). Scale bars are 2 μm in (a–d).

Interestingly, the dissolution of the freshly deposited Li spheres in the interstices, such as those marked by the yellow circles in Fig. 5b, leads to wrinkled SEIs (Fig. 5c and d), which are remarkably different from the open-shelled SEIs. It is noteworthy that such Li deposits are supposed to have the SEI-Li shell on the entire surface, and we denote such an SEI shell as the "integrated-SEI" shell, which will be the subject of discussion in the following sections.

In situ AFM monitoring of the effects of SEI-Cu-step and SEI-Li

Upon potential stepping, *e.g.* to −100 mV, Li nucleation and growth occurs with the simultaneous reduction of the electrolyte, which leads to the formation of SEIs on Li nuclei (SEI-Li) and on the sites not yet taken for nucleation (SEI-Cu-step). As shown in Fig. 6, nuclei grow in a three-dimensional manner with a sphere-like shape and can reach a size of 1.4 μm at *ca.* 0.1 mA h cm^{-2} for the red-circled nucleus (Fig. S5†), similar to the case of SEI-Cu-lsv. However, in the initial stage, newly grown nuclei are easily scratched away by the scanning tip, which is likely due to the SEI-Cu-step not being completely formed and, thus, it could not accommodate nucleation underneath, which is unlike the situation with pre-formed SEI-Cu-lsv. The challenge in the successful *in situ* AFM imaging of the initial stage of Li deposition after potential stepping makes it difficult, if not impossible, to distinguish between SEI-Cu-step and SEI-Li on a specific Li deposit according to their sequence of nucleation. Nevertheless, this may be partially recognized upon dissolution, as nuclei with different SEI features are expected to exhibit different dissolution behaviors.

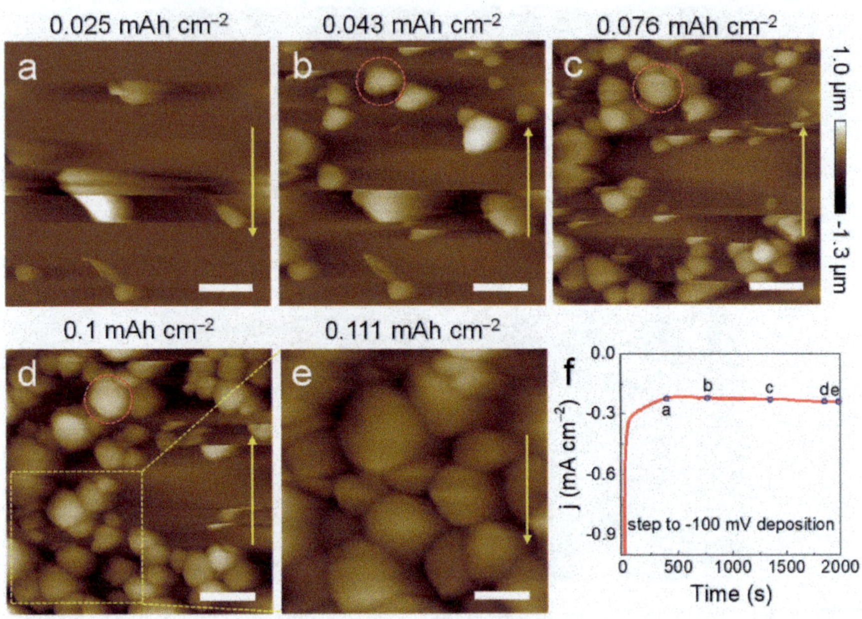

Fig. 6 (a–e) *In situ* AFM monitoring of Li deposition on the Cu(100) surface at various time points after potential stepping from 2 V to −100 mV. (f) Current profile during Li deposition at −100 mV. Scale bars are 2 μm in (a–d) and 1 μm in (e).

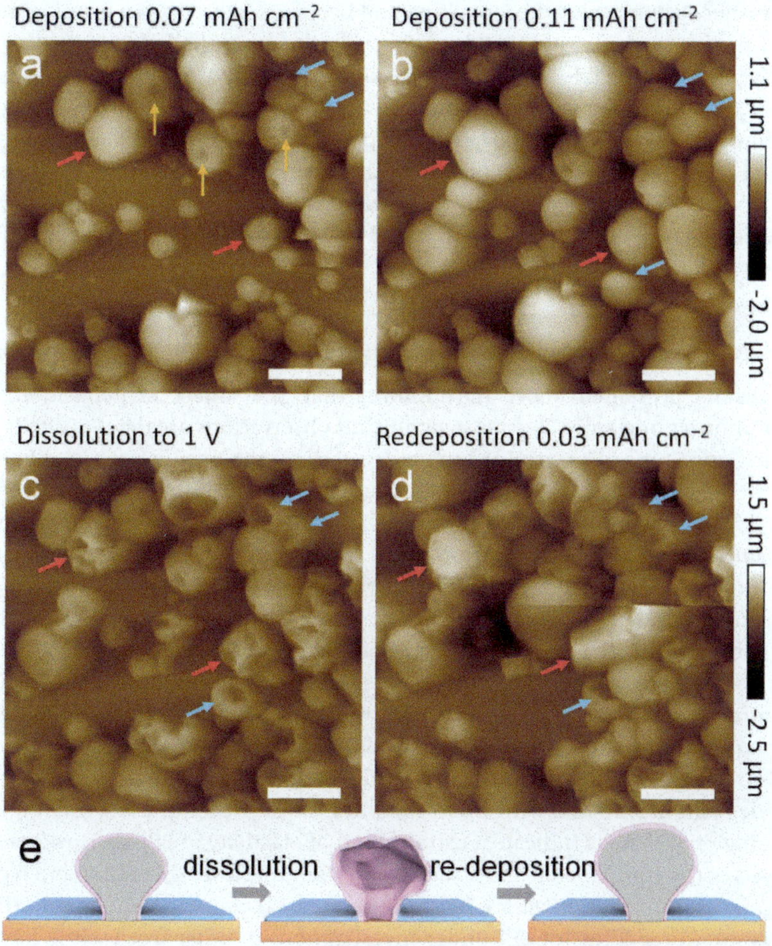

Fig. 7 *In situ* AFM images showing Li deposition, dissolution and re-deposition on Cu(100) after potential stepping from 2 V to −100 mV. (a and b) After Li deposition at −100 mV for 2400 and 3730 s, respectively. (c) Dissolution at 0.2 mA cm^{-2} during the potential rising to 1 V after (b). (d) Li re-deposition for 0.03 mA h cm^{-2} after (c). (e) Schematic diagram of the Li deposition, dissolution and re-deposition processes. Scale bars are 2 µm in (a–d).

Fig. 7 shows a set of *in situ* AFM images, along with the deposition, dissolution and re-deposition sequence. Fig. 7a and b were recorded after Li deposition at −100 mV for 0.07 mA h cm^{-2} (2400 s) and 0.11 mA h cm^{-2} (3730 s), respectively. During deposition, the nuclei indicated by red arrows and blue arrows do not show much difference from each other except for the size and deformation of the spherical shape. However, differences appear after anodic dissolution during the potential rising to 1 V (Fig. 7c). The nuclei marked with red arrows are associated with wrinkled SEI, implying a mechanism of multi-site Li dissolution, while those marked by blue arrows are open-shelled SEIs. The red marked nuclei have a larger size compared to that of the blue ones, implying the earlier nucleation of the former and also confirming the coexistence of SEI-Li and SEI-Cu-step on different nuclei. More interestingly and importantly, the wrinkled SEIs-Li are active and can further accommodate Li deposition and become restored, even forming much larger nuclei, as shown in Fig. 7d. In contrast, the open-shelled SEIs-Cu-step are inactive for further Li deposition, and become "dead-SEI", like those of the open-shelled SEIs-Cu-lsv. In Fig. 7c and d, some nuclei do not show changes during the dissolution and re-deposition cycle, implying dead SEIs which had already formed in the first cycle of deposition. As shown in the schematic diagram in Fig. 7e, the integrated-SEI shell only wrinkles upon dissolution and may be recovered after re-deposition, which could be essential for the high-cycling life of Li metal based batteries.

Segmented-SEI shells *vs.* integrated-SEI shells

The results presented above have shown that SEIs affect Li deposition and dissolution significantly. The most significant observations are that SEI shells can become open-topped or wrinkled upon dissolution; the opening of the SEI shells is irreversible and the open-shelled SEIs become "dead SEI", hindering further Li deposition, while the wrinkled SEIs can be restored and accommodate further deposition. Several pieces of information are worth noting for further understanding the origins leading to the different fates of the SEI-shells and the Li deposition–dissolution cycling behavior. First, the shell-opening and shell-wrinkling phenomena are most clearly observed in the dissolution stage for the systems with preformation of SEI-Cu-lsv and with fresh formation of SEI-Li *via* potential stepping, respectively. However, the shell-opening can also occur in the dissolution stage after Li deposition initiated by potential stepping, while the shell-wrinkling can appear in the dissolution stage after the deposition–dissolution–deposition initiated with preformed SEI-Cu-lsv. Second, the wrinkling and restoring of SEI shells indicate a certain kind of elasticity of SEI-Li. However, the analysis of the components and structures already discloses that SEI-Li is a typical inorganic-dominated SEI with a heavy amount of LiF and Li_2O and is the stiffest among the three kinds of SEI under investigation. In other words, it is hard to solely attribute the shell-wrinkling to the elasticity and the shell-opening to the lack of elasticity. Third, only the very initially nucleated Li nuclei after potential stepping, when the SEI-Cu-step has not been effectively formed, can be covered ideally by the SEI-Li shell. In many cases, however, new nuclei with delayed arrival have to break out of the preformed SEI on Cu and grow towards the solution side, when SEI-Li can be almost immediately formed (see Fig. 7e). In this situation, the coexistence of the SEI-Li and SEI-Cu on a single nucleus is inevitable sooner or later.

The above analysis further explains that, for a Li nucleus grown beneath and then out of the SEI-Cu protective layers (either SEI-Cu-lsv or SEI-Cu-step), its body is covered by SEI-Li, but its top, big or small, is always almost covered by the SEI-Cu type of SEI, forming the segmented-SEI with junctions of the SEI-Cu and SEI-Li. On the other hand, for a nucleus formed immediately after potential stepping, an integrated-SEI, ideally without clear junctions of different types of SEIs, is expected to cover the entire surface of the Li nucleus. It should be mentioned that the integrated-SEIs are not so "pure" due to interference of the process of SEI-Cu-step formation in the very initial stage, and a number of tiny pinhole-like defects, despite being very small on the tops (yellow arrows in Fig. 7a), may be an indication of the coexistence of the two types of SEIs.

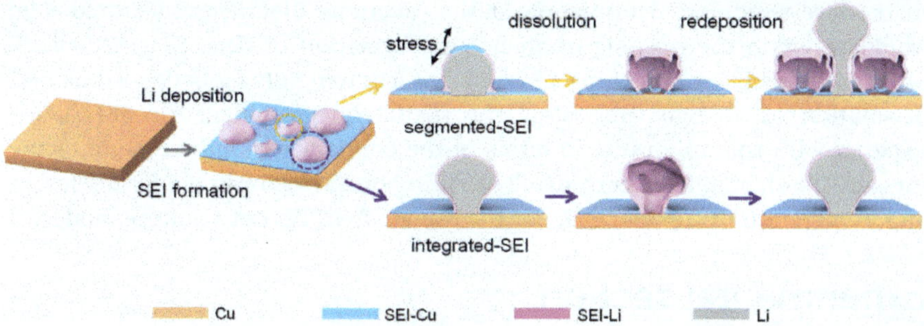

Fig. 8 Schematic diagram of Li deposition, dissolution and re-deposition on Cu associated with segmented-SEI shells and integrated-SEI shells.

Fig. 8 further illustrates the scenarios of Li deposition and dissolution on Cu associated with segmented-SEI and integrated-SEI. The sequence of SEI formation with respect to Li nucleation and growth determines the configuration of the SEI, *i.e.* segmented or integrated, at the surface of the Li deposits in a later stage. In the case of segmented-SEI where junctions exist between the SEI-Cu and SEI-Li, which have different mechanical properties, failure to adapt to the stress-induced deformation at the junction would result in the opening of the shell with the broken shell top remnant falling inside the shell. This makes the SEI dead, so that further Li deposition could only proceed at the interstice of the dead SEIs. Accumulation of dead Li, dead SEIs and the resultant forced interstice deposition upon repetitive cycling may severely shorten the cycling life of an "anode-free" type of Li metal anode. In contrast, in the case of the integrated-SEI, the thin I–O hybrid shell on Li is spatially integrated without distinct junctions of different SEIs, preventing the SEI from complete damage upon dissolution. Rather, stress is released through wrinkling, which can be recovered in the subsequent deposition, facilitating the long-term deposition and dissolution cycling of Li metal anodes.

Conclusions

In conclusion, we have demonstrated that the formation of two different types of SEIs, namely with and without the participation of Li by OPD, respectively, can occur in the initial stage of Li deposition on a Cu substrate. Both types of SEIs formed in the pyrrolidinium-based ionic liquid are inorganic-rich, with the one formed on Cu thicker and less stiff than the one formed on freshly grown Li nuclei. More remarkably, integrated-SEI shells composed of entirely SEI-Li and segmented-SEI shells of two different-featured SEIs may be formed on different Li deposits in the later stage of deposition, causing different fates for the subsequent dissolution–deposition cycles. The integrated-SEI shells only show wrinkles, which can be recovered upon subsequent Li deposition, but the segmented-SEI shells are top-opened at the junctions of different SEIs upon dissolution and become "dead" SEIs that are no longer usable for accommodating further Li deposition. The present work reveals that SEIs are more influential for Li dissolution, and the spatial integration of SEI shells on Li deposits is important for maintaining reversible deposition and dissolution cycling, while failure to adapt to the stress-induced deformation at the junctions of different-featured SEIs would result in the opening of shells and formation of dead SEIs. Our work provides insights into the impact of different-featured SEIs on Li deposition and dissolution on foreign substrates. The spatial integration of the SEI shells, together with the uniformity in terms of the components and structure of the integrated-SEI, should be considered as an important factor in the future design and preparation of superior SEIs for the "anode-free" type of Li metal anodes.

Experimental section

Electrodes and electrolytes

Polycrystalline Cu disk electrodes ($\Phi = 1$ mm) and a Cu(100) single crystal ($\Phi = 18$ mm, Ma Teck) were used as working electrodes for the electrochemical and *in situ* AFM measurements, respectively. The Cu disk electrodes were mechanical polished by alumina power down to 0.05 μm, while the Cu(100) single crystal was chemically polished in H_3PO_4 (85%, Sinopharm) prior to each experiment. For preparing the samples for XPS measurements, copper foils (≥99.9%) were used after washing by immersing in 0.1 M HCl for 10 min, then rinsing with deionized water and ethanol. Li metal foils (≥99.9%) were used as received and served as the counter and reference electrodes. Ionic liquids composed of *N*-butyl-*N*-methyl-pyrrolidinium (Py_{14}) with bis(trifluoromethanesulfonyl)imide (TFSI) or with bis(fluorosulfonyl)imide (FSI) ($Py_{14}FSI$, Solvionic 99.5%) ($Py_{14}TFSI$, Solvionic 99.5%) were vacuumed and stirred for at least ten hours. LiTFSI (≥99.95%, Sigma-Aldrich) was dried under vacuum at 100 °C for two days. An electrolyte of 0.5 M LiTFSI/Py_{14}TFSI-Py_{14}FSI (1 : 1, v/v) with addition of 1300 ppm H_2O was used in this work.

Electrochemical measurements and preparation of SEIs

Cyclic voltammograms (CVs) and transient *i*–*t* curves were measured on the Cu disk electrode in a homemade three-electrode cell on a CHI750E electrochemical workstation (Chenhua Instruments). Different SEIs were prepared by linear sweep voltammetry (LSV) and potential stepping techniques. The LSV method involves a negative scanning potential from 2 V to 0.1 V at a sweep rate of 10 mV s^{-1} on a Cu electrode to obtain SEI-Cu-lsv, while the potential stepping method involves the direct application of a potential step from 2 V to −0.1 V for 100 s and then positive scanning to 1 V was conducted to obtain Cu-SEI-step. For the SEI formation on the Li surface, a time period of 1800 s was used after the potential stepping from 2 V to −0.1 V for obtaining SEI-Li. All electrochemical measurements were conducted in an argon-filled glove box.

EC-AFM imaging and nanomechanical measurements

All AFM measurements were accomplished on a Dimension Icon scanning probe microscope (Bruker Corp.) equipped with a CHI660E electrochemical workstation (Chenhua Instruments) in an Ar-filled glove box (MIKROUNA, $H_2O < 0.1$ ppm, $O_2 < 0.1$ ppm). *In situ* AFM images were obtained in the peak force tapping mode using a NSG10 tip (TipsNano) with a spring constant of 11.8 N m^{-1} and at a scanning rate of 1 Hz. In a homemade EC-AFM cell, a freshly polished Cu(100) single crystal was used as the working electrode, and lithium wires were the counter and reference electrodes. In order to follow the formation of the SEI, a scanning rate of 1 mV s^{-1} was applied during LSV from 2 V to 0.1 V.

XPS characterizations

PHI 5000 Versa Probe II (UIVAC-PHI) was employed for *ex situ* XPS measurements on the two SEI-Cu using a monochromatic Al Kα (1486.6 eV) X-ray source at 16 kV with a beam spot size of 100 μm. EscaLab Xi+ (Thermo Fisher Scientific) was applied to perform *ex situ* XPS measurements of the SEI on Li using a monochromatic Al Kα (1486.7 eV) X-ray source at 15 kV with a beam spot size of 650 μm. The binding energies of the C 1s line from adventitious carbon at 284.8 eV was used as a reference. Depth profiling was achieved using Ar ion sputtering over an area of 2 × 2 mm^2 in the *x*–*y* scan mode at an ion acceleration of 2 V and an ion beam current of 2 μA. The thicknesses of the SEIs were estimated based on the calibrated sputtering rate of 5 nm per minute for Si. All samples were rinsed with DME solvent to remove the remaining electrolyte and then sealed in commercial air-isolating containers in the glove box, and then quickly transferred into the XPS chamber.

Conflicts of interest

There are no conflicts of interest to declare.

Acknowledgements

This work was supported by the NSFC projects (21972119, 22002129, 22102137, 22072123) and CPSF (2019TQ0177). We thank Dr Xiao-Bin Lian at Quanzhou Normal University for technical assistance with the XPS measurements. We also thank Profs. Quan-Feng Dong and Min-Seng Zheng at Xiamen University for their various support.

References

1 X. B. Cheng, R. Zhang, C. Z. Zhao and Q. Zhang, *Chem. Rev.*, 2017, **117**, 10403–10473.
2 Y. Guo, H. Li and T. Zhai, *Adv. Mater.*, 2017, **29**, 1700007.
3 D. Lin, Y. Liu and Y. Cui, *Nat. Nanotechnol.*, 2017, **12**, 194–206.
4 D. G. E. Peled and G. Ardel, *J. Electrochem. Soc.*, 1997, **144**, L208–L210.
5 E. Peled and S. Menkin, *J. Electrochem. Soc.*, 2017, **164**, A1703–A1719.
6 Y. Li, W. Huang, Y. Li, A. Pei, D. T. Boyle and Y. Cui, *Joule*, 2018, **2**, 1–11.
7 Y. Gu, W.-W. Wang, J.-W. He, S. Tang, H.-Y. Xu, J.-W. Yan, Q.-H. Wu, X.-B. Lian, M.-S. Zheng, Q.-F. Dong and B.-W. Mao, *ChemElectroChem*, 2019, **6**, 181–188.
8 N.-W. Li, Y. Shi, Y.-X. Yin, X.-X. Zeng, J.-Y. Li, C.-J. Li, L.-J. Wan, R. Wen and Y.-G. Guo, *Angew. Chem., Int. Ed.*, 2018, **57**, 1505–1509.
9 C. Yan, H.-R. Li, X. Chen, X.-Q. Zhang, X.-B. Cheng, R. Xu, J.-Q. Huang and Q. Zhang, *J. Am. Chem. Soc.*, 2019, **141**, 9422–9429.
10 X. Gao, Y.-N. Zhou, D. Han, J. Zhou, D. Zhou, W. Tang and J. B. Goodenough, *Joule*, 2020, **4**, 1864–1879.

11 J. W. He, Y. Gu, W. W. Wang, J. H. Wang, Z. B. Chen, H. Y. He, Q. H. Wu, J. W. Yan and B. W. Mao, *ChemElectroChem*, 2021, **8**, 62–69.

12 J. Zheng, M. S. Kim, Z. Tu, S. Choudhury, T. Tang and L. A. Archer, *Chem. Soc. Rev.*, 2020, **49**, 2701–2750.

13 K. N. Wood, M. Noked and N. P. Dasgupta, *ACS Energy Lett.*, 2017, **2**, 664–672.

14 Y. Zhang, T.-T. Zuo, J. Popovic, K. Lim, Y.-X. Yin, J. Maier and Y.-G. Guo, *Mater. Today*, 2020, **33**, 56–74.

15 W.-W. Wang, Y. Gu, H. Yan, S. Li, J.-W. He, H.-Y. Xu, Q.-H. Wu, J.-W. Yan and B.-W. Mao, *Chem*, 2020, **6**, 2728–2745.

16 X. Guan, A. Wang, S. Liu, G. Li, F. Liang, Y.-W. Yang, X. Liu and J. Luo, *Small*, 2018, **14**, 1801423.

17 P. Zhai, L. Liu, X. Gu, T. Wang and Y. Gong, *Adv. Energy Mater.*, 2020, **10**, 2001257.

18 P. Bai, J. Li, F. R. Brushett and M. Z. Bazant, *Energy Environ. Sci.*, 2016, **9**, 3221–3229.

19 J. Yamaki, S. Tobishima, K. Hayashi, K. Saito, Y. Nemoto and M. Arakawa, *J. Power Sources*, 1998, **74**, 219–227.

20 J.-N. Chazalviel, *Phys. Rev. A*, 1990, **42**, 7355–7367.

21 Y. S. Hu and Y. X. Lu, *ACS Energy Lett.*, 2020, **5**, 3633–3636.

22 K. Xu, *Chem. Rev.*, 2014, **114**, 11503–11618.

23 J. Qian, W. A. Henderson, W. Xu, P. Bhattacharya, M. Engelhard, O. Borodin and J. G. Zhang, *Nat. Commun.*, 2015, **6**, 6362.

24 L. Suo, Y.-S. Hu, H. Li, M. Armand and L. Chen, *Nat. Commun.*, 2013, **4**, 1481.

25 S. Chen, J. Zheng, D. Mei, K. S. Han, M. H. Engelhard, W. Zhao, W. Xu, J. Liu and J.-G. Zhang, *Adv. Mater.*, 2018, **30**, 1706102.

26 H. Wang, W. Huang, Z. Yu, W. Huang, R. Xu, Z. Zhang, Z. Bao and Y. Cui, *ACS Energy Lett.*, 2021, **6**, 816–825.

27 H. Wu, H. Jia, C. Wang, J. G. Zhang and W. Xu, *Adv. Energy Mater.*, 2021, **11**, 2003092.

28 J.-G. Zhang, W. Xu, J. Xiao, X. Cao and J. Liu, *Chem. Rev.*, 2020, **120**, 13312–13348.

29 C. Yan, L. L. Jiang, Y. X. Yao, Y. Lu, J. Q. Huang and Q. Zhang, *Angew. Chem., Int. Ed.*, 2021, **60**, 8521–8525.

30 A. Basile, A. I. Bhatt and A. P. O'Mullane, *Nat. Commun.*, 2016, **7**, 11794.

31 X.-B. Cheng, C. Yan, X. Chen, C. Guan, J.-Q. Huang, H.-J. Peng, R. Zhang, S.-T. Yang and Q. Zhang, *Chem*, 2017, **2**, 258–270.

32 Y. Gu, W. W. Wang, Y. J. Li, Q. H. Wu, S. Tang, J. W. Yan, M. S. Zheng, D. Y. Wu, C. H. Fan, W. Q. Hu, Z. B. Chen, Y. Fang, Q. H. Zhang, Q. F. Dong and B. W. Mao, *Nat. Commun.*, 2018, **9**, 1339.

33 B. J. Neudecker, N. J. Dudney and J. B. Bates, *J. Electrochem. Soc.*, 2000, **147**, 517–523.

34 J. Qian, B. D. Adams, J. Zheng, W. Xu, W. A. Henderson, J. Wang, M. E. Bowden, S. Xu, J. Hu and J.-G. Zhang, *Adv. Funct. Mater.*, 2016, **26**, 7094–7102.

35 Z. Xie, Z. Wu, X. An, X. Yue, J. Wang, A. Abudula and G. Guan, *Energy Storage Materials*, 2020, **32**, 386–401.

36 Y. Gu, H.-Y. Xu, X.-G. Zhang, W.-W. Wang, J.-W. He, S. Tang, J.-W. Yan, D.-Y. Wu, M.-S. Zheng, Q.-F. Dong and B.-W. Mao, *Angew. Chem., Int. Ed.*, 2019, **58**, 3092–3096.

37 A. Pei, G. Y. Zheng, F. F. Shi, Y. Z. Li and Y. Cui, *Nano Lett.*, 2017, **17**, 1132–1139.

38 A. M. Tripathi, W.-N. Su and B. J. Hwang, *Chem. Soc. Rev.*, 2018, **47**, 736–851.

39 R. Kempaiah, G. Vasudevamurthy and A. Subramanian, *Nano Energy*, 2019, **65**, 103925.

40 W. D. Zhao, W. T. Song, L. Z. Cheong, D. Y. Wang, H. Li, F. Besenbacher, F. Q. Huang and C. Shen, *Ultramicroscopy*, 2019, **204**, 34–48.

41 S. W. Wang, Q. Liu, C. L. Zhao, F. Z. Lv, X. Y. Qin, H. D. Du, F. Y. Kang and B. H. Li, *Energy Environ. Mater.*, 2018, **1**, 28–40.

42 X. Chen, J. Lai, Y. Shen, Q. Chen and L. Chen, *Adv. Mater.*, 2018, 1802490.

43 D. Aurbach and Y. Cohen, *J. Electrochem. Soc.*, 1996, **143**, 3525–3532.

44 K. A. Hirasawa, T. Sato, H. Asahina, S. Yamaguchi and S. Mori, *J. Electrochem. Soc.*, 1997, **144**, L81–L84.

45 R. Mogi, M. Inaba, Y. Iriyama, T. Abe and Z. Ogumi, *J. Electrochem. Soc.*, 2002, **149**, A385–A390.

46 S. Shiraishi, K. Kanamura and Z. Takehara, *J. Phys. Chem. B*, 2001, **105**, 123–134.

47 C. Shen, S. Wang, Y. Jin and W.-Q. Han, *ACS Appl. Mater. Interfaces*, 2015, **7**, 25441–25447.

48 J. Wan, Y. Hao, Y. Shi, Y.-X. Song, H.-J. Yan, J. Zheng, R. Wen and L.-J. Wan, *Nat. Commun.*, 2019, **10**, 3265.

49 Z. Y. Zhang, K. Smith, R. Jervis, P. R. Shearing, T. S. Miller and D. J. L. Brett, *ACS Appl. Mater. Interfaces*, 2020, **12**, 35132–35141.

50 M. Kitta and H. Sano, *Langmuir*, 2017, **33**, 1861–1866.

51 S.-Y. Lang, Z.-Z. Shen, X.-C. Hu, Y. Shi, Y.-G. Guo, F.-F. Jia, F.-Y. Wang, R. Wen and L.-J. Wan, *Nano Energy*, 2020, **75**, 104967.

52 S. Wang, X. Yin, D. Liu, Y. Liu, X. Qin, W. Wang, R. Zhao, X. Zeng and B. Li, *J. Mater. Chem. A*, 2020, **8**, 18348–18357.

53 Y. Shi, G.-X. Liu, J. Wan, R. Wen and L.-J. Wan, *Sci. China: Chem.*, 2021, **64**, 734–738.

54 M. Han, C. Zhu, T. Ma, Z. Pan, Z. Tao and J. Chen, *Chem. Commun.*, 2018, **54**, 2381–2384.

55 F. Shi, A. Pei, D. T. Boyle, J. Xie, X. Yu, X. Zhang and Y. Cui, *Proc. Natl. Acad. Sci. U. S. A.*, 2018, **115**, 8529–8534.

56 J. Qian, W. Xu, P. Bhattacharya, M. Engelhard, W. A. Henderson, Y. Zhang and J.-G. Zhang, *Nano Energy*, 2015, **15**, 135–144.

57 N. Togasaki, T. Momma and T. Osaka, *J. Power Sources*, 2014, **261**, 23–27.

58 B. Thirumalraj, T. T. Hagos, C. J. Huang, M. A. Teshager, J. H. Cheng, W. N. Su and B. J. Hwang, *J. Am. Chem. Soc.*, 2019, **141**, 18612–18623.

59 G. Gunawardena, G. Hills, I. Montenegro and B. Scharifker, *J. Electroanal. Chem.*, 1982, **138**, 225–239.

60 M. Watanabe, M. L. Thomas, S. Zhang, K. Ueno, T. Yasuda and K. Dokko, *Chem. Rev.*, 2017, **117**, 7190–7239.

61 S. Zhang, N. Sun, X. He, X. Lu and X. Zhang, *J. Phys. Chem. Ref. Data*, 2006, **35**, 1475–1517.

Faraday Discussions

PAPER

Nano-confined electrochemical reaction studied by electrochemical surface forces apparatus†

Motohiro Kasuya,[a] Daiki Kubota,[a] Sho Fujii[b] and Kazue Kurihara*[abc]

Received 29th August 2021, Accepted 21st September 2021
DOI: 10.1039/d1fd00060h

Electrochemical reactions in a nano-space are different from those in bulk solutions due to structuring of the liquid molecules and peculiar ion behavior at the electric double layer and are important for applications involving sensors and energy devices. The electrochemical surface forces apparatus (EC-SFA) we developed enabled us to study the electrochemical reactions in a solution nano-confined between the electrodes with varying distance (D) at nm resolution. We recorded measurements of the current–distance profiles due to the electrochemical reaction of the redox couples in the electrolyte nano-confined between Pt electrodes using our EC-SFA. We observed a long-range feedback current due to redox cycling and the sudden current increase at a short distance, the latter for the first time. This sudden current increase was two orders greater than the conventional feedback current and was observed at $D < 5$ nm when the electrodes were approaching and $D < 200$ nm on separation. We simultaneously measured the electric double layer force and the current between the electrodes in the solution to study the mechanisms of this sudden current increase in the short distance range. The results revealed a molecular insight as to how the redox species affect the current between two electrodes under nano-confinement. This study demonstrated that EC-SFA is a powerful tool for obtaining fundamental knowledge about the nano-confined electrochemical reactions for nanoelectrodes which can be applied to sensors and energy devices.

1. Introduction

Electrolyte solutions confined in nano-spaces exhibit different properties from those of bulk solutions due to structuring of the liquid and restricted molecular

[a]Institute of Multidisciplinary Research for Advanced Materials, Tohoku University, Sendai 980-8577, Japan. E-mail: kazue.kurihara.b7@tohoku.ac.jp
[b]Advanced Institute for Materials Research (AIMR), Tohoku University, Sendai 980-8577, Japan
[c]New Industry Creation Hatchery Center, Tohoku University, Sendai 980-8578, Japan

† Electronic supplementary information (ESI) available. See DOI: 10.1039/d1fd00060h

motions. For example, when ionic liquids are under confinement, they form layered structures[1] and their viscosities increase.[2] These viscosity increases were applied to the gelation of the electrolyte for the quasi-solid battery.[3] Another example is the pH under confinement. When water is confined between negative surfaces, the pH increases. We reported that such interfacial pH changes are due to both concentrating of the proton in the electric double layer and hydration of the solid surfaces and absorbed ions.[4]

For nano-electrochemical devices, such as sensors and energy devices, their peculiar properties, fast time responses, small capacitive background, and enhanced voltammetric responses are known in practice and considered as advantages.[5–7] Therefore, various electrode configurations in small-spaces, such as the ring-disk electrode,[8] interdigitated array electrode,[9,10] ultrathin layer cell[11] and confined bipolar electrode,[12,13] were developed and applied to those electrochemical devices for achieving a highly sensitive detection in simultaneous multi-sample analyses for small sample volumes. However, the mechanisms of these advantageous properties have not been well elucidated. Electrochemical reaction behaviors in nano-structured electrodes were characterized by only limited techniques such as model analysis and spatially resolved characterization methods of scanning electrochemical microscopy (SECM) and electrochemical atomic force microscopy.[14–17] They have been used to monitor 2D images of electrochemical reactions and/or electrodes with a nanometer resolution, however, there is a limitation of low vertical spatial resolution, *i.e.*, several tens nm, between electrodes because typical confinement effects appear at separation distances below *ca.* 20 nm. Therefore, a suitable characterization method for studying the confinement effects on electrochemical reactions is desired.

Another potentially powerful technique can be surface forces measurement. This technique provides a unique tool for studying confined liquids of varying thicknesses (surface separation, D) at a 0.1 nm resolution.[18–24] By employing a surface force apparatus (SFA), we have developed novel techniques, such as the resonance shear measurement (RSM),[23–25] X-ray diffraction apparatus,[26] and SFA fluorescence spectroscopy,[4,27] and studied the behavior of confined liquids.

Electrochemical (EC)-SFA has been another direction for the development of SFA. The EC-SFA can directly monitor the electric double layer forces as a function of the surface separation and quantitatively estimate the surface potential and charges of the electrodes.[28–30] Naturally, there was interest in developing EC-SFA. Conner and Horn succeeded in measuring the interactions between mica and a mercury droplet which was connected to a potentiostat.[29] Vanderlick measured the interactions between mica and a gold electrode.[30] However, these EC-SFAs used the conventional SFA with FECO (fringes of equal chromatic order) for the distance determination, and at least one transparent substrate (typically mica) was required.[29,30] This condition renders the analysis of interactions more difficult due to asymmetric configurations, and it is impossible to study two electrodes under various applied potentials. We recently developed a new-type of surface forces apparatus, *i.e.*, the twin-path SFA, using modified two-beam (twin-path) interferometry for measuring the interactions between non-transparent substrates.[31] Based on this twin-path SFA, we developed a new type of EC-SFA which enables us to measure the interaction between non-transparent electrodes in a symmetric configuration.[28] We applied this EC-SFA (twin-path type) for characterizing non-transparent electrodes such as gold and platinum.[32–34] These studies revealed how the surface potentials and charge density were affected by the various selectively adsorbed anions on the gold electrode and the hydrogen adsorption on the platinum electrodes.[32,33] We also studied charge neutralization of the ferrocene modified electrode by counterions and quantitatively evaluated the effective charges on the electrode and the ratio of the strongly bound counterions and the loosely bound counterions in the Stern layer.[34]

Taking advantage of the precise control of the separation distance of SFA, Fan and Bard studied the currents between two very thin Pt electrodes deposited on mica by changing the separation.[35] However, their instrument could not determine distances shorter than 500 nm, and therefore, assumed the zero distance from the point which showed the sudden current increase and estimated the distance from the approaching speed of the electrodes. They also did not measure the forces between the electrodes. The EC-SFA we developed enabled us to study both the force and the current between the electrodes by varying their distances with a nm resolution. The measured force profiles provided important information, such as surface potentials, charge densities and ion adsorption of the electrodes, which could be useful for understanding the detailed mechanisms of the observed electrochemical events under nano-confinement.

In this study using our EC-SFA, we measured the current–distance profiles due to the electrochemical reaction of redox couples in an electrolyte confined between Pt electrodes. We observed, for the first time, a sudden current increase of three orders of magnitude greater than the conventional feedback current at short distances on a nm level. We discussed the mechanisms of this sudden current increase at the short distance based on the measured double layer interactions. This study indicated that EC-SFA is useful for characterizing the electrochemical events under nano-confinement, which should be useful for the design of electrochemical systems such as sensors and energy devices.

2. Experimental

2.1. Materials

$K_3[Fe(CN)_6]$ (99.5%, Nacalai Tesque), $K_4[Fe(CN)_6]\cdot 3H_2O$ (99%, Nacalai Tesque), and $KClO_4$ (99.99%, Aldrich) were used as received. The Pt plate (99.99%) from Toyoshima Manufacturing was used as a sputtering source. The Pt wire (99.9999, $\phi = 0.2$ mm) was purchased from Tanaka Kikinzoku. Ultrapure water (NANOpure II, Barnstead, 18 MΩ cm^{-1} resistance) was used after double distillation.

2.2. Preparation of sample surfaces

The Pt electrode surfaces were prepared by the mica-stripping method.[36] A Pt film of *ca.* 100 nm in thickness was deposited on freshly-cleaved mica by a sputtering deposition system (SPV-200, Toei Scientific Industrial) in an argon atmosphere as previously reported.[33] The as-deposited Pt films were glued on cylindrical quartz disks (curvature radius, $R = 20$ mm) by a hot-melt epoxy resin (E1004, Shell). For the EC-SFA measurements, a conductive wire was connected to the Pt surface using a conductive epoxy (CW2400, ITW Chemtronics), then covered by an epoxy resin. After removal of the mica template, a hydrophilic Pt surface was prepared.

2.3. Electrochemical surface forces apparatus (EC-SFA)

A schematic illustration of the EC-SFA measurement system[31] is shown in Fig. 1(a). Two prepared Pt electrodes were used as the working electrodes using a bipotentiostat (Model 2325, ALS) for controlling the electrochemical potential (E) and measuring the current (I). The counter electrode was a Pt-wire, and the reference electrode was a Ag/AgCl (saturated KCl) electrode (BAS). The electrochemical measurements were performed inside the chamber of the twin-path SFA. The SFA chamber was filled with the aqueous electrolytes. Before all the measurements, argon (99.9999%) was bubbled through the solutions for more than 30 min for deaeration, and the experiments were done under an argon atmosphere.

Force measurements were carried out by a twin-path SFA (RSM-1, Advance Riko) which determined the distance between the two surfaces by the modified two-laser beam interferometry technique.[31] The interaction force (F) between the Pt electrode surfaces was measured as a function of the surface separation (D) in an aqueous solution. The spring constant of the cantilever was in the range of 160–220 N m^{-1}. We continuously changed the surface separation between the Pt surfaces at a constant approaching rate (20 nm s^{-1}) in order to obtain the force–distance profiles. The obtained force F was normalized by the radius R of the surface curvature using the Derjaguin approximation,[18] $F/R = 2\pi G_f$, where G_f is the interaction free energy per unit area between two flat surfaces. The observation of the electric double layer repulsion and hardwall contact (see Results parts) confirmed the cleanliness of the electrode surfaces and the accuracy of the distance control with a nanometer resolution.

The measurement of the current–distance profiles was simultaneously performed along with the forces measurement. The potentials E of the upper and lower electrodes were $E = E_{redox} + \Delta V$ and $E = E_{redox} - \Delta V$, respectively. We set $E_{redox} = 220$ mV, obtained from the cyclic voltammogram shown in Fig. S1† for the 0.1 mM aqueous solution of the $Fe(CN)_6{}^{3-/4-}$ redox couple. We waited *ca.* 200 s before the current reached a constant value after setting the E values. The two electrodes were then approached from a distance greater than 2000 nm while measuring I and D by the bipotentiostat and twinpath interferometer, respectively, in order to obtain the current–distance profiles. The current–distance

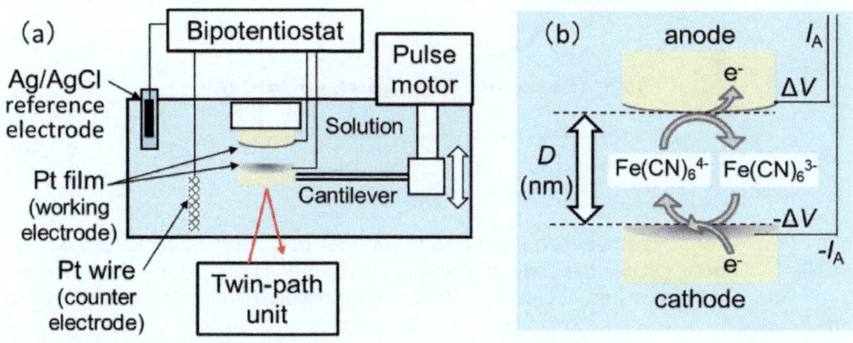

Fig. 1 (a) Schematic illustration of EC-SFA. (b) Schematic illustration of nano-confined electrochemical reaction between the electrode surfaces.

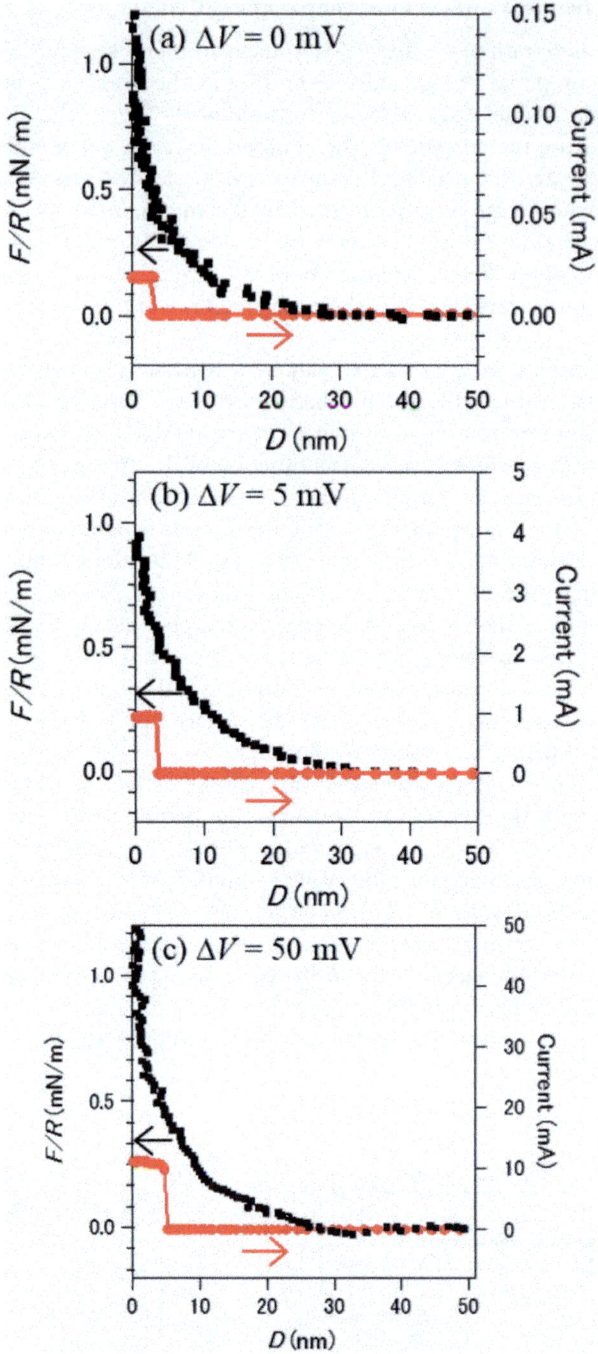

Fig. 2 Profiles of forces between Pt electrode surfaces in 0.1 mM aqueous solution of $Fe(CN)_6{}^{3-/4-}$ redox couple depending on the applied potentials: (a) 0 mV, (b) 5 mV, (c) 50 mV. Current profiles (red circles) simultaneously measured were also plotted in the same figure.

profiles of the anode and cathode of which the potentials were set at $E_{redox} + \Delta V$ and $E_{redox} - \Delta V$, respectively, showed the same absolute values of positive and negative currents as shown in Fig. S2,† which supported the fact that the observed current should be due to the current between the two electrodes as illustrated in Fig. 1(b). All measurements were done at room temperature (22 ± 2 °C).

3. Results and discussion

3.1. Forces between Pt surfaces in various electrolytes

Fig. 2(a) shows a force–distance profile for the Pt electrode surfaces upon approach in a 0.1 mM aqueous solution of the $Fe(CN)_6{}^{3-/4-}$ redox couple at $\Delta V =$ 0 mV. The forces consisting of a long-range repulsion at $D < 30$ nm, which followed an exponential function, and the hardwall contact were observed. The decay lengths of the repulsion were 6.5 ± 0.3 nm and in good agreement with the theoretical Debye length ($1/\kappa$) for the current concentration of the electrolyte ($1/\kappa = 5.6$ nm), indicating that the observed repulsion was attributed to the electric double layer repulsion. It was reported that the potential of zero charge (pzc) of this electrode was *ca.* 0.2 V *vs.* Ag/AgCl.[33] Therefore, both the cathode and anode should be positively charged because the potentials of these electrodes were 0.22 V. The surface potentials (ψ_0) and charge density (σ) could be evaluated by fitting the profile to the Poisson–Boltzmann equation of the electric double layer force for the constant potential and charge model (see Fig. S3†), respectively.[18,37] The evaluated ψ_0 and σ of the Pt electrode at $\Delta V = 0$ mV were 25.0 ± 2.2 mV and 0.30 ± 0.02 μC cm^{-2}, respectively, as summarized in Table 1.

At $\Delta V = 5$ and 50 mV, a similar long-range repulsion was observed, as shown in Fig. 2(b) and (c), respectively. The decay lengths of the repulsion (6.3 ± 1.0 nm ($\Delta V = 5$ mV) and 5.7 ± 0.2 nm ($\Delta V = 50$ mV)) were similar to that at $\Delta V = 0$ mV and were in good agreement with $1/\kappa$ (5.6 nm), indicating that the observed repulsion was also attributed to the double layer repulsion. The appearance of the double layer repulsion showed that the sign of the two charged electrodes was the same perhaps due to ion adsorption even when the applied voltages were different in their signs. Assuming a symmetrical configuration, the σ values were obtained by fitting to the Poisson–Boltzmann equation to be 0.35 and 0.37 μC cm^{-2} at $\Delta V = 5$ and 50 mV, respectively, as summarized in Table 1 together with the ψ_0 values. The force–distance profiles upon separation under the same conditions were also

Table 1 Surface potential, ψ_0, and charge density, σ, of Pt electrodes in aqueous solution of $Fe(CN)_6{}^{3-/4-}$ redox couple obtained from fitting the force–distance profiles by the Poisson–Boltzmann equation

Concentration	ΔV [mV]	ψ_0 [mV]	σ [μC cm^{-2}]	Decay length [nm]	$1/\kappa$ [nm]
0.1 mM	0	25.0 ± 2.2	0.30 ± 0.02	6.2 ± 0.9	5.6
0.1 mM	5	28.3 ± 3.0	0.35 ± 0.03	6.3 ± 1.0	5.6
0.1 mM	50	33.3 ± 2.9	0.37 ± 0.03	5.7 ± 0.2	5.6
1.0 mM	0	6.6 ± 1.5	0.35 ± 0.15	2.5 ± 0.2	1.7
1.0 mM	5	8.5 ± 2.2	0.48 ± 0.13	2.2 ± 0.3	1.7
1.0 mM	50	9.8 ± 2.0	0.53 ± 0.10	2.3 ± 0.8	1.7

measured as shown in Fig. S4.† No adhesion was observed at $\Delta V = 0$ mV and 5 mV, and a small adhesion force was observed at $\Delta V = 50$ mV.

We performed the same measurement for the Pt electrodes in the 1 mM aqueous $Fe(CN)_6^{3-/4-}$ and 10 mM $KClO_4$ at various ΔV's. The electric double layer repulsion and the hardwall contact were observed in all cases (see ESI†). A small adhesion force was observed in the force profiles upon separation at $\Delta V = 5$ mV and 50 mV, while no adhesion was observed at $\Delta V = 0$ mV in the 1 mM $Fe(CN)_6^{3-/4-}$. The σ values obtained by fitting to the Poisson–Boltzmann equation are summarized in Table 1.

3.2. Currents between Pt surfaces in $Fe(CN)_6^{3-/4-}$ solutions

We simultaneously performed current–distance measurements for the Pt electrodes with the forces measurement using the setup shown in Fig. 1(b). Fig. 3(a)

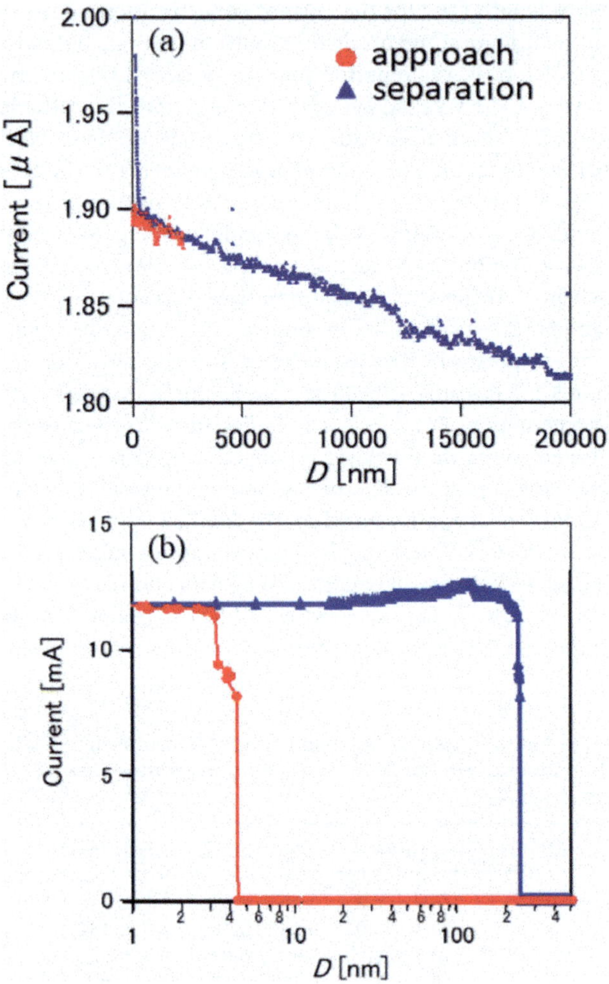

Fig. 3 Current profiles for the Pt electrode surfaces in 0.1 mM $Fe(CN)_6^{3-/4-}$ solution during approach (red circles) and separation (blue triangles) in μm range (a) and nm range (b). $\Delta V = 50$ mV.

and (b) show a current–distance profile (an enlarged one in (b)) upon approach for the Pt electrodes in the 0.1 mM aqueous $Fe(CN)_6^{3-/4-}$ redox couple at $\Delta V = 50$ mV. A gradual current increase of 10 nA (from 1.88 to 1.89 μA) was observed from D = 2000 nm to D = 4.4 nm. A sudden current increase from 1.89 μA to 12 mA was observed at D = 4.4 nm when D was further reduced as shown in Fig. 3(b). This high current was constant from a separation of 4.4 nm to the hardwall contact. When the electrodes were separated from the hardwall, the current was constant, *ca.* 12 mA, and suddenly decreased to 1.9 μA at D = *ca.* 250 nm and gradually decreased further to 1.81 μA at 20 000 nm. The latter gradual current decrease was identical to that observed in the approaching process.

For understanding the detailed behavior of the sudden current increase at a short distance, the current–distance profiles for the Pt electrodes in the same solution at ΔV = 5 and 0 mV were measured upon approach and separation as shown in Fig. 4(a) and (b), respectively. At ΔV = 5 mV, the sudden increase to 1 mA was observed at D = 3.4 nm when the electrodes were approaching, as shown in Fig. 4(a). This current was constant from D = 3.4 nm to the hardwall contact. When the surfaces were separated from the hardwall contact, an identical constant current was observed as for the approaching process until D = 3.3 nm, where a sudden current decrease was observed (Fig. 4(b)). At ΔV = 0 mV, we observed a similar sudden current increase and decrease at D = 2.0 nm during the approach and separation processes, respectively.[38] The amplitude of this current was 0.2 mA.

The effect of the $Fe(CN)_6^{3-/4-}$ concentration was also studied. When the $Fe(CN)_6^{3-/4-}$ concentration was increased to 1 mM, a gradual current increase of 400 nA from D = 2000 nm to 5 nm was observed at ΔV = 50 mV during the approach, and this current was 40 times in magnitude greater than that for the 0.1 mM solution as shown in Fig. 5(a). When the surfaces further approached, a sudden current increase to 0.4 mA was observed at D = 4.5 nm (Fig. 5(b)). This high current was constant until the hardwall contact. On the other hand, when the electrodes were separated from the hardwall contact at ΔV = 50 mV, the current was identical in magnitude as for the approach process but the distance range was further extended to D = 100 nm (Fig. 5(c)) which was much longer than the 5 nm upon approach. At ΔV = 5 mV, the sudden increase to 0.4 mA was observed at D = 3.6 nm when the electrodes approached. This current was constant from D = 3.6 nm to the hardwall contact. When the surfaces were separated from the hardwall contact, the current was identical in magnitude for the approach but the distance range was further extended to D = 50 nm. At ΔV = 0 mV, we observed a slight sudden current increase and decrease (0.01 mA) at D = *ca.* 2 nm in the approach and separation processes, respectively.

We performed the same measurements in the 10 mM aqueous $KClO_4$ for studying the influence of the redox species on the current profile because this electrolyte showed no redox reaction over the entire range of E used in this study (0.17–0.27 V). There was no current change when the surfaces approached at D > 2.5 nm as shown in Fig. 6(a), which was different from the case of the $Fe(CN)_6^{3-/4-}$ solutions (Fig. 4 and 5). A sudden current increase to 3.8 mA at D = 2.5 nm was observed upon the approach when ΔV = 50 mV (Fig. 6(b)). The same sudden current increases were observed in the cases of ΔV = 0 and 5 mV at D = 2.5 and 2.3 nm, respectively, which were practically identical to that at ΔV = 50 mV. The same current profiles were observed during the separating process as those

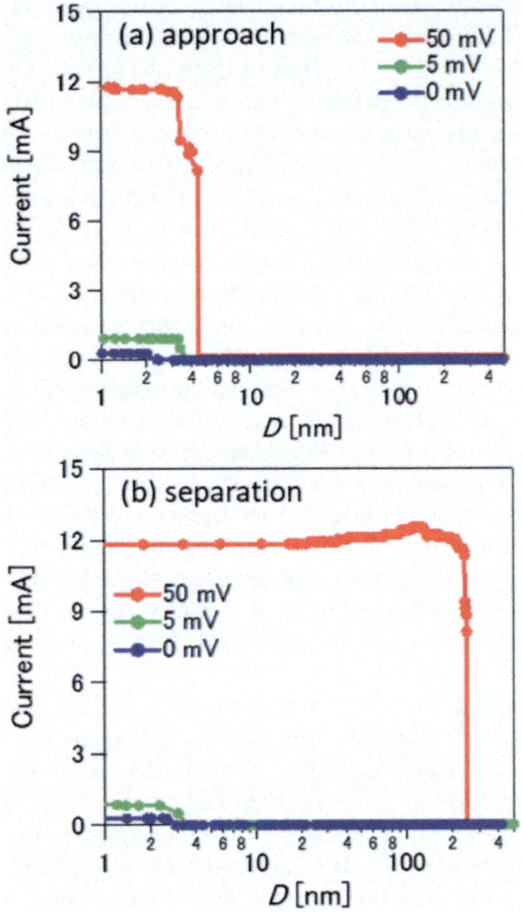

Fig. 4 Current profiles for the Pt electrode surfaces in 0.1 mM aqueous $Fe(CN)_6^{3-/4-}$ at various ΔV values during approach (a) and separation (b) in the nm range.

during the approaching process for all three ΔV cases, as shown in Fig. 6(b). On the other hand, the D values increased with the increasing ΔV in the case of the 0.1 mM $Fe(CN)_6^{3-/4-}$ solution (2.0, 3.4 and 4.4 nm at 0, 5, and 50 mV, respectively). The obtained results from the 10 mM aqueous $KClO_4$ (shorter D_c than that of the $Fe(CN)_6^{3-/4-}$ solution, no dependence of D_c on ΔV and no hysteresis between approaching and separating processes) were different from those in the case of the $Fe(CN)_6^{3-/4-}$ solution.

3.3. Discussion

As described in Section 3.2, the gradual current increase at the longer D and sudden increase at the shorter D were observed in the current profiles for the Pt electrodes in the $Fe(CN)_6^{3-/4-}$ solutions. In this section, the origin of these two current increases is discussed.

First, we discuss the origin of the gradual current increase at the longer D. The amplitude of the current change in the 1 mM $Fe(CN)_6^{3-/4-}$ solution was 40 times

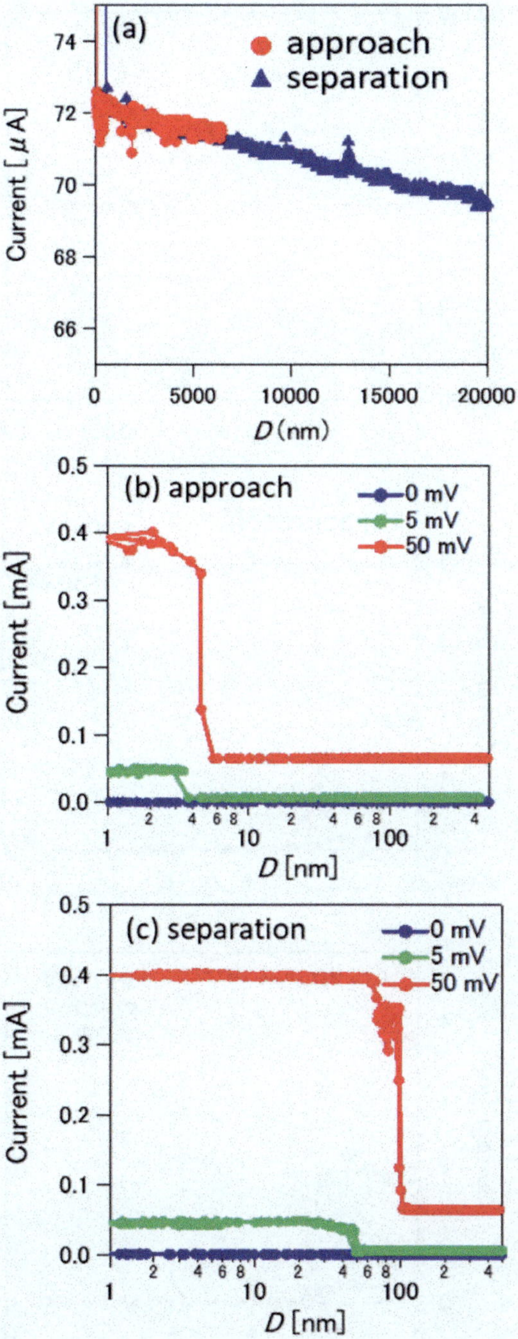

Fig. 5 (a) Current profiles for the Pt electrode surfaces in 1 mM $Fe(CN)_6^{3-/4-}$ solution during approach (red circles) and separation (blue triangles) in μm range ($\Delta V = 50$ mV). Same current profiles at various ΔV values in nm range during approach (b) and separation (c) are shown.

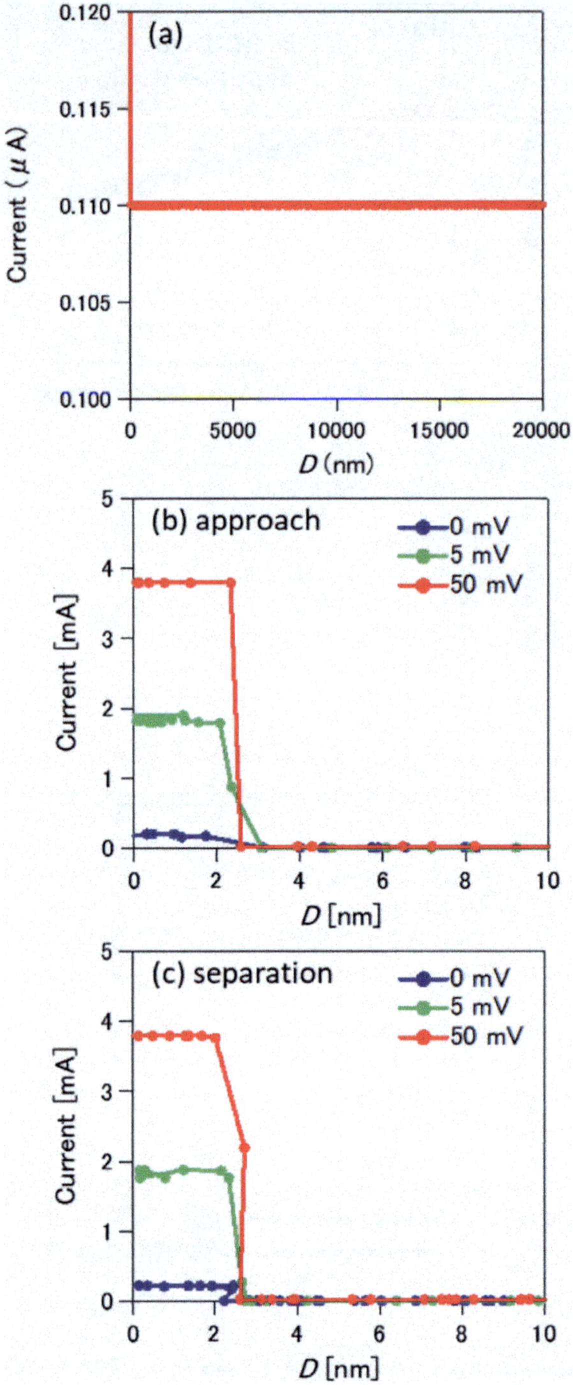

Fig. 6 (a) Current profile for the Pt electrode surfaces in 10 mM $KClO_4$ solution during approach in μm range ($\Delta V = 50$ mV). Same current profiles at various ΔV values in nm range during approach (b) and separation (c) are shown.

in magnitude greater than that for the 0.1 mM $Fe(CN)_6^{3-/4-}$ solution as shown in Fig. 3(a) and 5(a). Such a distance dependent increase was not observed in the case of the 10 mM $KClO_4$ solution as shown in Fig. 6(a). These results indicated that the current change in the $Fe(CN)_6^{3-/4-}$ solutions can be attributed to the feedback current due to the redox cycle reaction which resulted in the gradual current amplification observed when the cathode and anode were brought closer at a μm level distance in the redox couple solution.[9] On the other hand, the sudden current increase occurred at a nm distance, and its amplification reached 10^6 times. This phenomenon can be at least different in origin from the already known feedback current.

Second, we discuss the origin of the behavior in the short distance range. In order to analyze the behavior more in detail, the amplitude of the sudden current change (I_c) and the critical distance (D_c) are plotted *vs.* ΔV in Fig. 7(a) and (b), respectively. I_c linearly increased when the ΔV was increased in both of the 0.1 and 1 mM $Fe(CN)_6^{3-/4-}$ solutions as shown in Fig. 7(a). The I_c values of the 0.1 mM solution were 3 times higher than those of the 1 mM $Fe(CN)_6^{3-/4-}$ solution, which was the opposite trend of the conventional feedback current observed in the long D range. Concerning D_c, the current hysteresis was observed between the approach and the separation processes in both of the 0.1 and 1 mM $Fe(CN)_6^{3-/4-}$ solutions except for those at $\Delta V = 5$ mV in the 0.1 mM solution which exhibited the same D_c values in both the approach and separation processes (Fig. 7(b) and (c)). Such a current hysteresis was not found at 0 mV, which showed only a minor current increase (Fig. 5(c)). For the $KClO_4$ solution, there was no current hysteresis as shown in Fig. 6(b) and (c).

A possible mechanism of the sudden current change at the short D is proposed based on the simultaneously measured force profiles. The double layer repulsion forces observed at D_c in the 0.1 mM $Fe(CN)_6^{3-/4-}$ solution at 0, 5, and 50 mV were 0.49 ± 0.08 mN m^{-1}, 0.51 ± 0.08 mN m^{-1} and 0.49 ± 0.03 mN m^{-1}, respectively (see Fig. 2(a)–(c)). It is interesting that the double layer repulsion observed at D_c was nearly identical for all the cases. The double layer repulsion was produced by the osmotic pressure of the counterions in the electric double layer, thus the amplitude of the repulsion was determined by the concentration of the total confined electrolyte species. The sudden current increase could be correlated with the identical double layer repulsion at D_c, which could provide the threshold concentration of the redox species in the double layer. Therefore, we tried to evaluate the threshold concentration of the redox species at D_c using a previously reported model for calculating the confined ion concentration in the electric double layer.[4] This model can calculate the ion concentration between two electrodes (Fig. S8†) by integrating the ion distribution in the double layer estimated from the ψ_0 and $1/\kappa$ values obtained using the measured force profiles. The calculated concentration was 8.3 ± 0.3 mM (80 times more concentrated than the bulk value) for the 0.1 mM aqueous $Fe(CN)_6^{3-/4-}$ at 50 mV. The average distance between the redox molecules at this concentration was 2.2 nm (9.7 nm in the bulk solution). The threshold concentration for the 1 mM aqueous $Fe(CN)_6^{3-/4-}$ at 50 mV was calculated to be 18.5 ± 5.4 mM (18 times more concentrated than the bulk values). This value was similar to that for the 0.1 mM solution because the error at this concentration was high due to the short Debye length of 1.7 nm.

We may discuss how the redox reactions proceeded in such concentrated solutions of redox species confined in the nanospace, which resulted in the

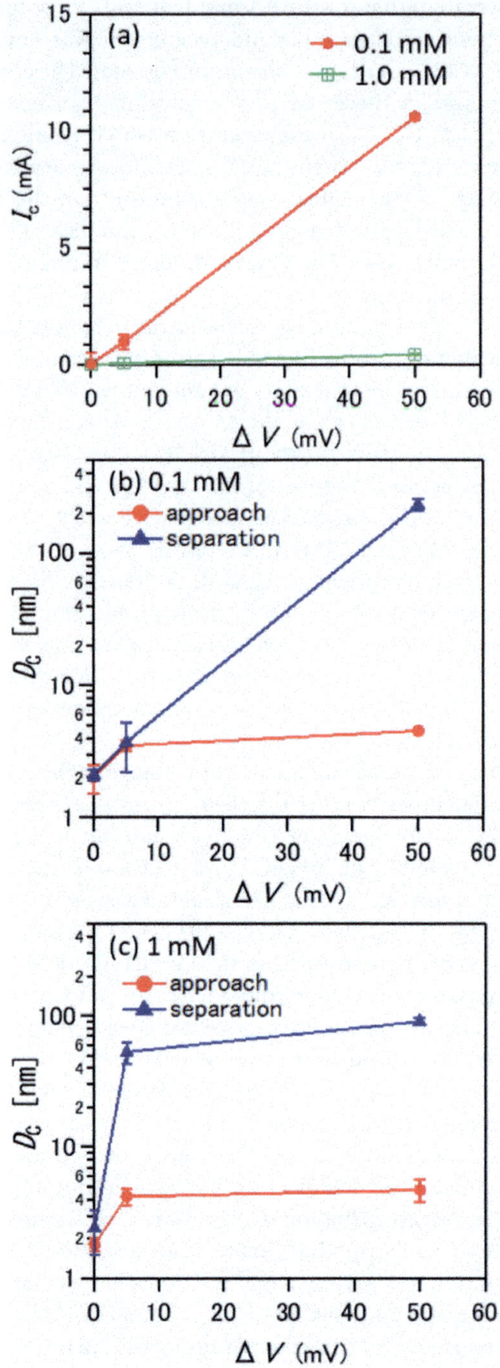

Fig. 7 Amplitude, I_c, (a) and distance range, D_c, (b and c) of the sudden current increases at a short distance depending on ΔV, obtained from the current–distance profiles in aqueous $Fe(CN)_6^{3-/4-}$.

sudden current increase at D_c. Such a long distance of the electron transfer current between the electrodes suggested that molecular organization of $Fe(CN)_6^{3-/4-}$ formed under confinement, which enhanced the collisions and/or electron transfer (or hopping) between the redox species. The current hysteresis, longer D_c upon separation, accompanied observation of adhesion force at $\Delta V = 50$ mV in the 0.1 mM $Fe(CN)_6^{3-/4-}$ solution and at $\Delta V = 5$ and 10 mV in the 1 mM $Fe(CN)_6^{3-/4-}$ solution. This supported the presence of assembled redox species under confinement.

4. Conclusion

We studied the electrochemical current between electrodes under nano-confinement in the presence of redox species using our EC-SFA. Our results are summarized as follows:

(1) Based on the current–distance profiles for two Pt electrodes in aqueous $Fe(CN)_6^{3-/4-}$, gradual current increases at a long distance and sudden current increase at a short distance (~5 nm on approach and ~200 nm on separation) were observed, the latter for the first time.

(2) The amplitude of the gradual current increases (~400 nA with 2000 nm D changes) at a long distance increased with the increasing $Fe(CN)_6^{3-/4-}$ concentration. In contrast, we did not find this current change in the case of the aqueous $KClO_4$. These results indicated that the current increase in the greater distance range was due to the conventional feedback current of the redox cycling.

(3) The amplitude of the sudden current increase at short distances decreased with the increasing $Fe(CN)_6^{3-/4-}$ concentration. The current suddenly increased at D_c which was longer for the greater ΔV. In addition, we found greater D_c values upon separation than those on approach when ΔV was high ($\Delta V = 50$ mV for 0.1 mM solution and $\Delta V \gtrsim 5$ mV for 1 mM solution). In the case of the aqueous $KClO_4$, the distance range of the sudden current change, D_c, was smaller than that of the $Fe(CN)_6^{3-/4-}$ solution with no dependence on ΔV. There was also no current hysteresis between the approaching and separating processes in the aqueous $KClO_4$. These results indicated that the presence of redox species strongly affect this sudden current change.

(4) The force profiles simultaneously obtained with the current–distance profiles showed that the same amplitude of the double layer repulsion was observed at D_c for all three ΔV values on approach, suggesting that the concentrated and/or assembled redox species induced the sudden current increase due to a chain reaction of electron transfer between the redox species. There was a threshold concentration of the redox species for this phenomenon.

These results revealed an insight into electrochemical reactions under nano-confinement. We believe that the EC-twin-path SFA we developed will become a powerful tool for obtaining fundamental knowledge about the nano-confined electrochemical reactions for nanoelectrodes which can be applied to sensors and energy devices.

Conflicts of interest

There are no conflicts to declare.

Acknowledgements

This work was supported by the CREST program of the Japan Science and Technology Agency (JST), the Regional Innovation Ecosystem Program (MEXT), and JSPS KAKENHI Grant Numbers/26248002.

References

1 C. Aliaga, C. S. Santos and S. Baldeli, *Phys. Chem. Chem. Phys.*, 2007, **9**, 3683–3700.
2 K. Ueno, M. Kasuya, M. Watanabe, M. Mizukami and K. Kurihara, *Phys. Chem. Chem. Phys.*, 2010, **12**, 4066–4071.
3 S. Ito, A. Unemoto, H. Ogawa, T. Tomai and I. Honma, *J. Power Sources*, 2012, **208**, 271–275.
4 M. Kasuya, Y. Sano, M. Kawashima and K. Kurihara, *Langmuir*, 2021, **37**, 5073–5080.
5 D. W. M. Arrigan, *Analyst*, 2004, **129**, 1157–1165.
6 S. Pennathur, J. C. T. Eijkel and A. van der Berg, *Lab Chip*, 2007, **7**, 1234–1237.
7 F.-R. F. Fan and A. J. Bard, *Science*, 1995, **267**, 871–874.
8 C. Mao, N. M. Contento, L. R. Gibson and P. W. Bohn, *ACS Nano*, 2013, **7**, 5483–5490.
9 A. J. Bard, J. A. Crayston, G. P. Kittlesen, T. V. Shea and M. S. Wrighton, *Anal. Chem.*, 1986, **58**, 2321–2331.
10 K. Ueno, M. Hayashida, J. Y. Ye and H. Misawa, *Electrochem. Commun.*, 2005, **7**, 161–165.
11 P. S. Singh, H. S. M. Chan, S. Kang and S. G. Lemay, *J. Am. Chem. Soc.*, 2011, **133**, 18289–18295.
12 S. Pennathur and J. G. Santiago, *Anal. Chem.*, 2005, **77**, 6782–6789.
13 A. Eden, K. Scida, N. Arroyo-Curras, J. C. T. Eijkel, C. D. Meinhart and S. Pennathur, *Electrochim. Acta*, 2020, **330**, 135275.
14 A. Schulte and W. Schumann, *Angew. Chem., Int. Ed.*, 2007, **46**, 8760–8777.
15 Y. Takahashi, A. I. Shevchuk, P. Novak, Y. Zhang, N. Ebejer, J. V. Macpherson, P. R. Unwin, A. J. Pollard, D. Roy, C. A. Clifford, H. Shiku, T. Matsue, D. Klenerman and Y. E. Korchev, *Angew. Chem., Int. Ed.*, 2011, **50**, 9638–9642.
16 J. V. Macpherson, P. R. Unwin, A. C. Hillier and A. J. Bard, *J. Am. Chem. Soc.*, 1996, **118**, 6445–6452.
17 A. Anne, E. Cambril, A. Chovin, C. Bemille and C. Goyer, *ACS Nano*, 2009, **3**, 2927–2940.
18 J. N. Israelachvili, *Intermolecular and Surface Forces*, 3rd edn, Academic Press, Ltd., New York, 2010.
19 R. G. Horn and J. N. Israelachvili, *J. Chem. Phys.*, 1981, **75**, 1400–1411.
20 J. N. Israelachvili, P. M. McGuiggan and A. M. Homola, *Science*, 1988, **240**, 189–191.
21 J. V. Alsten and S. Granick, *Phys. Rev. Lett.*, 1988, **61**, 2570–2573.
22 J. Klein and E. Kumacheva, *Science*, 1995, **269**, 816–818.
23 C. D. Dushkin and K. Kurihara, *Colloids Surf., A*, 1997, **129–130**, 131–139.
24 M. Mizukami and K. Kurihara, *Rev. Sci. Instrum.*, 2008, **79**, 113705.
25 H. Sakuma, K. Otsuki and K. Kurihara, *Phys. Rev. Lett.*, 2006, **96**, 046104.

26 S. Nakano, M. Mizukami, N. Ohta, N. Yagi, I. Hatta and K. Kurihara, *Jpn. J. Appl. Phys.*, 2013, **52**, 035002.
27 D. Fukushi, M. Kasuya, H. Sakuma and K. Kurihara, *Chem. Lett.*, 2011, **40**, 776–778.
28 T. Kamijo, M. Kasuya, M. Mizukami and K. Kurihara, *Chem. Lett.*, 2011, **40**, 674–675.
29 J. N. Connor and R. G. Horn, *Langmuir*, 2001, **17**, 7194–7197.
30 J. Fréchette and T. K. Vanderlick, *Langmuir*, 2001, **17**, 7620–7627.
31 H. Kawai, H. Sakuma, M. Mizukami, T. Abe, Y. Fukao, H. Tajima and K. Kurihara, *Rev. Sci. Instrum.*, 2008, **79**, 043701.
32 M. Kasuya, T. Sogawa, T. Masuda, T. Kamijo, K. Uosaki and K. Kurihara, *J. Phys. Chem. C*, 2016, **120**, 15986–15992.
33 S. Fujii, M. Kasuya and K. Kurihara, *J. Phys. Chem. C*, 2017, **121**, 26406–26413.
34 M. Kasuya and K. Kurihara, *Langmuir*, 2014, **30**, 7093–7097.
35 F.-R. F. Fan and A. J. Bard, *J. Am. Chem. Soc.*, 1987, **109**, 6262–6268.
36 M. Hegner, P. Wagner and G. Semenza, *Surf. Sci.*, 1993, **291**, 39–46.
37 D. Y. C. Chan, R. M. Pashley and L. R. A. White, *J. Colloid Interface Sci.*, 1980, **77**, 283–285.
38 One may wonder why a current was observed even at $\Delta V = 0$ mV. We supposed that it may be due to a tiny difference in the electrode potentials (<1 mV) between the two electrodes.

Faraday Discussions

ROYAL SOCIETY OF CHEMISTRY

PAPER

Nanocollision mediated electrochemical sensing of host–guest chemistry at a nanoelectrode surface†

Ruixia Wang,[a] Gongming Qian,[a] Jing Guo,[a] Qiushuang Ai,[c] Simin Liu,[b] Yichong Liu,*[a] Feng Liang*[b] and Shuai Chang*[a]

Received 16th August 2021, Accepted 5th October 2021

DOI: 10.1039/d1fd00054c

Electrochemical (EC) measurements of dynamic nanoparticle collisions on a support electrode provide a powerful approach to study the electrical properties of interfacial molecules self-assembled on the electrode surface. By introducing a special cage-shaped macrocyclic molecule, cucurbit[7]uril (CB7), onto a gold nanoelectrode surface, we show that the dynamic interactions between CB7 and the colliding nanoparticles can be real-time monitored *via* the appearance of distinct EC current switching signals. When a guest molecule is included in the CB7 cavity, the changed host–guest chemistry can be probed *via* the amplitude change of the EC current signals. In addition, different guest molecules can be recognized by CB7 on the nanoelectrode surface, giving rise to distinguishable current jump signals for different host–guest systems. Remarkably, two well-defined current states are observed in the EC measurements of the CB7–ferrocene complex, indicating two orientation geometries of ferrocene inside the CB7 cavity can be resolved in this EC sensing platform. This work demonstrates an effective approach for studying the dynamics of host–guest chemistry at the liquid–solid interface and sheds light on a convenient EC sensor for the recognition of target molecules with the aid of CB7.

Introduction

Nanocollision electrochemistry provides a powerful and convenient analysis method for investigating and characterizing the structural and dynamic behavior of single nanoparticles.[1] By performing nanoparticle (NP) impact experiments at

[a]*The State Key Laboratory of Refractories and Metallurgy, The Institute of Advanced Materials and Nanotechnology, Wuhan University of Science and Technology, Wuhan, Hubei, 430081, China. E-mail: schang23@wust.edu.cn*

[b]*The State Key Laboratory for Refractories and Metallurgy, School of Chemistry and Chemical Engineering, Wuhan University of Science and Technology, Wuhan 430081, China*

[c]*Institute for Quality & Safety and Standards of Agricultural Products Research, Jiangxi Academy of Agricultural Sciences, Nanchang, Jiangxi 330200, China*

† Electronic supplementary information (ESI) available. See DOI: 10.1039/d1fd00054c

a metal electrode, the intrinsic properties of the NPs and their catalytic performance can be revealed through analyzing the EC current response upon collisions.[2–6] When the electrode is chemically functionalized, the surface chemistry on the electrode can be sensitively probed in each NP collision event and the electrical properties of the surface molecule can be traced from its behavior in affecting the electron transfer between the NP and electrode.[7–9] A few types of self-assembly chemistry have been employed to functionalize the electrode, and the effect of the molecular properties, such as length, terminal groups and the molecular conductance, on the EC current features (*e.g.*, peak current or switching frequency) are unambiguously uncovered both experimentally[10–14] and theoretically,[15,16] rendering the NP collision method a sensitive means of probing surface chemistry.

Here we use this technique to study a new type of surface chemistry, CB7-based host–guest chemistry that is self-assembled on a gold nanoelectrode (GNE) surface. CB7 is a barrel-shaped macrocyclic molecule with an inner cavity and two identical open portals fringed with 7 carbonyl groups.[17] It is well known for its prominent ability to facilitate noncovalent binding with a large variety of targets and form stable host–guest geometry. In our previous study,[18–20] we have shown that CB7 can be conveniently modified onto the gold surface through collective interactions between gold and the 7 carbonyl groups. When a guest molecule is encapsulated by the CB7 cavity, it can be confined to the gold surface and characterized at single-molecule scale using scanning tunneling microscopy (STM) break junction[19] and fixed junction techniques[18,21] or surface enhanced Raman spectroscopy.[20] In this work, we employ the same setup by modifying the host–guest chemistry onto a GNE surface and explore its EC current response on collisions with gold nanoparticles (GNP). By introducing two model guest molecules, ferrocene (FC) and aminoferrocene (AFC), into the system, we show that distinctive electronic features of different host–guest chemistry on a GNE surface can be revealed in this platform.

Methods

Chemicals

CB7 and gold nanoparticles were prepared by following reported procedure.[22,23] Gold wire, potassium ferrocyanide, potassium chloride, potassium dihydrogen phosphate, dibasic sodium phosphate, and absolute ethanol were purchased from Sinopharm Chemical Reagent Co., Ltd. All of the purchased chemicals were used directly without further purification. Deionized water (18.2 MΩ) from a Milli-Q water purifying system was used to prepare all of the solutions.

GNE fabrication and modification

CB7 solution was prepared by dissolving CB7 powders into DI water to a final concentration of 0.5 mM. The host–guest complex solution was prepared by mixing the host (CB7) and guest molecule (ferrocene (FC) and aminoferrocene (AFC)) at 1 : 2 stoichiometry in DI water, followed by a treatment of sonication for ~30 minutes. The concentration of the complexed host–guest solution was 0.5 mM.

The fabrication of GNEs was derived from previous reports.[20] First, a gold wire (Electron Microscopy Sciences, 99.99%) with a 0.2 mm diameter was electrochemically etched and cleaned by immersion in piranha solution ($H_2SO_4 : H_2O_2 = 3 : 1$, by volume), rinsed with copious amounts of deionized water, and blown dry with argon. Second, the gold tip was insulated with high-density polyethylene (HDPE) to expose only the tip apex.

Fresh GNEs were immersed in host or host–guest molecule solution overnight at room temperature. The modified GNEs were washed with copious amounts of ethanol and deionized water, dried with argon gas, and used immediately.

I–t measurements

Electrochemical current measurements were achieved using the current recording instrument Axon Digidata 1440B. An Axon 200B patch-clamp amplifier (Molecular Devices Inc., CA) in the voltage-clamp mode was used to supply the bias and amplify the current. A 1 kHz Bessel low-pass filter and 10 kHz sampling rate were typically used for current measurements. The GNE was used as the working electrode. The Ag/AgCl electrode was grounded and used as the quasi-reference electrode. The GNE was placed in a 5 mM phosphate buffer solution (PBS, pH = 7.4) with 3 mM potassium ferrocyanide. The concentration of GNPs was 30 pM. All of the measurements were performed in a Faraday cage at room temperature. All the curves were collected at 800 mV in 5×10^{-3} M phosphate-buffered saline (PBS) with pH = 7.4. The redox mediator ($Fe(CN)_6{}^{4-}$) concentration was 3×10^{-3} M.

Results and discussion

Fig. 1A shows a schematic layout of the NP-impact measurement on a GNE surface. A gold wire probe, etched to a sharp tip and insulated to expose only the tip apex, was used in an electrochemical setup as described elsewhere.[11,20] The exposed apex has a typical radius of ~300 nm (inset of Fig. 1A) and it was functionalized with CB7 (see CV characterization in Fig. S1†). The measurements were carried out in 5 mM phosphate buffer (pH = 7.4) containing 3 mM potassium ferrocyanide. The potential of the GNE was controlled at a voltage $V = 800$ mV with respect to a Ag/AgCl reference and the electrochemical current *versus* time was recorded with an Axon Digidata 1440B. Fig. 1C shows typical current–time traces collected before and after adding gold nanoparticles in the solution. In the blank control experiment without GNPs, no current features were observed beyond the shot noise and system electronic noise (trace I). When 30 pM GNPs with a diameter of 40 nm are added, distinctive current jumps occur, indicating the occurrence of electrochemical reactions at the GNE surface caused by the GNP collisions (trace II).[24]

Fig. 1B depicts the mechanism of the electrochemical reaction process. The CB7 molecules were modified on the GNE with an open portal consisting of seven carbonyl groups.[18] When a freely moving GNP collides with the molecule, a metal–molecule–metal junction is transiently formed and electron tunneling occurs from the GNP to the electrode,[25,26] giving rise to the current pulses. It is worth noting that a high positive potential applied to the GNE is generally required for the current signals to be observed in our system. We have performed the

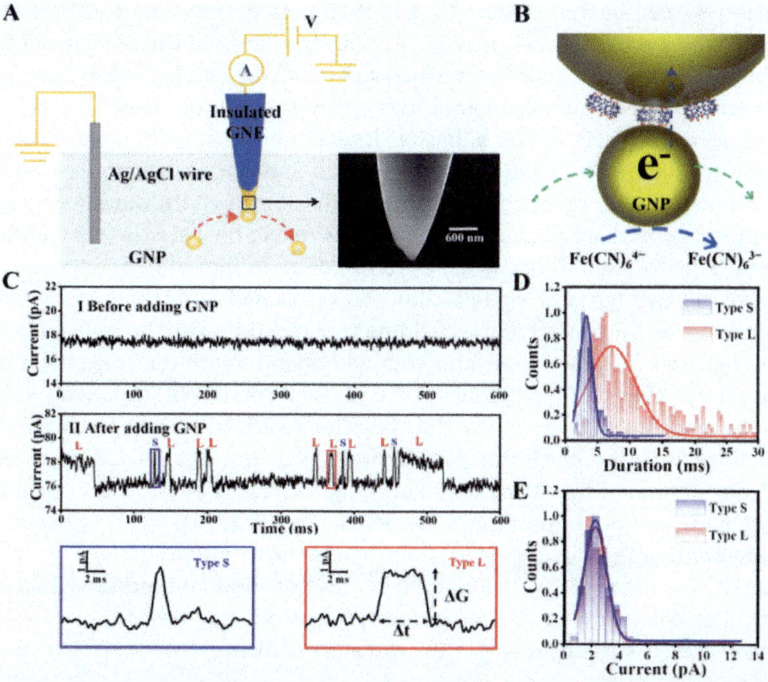

Fig. 1 (A) Schematic layout of the NP-impact measurement on a GNE. The inset to the right shows the SEM image of a GNE apex (non-insulated) with a tip radius of about 300 nm. (B) A schematic representation to show the mechanism of the electrochemical reaction process. (C) Typical current–time traces collected from electrochemical current measurement at CB7 modified GNEs before (I) and after (II) adding GNPs. Expanded views of two types of signals are shown in the lower panel. The first type is a fast event (in blue frame) and the second type of signal lasts much longer, with a current plateau shown (in red frame). (D) Statistical distribution of the lifetime of the two types of current signals shown in (C). (E) Statistical distribution of the peak current amplitude of the two types of current signals. The histograms are built from 1000 events. The solid lines are the Gaussian fittings to the histograms. All the curves were collected at 800 mV in 5 mM phosphate buffer (pH = 7.4) with 3 mM redox mediator ($Fe(CN)_6^{4-}$).

experiments at different applied bias, finding that no current switching events can be detected below a threshold bias of 500 mV (see details in Fig. S2†). This might be explained by the fact that the electron rich carbonyl portal of CB7 exposed to the solution could exert an electrostatic repulsion to the negatively charged NPs, which suppresses the NP collisions onto the GNE surface. While an insufficient redox turnover on the NP surface may also occur at low potential bias.

Two types of signals are obvious in each typical run of current–time trace. The first type is a fast event (marked "S" in Fig. 1C), where the current jumps up to a certain point and declines immediately. While a second type of signal lasts much longer, with a current plateau appearing on top of the signal (marked "L"). The zoomed in plots of the two types of events are displayed in the lower panel of Fig. 1C. The amplitudes of the current jumps (ΔG) above the baseline current are similar between the two types of events, while their lifetimes (Δt) are quite different. Statistical histograms of combined events show a Δt of ~3 ms for type S event and ~7 ms for type L (Fig. 1D), and their current amplitudes ΔG are about identical at ~2 pA, as shown in Fig. 1E. Similar results of the two types of NP-impact signals are also reported for measurements at Au electrodes functionalized with 11-mercaptoundecanoic acid,[24] where the authors attribute the long-lived "plateau" signals to the adhesion interaction arising from the hydrogen bonding between the –COOH functional group and the Au electrode, while the short "spike" signals reflect cases with no adhesion force. In our situation, the same process could apply. The type "S" signals reflect the "elastic" NP collisions, while the type "L" signals imply the existence of molecule–gold interactions. The origin of the two types of signals could be associated with the heterogeneous nanoscale surface geometry at the colliding spot together with the local molecular conditions. CB7 is known to be able to bind GNPs through the collective contributions from seven carbonyl groups.[27–29] In a given condition, the colliding GNPs could be transiently trapped on the molecular surface *via* the carbonyl–gold interactions and their dwell time is thus prolonged. Some GNPs, although fewer, could stay permanently on the GNE surface, as revealed by the SEM image performed after the NP-impact measurements (Fig. S1F†). It also suggests that a strong binding interaction could occur between CB7 and GNPs.

One unique asset of CB7 is that it facilitates noncovalent binding with a large variety of guests by molecular recognition. By introducing target molecules that form complexed structures with CB7, the surface chemistry at the GNE can be easily tuned and the electrical properties of the targets of interest could be potentially probed in this facile NP-impact platform. As a first step in exploring this, we tested two model guest molecules FC and AFC, which are known to bind CB7 with a strong affinity.[18] Fig. 2A shows the complexed structures of the two guests included in the CB7 cavity. They are prepared by mixing the guest molecules into CB7 solution to form the host–guest complex, followed by the GNE deposition. Following the same procedure, the electrochemical measurements of NP collisions at CB–FC and CB–AFC modified GNE are performed at an applied bias of 800 mV, and similar current fluctuation signals are frequently seen in the typical current–time traces, as shown in Fig. 2B. Two types of signals, short spikes and long-lived events (marked as S and L) are also observed for these complexed systems, as observed in the case of CB7 alone. The statistical lifetime for both S and L types of signals recorded for both CB7–FC and CB7–AFC are about identical to the results of CB7 (Fig. S3†), indicating the same origin of the current pulses involving the collisions between CB7 and GNPs. However, the peak current amplitude is quite different for measurements on different surface chemistry. We have separately analyzed the S and L types of signals for CB7, CB7–FC and CB7–AFC, with their results summarized in Fig. 2C and D. For both types of signals, the peak current amplitude shows a clear increase upon addition of guest molecules. In addition, different host–guest chemistry gives rise to distinguishable current features, as shown in the well-separated current amplitude histograms between CB–FC and CB–AFC.

It has been reported that the peak current induced by NP collisions is influenced by three sequential factors, the mass transport of redox species toward the GNP, the electron transfer from the redox molecule to GNP and electron tunneling through the molecular junction (illustrated in Fig. 1B).[15,30] In our previous study of the amine-terminated molecules in the same setup, we find that the amplitude of the current jump signals is determined by the electron transfer

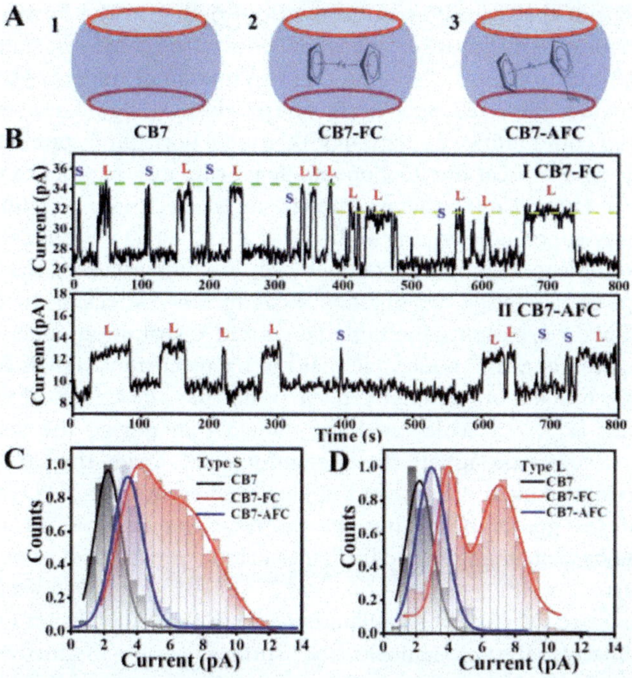

Fig. 2 EC measurements of CB7 inclusion complexes. (A) Illustration images of different host–guest complexes including CB7, CB7–FC and CB7–AFC. (B) Typical current–time traces obtained from the EC measurements of NP collisions at CB–FC and CB–AFC modified GNE at an applied bias of 800 mV. Statistical histograms of the peak current amplitude of the type S (C) and type L (D) signals for three different molecular systems. The histograms are constructed from 1000 events.

kinetics occurring at the GNP–solution interface and through the molecular junctions.[11] The electron transfer from the redox molecule to the GNP is limited by the electrochemical resistance R_{EC} and the electron transfer behavior through the molecular layer is associated with the metal–molecule–metal junction conductance. In the current study, R_{EC} at the GNP surface is expectedly unchanged in the same EC setup for measurements of three molecular systems, thus the measured peak current change between different surface chemistries should be reasonably explained by the different conductance of the molecules binding between two gold surfaces. In Fig. 2C and D, we see that both CB[7]–FC (red histograms) and CB7–AFC complexes (histograms in blue) give a higher current amplitude distribution than that of CB7, suggesting the existence of a more efficient conducting channel involving FC and AFC in the tunneling path. In addition, the histogrammed peak current amplitude for CB7–FC is obviously larger than that of CB–AFC, indicating a larger conductance for CB–FC in the tunneling junction. This conductance trend is consistent with the results of our single-molecule conductance measurements of the same molecular systems between two gold electrodes.[19] Surprisingly, the current signals for CB7–FC show a bimodal distribution (also see the typical trace in Fig. 2B), indicating the observation of two conductance states of CB7–FC in the tunneling junction. This is reminiscent of the two different geometries existing in the formation of the CB7–FC complex. In our previous measurements of the same system using electrical and optical methods,[19,20] we find that two rotational states of FC inside the CB7 cavity could be detected. Specifically, the proximity of a gold nanoelectrode to the CB7 portal could induce a rotation of FC from a horizontal state (FC lying flat in CB cavity) to a vertical one (cyclopentadienyl ring axis parallel to CB7 portal axis) due to a favored total energy. The enhanced electronic coupling between gold and the cyclopentadienyl ring in the vertical CB7–FC state could significantly increase the molecular junction conductance, in comparison to the horizontal state. This gives rise to the two distinctive conductance states of the complexed system, and likely explaining the bimodal current distribution obtained here in the EC measurement. It is noticeable that the L type current signals for CB7–FC (Fig. 2D) show much more well-defined two peaks than the statistical result of the S type signals (Fig. 2C). This can be explained that during the fast "elastic" collision process (S type signal), the orientation of FC inside the CB7 cavity may not be as stabilized as in the case of the long-lived events where FC could stably interact with the gold surface, thus the two rotational states of FC in the CB7 cavity cannot be well distinguished from the current amplitude distribution of the S type signals.

The distinctive EC current signals arising from the NP collision on different surface chemistry at the gold nanoelectrode surface reveals a distinctive feature of the surface molecules. It also sheds light on an effective approach for molecular recognition based on the dynamic interaction property between CB7 and the guest molecules, which could possibly provide an EC sensor toward targets in ambient conditions. To prove the principle of such an EC sensing platform, we place a CB7-modified GNE in a testing solution (5 mM phosphate buffer and 3 mM potassium ferrocyanide, pH = 7.4) and monitor the EC current signals in the presence of GNPs. Fig. 3 shows the time resolved distribution of EC current signals statistically analyzed every five minutes (see Fig. S4† for detailed analysis). In the first 20 minutes, no guest molecules are added and the switching current signals have a stable peak amplitude around 2 pA (see the scatter points located in the grey shaded area in Fig. 3, the error bars are ±1 SD of the amplitude distribution of current jump signals recorded every five minutes). At 20 min, 1 mM FC was added into the solution, the current signals show an immediate increase in the peak current amplitude (at 25 min) and two distinctive current features with well-separated peak values appear at 30 min (see data points in the red shaded area). The values of the two current states are consistent with the results of the NP-impact measurements of CB7–FC, indicating that a new surface chemistry of CB7–FC is formed spontaneously (and instantly) on the gold nanoelectrode surface. The signal amplitudes of the two current states stay unchanged for more than 10 minutes until a new guest molecule AFC was added at 40 min. From this point, two current peaks are detected at the beginning (at 45 min), which change to a single current peak (statistical points in blue shaded area) with the statistical value identical to the amplitude of the measured EC current signals of CB7–AFC. Because AFC has a much larger binding affinity toward CB7 than FC, the changed current signal after adding AFC is indicative of a dynamically changed surface chemistry where FC is replaced by AFC in the CB7 cavity. The instant detection and selective recognition of different guest molecules in this CB7 mediated system exemplify a remarkable sensing technology, establishing a versatile

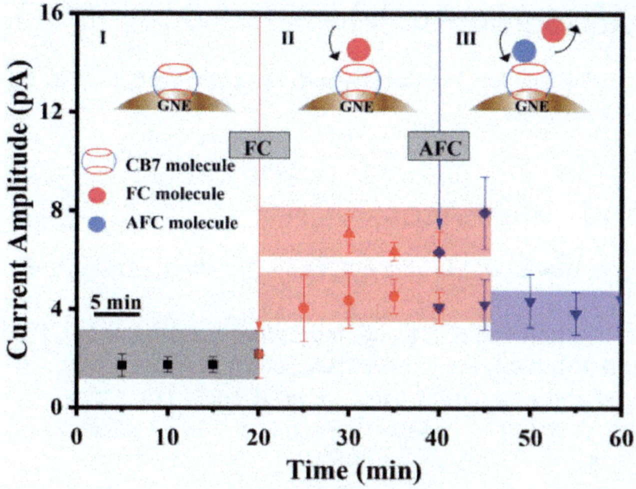

Fig. 3 The recognition of different guest molecules. Statistical results of the amplitude of collected EC current signals recorded and analyzed every 5 minutes in a testing solution (5 mM phosphate buffer and 3 mM potassium ferrocyanide, pH = 7.4). 1 mM FC is added at 20 min to the solution and 1 mM AFC is added to the same solution at 40 min.

electronic technique of analytical, mechanistic, and potentially pharmacological interest that could be further extended to screen important biomacromolecules, such as peptides and proteins.

Conclusions

In conclusion, we have shown that NP collisions at a CB7-modified GNE surface can generate two types of current signals (short spikes and long-lived current jump events) which arise from the different interactions between CB7 and GNPs. By introducing guest molecules into the CB7 cavity, the electrical properties of the complexed host–guest chemistry can be sensitively probed from the NP-impact measurements, and the generated EC current signals show clear increase in the peak current amplitude, indicating effective tunneling occurs through the guest molecule in the junction between GNP and GNE. Remarkably, measurements of different guest molecules (FC and AFC) included in the CB7 cavity show different peak current values, attributed to their different molecular conductance in the tunneling junction. This enables an EC sensor which could distinguish different molecules at ambient conditions. More interestingly, we observed two distinctive peak current states in EC measurement of CB7–FC, which is likely associated with the two different orientations of FC inside the CB7 cavity, as suggested by our single molecule study of the host–guest chemistry using both electrical and optical methods. This work demonstrates a feasible method for studying the host–guest chemistry at the solid–liquid interface and might pave the way toward a novel EC sensor for the next generation.

Conflicts of interest

There are no conflicts to declare.

Acknowledgements

The authors acknowledge helpful discussions with Prof. Jin He from Florida International University.

Notes and references

1 Z. Sun, M. E. Hafez, W. Ma and Y.-T. Long, *Sci. China: Chem.*, 2019, **62**, 1588–1600.

2 A. J. Bard, H. Zhou and S. J. Kwon, *Isr. J. Chem.*, 2010, **50**, 267–276.

3 X. Xiao and A. J. Bard, *J. Am. Chem. Soc.*, 2007, **129**, 9610–9612.

4 R. Tel-Vered and A. J. Bard, *J. Phys. Chem. B*, 2006, **110**, 25279–25287.

5 M. E. Hafez, H. Ma, Y.-Y. Peng, W. Ma and Y.-T. Long, *J. Phys. Chem. Lett.*, 2019, **10**, 3276–3281.

6 Y.-Y. Peng, D. Guo, W. Ma and Y.-T. Long, *ChemElectroChem*, 2018, **5**, 2982–2985.

7 X. Xiao, S. Pan, J. S. Jang, F.-R. F. Fan and A. J. Bard, *J. Phys. Chem. C*, 2009, **113**, 14978–14982.

8 J. Kim, B.-K. Kim, S. K. Cho and A. J. Bard, *J. Am. Chem. Soc.*, 2014, **136**, 8173–8176.

9 A. D. Castañeda, T. M. Alligrant, J. A. Loussaert and R. M. Crooks, *Langmuir*, 2015, **31**, 876–885.

10 C.-H. Chen, E. R. Ravenhill, D. Momotenko, Y.-R. Kim, S. C. Lai and P. R. Unwin, *Langmuir*, 2015, **31**, 11932–11942.

11 J. Guo, J. Pan, S. Chang, X. Wang, N. Kong, W. Yang and J. He, *Small*, 2018, **14**, e1704164.

12 J. Zhao, C. R. Bradbury, S. Huclova, I. Potapova, M. Carrara and D. J. Fermín, *J. Phys. Chem. B*, 2005, **109**, 22985–22994.

13 J. B. Shein, L. M. Lai, P. K. Eggers, M. N. Paddon-Row and J. J. Gooding, *Langmuir*, 2009, **25**, 11121–11128.

14 B. Zhang, L. Fan, H. Zhong, Y. Liu and S. Chen, *J. Am. Chem. Soc.*, 2013, **135**, 10073–10080.

15 J.-N. Chazalviel and P. Allongue, *J. Am. Chem. Soc.*, 2011, **133**, 762–764.

16 B. Zhang, L. Fan, H. Zhong, *et al.*, *J. Am. Chem. Soc.*, 2013, **135**(27), 10073–10080.

17 K. I. Assaf and W. M. Nau, *Chem. Soc. Rev.*, 2015, **44**, 394–418.

18 B. Xiao, F. Liang, S. Liu, J. Im, Y. Li, J. Liu, B. Zhang, J. Zhou, J. He and S. Chang, *Nanotechnology*, 2018, **29**, 365501.

19 M. Huang, M. Sun, X. Yu, S. He, S. Liu, W. M. Nau, Y. Li, T. Wu, Y. Wang, S. Chang and J. He, *J. Phys. Chem. C*, 2020, **124**, 16143–16148.

20 Q. Ai, J. Zhou, J. Guo, P. Pandey, S. Liu, Q. Fu, Y. Liu, C. Deng, S. Chang, F. Liang and J. He, *Nanoscale*, 2020, **12**, 17103–17112.

21 A. Day, A. P. Arnold, R. J. Blanch, *et al.*, *J. Org. Chem.*, 2001, **66**(24), 8094–8100.

22 A. Day, A. P. Arnold, R. J. Blanch and B. Snushall, *J. Org. Chem.*, 2001, **66**, 8094–8100.

23 Q. Ai, L. Jin, Z. Gong and F. Liang, *ACS Omega*, 2020, **5**, 10581–10585.

24 C. H. Chen, E. R. Ravenhill, D. Momotenko, Y. R. Kim, S. C. Lai and P. R. Unwin, *Langmuir*, 2015, **31**, 11932–11942.

25 J. Kim, B. K. Kim, S. K. Cho and A. J. Bard, *J. Am. Chem. Soc.*, 2014, **136**, 8173–8176.
26 J. N. Chazalviel and P. Allongue, *J. Am. Chem. Soc.*, 2011, **133**, 762–764.
27 S. T. Jones, R. W. Taylor, R. Esteban, E. K. Abo-Hamed, P. H. Bomans, N. A. Sommerdijk, J. Aizpurua, J. J. Baumberg and O. A. Scherman, *Small*, 2014, **10**, 4298–4303.
28 C. A. Tao, Q. An, W. Zhu, H. Yang, W. Li, C. Lin, D. Xu and G. Li, *Chem. Commun.*, 2011, **47**, 9867–9869.
29 R. W. Taylor, T. C. Lee, O. A. Scherman, R. Esteban, J. Aizpurua, F. M. Huang, J. J. Baumberg and S. Mahajan, *ACS Nano*, 2011, **5**, 3878–3887.
30 C. M. Hill, J. Kim and A. J. Bard, *J. Am. Chem. Soc.*, 2015, **137**, 11321–11326.

Faraday Discussions

ROYAL SOCIETY OF CHEMISTRY

PAPER

Engineering PtCu nanoparticles for a highly efficient methanol electro-oxidation reaction†

Pengfei Yao,[ab] Jing Cao,[ab] Mingbo Ruan,[a] Ping Song,[a] Xue Gong,[a] Ce Han[a] and Weilin Xu *[ab]

Received 1st August 2021, Accepted 31st August 2021
DOI: 10.1039/d1fd00047k

Achieving a highly efficient and durable methanol electro-oxidation catalyst in acid media is critical for the practical utilization of direct methanol fuel cells (DMFCs) at the commercial scale. Herein, we report a facile and effective one-pot strategy for the synthesis of carbon-supported PtCu alloy nanoparticles (PtCu NPs) with a Pt-rich surface, small particle size and uniform dispersion. The as-prepared PtCu NPs with the optimal alloy composition (Pt_2Cu) exhibit a significantly improved electrochemical methanol oxidation reaction performance in terms of a high activity, superior CO tolerance and remarkable durability, in contrast to those of commercial Pt/C catalysts in acid media. Particularly, the Pt_2Cu/C catalyst exerts a 4.5 times enhancement in the mass activity and a larger I_f/I_b value compared to those of commercial Pt/C (Pt/C_{comm}). The enhanced catalytic activities can be ascribed to the high utilization of Pt and the high index facets of the surface. Also, the addition of Cu downshifts the d-band center of Pt and improves the CO tolerance during the methanol oxidation reaction process. This work provides an efficient strategy for designing desired Pt-based alloys for various catalytic reactions.

1. Introduction

Direct methanol fuel cells (DMFCs), being promising next-generation electronic devices thanks to their high energy density, convenient fuel storage and transportation and low environmental pollution,[1–4] have become of utmost importance all over the world in recent years. Nevertheless, there still exists some performance-hindering critical drawbacks impeding the widespread commercialization of DMFC technology,[5–7] including the sluggish anode reaction kinetics

[a]*State Key Laboratory of Electroanalytical Chemistry, Jilin Province Key Laboratory of Low Carbon Chemical Power, Changchun Institute of Applied Chemistry, Chinese Academy of Sciences, 5625 Renmin Street, Changchun, Jilin, 130022, P. R. China. E-mail: weilinxu@ciac.ac.cn*

[b]*University of Science and Technology of China, Hefei, Anhui, 230026, P. R. China*

† Electronic supplementary information (ESI) available: Full experimental details, analysis method description and supplementary figures. See DOI: 10.1039/d1fd00047k

and the poisoning caused by carbonaceous intermediates (*e.g.*, CH_3O, CH_2O, CHO and CO_{ads}). A scheme of the overall reaction process occurring at the anode in a DMFC equipped with a proton conducting electrolyte is outlined below:

$$CH_3OH + H_2O \rightarrow CO_2 + 6H^+ + 6e^-,\ E^0 = 0.016\ \text{V}\ (vs.\ \text{SHE})$$

By far, Pt has been recognized as the most active metal-based electrocatalyst for the MOR at the anode.[8–10] However, its scarcity and high cost necessitate the development of catalytic systems with reduced Pt loadings, improved activity and durability.[11–13] Theoretical and experimental investigations have also demonstrated that the CO intermediate tends to adsorb on the surface of Pt catalysts, then it takes up the MOR active sites and causes the dramatic poisoning/deactivation of the catalysts.[14,15]

Considering the above issues, alloying Pt with less expensive 3d-transition metals (M), such as Ni,[16–18] Co,[19–21] Cu,[22–24] Fe,[25,26] and so on, to form PtM alloys is an effective way to improve the utilization efficiency of Pt and regulate the MOR performance (including the activity, durability, CO resistance, *etc.*). By partially replacing Pt atoms, the alloying components not only reduce the usage of Pt, but also boost the catalytic activity through synergetic, geometric or electronic effects, which originate from the rearrangement of platinum atoms and the shift of the d-band center of Pt concerning the Fermi level in the bimetallic structures.[27–31] Accordingly, substantial research efforts have been devoted to the controlled synthesis of PtM alloys with advanced MOR performance. Among them, PtCu catalysts have attracted much attention due to their lower cost, enhanced activity and high anti-poisoning ability. The introduced Cu can considerably decrease the adsorption energy of intermediate species on platinum and reduce the oxidation potential of carbon monoxide.[32–34] Furthermore, PtCu NPs with various morphologies, such as cubes, cages, dendrites, wires, octahedrons, frames and mesoporous structures,[35–41] have been synthesized to modulate the exposed crystal planes and atomic arrangements and, hence, to influence their performance for the MOR. However, it is noteworthy that the oxidation and dissolution of the Cu in PtCu nanostructures in acid media under corrosive MOR conditions are great challenges.[42,43] In this respect, nanoparticles with core–shell structures, composed of a copper-rich core coated with a thin platinum-rich shell, are a promising platform that offers both design flexibility and reduced Pt loadings, and prevents the dissolution of the transition metals.[31,44,45]

We present here a simple one-pot synthetic strategy to fabricate PtCu alloy nanoparticles with a Pt-rich surface, which are loaded *in situ* on a commercial XC-72 carbon support. The as-prepared PtCu NPs with an optimal Pt/Cu molar ratio of 2/1 (named Pt_2Cu/C) exhibit the best catalytic performance toward the MOR under acidic conditions. The obtained Pt_2Cu/C shows a mass activity of 4.1 A mg_{Pt}^{-1} at 0.7 V, a 4.5-fold enhancement relative to that of the commercial Pt/C (named Pt/C_{comm}). The negative shift of the XPS peak of Pt predicts weaker adsorption energies of the intermediates on Pt_2Cu/C. Furthermore, the high peak current ratio of the forward scan (I_f) to the backward scan (I_b), I_f/I_b, of the Pt_2Cu/C indicates its strong resistance to the poisoning by carbonaceous species. At the same time, the durability of Pt_2Cu/C is substantially improved.

2. Results and discussion

2.1. Characterization

The PtCu alloy NPs with a Pt-rich surface were synthesized by the contemporaneous reduction of $CuCl_2 \cdot 2H_2O$ and $H_2PtCl_6 \cdot 6H_2O$ with OAm as the solvent, $Mo(CO)_6$ as the reducing agent and CTAB as the surfactant. With the same Pt loadings, the PtCu nanoparticles with different molar ratios of Pt/Cu were synthesized by adjusting the amount of $CuCl_2 \cdot 2H_2O$ and named Pt_3Cu/C, Pt_2Cu/C and PtCu/C on the basis of the average molar ratio of Pt/Cu, determined by inductively coupled plasma atomic emission spectroscopy (ICP-AES).

The morphology and structure of the as-synthesized PtCu NPs with the optimum molar ratio of Pt/Cu (Pt_2Cu/C) were characterized by transmission electron microscopy (TEM). As shown in Fig. 1a, the Pt_2Cu NPs, with an average diameter of 6.4 nm, are dispersed on the surface of the carbon support uniformly. Meanwhile, the particle size (with an average of 9.8 nm) of PtCu/C is much larger than that of Pt_2Cu/C and the sizes of the NPs in Pt_3Cu/C vary in a much wider range (Fig. S1†). The larger and heterogeneous nanoparticles are not beneficial for the efficient utilization of Pt and modulation of the catalyst

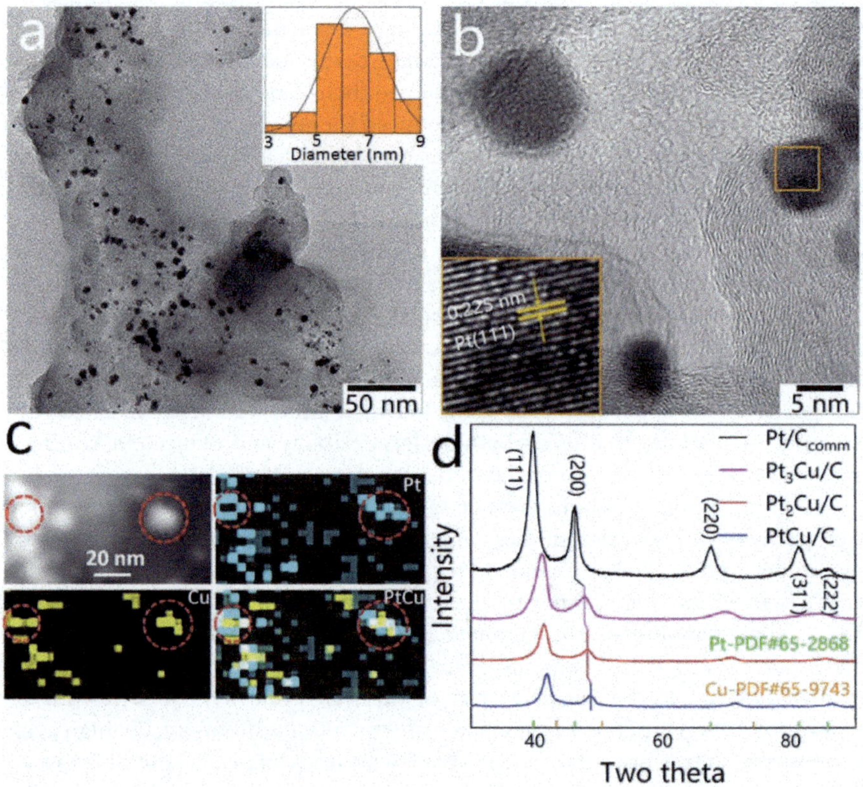

Fig. 1 (a) TEM image and (b) HRTEM image of Pt_2Cu/C; (c) HAADF-SEM image and the corresponding elemental mapping images of Pt_2Cu/C; (d) XRD patterns of Pt/C_{comm}, Pt_3Cu/C, Pt_2Cu/C and PtCu/C.

structure, respectively. The high-resolution TEM (HRTEM) image (Fig. 1b) shows well-resolved lattice fringes with an interplanar spacing of 0.225 nm, implying that the NPs are enclosed by (111) facets of Pt.[46] The high-angle annular dark-field scanning TEM (HAADF-STEM) image (Fig. 1c) shows that the Cu atoms are mainly distributed in the centers of the nanoparticles, while the Pt atoms are mainly on their surface skin, indicating a core/shell nanostructure with a Pt skin.[47] The powder XRD patterns (Fig. 1d) of these NPs display the typical face-centered cubic (fcc) structure associated with the alloyed PtCu nanomaterials. The typical peaks can be indexed to the (111), (200), (220), (311) and (222) planes of the fcc PtCu alloy. It is obvious that the platinum peaks of Pt_3Cu/C, Pt_2Cu/C and PtCu/C shift gradually to higher 2θ values with the addition of more and more Cu. Significantly, no diffraction peak is observed here for pure Pt, Cu or CuO, suggesting the dissolution of Cu atoms into the Pt lattice and the successful fabrication of PtCu alloy nanostructures in Pt_3Cu/C, Pt_2Cu/C and PtCu/C.

In addition, the electronic structure and surface chemical states of the PtCu NPs with different molar ratios were characterized with X-ray photoelectron spectroscopy (XPS). The XPS patterns reveal that both Pt (Fig. 2a) and Cu (Fig. 2b) are mainly in their metallic states. The Pt 4f spectrum can be deconvoluted into the doublets of $4f_{7/2}$ and $4f_{5/2}$ and each doublet consists of Pt^0 and Pt^{2+} species. The peak at 71.0 eV corresponds to metallic Pt $4f_{7/2}$. With the addition of more and more Cu, the binding energy of metallic Pt $4f_{7/2}$ shifts from 71.0 eV to 70.8 eV (Pt_3Cu/C), 70.5 eV (Pt_2Cu/C) and 70.3 eV (PtCu/C), respectively, indicating that Cu has a remarkable influence on the electronic structure of Pt in the alloy. The downshifted binding energy of Pt indicates that Cu donates electrons to Pt, leading to a substantial increase in the electron density around the Pt sites and a downshift of the d-band center of Pt.[22,30,48] Theoretically, a lower d-band center is usually related to a weaker adsorbate binding to the metal atoms and improved CO tolerance during the MOR.[49] On the other hand, as shown in Fig. 2b, the Cu 2p spectra can be fitted to pure metallic Cu very well, indicating firmly that the Cu atoms are mainly distributed in the cores of the PtCu alloy covered by a thin Pt skin, otherwise the surface Cu atoms would tend to be oxidized and then detected by XPS.

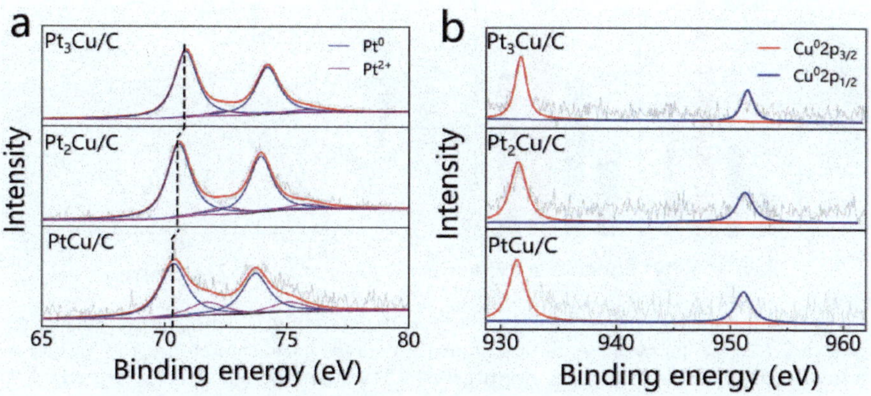

Fig. 2 XPS spectra: (a) Pt 4f and (b) Cu 2p spectra of PtCu/C, Pt_2Cu/C and Pt_3Cu/C.

2.2. Electrochemical measurements

Fig. 3a shows the cyclic voltammograms (CVs) of Pt_3Cu/C, Pt_2Cu/C, PtCu/C and Pt/C_{comm} in 0.1 M $HClO_4$ aqueous solution. The electrochemical surface area (ECSA) was evaluated for each catalyst by integrating the coulombic charge for hydrogen desorption, assuming that the adsorbed hydrogen monolayer on the metal surface is 210 μC cm^{-2}.[50] After double-layer charging correction, the specific ECSAs of Pt_3Cu/C, Pt_2Cu/C, PtCu/C and Pt/C_{comm} are estimated to be 71.43 m^2 g_{Pt}^{-1}, 95.24 m^2 g_{Pt}^{-1}, 42.86 m^2 g_{Pt}^{-1} and 48.57 m^2 g_{Pt}^{-1}, respectively (Fig. S2†). The Pt_2Cu/C NPs present the largest ECSA, which is probably due to them having the smallest size, as shown in Fig. 1 and S1,† leading to a high Pt utilization. Furthermore, as shown in Fig. 3b, the MOR catalytic activities of these catalysts were measured in 0.1 M $HClO_4$ aqueous solution containing 0.5 M CH_3OH. Interestingly, Pt_2Cu/C also shows a higher methanol oxidation current density, a larger I_f/I_b ratio (where I_f is the forward current density and I_b is the backward current density) and a lower onset potential than any other catalyst studied here, all in the same order as that of the ECSA, shown in Fig. S2.† The lowest MOR onset potential on Pt_2Cu/C NPs indicates that the MOR can occur easily under the smallest driving force, and the highest I_f/I_b ratio (Fig. S3†) on Pt_2Cu/C NPs implies

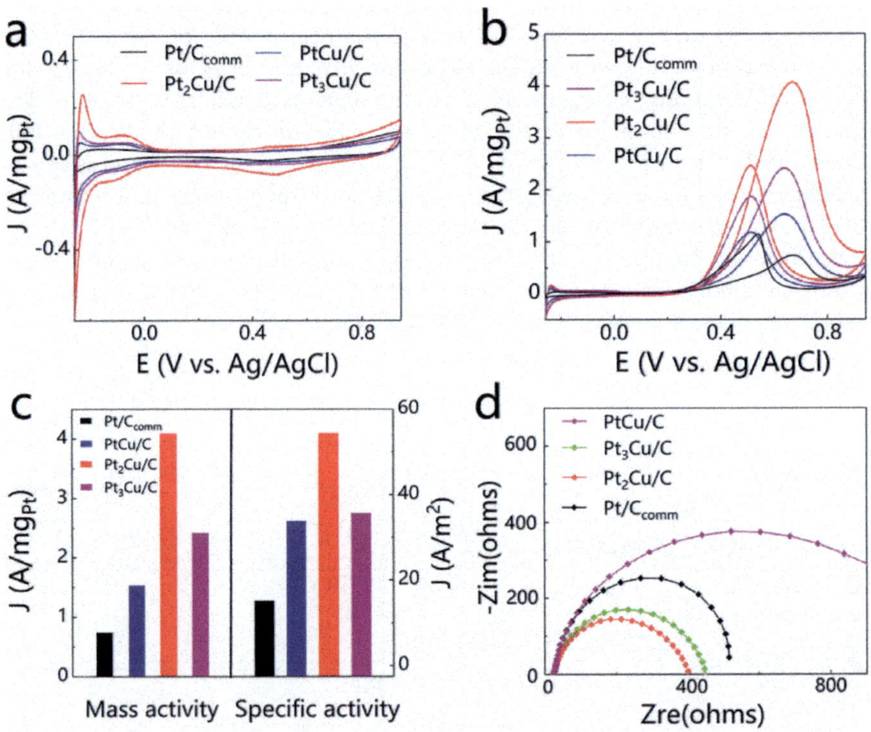

Fig. 3 Performance of the PtCu NPs catalysts with different Pt/Cu molar ratios relative to that of Pt/C_{comm}: (a) cyclic voltammogram (CV) curves in Ar-saturated 0.1 M $HClO_4$ at a scan rate of 50 mV s^{-1}; (b) mass-normalized MOR voltammograms in Ar-saturated 0.1 M $HClO_4$ and 0.5 M CH_3OH at a scan rate of 50 mV s^{-1}; (c) mass activity and specific activity of the catalysts; (d) electrochemical impedance spectroscopy (EIS) of the catalysts.

that the methanol could be oxidized most thoroughly with the least formation of poisonous CO species, thereby it possesses the highest tolerance toward CO poisoning.[51,52] Notably, all of the I_f/I_b ratios of the PtCu alloy NPs are much larger than that of Pt/C_{comm} (0.64), further confirming that the addition of Cu can indeed speed up the MOR with less CO poisoning.[52,53]

The MOR mass activities and specific activities for these catalysts were further calculated by normalizing the current densities with the corresponding Pt loading amount and ECSA, respectively. As shown in Fig. 3c, Pt_2Cu/C exhibits the highest mass activity of 4.1 A mg_{Pt}^{-1} at 0.7 V, which is much larger than those of any of the other catalysts (Pt_3Cu/C: 2.4 A mg_{Pt}^{-1}, PtCu/C: 1.5 A mg_{Pt}^{-1} and Pt/C_{comm}: 0.7 A mg_{Pt}^{-1}) studied here. Meanwhile, Pt_2Cu/C also exhibits the highest specific activity among these catalysts of 54.3 A m^{-2} (Pt_3Cu/C: 35.7 A m^{-2}, PtCu/C: 33.8 A m^{-2} and Pt/C_{comm}: 15.2 A m^{-2}). Significantly, the MOR performance of Pt_2Cu/C obtained here has also been compared with the best PtCu catalysts reported before by others (Table S1†).

It is known that the electrochemical impedance spectra (EIS) can reflect the reaction resistance of some intermediate processes and reveal the kinetic information of the MOR to some extent.[54] The Nyquist plots for these catalysts (Fig. 3d) were obtained in a mixture of 0.1 M $HClO_4$ and 0.5 M CH_3OH at 0.4 V. The electrochemical impedance contains an ohm resistance, which is attributed to the electrode and half-cell components (including the bulk resistance of the electrocatalysts), the charge transfer resistance and the inductance of each elementary reaction.[54] The minimal impedance arc of Pt_2Cu/C proves that the intermediates on its surface can be converted fastest during the MOR process, leading to the acceleration of the rate-determining step to reduce the accumulation of poisonous intermediates on the catalyst surface, which then greatly facilitates the electro-oxidation of methanol.

The enhanced activities of the core/shell structured PtCu alloy NPs can be attributed to the following two reasons. First, the high index surfaces of the Pt skin can boost the MOR activity, thanks to the low coordinated atoms on the surface.[55,56] Second, the regulated electronic structure of Pt by Cu alloying also can promote the MOR process.[53,57] For Pt catalysts, the indirect MOR route occurring on Pt NPs usually means that methanol is oxidized first to CO or other carbonaceous intermediates and consequently causes the poisoning of the Pt/C catalysts.[58] For the case here, the alloying with Cu downshifts the d-band center of Pt, as indicated by the XPS data shown in Fig. 2,[22,30] thus weakening the adsorption of intermediate species (such as CO) and, hence, promoting the direct oxidation of methanol. Furthermore, the CO stripping experiment was further done to evaluate the tolerance of the catalysts toward CO poisoning. As shown in Fig. 4, it can be seen that the peak potentials of the CO stripping curves of Pt_2Cu/C are negatively shifted compared to those of Pt/C_{comm}, indicating that the CO molecules are more easily oxidized/removed from the surface of Pt_2Cu/C, further demonstrating the enhanced remarkable CO tolerance of Pt_2Cu/C, consistent with that shown in Fig. 3.

The accelerated durability experiment was also done by cycling the potentials between −0.26 and 0.94 V for both Pt_2Cu/C and Pt/C_{comm}. As shown in Fig. 5a, the CV measurement of the MOR shows an activity loss of 53% after 1000 CV cycles. To evaluate whether such a performance decay is from MOR-induced poisoning or something else, 10 CV cycles are carried out in 0.1 M $HClO_4$ solution without

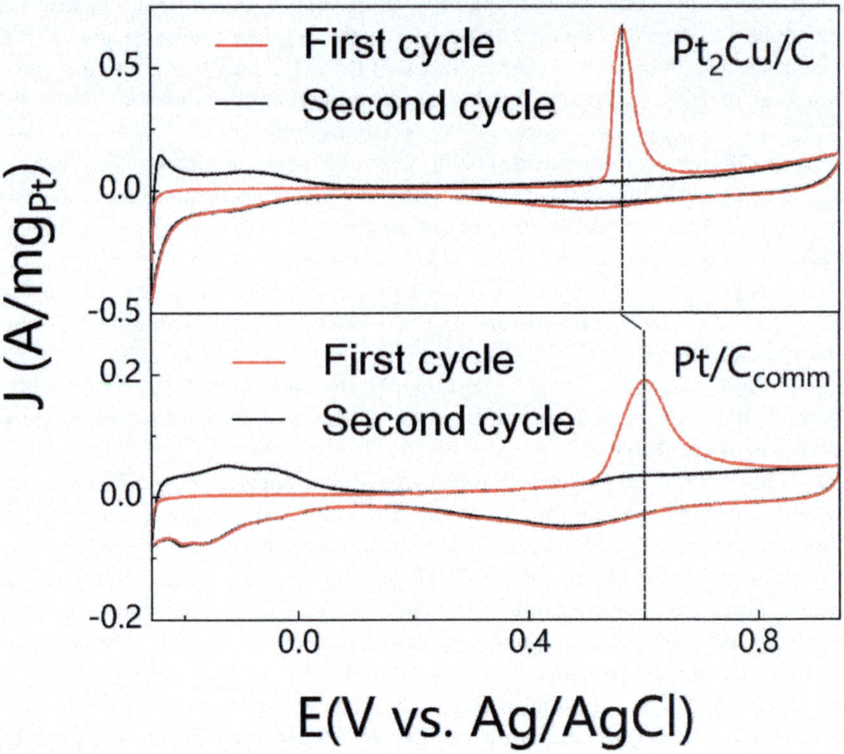

Fig. 4 CO-stripping tests of Pt_2Cu/C and Pt/C_{comm}.

methanol to remove the possible surface species formed from the MOR on Pt_2Cu/C. Surprisingly, a huge activity regeneration is observed, as indicated by the increase of the peak current. It should be noted here that the regenerated I_f is slightly lower than the fresh I_f, while the regenerated I_b is approximately the same as the fresh I_b, indicating that besides the reversible poisoning-induced performance decay, some other possible reasons, such as the CV-induced etching of the Pt surface, lead to a slow irreversible performance decay. On the contrary, the activity loss of Pt/C_{comm} is dramatic, as shown in Fig. 5b. It shows clearly that, due to the strong poisoning and etching at high potential, the MOR activity of Pt/C_{comm} almost disappears completely after 1000 CV cycles and, even after the activation regeneration treatment, the activity of the regenerated Pt/C_{comm} is only 27% of the original (Fig. 5c). The stability of Pt_2Cu/C was further evaluated by a chronoamperometry (CA) test. Fig. 5d shows the CA curves of MOR by holding the potential at 0.65 V, which are normalized with the initial current as 100%. It shows clearly that the current density of Pt/C_{comm} remains at only 1% after 6000 s, while the residual current density of Pt_2Cu/C is about 6% of the initial, further confirming that Pt_2Cu/C possesses better durability than Pt/C_{comm}. The activation regeneration is also observed during the CA test, which shows that the current density of Pt_2Cu/C reverts to 85% of the original after 10 CV cycles in 0.1 M $HClO_4$ (more than that of Pt/C_{comm}, which is 67%).

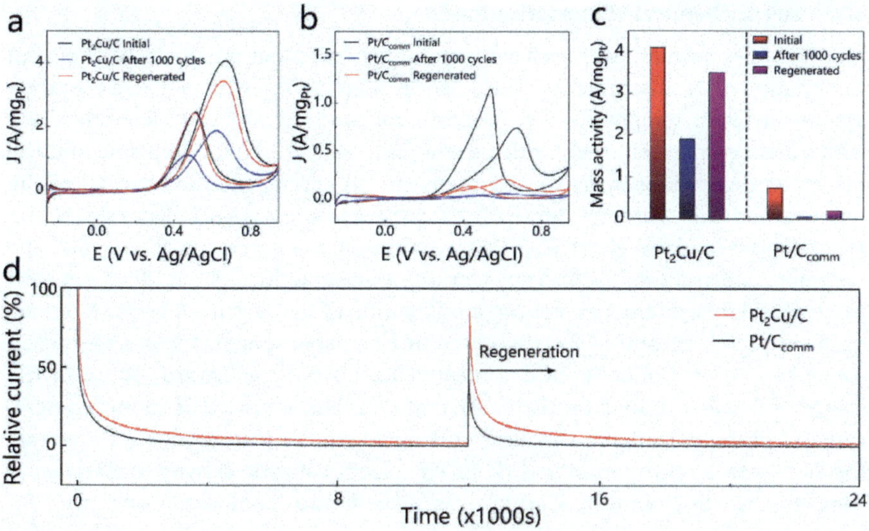

Fig. 5 Mass-normalized MOR voltammograms of (a) Pt_2Cu/C and (b) Pt/C_{comm} before and after 1000 cycles and regeneration in Ar-saturated 0.1 M $HClO_4$ and 0.5 M CH_3OH at a scan rate of 50 mV s^{-1}; (c) the mass activity of Pt_2Cu/C and Pt/C_{comm} after stability cycling and regeneration in acidic media; (d) the chronoamperometric test of Pt_2Cu/C and Pt/C_{comm} before and after regeneration at 0.65 V in Ar-saturated 0.1 M $HClO_4$ + 0.5 M CH_3OH solution for 12 000 s.

3. Experimental

3.1. Chemicals

Molybdenumhexacarbonyl ($Mo(CO)_6$), cetyltrimethylammonium bromide (CTAB), oleylamine (OAm) and chloroplatinic acid hexahydrate ($H_2PtCl_6 \cdot 6H_2O$) were purchased from Aladdin and copper(II) chloride dehydrate ($CuCl_2 \cdot 2H_2O$) was obtained from Beijing Chemical Works. All of the chemicals were used as delivered without further treatment.

3.2. Preparation of PtCu NPs with different Pt/Cu molar ratios loaded on XC-72 carbon

In a typical synthesis, XC-72 carbon (47.5 mg), CTAB (120 mg), $CuCl_2 \cdot 2H_2O$ (1 mg) and $H_2PtCl_6 \cdot 6H_2O$ (1.5 mg) were placed into a flask with 30 mL OAm and then the flocculent mixture was ultrasonicated for 30 min. Next, $Mo(CO)_6$ (200 mg) was added to the above solution. After stirring for 1 h, the mixture was moved into a Teflon vessel, which was sealed in a stainless steel reactor and heated from room temperature to 180 °C, then kept at 180 °C for 12 h in an oven before it was cooled to room temperature. The resulting product was collected by repeated centrifugation and washed several times with an ethanol/cyclohexane mixture. Finally, the collected products were dried at 50 °C in a vacuum oven overnight. The contents of Pt and Cu in the catalyst were determined by ICP-AES. The result indicated that the average molar ratio of Pt/Cu was 2 : 1. We accordingly named this product Pt_2Cu/C. With the same Pt loading, the other PtCu nanoparticles were synthesized by adjusting the amount of $CuCl_2 \cdot 2H_2O$.

3.3. Electrochemical characterization

All of the electrochemical measurements were carried out on an electrochemical workstation (CH Instruments, Inc., Model CHI 750E) with a three-electrode system at room temperature. A graphite rod and an Ag/AgCl electrode filled with saturated potassium chloride were chosen as the counter electrode and the reference electrode, respectively. If there are no special instructions, all of the potentials in this work are quoted with respect to the Ag/AgCl electrode. A pretreated glass carbon electrode (GCE, 5 mm in diameter) modified with the prepared catalysts was used as the working electrode throughout all of the electrochemical measurements. For the modification of the working electrode, 6.0 mg of the as-prepared catalyst was dispersed in 1 mL isopropanol, 1.95 mL deionized water and 50 μL Nafion™, and sonicated for 30 min. Afterward, 10 μL of the catalyst ink was dropped on the surface of GCE and air-dried at room temperature. All of the currents in the electrochemical studies were normalized against the Pt contents on the modified GCE. The electrocatalytic activity of these catalysts toward methanol oxidation was tested and compared with that of a commercial 20 wt% Pt/C catalyst. The cyclic voltammograms (CV) were recorded in Ar-purged 0.1 M $HClO_4$ solution at a scan rate of 50 mV s^{-1}. The methanol oxidation reaction was performed in Ar-purged 0.1 M $HClO_4$ + 0.5 M CH_3OH solution by potential cycling at a 50 mV s^{-1} potential sweep rate and in a potential window between −0.26 and 0.94 V. Chronoamperometry (CA) curves were recorded in 0.1 M $HClO_4$ + 0.5 M CH_3OH aqueous solution at 0.65 V for 12 000 s. The accelerated durability test was performed by continuous cycling between −0.26 and 0.94 V in 0.1 M $HClO_4$ + 0.5 M CH_3OH solution.

The ECSA was calculated using the equation:

$$\text{ECSA}\ \left(\text{m}^2\ \text{g}^{-1}\right) = \frac{Q\ (\text{C})}{210\ (\mu\text{C cm}^{-2}) \times w_{\text{Pt}}\ (\text{g})} \times 100$$

For the CO stripping voltammograms: the GCE modified with the prepared catalyst was immersed in CO-purged 0.1 M $HClO_4$ solution and CO was strongly adsorbed onto the surface of Pt to form a monolayer. After that, two CVs were tested in Ar-purged 0.1 M $HClO_4$ solution. CO was removed quickly and completely by electrochemical oxidation in the first cycle of the CV.

The electrochemical impedance spectra for the catalysts were recorded in a mixture of 0.1 M $HClO_4$ and 0.5 M CH_3OH at 0.4 V.

4. Conclusion

In conclusion, PtCu alloy nanoparticles with a Pt-rich surface have been synthesized. The one-pot synthetic approach of loading metal on carbon *in situ* can be used to synthesize other metal nanoparticles with adjustable morphologies. Compared to the commercial Pt/C, the optimum Pt_2Cu/C exhibits a much enhanced catalytic performance for the MOR. The mechanism of the enhancement is ascribed to the high index facets of the Pt-rich surface and the unique electronic structure of Pt induced by Cu alloying. The downshift of the d-band center improves the CO tolerance ability of Pt_2Cu/C, contributing to the boost of the MOR activity and durability.

Author contributions

Pengfei Yao: methodology, validation, formal analysis, investigation, data curation, writing – original draft, visualization. Jing Cao: investigation. Mingbo Ruan: investigation, resources, project administration. Ping Song: conceptualization, supervision, funding acquisition. Xue Gong: resources. Ce Han: supervision. Weilin Xu: conceptualization, resources, writing – review and editing, supervision, funding acquisition.

Conflicts of interest

There are no conflicts to declare.

Acknowledgements

The work was funded by the National Natural Science Foundation of China (21733004, 21925205, 22072145, 22005294, 21633008, 21721003) and the National Key R&D Program of China (2018YFB1502302, 2017YFE0197900).

Notes and references

1 S. S. Munjewar, S. B. Thombre and R. K. Mallick, *Renewable Sustainable Energy Rev.*, 2017, **67**, 1087–1104.
2 R. K. Mallick, S. B. Thombre and N. K. Shrivastava, *J. Power Sources*, 2015, **285**, 510–529.
3 Y. Tong, X. Yan, J. Liang and S. X. Dou, *Small*, 2021, **17**, 1904126.
4 Z. Xia, X. Zhang, H. Sun, S. Wang and G. Sun, *Nano Energy*, 2019, **65**, 104048.
5 H. J. Huang and X. Wang, *J. Mater. Chem. A*, 2014, **2**, 6266–6291.
6 M. K. Debe, *Nature*, 2012, **486**, 43–51.
7 J. N. Tiwari, R. N. Tiwari, G. Singh and K. S. Kim, *Nano Energy*, 2013, **2**, 553–578.
8 Z. Z. Yang, Y. Shi, X. S. Wang, G. L. Zhang and P. Cui, *J. Power Sources*, 2019, **431**, 125–134.
9 Y. J. Wang, W. Y. Long, L. L. Wang, R. S. Yuan, A. Ignaszak, B. Z. Fang and D. P. Wilkinson, *Energy Environ. Sci.*, 2018, **11**, 258–275.
10 Z. Dasdelen, Y. Yildiz, S. Eris and F. Sen, *Appl. Catal., B*, 2017, **219**, 511–516.
11 N. Kakati, J. Maiti, S. H. Lee, S. H. Jee, B. Viswanathan and Y. S. Yoon, *Chem. Rev.*, 2014, **114**, 12397–12429.
12 R. N. Singh, R. Awasthi and C. S. Sharma, *Int. J. Electrochem. Sci.*, 2014, **9**, 5607–5639.
13 C. S. Shang and E. K. Wang, *Phys. Chem. Chem. Phys.*, 2019, **21**, 21185–21199.
14 L. Yaqoob, T. Noor and N. Iqbal, *Int. J. Energy Res.*, 2021, **45**, 6550–6583.
15 M. Mansor, S. N. Timmiati, K. L. Lim, W. Y. Wong, S. K. Kamarudin and N. H. Nazirah Kamarudin, *Int. J. Hydrogen Energy*, 2019, **44**, 14744–14769.
16 Y. Wang, J. H. Yang, S. Sun, L. N. Wang, T. Guo, D. X. Zhang, Z. H. Xue and X. B. Zhou, *Chem. Phys. Lett.*, 2019, **730**, 575–581.
17 Y. Kwon, Y. Kim, J. W. Hong, Y. Whang, S. Kim, D. H. Wi, H. R. Byon and S. W. Han, *J. Mater. Chem. A*, 2020, **8**, 25842–25849.
18 F. X. Wu, K. Eid, A. M. Abdullah, W. X. Niu, C. Wang, Y. X. Lan, A. A. Elzatahry and G. B. Xu, *ACS Appl. Mater. Interfaces*, 2020, **12**, 31309–31318.

19 J. T. Ding, S. Ji, H. Wang, B. G. Pollet and R. F. Wang, *Electrochim. Acta*, 2017, **255**, 55–62.
20 Y. W. Liu, Z. L. Chen, C. Liu, J. F. Zhang, X. P. Han, C. Zhong, D. W. Rao, Y. S. Wang, W. B. Hu and Y. D. Deng, *ACS Appl. Energy Mater.*, 2019, **2**, 1588–1593.
21 R. Baronia, J. Goel and S. K. Singhal, *J. Nanosci. Nanotechnol.*, 2019, **19**, 4315–4322.
22 Y. Shi, Y. Fang, G. Zhang, X. Wang, P. Cui, Q. Wang and Y. Wang, *J. Mater. Chem. A*, 2020, **8**, 3795–3802.
23 J. P. Zhong, L. L. Li, M. Waqas, X. Q. Wang, Y. J. Fan, J. H. Qi, B. Yang, C. Y. Rong, W. Chen and S. G. Sun, *Electrochim. Acta*, 2019, **322**, 134677.
24 H. H. Li, Q. Q. Fu, L. Xu, S. Y. Ma, Y. R. Zheng, X. J. Liu and S. H. Yu, *Energy Environ. Sci.*, 2017, **10**, 1751–1756.
25 M. C. Luo, Y. J. Sun, X. Zhang, Y. N. Qin, M. Q. Li, Y. J. Li, C. J. Li, Y. Yang, L. Wang, P. Gao, G. Lu and S. J. Guo, *Adv. Mater.*, 2018, **30**, 1705515.
26 Q. Lv, Y. Xiao, M. Yin, J. J. Ge, W. Xing and C. P. Liu, *Electrochim. Acta*, 2014, **139**, 61–68.
27 J. B. Wu, L. Qi, H. J. You, A. Gross, J. Li and H. Yang, *J. Am. Chem. Soc.*, 2012, **134**, 11880–11883.
28 P. Strasser, S. Koh, T. Anniyev, J. Greeley, K. More, C. F. Yu, Z. C. Liu, S. Kaya, D. Nordlund, H. Ogasawara, M. F. Toney and A. Nilsson, *Nat. Chem.*, 2010, **2**, 454–460.
29 I. Katsounaros, S. Cherevko, A. R. Zeradjanin and K. J. J. Mayrhofer, *Angew. Chem., Int. Ed.*, 2014, **53**, 102–121.
30 L. Guo, L. B. Huang, W. J. Jiang, Z. D. Wei, L. J. Wan and J. S. Hu, *J. Mater. Chem. A*, 2017, **5**, 9014–9021.
31 A. Glusen, F. Dionigi, P. Paciok, M. Heggen, M. Muller, L. Gan, P. Strasser, R. E. Dunin-Borkowsk and D. Stolten, *ACS Catal.*, 2019, **9**, 3764–3772.
32 H. Xiang, Y. Zheng, Y. Sun, T. Guo, P. Zhang, W. Li, S. Kong, M. Ouzounian, H. Chen, H. Li, T. S. Hu, G. Yu, Y. Feng and S. Liu, *Nanoscale Adv.*, 2020, **2**, 1603–1612.
33 C. Li, T. Liu, T. He, B. Ni, Q. Yuan and X. Wang, *Nanoscale*, 2018, **10**, 4670–4674.
34 Z. Zhang, Z. Luo, B. Chen, C. Wei, J. Zhao, J. Chen, X. Zhang, Z. Lai, Z. Fan, C. Tan, M. Zhao, Q. Lu, B. Li, Y. Zong, C. Yan, G. Wang, Z. J. Xu and H. Zhang, *Adv. Mater.*, 2016, **28**, 8712–8717.
35 L. Huang, X. Zhang, Q. Wang, Y. Han, Y. Fang and S. Dong, *J. Am. Chem. Soc.*, 2018, **140**, 1142–1147.
36 X. W. Zhou, R. H. Zhang, Z. Y. Zhou and S. G. Sun, *J. Power Sources*, 2011, **196**, 5844–5848.
37 Y. Y. Zhang, R. Shi, J. Ren, Y. Dai, Y. J. Yuan and Z. H. Wang, *Langmuir*, 2019, **35**, 16752–16760.
38 X. L. Li, Y. S. Zhou, Y. Y. Du, J. Xu, W. C. Wang, Z. D. Chen and J. Y. Cao, *Int. J. Hydrogen Energy*, 2019, **44**, 18050–18057.
39 J. Liu, G. Xu, B. Liu and J. Zhang, *Chem. Commun.*, 2017, **53**, 7457–7460.
40 T. Bian, B. Sun, S. Luo, L. Huang, S. Su, C. Meng, S. Su, A. Yuan and H. Zhang, *RSC Adv.*, 2019, **9**, 35887–35894.
41 Z. J. Li, X. Jiang, X. R. Wang, J. R. Hu, Y. Y. Liu, G. T. Fu and Y. W. Tang, *Appl. Catal., B*, 2020, **277**, 119135.

42 Q. Xue, H. Huang, J. Y. Zhu, Y. Zhao, F. M. Li, P. Chen and Y. Chen, *Appl. Catal., B*, 2020, **278**, 119269.
43 E. Antolini, J. R. C. Salgado and E. R. Gonzalez, *Appl. Catal., B*, 2006, **63**, 137–149.
44 X. Xiao, E. Jung, S. Yu, H. Kim, H. K. Kim, K. Y. Lee, J. P. Ahn, T. Lim, J. Kim and T. Yu, *Catalysts*, 2021, **11**, 130.
45 X. Yuan, X. Jiang, M. Cao, L. Chen, K. Nie, Y. Zhang, Y. Xu, X. Sun, Y. Li and Q. Zhang, *Nano Res.*, 2019, **12**, 429–436.
46 G. Chen, C. Xu, X. Huang, J. Ye, L. Gu, G. Li, Z. Tang, B. Wu, H. Yang, Z. Zhao, Z. Zhou, G. Fu and N. Zheng, *Nat. Mater.*, 2016, **15**, 564–569.
47 K. Jiang, D. Zhao, S. Guo, X. Zhang, X. Zhu, J. Guo, G. Lu and X. Huang, *Sci. Adv.*, 2017, **3**, e1601705.
48 C. Z. Li, T. Y. Liu, T. He, B. Ni, Q. Yuan and X. Wang, *Nanoscale*, 2018, **10**, 4670–4674.
49 Z. Yang, S. Pedireddy, H. K. Lee, Y. J. Liu, W. W. Tjiu, I. Y. Phang and X. Y. Ling, *Chem. Mater.*, 2016, **28**, 5080–5086.
50 W. Q. Yao, X. Jiang, M. Li, Y. L. Li, Y. Y. Liu, X. Zhan, G. T. Fu and Y. W. Tang, *Appl. Catal., B*, 2021, **282**, 119595–119605.
51 M. X. Gong, G. T. Fu, Y. Chen, Y. W. Tang and T. H. Lu, *ACS Appl. Mater. Interfaces*, 2014, **6**, 7301–7308.
52 Y. Y. Jia, J. Y. Su, Z. B. Chen, K. Tan, Q. L. Chen, Z. M. Cao, Y. Q. Jiang, Z. X. Xie and L. S. Zheng, *RSC Adv.*, 2015, **5**, 18153–18158.
53 T. Song, H. Xue, N. Guo, J. Sun, L. Qin, L. Guo, K. Huang, F. He and Q. Wang, *Chem. Commun.*, 2020, **56**, 7136–7139.
54 I. M. Hsing, X. Wang and Y. J. Leng, *J. Electrochem. Soc.*, 2002, **149**, A615–A621.
55 A. Kuzume, E. Herrero and J. M. Feliu, *J. Electroanal. Chem.*, 2007, **599**, 333–343.
56 K. M. Yeo, S. Choi, R. M. Anisur, J. Kim and I. S. Lee, *Angew. Chem., Int. Ed.*, 2011, **50**, 745–748.
57 S.-H. Ye, X.-J. He, L.-X. Ding, Z.-W. Pan, Y.-X. Tong, M. Wu and G.-R. Li, *Chem. Commun.*, 2020, **56**, 11142.
58 X. Zhao, M. Yin, L. Ma, L. Liang, C. P. Liu, J. H. Liao, T. H. Lu and W. Xing, *Energy Environ. Sci.*, 2011, **4**, 2736–2753.

Faraday Discussions

ROYAL SOCIETY OF CHEMISTRY

PAPER

Lateral voltage as a new input for artificial lipid bilayer systems

Teng Ma,[ab] Madoka Sato,[cd] Maki Komiya,[c] Kensaku Kanomata,[e] Takaya Watanabe,[cd] Xingyao Feng,[c] Ryusuke Miyata,[cd] Daisuke Tadaki,[c] Fumihiko Hirose,[e] Yuzuru Tozawa[f] and Ayumi Hirano-Iwata *[abc]

Received 27th July 2021, Accepted 2nd September 2021

DOI: 10.1039/d1fd00045d

In this work, we propose lateral voltage as a new input for use in artificial lipid bilayer systems in addition to the commonly used transmembrane voltage. To apply a lateral voltage to bilayer lipid membranes, we fabricated electrode-equipped silicon and Teflon chips. The Si chips could be used for photodetector devices based on fullerene-doped lipid bilayers, and the Teflon chips were used in a study of the ion channel functions in the lipid bilayer. The findings indicate that the lateral voltage effectively regulates the transmembrane current, in both ion-channel-incorporated and fullerene-incorporated lipid bilayer systems, suggesting that the lateral voltage is a practicable and useful additional input for use in lipid bilayer systems.

Introduction

The bilayer lipid membrane (BLM) is an essential component of the cell membrane. It is formed *via* the self-assembly of amphiphilic lipid molecules at the water/water interface. The first artificial free-standing BLM was reported by Mueller *et al.*, who painted a solution containing lipids on a hydrophobic aperture.[1] Since then, artificial BLMs have been widely accepted as a feasible platform for studies of the biological functions of cell membranes.[2–5] After 60 years of intensive research on artificial BLM systems, techniques for preparing BLMs have greatly evolved. For example, the "folding" method was proposed to reduce the oil (non-volatile solvent) residue that is present in free-standing

[a]*Advanced Institute for Materials Research (WPI-AIMR), Tohoku University, 2-1-1 Katahira, Aoba-ku, Sendai 980-8577, Japan. E-mail: ayumi.hirano.a5@tohoku.ac.jp*

[b]*Core Research Cluster, Tohoku University, 2-1-1 Katahirano, Aoba-ku, Sendai 980-8577, Japan*

[c]*Research Institute of Electrical Communication, Tohoku University, 2-1-1 Katahira, Aoba-ku, Sendai 980-8577, Japan*

[d]*Graduate School of Biomedical Engineering, Tohoku University, 6-6 Aoba, Aramaki, Aoba-ku, Sendai 980-8579, Japan*

[e]*Graduate School of Science and Engineering, Yamagata University, 4-3-16 Jonan, Yonezawa 992-8510, Japan*

[f]*Graduate School of Science and Engineering, Saitama University, 255 Shimo-Okubo, Sakura-ku, Saitama-shi, Saitama 338-8570, Japan*

membranes;[6] and the droplet interface bilayer (DIB) method was developed to form a stable BLM in an oil bath.[7,8] In our group, we also adopted the well-established micro-fabrication process used in semiconductor technology to form silicon (Si) chips with a nano-tapered aperture, on which stable and strong oil-free BLMs could be formed.[9,10] Based on the Si chip, we incorporated various ion channel proteins into BLMs and investigated drug-related side effects on cardiac ion channels.[11,12] Such BLM systems are now being applied to a wider range of fields. For instance, nanomaterial-doped BLMs have been demonstrated to have great potential in studies related to self-assembled devices for different applications, such as light sensing devices and water desalination.[13–16] In previous studies, we also doped BLMs with fullerene derivatives, and fabricated self-assembled devices based on hybrid membranes composed of lipids and fullerene derivatives.[17,18] The devices demonstrated reversible and repetitive responses to light illumination.

In contrast to versatile progress in BLM platforms, the measurement approach for artificial BLM systems has not changed substantially in the past several decades. In a typical BLM system, both sides of a BLM are filled with buffer solutions; and a transmembrane voltage is applied to the BLM through two electrodes that are immersed in the buffer solution. The transmembrane voltage is usually used as the input, and the measured transmembrane current is used as the output of the system. This simple single-input, single-output system can provide valuable information in terms of understanding the basic functions of membrane proteins, and for developing functional biosensors and nano–bio hybrid devices.[2,19,20]

Inspired by the evolution from two-terminal diodes to three-terminal transistors in semiconductor technology, we expected that introducing an additional input to a BLM system might allow better control over the system and additional information could be obtained from the system, which may help to create new types of lipid membrane platforms that do not exist in nature. In this work, we propose a method to improve the control of the BLM system by applying a voltage parallel to the BLM (lateral voltage) in addition to the traditional transmembrane voltage (Fig. 1). To apply the lateral voltage to

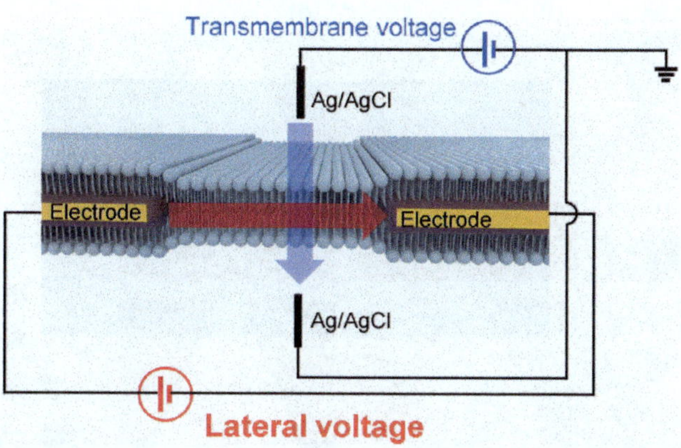

Fig. 1 Conceptual scheme for the two inputs (lateral and transmembrane voltages) BLM system.

artificial BLM systems, we fabricated two different types of electrode-equipped membrane supports based on Teflon films and Si chips. The resulting BLMs were then functionalized with ion channel proteins and nanomaterials. The effects of the lateral voltage on both the ion-channel-incorporated and the nanomaterial-doped BLM systems were evaluated.

Experimental

In order to apply a lateral voltage to the BLM system, we needed to fabricate an electrode-equipped membrane support for the free-standing BLM. In previous studies, we used micro-fabricated Si chips as membrane supports to form BLMs and applied them to producing biosensors and light sensing devices.[10,17,21] On the other hand, Teflon films have traditionally been used as a support for BLMs in functional analyses of ion channel activities.[6,22,23] In this study, we developed suitable processes for fabricating electrodes on these Teflon films and Si chips.

For the Si chips, we previously reported on the use of mask deposition to form electrodes for the lateral voltage applications.[18] The simplified fabrication process is shown in Fig. 2: a single crystal silicon substrate with a Si_3N_4 coating was first thermally oxidized to form a SiO_2 layer (Fig. 2(a and b)); Si was then selectively etched by a photolithography process (Fig. 2(c and d)); after which, the chip was thermally oxidized to form a protective SiO_2 coating (Fig. 2(e)); photolithography was then used to selectively remove the SiO_2 to form a circular window at the center of the Si chip (Fig. 2(f)); the underlying Si_3N_4 and SiO_2 were then sequentially etched to form a center aperture in the Si chip (Fig. 2(g and h)); we then used thermal evaporation to form aluminium (Al) electrodes through a metal

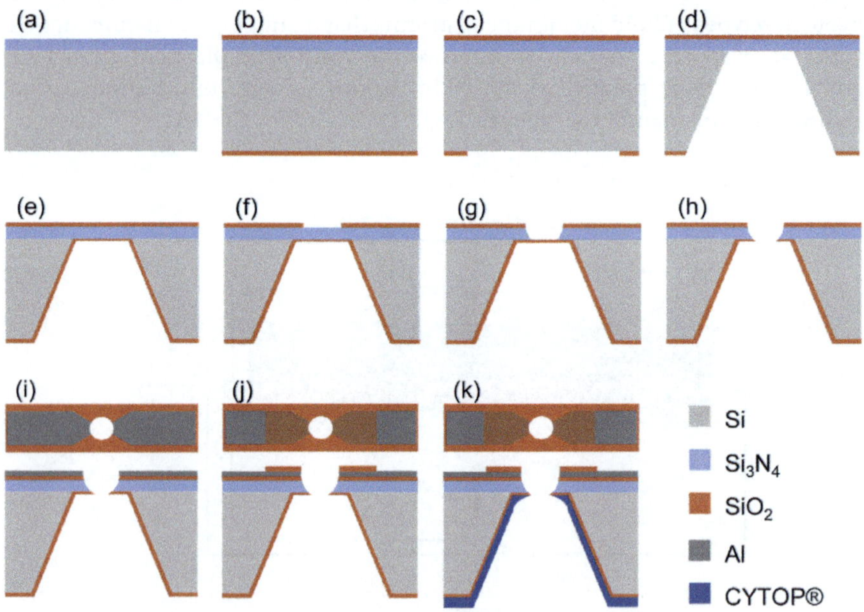

Fig. 2 Process used to fabricate the electrode-equipped Si chip for the application of lateral voltages.

mask (Fig. 2(i)); to prevent the buffer from causing the corrosion of the electrodes, we formed a SiO_2 protective layer on the electrodes (Fig. 2(j)). Note that not all of the Al surfaces were covered by SiO_2, the uncovered parts of the Al electrodes were used to connect to external circuits for the application of a lateral voltage. Finally, the back side of the chip was coated with CYTOP® (AGC) to reduce the capacitance of the Si chip system and to reduce system noise (Fig. 2(k)).[10] Before BLM formation, the surface of the electrode-equipped Si chip was subjected to treatment with a long-chain perfluorocarbon, (tridecafluoro-1,1,2,2-tetrahydrooctyl) dimethylchlorosilane (PFDS).[24]

To fabricate an electrode-equipped Teflon chip, we formed a circular aperture (~100 μm diameter) on a 12 μm-thick Teflon film by an electrical discharge method (Fig. 3(a and b)).[25] On the top of the film, an electron-beam (EB) evaporation technique was used to deposit titanium (Ti) electrodes (Fig. 3(c)). To ensure that the lateral voltage is effectively applied to the BLM, the Ti electrodes extended to the very edge of the aperture. Similar to a Si chip, to protect the electrodes from buffer corrosion, we formed a protective SiO_2 layer on the surface of the electrodes *via* EB evaporation (Fig. 3(d)). Finally, to prevent the oxidation of the Ti surfaces that were exposed to the atmosphere, we deposited a layer of Pt on the exposed Ti electrodes (Fig. 3(e)). The resulting Teflon chip is shown in Fig. 3(f). Before BLM formation, the SiO_2 surface was subjected to silanization with PFDS in the same manner as was applied to Si chips.

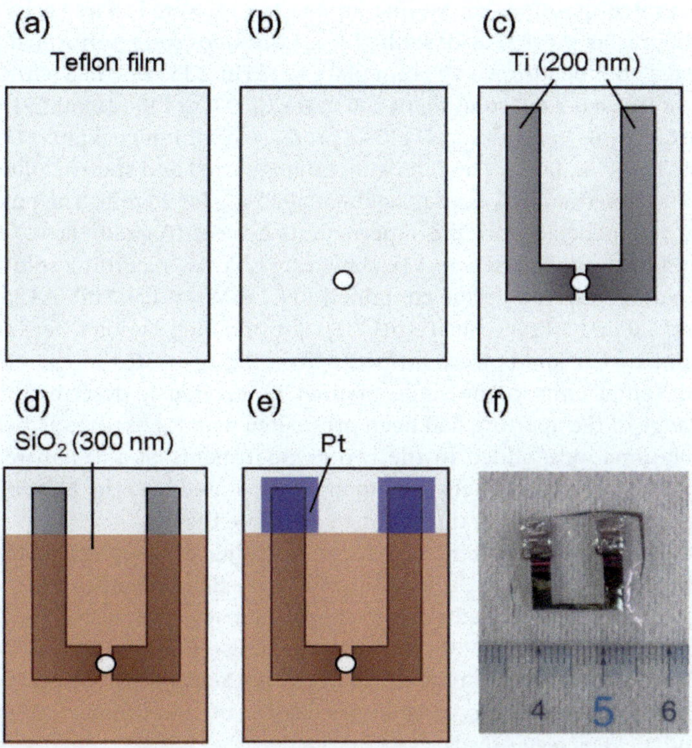

Fig. 3 Process used in the fabrication of the electrode-equipped Teflon chip for the application of lateral voltages.

Both the DC and AC voltage sources that were used for the application of lateral voltages were generated by home-made circuits with batteries as power sources to minimize hum noise. The AC voltage source has a sinusoidal waveform with a frequency of 11.8 kHz.[18]

The electrode-equipped Si chip was used for producing BLMs doped with a fullerene derivative as a light-sensitive membrane. 1,2-Diphytanoyl-*sn*-glycero-3-phosphocholine (DPhPC) and phenyl-C_{61}-butyric acid methyl ester (PCBM) were dissolved in chloroform/*n*-hexane to form a solution with a molar ratio of 5 : 1 (DPhPC : PCBM). The concentration of the DPhPC was 10 mg mL^{-1}. Solvent-free BLMs doped with PCBM were formed by the folding method.[17,18] Two electrodes on the Si chip, used as the lateral electrodes, were connected to the AC voltage source. Two Ag/AgCl electrodes were placed in buffer solution surrounding BLMs, on which transmembrane voltages were applied. Lights from a light emitting diode (525 nm, Solis High Power LED, Thorlabs) were used to illuminate the hybrid membrane.

The electrode-equipped Teflon chip was used for BLMs that were incorporated with human voltage-gated sodium channel Na_V1.5 that was prepared from human embryonic kidney (HEK) 293T cells expressing the channel. HEK 293T cells (DS Pharma Biomedical Co., Ltd.) were cultured in Dulbecco's modified Eagle's medium/GlutaMAX medium supplemented with 10% fetal bovine serum (GIBCO). One to two days before the protein extraction, the cells were transfected with a plasmid, Nav1.5(SCN5A) (NM_198056) Human Untagged Clone (ORIGENE), using PEI MAX reagent (Polysciences, Inc.) and OPTI-MEM medium (GIBCO). Proteoliposomes containing human Na_V1.5 were isolated according to the procedures as previously described.[11] All the steps were performed at 4 °C. Briefly, cells obtained from a 100 mm plate were rinsed with HBSS (GIBCO) and scraped off into a 0.9 mL solution of 200 mM KCl, 33 mM KF, 10 mM EDTA, and 50 mM HEPES (pH 7.4 with KOH) plus a protease inhibitor cocktail (EDTA-free) (NACALAI TESQUE, INC.). The cells were homogenized and spun at 2000 × *g* for 10 min. The supernatant was spun again at 4000 × *g* for 20 min. The membrane fractions were pelleted from the supernatants by centrifugation at 157 000 × *g* for 1 h and resuspended in a 4 : 1 (v/v) mixture of a Na_V recording solution and glycerol. Na_V recording solution contained 149.2 mM NaCl, 4.7 mM KCl, 2.5 mM $CaCl_2$, and 5.0 mM HEPES–NaOH (pH 7.3). The obtained samples were stored at −30 °C until used. BLMs were formed across the apertures in the electrode-equipped Teflon chip in the same manner as previously described,[10] except that the edge of the aperture had been pre-coated with *n*-hexadecane before the buffer solutions were added to the two compartments of a Teflon recording chamber. The Na_V1.5 channels were then incorporated into the BLM by fusing proteoliposomes containing the channels with the BLMs.[12]

The transmembrane current was recorded using an Axopatch 200B patch-clamp amplifier (Molecular Devices). Signals were filtered with a low-pass Bessel filter, digitized, and stored online using a data acquisition system (Digidata 1440A and 1550B, and pCLAMP, Molecular Devices). For the BLMs doped with PCBM, the cut-off and sampling frequencies were 1 and 10 kHz, respectively. For the BLMs integrated with Na_V1.5 channels, the cut-off and sampling frequencies were 2 and 25 kHz, respectively.

Results and discussion

Construction of lateral-voltage-equipped BLM system

To investigate the influence of additive layers (metal electrodes, SiO_2) of the BLM support on BLM quality, we formed free-standing BLMs on electrode-equipped Si chips and Teflon chips and evaluated their electrical characteristics. The resistance of the BLMs formed on the Si chip and the Teflon chip both exceeded the upper limit of the measuring instrument (200 GΩ), indicating that the additional electrodes and SiO_2 had no measurable effect on the BLM formation. We also confirmed that there was no leakage current observed between the two electrodes in buffer solutions and after BLM formation, indicating that the SiO_2 protective layer on the chip surface effectively protected the metal electrodes from the corrosion by the buffer solution.

The effect of lateral voltage, including DC and AC voltages, on BLM stability and transmembrane current noise were next investigated using membranes containing only lipids. Applying the lateral AC voltage to the Si chip, as previously reported, caused no noticeable effect on the stability and noise level of the system.[18] When a lateral voltage of 1 V (peak to peak) was applied, the resistance remained above 200 GΩ. Moreover, the current noise was as low (~3 pA) as that of BLMs in the absence of a lateral voltage (Fig. 4(a)). However, when applying a DC voltage to the Si chip, we observed that the resistance of the BLM was reduced to about 100 GΩ, and the noise level increased significantly from 3 pA to over 80 pA, which is not suitable for current recordings at a pA level (Fig. 4(a)). When a higher DC voltage (2 V) was applied to the lateral electrodes, electrochemical corrosion was observed at the edge of the lateral electrodes, indicating that the SiO_2 coating on the Si chip could not withstand such a high DC lateral voltage.

In the Teflon-chip-based system, it was observed that the BLM maintained a large resistance (>200 GΩ) and a small transmembrane current noise (~3 pA) when a DC lateral voltage was applied, as shown in Fig. 4(b). However, when we applied an AC voltage to the Teflon chip, we observed that even a small voltage

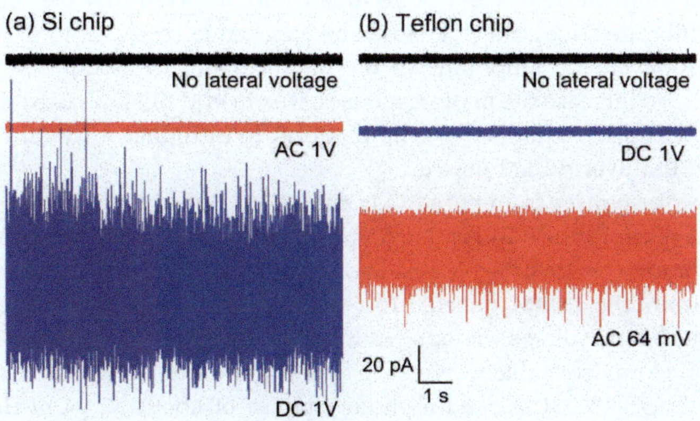

Fig. 4 Transmembrane current noise of BLM systems based on a Si chip (a) and a Teflon chip (b). Current noises without lateral voltage, with AC lateral voltage and DC lateral voltage are shown in black, red, and blue, respectively. The scale bar shown in (b) applies to all the figures. A low pass filter of 1 kHz was applied to all the recorded signals.

was sufficient to cause a drastic increase in noise level (>40 pA), and the AC voltage of 0.5 V (peak-to-peak) resulted in the rupture of the BLM. As shown above, the BLMs formed using the Teflon-chip and the Si-chip systems exhibited different tendency to generate noise when DC and AC lateral voltages were applied; to minimize transmembrane current noise induced by the application of the lateral voltage, DC voltages were preferable for BLMs in the Teflon chip systems, while AC voltages were preferable for BLMs in the case of Si chip systems.

Comparing these two systems, the pronounced differences lie in the flexibility of the membrane support (Teflon *vs.* Si_3N_4), and the necessity of *n*-hexadecane to be present around the aperture. For the present, we tentatively assume that the BLMs formed in the flexible Teflon films are likely to vibrate mechanically due to the Coulomb force from the AC lateral voltage between the two electrodes.[26,27] It is assumed that the mechanical vibration of the BLM would generate a displacement current near the BLM surface, which might result in the higher thermal noise in the transmembrane current in the case of the Teflon chip. On the other hand, we considered that the noise was suppressed in the Si chip due to its mechanical stability. At the same time, because it is necessary to apply *n*-hexadecane near the aperture to facilitate BLMs formation on Teflon chips, the edge of the lateral electrodes around the aperture would not be in direct contact with the BLM. A DC lateral voltage would not induce an overly large electric field around the edge of the lateral electrodes. On the other hand, *n*-hexadecane is not required to form BLMs on Si chips. At the edge of the Si chip aperture, the BLM is in direct contact with the SiO_2-protected electrodes. Therefore, we speculate that when the lateral DC voltage is applied to the Si chip, a larger electric field would be generated at the sharp edges of the lateral electrodes, resulting in a large increase in noise and/or the electric breakdown of the insulating layer.

Effect of lateral voltage on fullerene-derivative-doped lipid bilayer membranes

BLMs formed across the tapered apertures in the Si chips were sufficiently stable to withstand high transmembrane voltages, which is suitable for sensor applications.[10] In a previous study, we introduced a fullerene derivative, phenyl-C_{61}-butyric acid methyl ester (PCBM), into BLMs, and observed that the lipid/PCBM hybrid membrane showed a reversible and repeatable response to light illumination.[17] However, the amplitude of the photoresponse was quite small, only about 20 pA. To increase the photoresponse of the hybrid BLM, we used the abovementioned electrode-equipped Si chip platform to introduce a lateral voltage to the lipid/PCBM hybrid BLM device.

The photoresponses from pure BLMs and PCBM doped BLMs under different lateral voltages are shown in Fig. 5(a and b). In this experiment, a relatively high transmembrane voltage of −1 V was used to effectively extract photo-generated electrons from the PCBM doped BLM. To minimize transmembrane current noise, an AC lateral voltage was applied to the electrodes of the Si chip. The pure BLM exhibited no detectable response to light illumination or lateral voltage. The BLM doped with PCBM showed a photoresponse of about 22 pA at the lateral voltage of 0 V. This result is consistent with findings of our previous report.[17] As the lateral voltage was gradually increased, the amplitude of the photoresponse increased significantly (Fig. 5(b)). As shown in Fig. 5(c), when the lateral voltage was increased from 0 V to 3 V, the photoresponse increased by over 10 times (from

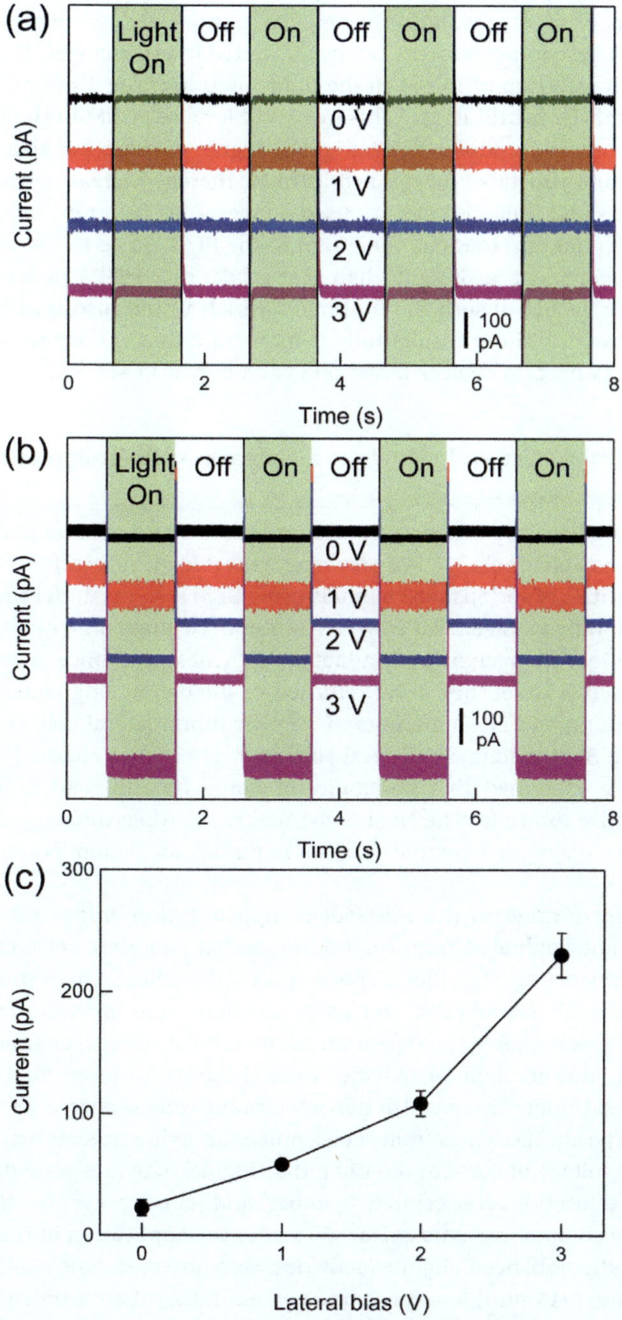

Fig. 5 Photoresponse from BLMs composed of pure DPhPC lipids (a) and PCBM-doped BLMs (b). (c) shows relationships between the photoresponse and applied lateral voltage. The statistical data are calculated based on the results of 3 separate experiments. The photoresponses were offset for clarity.

22 pA to 230 pA), indicating that the photoresponse can be effectively modulated and amplified by applying a lateral voltage to the BLM system. Compared with our previously reported results, both the amplitude and speed of the photoresponse increased.[18] These improvements can be attributed to two changes. First, we used a higher concentration of PCBM in the lipid solution (from 1 mg mL^{-1} to 2 mg mL^{-1}) to form the hybrid BLMs. This increased level of PCBM in the BLM would not only absorb more photons to generate a higher current, but also reduce the transmembrane resistance of the hybrid BLM, thereby increasing the response speed at the same time. Second, we used a green LED (525 nm) rather than an infrared LED (850 nm) to excite the PCBM in the BLM. Since PCBM has a higher absorption coefficient at 525 nm than that at 850 nm, PCBM molecules in the present BLMs would absorb more photons which would also lead to a higher photoresponse.[28,29] These results fully demonstrate that the lateral voltage has a modulating effect on devices based on PCBM-doped BLMs.

Effect of lateral voltage on lipid bilayer membranes with incorporated ion channels

Artificial BLMs have been used as an excellent platform for the functional analysis of biological ion channels.[2,4,19] We previously reported on successful recordings of channel activities of various ion channels using BLM systems under the control of a transmembrane voltage.[9–12,24] However, we noticed that some biological channels rapidly lost their channel conducting activities, and once they went into a non-conductive state, they never returned to the conducting states that show channel opening and closing events. If another input, lateral voltage, was to be added to the BLM system, the lateral voltage might trigger channel conducting activities. We examined this possibility by using human $Na_V1.5$ channels as a representative example. The $Na_V1.5$ channel is a cardiac voltage-gated sodium channel that plays an essential role in initiating an action potential in the heart.[30,31]

BLMs were formed in the electrode-equipped Teflon chips and fused with proteoliposomes prepared from $Na_V1.5$-transfected HEK293T cells. On applying a voltage protocol (Fig. 6(a)) that is commonly used for $Na_V1.5$ activation in patch-clamp studies,[32,33] rectangular and spiky currents were sporadically observed (Fig. 6(b)). However, after the repetition of the voltage sweeps, channel activities disappeared, and no channel activities were observed for more than 150 subsequent sweeps. Under the control of transmembrane voltage only, it was difficult to recover the channel activities from such a non-conducting state. When we applied a DC lateral voltage of 0.5 V to the BLM, the channel activities were dramatically enhanced, exhibiting consecutive opening and closing events. Because we recorded the channel currents at least 30 s after the application of the DC lateral voltage and the enhanced channel activities were observed over a relatively long period of time (>15 min), a steady-state process rather than a transient process (*e.g.* charging process of the lateral electrodes) is likely responsible for the enhanced channel activities. The enlarged view of the channel activities with lateral voltage was similar to that of the channel activities before entering non-conductive states (Fig. 6(c)). The channel activities were blocked by the addition of tetrodotoxin (TTX), a specific blocker of sodium channels.[34,35] When we applied a lateral voltage to a BLM containing no ion channels, the same transmembrane

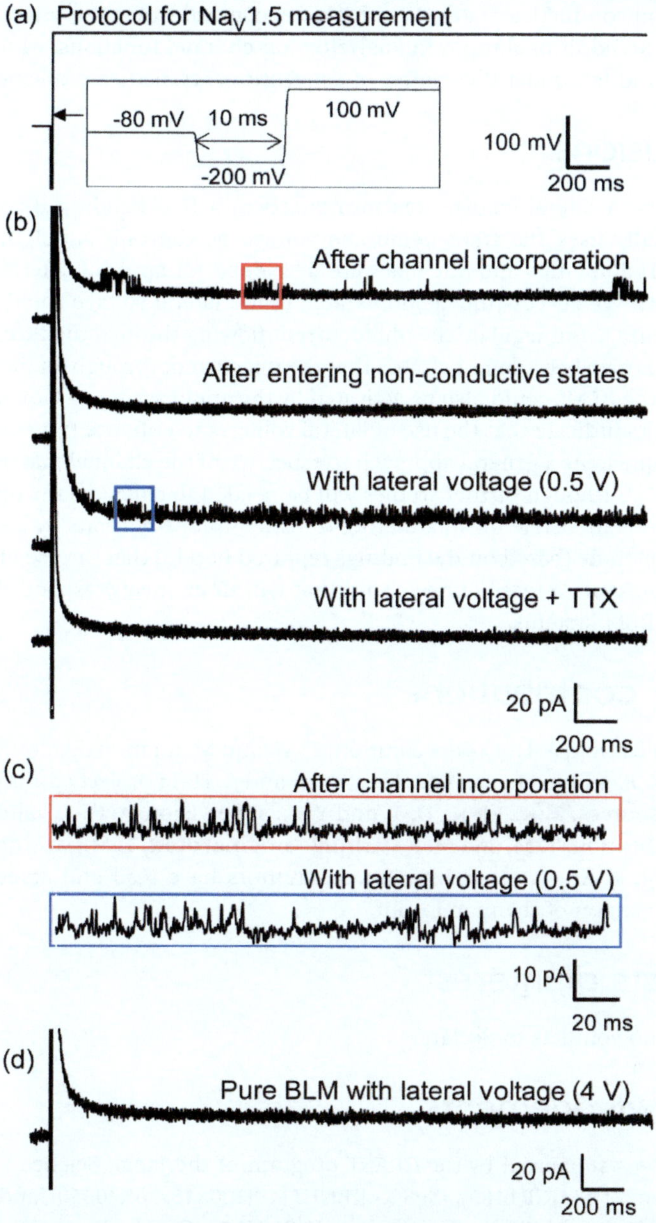

Fig. 6 (a) Transmembrane voltage protocol used for measuring Na_V1.5 channel currents. The inset shows the detail of the protocol. (b) Sequentially measured transmembrane currents using the voltage protocol: after the channel incorporation, after the channels lost their conductive activities, when a lateral voltage (0.5 V) was applied, after adding TTX to the buffer solution. The measured currents were offset for clarity. (c) Enlarged views of the channel activities indicated by red and blue squares in (b). (d) Transmembrane current of a pure BLM when a lateral voltage (4 V) was applied.

voltage protocol induced no significant currents even when a higher lateral voltage (4 V) was applied (Fig. 6(d)). These results indicate that the application of the lateral voltages effectively recovered the activities of the Na_V1.5 channel that were in non-conductive states. We therefore conclude that the lateral voltage represents an additional input in analyzing ion channel functions, which would have been hidden under the control of conventional transmembrane potentials.

Conclusions

In summary, a lateral voltage was introduced to a free-standing BLM system, which usually uses the transmembrane voltage as the only input. Electrode-equipped Teflon chips and Si chips were developed for applying lateral voltages to the BLMs. We successfully demonstrated that a lateral voltage could function as a "gate bias", and regulate the photocurrent flowing through the PCBM-doped BLMs. It has also been demonstrated that the ion current through an ion channel protein in the BLMs could also be activated by the application of a lateral voltage. These results indicate that the use of lateral voltages to enhance transmembrane currents represents a general approach irrespective of the channel materials used in the BLMs. Although further studies will be needed to elucidate the underlying mechanism responsible for modulating transmembrane currents by lateral voltages, we conclude (based on the findings reported herein) that the use of a lateral voltage represents a feasible new input that will dramatically expand the applicability of BLM systems.

Author contributions

Conceptualization, A. H.-I.; data curation, T. M. and M. K.; investigation, M. S., M. K., T. W., X. F. and R. M.; methodology, T. M. and A. H.-I.; project administration, A. H.-I.; resources, K. K., M. K., D. T. and Y. T.; supervision, A. H.-I.; validation, A. H.-I.; visualization, T. M. and M. K.; writing – original draft, T. M.; writing – review and editing, T. M., F. H. and A. H.-I. All authors have read and agreed to the published version of the manuscript.

Conflicts of interest

There are no conflicts to declare.

Acknowledgements

The work was supported by the CREST program of the Japan Science and Technology Agency (JPMJCR14F3), JSPS KAKENHI (19H00846, 20K20550), and partially supported by Asahi Glass Foundation, WPI-AIMR Fusion Research, and the Nation-wide Cooperative Research Project Program of the Research Institute of Electrical Communication at Tohoku University. Some of the equipment used in this research was manufactured by a technical staff member Kento Abe, at the machine shop division of Fundamental Technology Center, Research Institute of Electrical Communication, Tohoku University.

Notes and references

1 P. Mueller, D. O. Rudin, H. Ti Tien and W. C. Wescott, *Nature*, 1962, **194**, 979–980.

2 C. G. Siontorou, G. P. Nikoleli, D. P. Nikolelis and S. K. Karapetis, *Membranes*, 2017, **7**, 38.
3 M. Bally, K. Bailey, K. Sugihara, D. Grieshaber, J. Vörös and B. Städler, *Small*, 2010, **6**, 2481–2497.
4 H. Lee, Y. Lee, S. S. Oh and S. Q. Choi, *Small*, 2020, **16**, 2002541.
5 M. Zagnoni, *Lab Chip*, 2012, **12**, 1026.
6 M. Montal and P. Mueller, *Proc. Natl. Acad. Sci. U. S. A.*, 1972, **69**, 3561–3566.
7 K. Funakoshi, H. Suzuki and S. Takeuchi, *Anal. Chem.*, 2006, **78**, 8169–8174.
8 R. Syeda, M. A. Holden, W. L. Hwang and H. Bayley, *J. Am. Chem. Soc.*, 2008, **130**, 15543–15548.
9 M. Komiya, M. Kato, D. Tadaki, T. Ma, H. Yamamoto, R. Tero, Y. Tozawa, M. Niwano and A. Hirano-Iwata, *Chem. Rec.*, 2020, **20**, 730–742.
10 D. Tadaki, D. Yamaura, S. Araki, M. Yoshida, K. Arata, T. Ohori, K. I. Ishibashi, M. Kato, T. Ma, R. Miyata, Y. Tozawa, H. Yamamoto, M. Niwano and A. Hirano-Iwata, *Sci. Rep.*, 2017, **7**, 17736.
11 A. Hirano-Iwata, Y. Ishinari, M. Yoshida, S. Araki, D. Tadaki, R. Miyata, K. Ishibashi, H. Yamamoto, Y. Kimura and M. Niwano, *Biophys. J.*, 2016, **110**, 2207–2215.
12 A. Oshima, A. Hirano-Iwata, H. Mozumi, Y. Ishinari, Y. Kimura and M. Niwano, *Anal. Chem.*, 2013, **85**, 4363–4369.
13 R. H. Tunuguntla, R. Y. Henley, Y.-C. Yao, T. A. Pham, M. Wanunu and A. Noy, *Science*, 2017, **357**, 792–796.
14 P. Urban, S. R. Kirchner, C. Mühlbauer, T. Lohmüller and J. Feldmann, *Sci. Rep.*, 2016, **6**, 22686.
15 S. Niu and D. Mauzerall, *J. Am. Chem. Soc.*, 1996, **118**, 5791–5795.
16 K. C. Hwang and D. Mauzerall, *Nature*, 1993, **361**, 138–140.
17 K. Kanomata, T. Deguchi, T. Ma, T. Haseyama, M. Miura, D. Yamaura, D. Tadaki, M. Niwano, A. Hirano-Iwata and F. Hirose, *J. Electroanal. Chem.*, 2019, **832**, 55–58.
18 T. Ma, X. Feng, T. Ohori, R. Miyata, D. Tadaki, D. Yamaura, T. Deguchi, M. Komiya, K. Kanomata, F. Hirose, M. Niwano and A. Hirano-Iwata, *ACS Omega*, 2019, **4**, 18299–18303.
19 T. Ma, M. Sato, M. Komiya, X. Feng, D. Tadaki and A. Hirano-Iwata, *Chem.Lett.*, 2021, **50**, 418–425.
20 T. Osaki and S. Takeuchi, *Anal. Chem.*, 2017, **89**, 216–231.
21 R. Miyata, D. Tadaki, D. Yamaura, S. Araki, M. Sato, M. Komiya, T. Ma, H. Yamamoto, M. Niwano and A. Hirano-Iwata, *Micromachines*, 2021, **12**, 98.
22 M. Mayer, J. K. Kriebel, M. T. Tosteson and G. M. Whitesides, *Biophys. J.*, 2003, **85**, 2684–2695.
23 M. Kitta, H. Tanaka and T. Kawai, *Biosens. Bioelectron.*, 2009, **25**, 931–934.
24 D. Yamaura, D. Tadaki, S. Araki, M. Yoshida, K. Arata, T. Ohori, K. Ishibashi, M. Kato, T. Ma, R. Miyata, H. Yamamoto, R. Tero, M. Sakuraba, T. Ogino, M. Niwano and A. Hirano-Iwata, *Langmuir*, 2018, **34**, 5615–5622.
25 T. Gutsmann, T. Heimburg, U. Keyser, K. R. Mahendran and M. Winterhalter, *Nat. Protoc.*, 2015, **10**, 188–198.
26 B. Sajadi, F. Alijani, D. Davidovikj, J. Goosen, P. G. Steeneken and F. van Keulen, *J. Appl. Phys.*, 2017, **122**, 234302.
27 X. Zhang, K. Makles, L. Colombier, D. Metten, H. Majjad, P. Verlot and S. Berciaud, *Nat. Commun.*, 2020, **11**, 5526.

28 J. Y. Kim, K. Lee, N. E. Coates, D. Moses, T.-Q. Nguyen, M. Dante and A. J. Heeger, *Science*, 2007, **317**, 222–225.
29 L.-L. Deng, S.-L. Xie, C. Yuan, R.-F. Liu, J. Feng, L.-C. Sun, X. Lu, S.-Y. Xie, R.-B. Huang and L.-S. Zheng, *Sol. Energy Mater. Sol. Cells*, 2013, **111**, 193–199.
30 A. S. Amin, A. Asghari-Roodsari and H. L. Tan, *Pfluegers Arch.*, 2010, **460**, 223–237.
31 W. A. Catterall, *J. Physiol.*, 2012, **590**, 2577–2589.
32 G. Klein, A. Gardiwal, A. Schaefer, B. Panning and D. Breitmeier, *Forensic Sci. Int.*, 2007, **171**, 131–135.
33 A. Beyder, J. L. Rae, C. Bernard, P. R. Strege, F. Sachs and G. Farrugia, *J. Physiol.*, 2010, **588**, 4969–4985.
34 M. Stevens, S. Peigneur and J. Tytgat, *Front. Pharmacol.*, 2011, **2**, 71.
35 M. Kollarik, H. Sun, R. A. Herbstsomer, F. Ru, M. Kocmalova, S. N. Meeker and B. J. Undem, *J. Physiol.*, 2018, **596**, 1419–1432.

Faraday Discussions

ROYAL SOCIETY OF CHEMISTRY

DISCUSSIONS

Emerging electrochemical methods at the nanointerface: general discussion

Mark A. Buckingham, Xiangkun Elvis Cao, Shuai Chang, Hong-Yuan Chen, Qianjin Chen, Selvaraj Chinnathambi, Martin A. Edwards, Stefano Fornasaro, Justin Gooding, Caleb Hill, Ayumi Hirano-Iwata, Ali Reza Kamali, Frédéric Kanoufi, Steffi Krause, Kazue Kurihara, Serge G. Lemay, Steven Linfield, Xu Liu, Yi-Tao Long, Si-Min Lu, Hui Ma, Bing-Wei Mao, Gabriel N. Meloni, Svetlana Menkin, Shelley Minteer, Shane O'Neill, Popular Pandey, Hang Ren, Ben Slater, Zhongqun Tian, Patrick Unwin, Dimitrios Valavanis, Alain Walcarius, Katherine (Kallie) Willets, Yanfang Wu, Li Xiao, Weilin Xu, Wenrong Yang, Yi-Lun Ying and Zhu Zhang

DOI: 10.1039/d2fd90003c

Alain Walcarius opened a general discussion of the paper by Dimitrios Valavanis: Thank you for your nice talk. In relation to what you did to evidence electrochemically induced calcium carbonate precipitation, do you think that you could combine your system with a quartz crystal microbalance to follow mass variation simultaneously? (It would be interesting to follow the kinetics of other electrochemically induced growth processes).

Dimitrios Valavanis replied: Thank you for your question. While a scanning pipette probe method (and SECCM in particular) can be used to approach and scan over a quartz crystal microbalance (QCM) electrode, IRM is more restrictive to the type of media in the optical path. In a reflected light configuration, light from the objective would first reach a glass layer, and then the absorbing layer (in our case indium tin oxide).[1] A QCM electrode in this arrangement would allow the acquisition of IRM images.

1 S. Campidelli, R. Abou Khachfe, K. Jaouen, J. Monteiller, C. Amra, M. Zerrad, R. Cornut, V. Derycke and D. Ausserré, *Sci. Adv.*, 2017, **3**, e1601724.

Katherine (Kallie) Willets remarked: I noticed that you use ITO from SPI Supplies as your substrate for these studies. Our group and others have found significant heterogeneity in the performance of these substrates as supporting electrodes. Have you noticed this heterogeneity with your technique? In particular, have you seen inactive or "dead" regions within a sample or noted variations in the reaction kinetics across a single substrate or as you move from substrate-to-substrate?

Dimitrios Valavanis replied: Indeed, there are reports of this particular ITO substrate's heterogeneity from both our collaborator's group[1] and from our group (manuscript in preparation). During our work with the proposed SECCM-IRM technique, we did notice that substrates with an ITO thickness of 70 nm (rated 70–100 Ω sq.$^{-1}$ sheet resistance) offered a more uniform optical response, compared to substrates with an ITO thickness of 700 nm (rated 8–12 Ω sq.$^{-1}$ sheet resistance). In terms of the varied electrochemical activity, while this can affect local reaction rates, the precipitation, and the growth of crystals, we believe that the variation did not hinder our experiments at the scale investigated. A high frame rate IRM imaging of a large SECCM meniscus area (tens of μm in diameter) under controlled atmospheric conditions would allow us to closely follow the spatial distribution of events, and in turn record localised fluxes and surface heterogeneities.

1 P. Ciocci, J.-F. Lemineur, J.-M. Noël, C. Combellas and F. Kanoufi, *Electrochim. Acta*, 2021, **386**, 138498.

Qianjin Chen observed: The study demonstrated a powerful technique by hybridizing SECCM and IRM. I remember the study by Bo Zhang on the fluorescence microscopy of single gas nanobubbles on ITO during electrolysis in a neutral solution. They observed that a fluorescence signal arose ~500 mV before HER occurred. Did you ever observe a similar phenomenon in the KNO_3 solution using IRM? Were there any optical intensities which arose before HER on ITO?

Dimitrios Valavanis responded: Thank you, Qianjin. We did not observe any local change in the optical intensity before the electrochemical HER in KNO_3 solution. We did observe some change at more negative potentials ($E \approx -1.5$ V *vs.* Ag/AgCl), however this is more likely to be indium oxide reduction into metallic indium, considering that the centroid position of the features did not vary at all, as opposed to what is expected from nanobubble formation.[1]

1 P. Ciocci, J.-F. Lemineur, J.-M. Noël, C. Combellas and F. Kanoufi, *Electrochim. Acta*, 2021, **386**, 138498.

Popular Pandey remarked: Thank you Dimitrios for the nice presentation. How is the SECCM-IRM method advantageous over existing methods for tracking phase formation, as well as imaging the SECCM meniscus?

Dimitrios Valavanis answered: Thank you, Popular. In terms of imaging the SECCM meniscus, IRM offers a direct, *in situ* visualisation. Other methods perform indirect measurements,[1] *ex situ* imaging,[2] or require the addition of fluorescent dyes.[3] Here, the meniscus can be imaged directly during any possible SECCM experiment that can be conducted on optically appropriate substrates. Furthermore, SECCM-IRM uniquely allows the viewing of events within the confined meniscus cell (which is easily tuneable in size) and very close to the working substrate. These characteristics, along with the inherent capacity – owing to the scanning probe architecture – for effortless repeats of the same experiment, make the hybrid technique well-suited to tracking phase formation.

1 C. L. Bentley, M. Kang, F. M. Maddar, F. Li, M. Walker, J. Zhang and P. R. Unwin, *Chem. Sci.*, 2017, **8**, 6583–6593.
2 N. Ebejer, A. G. Güell, S. C. S. Lai, K. McKelvey, M. E. Snowden and P. R. Unwin, *Annu. Rev. Anal. Chem.*, 2013, **6**, 329–351.
3 M. E. Snowden, A. G. Güell, S. C. S. Lai, K. McKelvey, N. Ebejer, M. A. O'Connell, A. W. Colburn and P. R. Unwin, *Anal. Chem.*, 2012, **84**, 2483–2491.

Steven Linfield asked: Does the SECCM droplet experience any evaporation (or flow from within the pipette) during the course of the $CaCO_3$ crystallisation? If so, how much does flow within the droplet contribute to the observed precipitation near the meniscus edge? It seems from the COMSOL simulations that longer tip–substrate distances allow for more uniform O_2 saturation at the substrate surface, so perhaps longer precipitation times at further distances could reveal the contribution of internal flow to the meniscus deposition.

Dimitrios Valavanis replied: We do not expect that limited evaporation or flow from within the pipette would directly result in precipitation because the initial pH conditions are so far from the supersaturation point – indeed, significant currents are required to perturb the pH sufficiently to observe precipitation. We propose that the buffering effect on the pH due to diffusion from inside the pipette can explain the delayed precipitation around the centre of the wetted area; in combination with some effect from the atmosphere surrounding the meniscus (*e.g.*, CO_2 dissolution), contributing to the distinct display at the very edge of the wetted area. Both of those effects would hinder the precipitation. In terms of evaporation in general, we do not observe that taking place significantly in the $CaCO_3$ system, compared to other systems that we have investigated. Based on images just after pipette retraction and at later times, the meniscus residue is not decreasing by any considerable amount.

Yi-Tao Long opened a general discussion of the paper by Hang Ren: Have you considered the capacitance at the water (droplet)/substrate interface in the SECCM experiment? Could you include the effect of interface capacitance in your model? Could you please comment on the future spatial resolution that SECCM could achieve?

Hang Ren responded: In the SECCM experiment, the steady state is readily achieved within the timescale of a typical experiment. Therefore, in our paper (DOI: 10.1039/d1fd00046b), we focused on the steady-state behavior. Therefore, the capacitive behavior is not important.

We have previously estimated the capacitance of the water (droplet)/substrate interface in SECCM experiments and obtained a capacitance of $\sim$50 $\mu F\ cm^{-2}$ on a Pt electrode in 10 mM $HClO_4$.[1] The spatial resolution of the SECCM is increasing over time. It will be fundamentally limited by the smallest droplet that can be formed stably due to the capillary force. This has been theoretically estimated to be as wide as 5 water molecules (1.9 nm).[2] However, we are not there yet.

1 Y. Wang, E. Gordon and H. Ren, *Anal. Chem.*, 2020, **92**, 2859–2865.

2 J. Jang, G. C. Schatz and M. A. Ratner, *Phys. Rev. Lett.*, 2004, **92**, 085504.

Gabriel N. Meloni commented: Usually, we can control the electrolyte concentration and in your work these are very low (lower than the electroactive species). At a 1 : 2 ratio (electrolyte to electroactive species), there is almost no ohmic drop in the CVs, which is interesting, meaning that for the *IR*-drop's sake, there is no need for much electrolyte, but as mass transport changes a bit migration starts to be important. What is the reason for these concentrations? Is that due to the applications you are envisaging? Also, as the largest resistance in the system is the pipette opening, small changes in the solution composition due to electrochemical reactions for the system proposed (low electrolyte) would impact the *IR* drop in the system a lot.

Hang Ren said: You are correct that when you have sufficient electrolytes compared with the electroactive species, the *iR* drop is not significant. We have chosen redox couples with fast outer-sphere electron transfer rates, so that the deviation in the voltammogram can be associated with the *iR* drop. The application is towards electrocatalysis, where the determination of the Tafel slope has been widely used to probe electrode mechanisms, requiring a very accurate determination of the Tafel slope. Indeed, deviation from 30 mV dec^{-1} to 40 mV dec^{-1}, which can easily be caused by an uncompensated *iR* drop, can change the interpretation of the mechanism. There are scenarios in electrocatalysis where the electrolyte concentration is lower than the electroactive species, for example, the electrolysis of water, *i.e.* the oxygen evolution reaction and hydrogen evolution reaction from water, where the water concentration is 55 M.

Popular Pandey remarked: Thank you for the beautiful presentation. What is the contribution of the electric double layer at the inner wall of the nanopipette, as well as the ion concentration polarization near the pore opening, to the *iR* drop?

Hang Ren answered: It depends on the size of the nanopipette opening relative to the thickness of the electrical double layer (EDL), *i.e.*, the Debye length. We do not specifically consider the effect of EDL from the charged glass wall of the nanopipette. When the Debye length is much shorter than the size of the opening of the nanopipette, our model is expected to be accurate. This is often true in practical systems, in which the electrolyte concentration is typically >1 mM and the radius of the nanopipette is typically >100 nm. If the EDL at the opening of the nanopipette overlaps, the local conductance will change, which is observed as ion current rectification.

Xiangkun Elvis Cao asked: This is really interesting work to estimate and compensate for the *iR* drop. I was wondering whether you could provide a high-level summary of comparing your method with other local electrode kinetics measurement approaches?

Hang Ren replied: There are other local electrochemical kinetic measurements which have been developed, for example, some optical methods based on converting electrochemical measurements to an optical signal (*i.e.*, electron to photon). However, quantitative analysis of the transduction to reveal the true local electrode kinetics sometimes can be convoluted, and might not be generally applicable for all electrochemical systems. Scanning electrochemical microscopy (SECM), a popular scanning electrochemical probe technique, is also very good at measuring local electrode kinetics. Scanning electrochemical cell microscopy (SECCM), which our work is focused on, can isolate a small nanointerface to be interrogated, and the current is directly associated with the interface isolated by the nanodroplet.

Martin A. Edwards remarked: This is nice, clear work. Firstly, a comment. In your work, you assume uniform concentration. It is possible to analytically determine resistances with changes in concentration with ET and migration and/or diffusion. The poster by Kamsy Anderson from my group (#5 at this meeting), which we are presently writing up, details these expressions. I think this could be quite easily included in your model. I don't anticipate dramatic changes in the interpretation, but it would extend it to a wider range of scenarios, *e.g.*, insufficient supporting electrolyte.

Second, a question. Do you think that evaporation at the end of the pipette could cause enhanced concentration at/near the meniscus, altering the conductivity? Would you notice that? Or is mass transport so darn fast on those length scales that it wins out?

Hang Ren answered: Thanks. We considered integrating the concentration as a function of distance into the analytical expression, but decided to go with simplicity: a simple equation that captures the major effects from the geometry and the bulk electrolyte concentration. We look forward to your work. Evaporation is a good point. The concentration near the liquid–air interface could be higher, but we haven't experimentally verified whether it would dramatically change the conductivity. We do keep the experiments at a controlled humidity so that the evaporation rate in a given experiment is stable. The fact that the numeric model predicts the limiting current very well suggests that the role of evaporation is insignificant.

Mark A. Buckingham opened a general discussion of the paper by Caleb Hill: You use this technique to probe the charge carriers in photoelectrochemical excitation, can you also use this technique for thermo-excited carriers, as in thermoelectrics instead of photovoltaics? Do you imagine that there would be a difference in the results between the two excitation mechanisms? Both applications are obviously hugely topical for metal chalcogenides/dichalcogenides, so have you also looked at metal mono-chalcogenides?

Caleb Hill replied: We do suspect that the methodology would work similarly well, providing that a suitable probe can be used to locally heat the sample. As for the metal mono-chalcogenides, not yet, but we look forward to applying this approach to a wide variety of materials such as these!

Steffi Krause said: Could you please comment on the difference in the excitation profile between the bulk sample (Fig. 2b of the paper, DOI: 10.1039/d1fd00052g) and the bilayer (Fig. 3b of the paper)? It looks as though both samples display a narrow excitation profile at 0.45 V, while the profile for the bulk sample significantly expands at higher potentials of 0.60 V and 0.75 V. Do you have comparable data for the bilayer sample at higher potentials? As for the anomalously broad photocurrent profile in Fig. 4b of the paper, is it possible that light scattering at defects in the complex WSe_2 sample could have caused the photocurrents at a greater distance from the excitation point than expected?

Caleb Hill replied: Comparable data can be found in the accompanying photocurrent movies. Unlike in the bulk samples, the profile does not significantly expand at the stated potentials. We attribute the unique responses observed at bilayer systems to the formation of interlayer excitons which are effectively repelled from the pipet interface. As for the second question, the defects mentioned could certainly scatter light. However, the source in these experiments is a tightly focused (<1 μm spot size) laser, making it possible to excite carriers within a pristine region of the sample.

Yi-Lun Ying remarked: In bulk n-WSe_2, have you considered the carrier recombination, as the carriers seem to exhibit long diffusion lengths?

Caleb Hill responded: In the bulk measurements, we quantify the diffusion length, which itself depends on both the diffusion constants of the carriers, as well as their lifetimes:

$$\mathrm{L} = \sqrt{D\tau}$$

The carrier lifetime will be set by the bulk recombination rate. In order to determine the recombination rate independently, a time-resolved variant of the described technique would need to be employed.

Qianjin Chen asked: These are very elegant measurements using a carrier generation-tip collection mode SECCM technique. What is your laser power and beam size used to initiate the photoelectrochemical reaction? Will these cause any significant increases in the local temperature and background current noise? Would this be a issue?[1]

1 R. Hao, Y. Fan, M. D. Howard, J. C. Vaughan and B. Zhang, *Proc. Natl. Acad. Sci. U.S.A.*, 2018, **115**, 5878–5883.

Caleb Hill responded: The laser power was <1 μW and the beam spot size was <1 μm in diameter. While the resulting local intensity is high, the resulting temperature fluctuations one would expect are actually rather small (<0.1 K) due to efficient thermal transport away through the substrate.

Serge G. Lemay remarked: This is a rather open-ended question that was prompted by one of the answers during the discussion. In thin ion-sensitive field-effect transistors made with, *e.g.*, graphene, $1/f$ noise usually dominates the noise spectrum at low concentrations. This has been attributed to gate noise from the gating electrical double layer. I am curious as to whether the same mechanism dominates in your system, or if instrumental fluctuations (from either the light source or the readout circuitry) render this inaccessible.

Caleb Hill responded: Excellent question. Because these CG-TC experiments involve the detection of photogenerated carriers, the stability of the light source employed, with respect to both the intensity and spatial position, will dominate the apparent noise. This would not be associated with the measurement itself, but fluctuations in the concentration profile generated *via* excitation.

Ali Reza Kamali remarked: This is a very nice work. I am wondering how WSe_2 flakes could be characterised in order to identify the monolayer regions? What was the minimum thickness of the flakes? Thank you very much!

Caleb Hill replied: Transition metal dichalcogenides exhibit thickness-dependent changes in the band structure which make it possible to identify monolayers optically. In particular, monolayers possess direct bandgaps which enable strong photoluminescence (PL) to be generated. The strength of this PL, as well as its characteristic spectrum, allow monolayers to be easily identified.

Alain Walcarius opened a general discussion of the paper by Serge G. Lemay: Very nice work! What would be the best resolution (lowest microdroplet size) achievable with this technique? Would it be "simply" related to the characteristics of the nanoelectrode array (the nanoelectrodes' diameters and distance between them) or something else?

Serge G. Lemay responded: You are correct, as far as we understand it is that "simple". More specifically:

For *imaging* particles, the pitch of the nanocapacitor array is the defining parameter. This is, in turn, dictated by the CMOS process employed. Feature sizes have decreased exponentially with time, and a higher nanocapacitor density could be achieved today compared to the 90 nm technology that was used to fabricate the chips used in these measurements.

For *detecting* particles that are smaller than the array pitch, the key parameters are the analyte-to-electrode size ratio and the sensitivity (background noise) of the readout AD converters. As an example of what has been possible so far, we have previously published a measurement of single 40 nm dielectric particles.[1]

1 F. Widdershoven, A. Cossettini, C. Laborde, A. Bandiziol, P. P. van Swinderen, S. G. Lemay and L. Selmi, *IEEE Transactions on Biomedical Circuits and Systems*, 2018, **12**, 1369–1382.

Yi-Lun Ying asked: Is there any crosstalk between the nanoelectrodes? Is it possible to fabricate a nanoelectrode array with a smaller diameter? Can you comment on what the size limitation of the electrode is?

Serge G. Lemay replied: Cross-talk between the electrodes contributes to the background (stray) capacitance. This is not a problem *per se*, as neighboring electrodes already serve as the counter-electrode to the electrode row being actuated. In principle, changing the solution impedance at an electrode will also redistribute the coupling to neighboring electrodes, but we have no experimental evidence of this so far.

The electrodes are fabricated using a mostly standard CMOS process, the devices used in the present work being based on a 90 nm low-power node from

TSMC. Roughly speaking, the electrode diameter scales with the minimal metal half pitch, which means that substantial further decreases in size are already possible today. It is important to keep in mind, however, that in most sensing applications it is insufficient to decrease the size of the electrode. To retain an analytically relevant number of receptor sites, one must simultaneously scale up the number of nodes so as to retain sufficiently effective binding kinetics. We address this point further in ref. 1.

1 S. G. Lemay and T. Moazzenzade, *Anal. Chem.*, 2021, **93**, 9023–9031.

Ben Slater asked: (1) Please can you detail the method and elapsed time from the emulsion preparation with sonication to the measurement of an image? What is the process for incubation? What does successful incubation refer to? Some images in the paper (DOI: 10.1039/d1fd00044f) are recorded at $t = 0$; what is $t = 0$ defined as?

(2) Regarding the preparation of emulsions with more than one component, specifically referring to 3B – are nitrobenzene and silicone oil miscible? How can you be sure that the droplets are individual components?

(3) On page 9, you state "The FWHM of these Gaussian-like profiles is typically 5–10 mm", what is the significance of the FWHM of the lines?

Serge G. Lemay replied: (1) As stated in the article (DOI: 10.1039/d1fd0044f), the emulsions were prepared by sonication and the chip was incubated within less than a minute. The incubation process entailed loading the emulsion into a syringe and injecting it by hand into the microcavity above the chip through a narrow capillary. No in-line emulsion preparation was employed, although this could be implemented if desired.

(2) We make no specific claim about possible coalescence of the different components. These were exploratory experiments of the nanocapacitor array platform and we did not attempt to characterize the emulsions independently.

(3) The FWHM gives an idea of the size of the object being imaged. For micron-sized dielectric particles, this scales approximately with the size of the particle (see, for example, ref. 1). The situation is more complicated, however, as the sensitivity to the sub-parts of a larger object falls off nonlinearly with distance from the array. This (so far unresolved) complexity is highlighted by the discrepancy between the two numerical methods for particle sizing investigated in the present article. It is fair to say that a full solution to the problem of how best to determine particle size using this method remains to be determined.

1 C. Laborde, F. Pittino, H. A. Verhoeven, S. G. Lemay, L. Selmi, M. A. Jongsma and F. P. Widdershoven, *Nat. Nanotechnol.*, 2015, **10**, 791–795.

Xu Liu enquired: For the physicochemical characterization that you have mentioned, can you compare it with other similar methods, such as SEM?

Serge G. Lemay responded: There is no doubt that SEM and its derivatives, such as EDX, are far more powerful and versatile tools than impedance spectroscopy for material characterization, even using dense arrays as we do here. The one virtue of chip-based impedance methods is that they provide a route to rather sophisticated information in a very cheap way (if the chips are mass-produced).

The costs of a readout board are a fraction of those of a SEM. A somewhat oversimplified comparison (but certainly on the side of fair) is to think in terms of the relative advantages and disadvantages of an optical microscope *versus* a miniaturized photodiode array. The microscope is much better at imaging, but there are many experiments and applications where you would rather have the cheap, compact and versatile photodiodes.

Gabriel N. Meloni commented: I was curious if the lack of agreement in the silicone oil permittivity comes down to a frequency limitation due to the smaller capacitance of the system. Is the frequency limiting you to a few systems?

Serge G. Lemay answered: In principle no, insofar as the total charge that is transferred per cycle does not become much smaller than that from the background capacitance and leakage so as to become undetectable. That said, we do need to correct for the properties of the readout circuit when the effective capacitance and the corresponding average current deviate significantly from the conditions under which the system was calibrated. Such large differences appear with emulsions because of the large differences in the dielectric constants, and this might introduce systematic errors that we have not yet fully understood. These corrections were addressed in the context of measurements at different frequencies, which also lead to large changes in the average current, in ref. 1.

We, however, also consider the possibility that the discrepancy stems from the limitations of our modeling. At the moment, we do not model any electrical double layers and the associated relaxation when interpreting experimental data. But this is a crucial component in interpreting electroosmotic data, for example, and it is likely to be relevant here as well.

1 F. Widdershoven, A. Cossettini, C. Laborde, A. Bandiziol, P. P. van Swinderen, S. G. Lemay and L. Selmi, *IEEE Transactions on Biomedical Circuits and Systems*, 2018, **12**, 1369–1382.

Frédéric Kanoufi asked: To follow up on Gabriel N. Meloni's question, could the disagreement in the permittivity measurement when multiple emulsions are used be due to an uptake (partitioning) of the second organic solvent by the silicone droplets? I guess that the water used for the emulsion is saturated with the organic solvent, which may then be able to be captured by the silicone droplets. It is actually an interesting analytical platform, like liquid nanochromatography, where the silicone can as a stationary phase to concentrate and detect a wide range of analytes. As you also mentioned the dynamics aspect of this high throughput technique, would you be able to track the dynamics of whatever happens to the droplets? For example, would it enable a new picture of liquid–liquid interface electrochemistry dynamics, where ion transfer at these interfaces generally has faster kinetics than electron transfer?

Serge G. Lemay answered: I agree that partitioning may play a role. We have not attempted this analysis, as there are also some open questions about the overall calibration of the method when the effective capacitance varies over a large range (as mentioned in the article, DOI: 10.1039/d1fd00044f). It will be interesting to return to this question when we are absolutely confident about the absolute accuracy; until then we only claim a qualitative discrimination capability.

Furthermore, using the platform to monitor the dynamics of partitioning is an exciting suggestion. At the moment this would be difficult; our present time resolution for imaging is about 200 ms, which is barely comparable to the time scale for diffusive processes on the ~10 μm length scale. One could imagine achieving this in the future, but we are unfortunately not yet in a position to do so.

Yi-Tao Long commented: Single entity characterization *via* CMOS-based nanocapacitor arrays is amazing. I am wondering if this array could be used to measure redox nanoparticles?

Serge G. Lemay responded: In principle, absolutely, as we can measure the charge/discharge signals of a few electrons per cycle. But this would require fast electrochemical kinetics: the particle should be oxidized and reduced on the time scale of the cycling period, which is only 20 ns at the frequency where we normally operate.

Hang Ren said: What is the strategy for measuring such a small capacitance? To measure attofarad capacitance, is it because of much lower stray capacitance, or is it because it is measuring a difference and the stray capacitance is constant during the measurement? What can be done to measure even smaller capacitance changes?

Serge G. Lemay replied: The former. Implementing the readout circuit on a chip allows a decrease of the stray input capacitance to about 1 fF. Also important is that the $1/f$ noise from the modulating transistors does not contribute to the overall noise level; discharge is basically a passive process where the transistor is not actively gated, as would occur in an ISFET. This may seem like a detail, but note that the $1/f$ noise dominates at very high frequencies in nanoscale transistors. In the future, some gains are possible by improving the A/D converters in the readout circuitry.

Steffi Krause asked: Could you please show a schematic of your chip architecture and explain how you address individual electrodes? It would also be interesting to see how you extract the capacitance from the measurements, as the use of CMOS technology implies the use of an array of field effect capacitors, the capacitance of which would be mostly dependent on the insulator and the space charge layer in the semiconductor, but not necessarily dependent on what happens on the solution side.

Serge G. Lemay replied: This is a good question. Regarding addressing individual electrodes, this is explained in detail in ref. 1. Regarding the second part of your question, our use of the term 'effective capacitance' can lead to confusion. The effective capacitance is defined as the amount of charge induced on the nanocapacitor during one charge/discharge cycle, divided by the applied voltage step amplitude. At low frequencies or in situations where ion migration is negligible, this corresponds to the actual capacitance of the electrode in solution plus the stray capacitance. We believe that this is the case that applies in most of the *Faraday Discuss.* data (DOI: 10.1039/d1fd00044f). In situations where migration is relevant (at significant salt concentrations, such as in biological samples, for example), the effective 'capacitance' is in fact determined by the solution resistance. In addition, note that the transistor gate capacitance does not really enter the problem. This is because we do not use the gate charge as the signal, which is very different from the mode of operation of an ISFET.

1 F. Widdershoven, A. Cossettini, C. Laborde, A. Bandiziol, P. P. van Swinderen, S. G. Lemay and L. Selmi, *IEEE Transactions on Biomedical Circuits and Systems*, 2018, **12**, 1369–1382.

Zhu Zhang addressed Selvaraj Chinnathambi and Serge G. Lemay: Since the electrodes are very sharp, and the frequency of the applied potential is about 50 MHz, I was wondering if the nanoelectrode generates a dielectrophoresis force that can attract or repel single particles or molecules? If so, during the periodic switch on/off of the nanoelectrode rows, the particles could bounce between rows of electrodes, which would be very interesting.

Serge G. Lemay and Selvaraj Chinnathambi answered: We agree that dielectrophoresis can generate significant forces at nanoelectrodes. For example, under our operation conditions, we estimate a (negative) force of order 10 fN for a 10 nm radius particle at a distance of 100 nm from an electrode, assuming a purely dielectric particle with a low dielectric constant and a frequency above f_2. This corresponds to a drift velocity of $\sim$5 μm s^{-1} (!). The force, however, decreases to 10 aN for a 1 nm radius entity and falls off at the fifth power of distance, so the effect is mostly relevant for larger nanoparticles and very locally at the electrode, where the field gradient is strongest. Note, however, that these estimates are based on the conventional Clausius–Mossotti factor treatment, ignoring the electrical double layer of the particle, and are therefore coarse approximations. To our knowledge, the description of dielectrophoretic forces under conditions where the particle size, the electrode size and/or the electrical double layer thickness are comparable has not yet been fully elucidated (see, *e.g.*, ref. 1).

1 E.-S. Yu, H. Lee, S.-M. Lee, J. Kim, T. Kim, J. Lee, C. Kim, M. Seo, J. H. Kim, Y. T. Byun, S.-C. Park, S.-Y. Lee, S.-D. Lee and Y.-S. Ryu, *Nat. Commun.*, 2020, **11**, 2804.

Martin A. Edwards communicated: With your chips, is it possible to zoom-in, *i.e.*, on only a sample or subset of the electrodes? In doing so, it might be possible to gain higher temporal resolution while maintaining the spatial resolution. Hopefully, this would allow you to see more dynamics of your system. Obviously, this would limit the area you are looking at and so you would have to be lucky that the region that you are looking at contains a droplet (unless one wanted to take on the challenge of making some sort of real-time adaptive strategy!)

Serge G. Lemay communicated in reply: You are absolutely correct. Since the array is read out in a mostly sequential manner, it is possible to zoom-in on sub-regions so as to trade a smaller array size in exchange for higher speed. There are 8 AD converters on chip, so in principle one could push this idea as far as measuring only 8 electrodes with a concomitant 8192× increase in speed. However, implementing this zoom-in capability would require a substantial reprogramming of the readout board for the chip, which is not something our lab has delved into yet.

Hong-Yuan Chen commented: A very innovative concept, "Electronic Chemistry" (or "Electron Chemistry"), is presented for the topic of next generation nanoelectrochemistry, which is the study of the electrochemistry of single entities down to the nanometer or sub-nanometer scale. The fundamental theories and rules of classic "electrochemistry", such as the Cottrell equation and Coulomb's law, are appropriate to macroscopic chemical processes, but no longer apply in the limited conditions of single nanoparticle and single molecule measurements. These theories and practical measurements are confined to the macroscopic and continuous systems of study, and lose sight of the granularity on the meta-microscale. These discrete and stochastic processes have attracted much attention for the nanoelectrochemical analysis of single entities. This new understanding, "Electronic Chemistry", could expand the electrochemical theories approaching the limit of temporal and spatial resolutions and highlights the discontinuous granularity and multiscale nature of electrochemical study in the quantum regime. Guided by this science, next-generation nanoelectrochemistry and molecular electronics could be characterized more accurately.

Si-Min Lu replied: I totally agree with you that the electrochemical process is stochastic and discrete at the microscale. The ions and single entities cannot be regarded as the mass point in single entity electrochemistry. There is an urgent demand to develop high-bandwidth instrumentation and advanced data algorithms to track the transient electron transfer process. Also, a comprehensive theoretical modelling framework (a combination of Marcus theory, the random walk model, Simmons model, *etc.*...) is needed to simulate the mass transfer and electron transfer kinetics at the nanoconfined interface. And single entity electrochemistry at the nanoconfined interface is actually a interdiscipline with the combination of quantum physics, physical chemistry, organic chemistry...

Yi-Lun Ying responded: It is a very interesting concept; the transport of charge, electrons and ions at the nanointerface and underconfinement conditions needs to have new models.

Yi-Tao Long said: In this session, we look deep into the nanointerface under confined conditions, which has raised more questions than we expected but provides more chances to get close to the 'truth'. For example, the correlations should be well understand to bridge the characteristics we found in the nanointerface with the sensing/catalytic performance on a large scale interface. Moreover, massively parallel measurements are expected on the nanointerface. Any comments on the future development of nanoelectrochemical methods and their applications are welcome.

Weilin Xu answered: For the study of confined nanointerfaces, in the future, it probably will go deep into the interface on the electrode, especially the nanointerface between the electrode and single entities. Some questions, such as the ET process between them, are very interesting.

Stefano Fornasaro responded: I believe that in the future, research at the nanointerface will have to take into account the increasing use of simulations and data processing algorithms, including machine learning, in designing experiments to optimise performance or test hypotheses more effectively. These methods are currently reshaping how researchers collect, analyse, and interpret data. Their impact on nano-electrochemistry is still not fully realised.

Shelley Minteer replied: In the field of bioelectrocatalysis, there is a great interest in nanoconfinement for enzyme cascades (*i.e.* Fraser Armstrong's recent work[1]), but there are challenges in developing single entity measurements for evaluating these systems. I hope that as we see more advances in single enzyme measurement techniques, we will be able to learn more about nanoconfinement in enzyme cascades.

1 G. Morello, C. F. Megarity and F. A. Armstrong, *Nat. Commun.*, 2021, **12**, 340.

Li Xiao commented: I hope that the research and method development of nanoelectrochemistry can help us understand more problems encountered in practical devices. For example, we would like to see the transport path of ions in polymer electrolytes, and how ions transfer and participate in the reaction at the electrode/polyelectrolyte interface. I hope that this can be realized by nano-electrochemical methods.

Zhongqun Tian opened a general discussion of the paper by Bing-Wei Mao: *In situ* atomic force microscopy is a powerful surface characterization technique, and your work (DOI: 10.1039/d1fd00043h) has demonstrated that it can be used to study the initial stage of Li deposition and dissolution, and provide insights on the SEI influence on the reversibility of Li deposition–dissolution cycle. In your presentation, you gave a perspective, in which combined *in situ* AFM and SECM is recommended as a nanoelectrode-based scanning probe microscopic technique. Could you comment on what additional merits beyond the capability of ECAFM can be expected from the AFM-based SECM for the nanoscale study of Li deposition?

Bing-Wei Mao answered: Thank you for the good question. AFM can provide a multitude of nanoscale information based on its imaging and beyond imaging capabilities. And one of the most important merits of AFM in electrochemistry is that it can work in electrochemical environments. However, AFM itself lacks chemical and electrochemical identification ability, and not all measurements can be performed in liquids by conventional AFM. For example, measurements of electric properties, such as conductivity and surface potential, in a liquid environment are difficult if not impossible because faradaic processes occurring at a bare conducting probe would cause significant interferences. Incorporating nanoelectrodes into AFM probes to form AFM-based SECM can enable important issues to be studied under electrochemical control, which would otherwise be difficult by conventional AFM. Dedicated AFM-based SECM is already commercially available. By utilizing such an AFM-based SECM, many scientific problems in Li deposition can be explored. For example, the conductivity of the SEI could be measured during Li deposition, or the local electrochemical reactivity of a porous SEI could be probed. Also, the interface formed between the SEI and electrolyte could be studied by combined surface potential measurements at the SEI side and force curve probing of the structure at the electrolyte side. Such a study would not only provide important information about Li ion transport across the interface, but also extend our understanding of the electric double layer in an interface formed by two ionic conductors involving a SEI.

Yi-Lun Ying enquired: Would the type of solid electrolyte give different types of SEI formation? Could you please comment on the correlation and difference between nanoscale measurements and large scale performance in SEI studies?

Bing-Wei Mao responded: Thank you for your good questions. If the electrochemical window of a solid electrolyte is insufficiently large, SEIs will be formed at the interface. Their compositions are expected to be different from those formed from liquid organic electrolytes. Since AFM is a surface technique, it is unfavorable for studying the SEI formation taking place at the electrode and solid electrolyte interfaces. Nevertheless, it is possible to study the interface if a section of the interface can be exposed. Such investigations are still very limited.

Regarding the correlation and difference, nanoscale measurements by AFM, for example, are to characterize the formation and properties of SEIs from the microscopic point of view. The obtained information is helpful for understanding phenomena observed in the macroscopic performance measurements. The differences lie in that SEI-related issues are complex and the macroscopic performance is usually an averaged result of many influencing factors, while microscopic measurements are usually based on model systems and emphasize some discrete aspects. A thorough understanding of the related processes requires a multitude of *ex situ* and *in situ* characterizations by microscopic and spectroscopic techniques and, if possible, under *operando* conditions.

Frédéric Kanoufi remarked: It is a very interesting study showing what *in situ* AFM can bring to the study of the initial growth steps of the SEI during Li deposition. The diversity in the SEI behavior is very interesting. Looking at them quickly, it seems that they have similar chemical structures and comparable thicknesses; maybe the SEI-Li is the stiffest, even though it is the one that allows the more reversible Li dissolution/deposition. I wonder how you picture these deposition/dissolution processes? To be effective, Li ions have to enter or be expelled from these protected Li metal spheres. This suggests that quite large amounts of Li ions should be transported through the SEI. Is the SEI acting as an homogeneous ion-transport layer? Do you have an estimate of such ion transport through the different SEI layers?

One could also invoke a mechanism where small ion channels/pores within this SEI shell are allowing the growth/dissolution of Li metal. This is maybe more consistent with the evolution of the shape of the SEI shell during Li dissolution. As the SEI shape seems to change, does it mean that the SEI always adheres to the underlying Li metal or does it just behave as a deformable shell/scaffold (a deflating balloon)? The latter behavior is suggested in different NP conversion mechanisms implying inorganic shells. For example, the dissolution of nanoparticles (by galvanic exchange reactions) under precipitating conditions generates a scaffold SEI-like inorganic layer and the dissolution proceeds through multiple pitting channels.[1]

1 J. G. Smith, X. Zhang and P. K. Jain, *J. Mater. Chem. A.*, 2017, **5**, 11940–11948.

Bing-Wei Mao responded: Thank you for your critical questions. The differences of the two SEIs (SEI-Cu-step and SEI-Li) formed by potential stepping to the potential where Li overpotential deposition takes place are indeed small. But it is worth noting that their thickness difference is *ca.* 25%, which is not minor. Also, the small amount of Li_3N contained in the SEI-Li is not present in SEI-Cu. Nevertheless, I agree that the mechanism which explains the opening of segmented SEI-shells formed by the two kinds of SEI requires further explorations.

Because an ideal SEI is electron insulator and Li ion conductor, Li deposition and dissolution have to proceed underneath the SEI. It is right that Li ions have to transport in and out of the SEI layer during deposition and dissolution, and SEI does act as an ion-transport layer, but probably an inhomogeneous layer. The estimated diffusion coefficient of Li ions in SEIs is in the range of $\sim$ 5–8 $\times$ 10^{-8} cm^2 s^{-1}.[1] Regarding the mechanism of small ion channels/pores, the formation of the SEI is a self-limiting process; once a complete SEI is formed, it is expected to be an electron insulator, and pores and channels can only exist in the inner region of the SEI. So, it is possible that Li deposits fill in the pores and channels of the SEI, but this should be unlikely to cause changes of the shape of the SEI unless Li dendrite growth occurs. The evolution of the shape of the SEI is due to further formation of the SEI on the growing nucleus. Finally, for the preformed SEI-Cu-lsv, inorganic rich components dominate the thick inner layer. Such an SEI has insufficient elasticity to accommodate the growth of the Li nucleus by solely deforming itself. Even for the two SEIs (on Cu and on Li created after potential stepping) with inorganic–inorganic hybrid structures, it is unlikely that SEIs can be stretched to accommodate Li deposition. Rather, SEIs would always continually form on the growing nucleus.

1 J.-W. He, Y. Ge, W.-W. Wang, J.-H. Wang, Z.-B. Chen, H.-Y. He, Q.-H. Wu, J.-W. Yan and B.-W. Mao, *ChemElectroChem*, 2021, **8**, 62–69.

Svetlana Menkin enquired: How can the under-potential deposition of lithium be distinguished from the SEI formation?

Bing-Wei Mao responded: Thank you for the good question. Li underpotential deposition (UPD) can be distinguished from the SEI formation by comparing CVs of different cycles (which are not shown in this paper (DOI: 10.1039/d1fd00043h), but can be found in our previous paper[1]). The SEI formation is almost completed after the first cathodic linear sweep of potential, so that both the main peak at *ca.* 1.3 V and the background current in the lower potential region from the electrolyte reduction nearly disappear on subsequent CV curves, but the peak at *ca.* 0.5 V always persists. So, we can safely distinguish Li UPD from SEI formation. Nevertheless, the underpotential deposited Li layer may somehow interact with the partially formed SEI, as is mentioned in ref. 1, which could cause compositional rearrangement of the inner layer of the SEI. But we believe that such a kind of influence is different from that of the bulk Li deposit.

1 J.-W. He, Y. Ge, W.-W. Wang, J.-H. Wang, Z.-B. Chen, H.-Y. He, Q.-H. Wu, J.-W. Yan and B.-W. Mao, *ChemElectroChem*, 2021, **8**, 62–69.

Frédéric Kanoufi said: There is a further very nice concept that you are showing in your work: controlling the SEI chemistry and its formation dynamics allows the production of a self-limiting growth of the Li metal. Such homogeneous SEI formation actually explains the Li spherical shape growth rather than a dendritic one. However, this could also mean that the amount of deposited Li will also be limited. How far have you been able to grow the Li deposit? How thick the Li can be grown? In the extent of large capacity, do you end up with dendrites anyway or does this chemically-controlled SEI prevent the dendritic growth regime by a more controlled Li^+ ion release?

Bing-Wei Mao responded: Thank you for your good questions. In our paper (DOI: 10.1039/d1fd00043h), our efforts are to follow the nucleation and growth processes of the initial stage of Li deposition. This is partly because AFM is more suitable for working with a smoother surface, with several micrometers being the upper limit. The highest capacity we reached was 0.5 mA h cm^{-2} (*i.e.* 1.8 C cm^{-2}), which is equivalent of a 2.5 μm thick dense Li deposit. Since we did not perform experiments of long deposition and dissolution cycles and to sufficiently high capacitance, dendritic growth was not observed, even at the capacitance of 0.5 mA h cm^{-2}. But I wish to mention that the shell-opening phenomenon or creation of holes and defects have also been observed by the group of Wen and Wan,[1] as well as by Shiraishi and coworkers,[2] in PC-based electrolytes. So, it is reasonable to believe that the shell-opening phenomenon generally exists regardless of the solvent environment, and may be accounted for by the model of segmented SEI-shells. Such a phenomenon is different from dendritic growth, which is more associated with the inhomogeneity of the SEI and depletion of Li ions in a later stage, especially after long cycling, but it is also a negative effect that degrades the Li deposition and dissolution reversibility.

1 Y. Shi, G.-X. Liu, J. Wan, R. Wen and L.-J. Wan, *Sci. China Chem.*, 2021, **64**, 734–738.
2 S. Shiraishi and K. Kanamura, *Langmuir*, 1998, **14**, 7082–7086.

Svetlana Menkin asked: How is the "dead" SEI defined? Does it have different mechanical properties?

Bing-Wei Mao responded: Thank you for your good questions. "Dead" SEIs are defined as SEIs that are no longer usable as an ordinary ionic conductor for accommodating Li deposition. By nature, such SEIs are detached from the substrate so that Li deposition does not proceed by Li ions being transported through the SEI. In our paper (DOI: 10.1039/d1fd00043h), the dead SEIs arise from the opening of segmented SEI-shells during Li dissolution. We measured the mechanical properties of the SEIs and the results show that segmented SEI-shells do not necessarily bear very different Young's moduli from those of the integrated SEI-shells. It is the discontinuity of the SEI-shell kinds that is a more important factor leading to dead SEIs. Our work about the mechanical properties of SEIs will be published elsewhere.

Si-Min Lu opened the discussion of the paper by Kazue Kurihara: Could you please discuss the difference of the redox reaction in a confined space and bulk solution? Where does the difference come from? Is it from the aspects of mass transport or electrode kinetics?

Kazue Kurihara answered: In the bulk system, redox species diffuse from the bulk to the electrode, so we may say that this is the diffusion controlled reaction. When the redox species are in a confined space, they cannot diffuse far away, so are cycled between the cathode and anode, resulting in an enhanced redox reaction (an increase in the current with the decreasing separation distance between the electrodes). What we observe in this study is that the redox current sharply increases when the space becomes narrower. We think that the redox species are organized or linked like polymers in the narrower space to form redox chains, thus the current sharply increases.

Patrick Unwin commented: Thank you very much for your interesting paper. Evidently, the SFA has considerable potential (excuse the pun!) to probe the EDL at electrodes under operation. You discuss the approach curve (longer distance) before the sudden onset in current to redox cycling in the gap. Have you analysed the shape of the curve and magnitude of the current in terms of a simple thin layer cell model?

Kazue Kurihara responded: Thank you for your interest. We have not analyzed the current curve. Because the current in the longer distance range increased with the increasing redox concentration, we think that the current increase in the longer distance range is due to well-known redox cycling in the gap between electrodes. So, we should be able to analyze it by a thin layer cell model, although our interest was more in the short range component. Thank you for your suggestion.

Patrick Unwin said: In the very high current region at close distances, if the "chain reaction of electron transfer between redox species" you state in your paper (DOI: 10.1039/d1fd00060h) is operating, it would be expected to be sensitive to the particular redox species (homogeneous self-exchange rate constant). Have you tried these experiments with other redox couples?

Kazue Kurihara replied: We agree with this comment, although we have not tried other redox couples.

Justin Gooding asked: The hysteresis in this sudden current increase is very interesting. Is it possible that in the experiment Prussian blue is being formed, which then bridges between the two electrodes? Then, upon withdrawal, polymer strands extend and maintain the effective electrochemical short between the two electrodes?

Kazue Kurihara replied: When we stopped the separation of the electrode at the middle of the largest current range, the current gradually decreased. So we think that the bridging is a temporary phenomenon in our case. We do not know the stability of Prussian blue polymer chains.

Justin Gooding said: In light of the possibility of ferricyanide forming Prussian blue polymer chains, the suggestion of Professor Pat Unwin to try a different redox species that cannot form polymeric materials seems like a good idea.

Kazue Kurihara answered: We completely agree with this comment. Thank you.

Qianjin Chen asked: While the 10 mM $KClO_4$ solution gave a consistent sudden current increase between the approach and separation, the redox molecule $Fe(CN)_6{}^{3-/4-}$ gave a large hysteresis. Are you able to extract/obtain the cyclic voltammetry at each of the electrodes at different distances (redox cycling)? This may help understand the contribution from mass transport within the nanogap?

Kazue Kurihara responded: Yes, we can perform cyclic voltammetry at different distances, though we did not do such measurements for this case. I agree with this comment. The measurement will provide useful information. The observation which convinced us of the presence of organized redox molecules is that the current gradually decreased when we stopped separation at the middle of the separation. We think that the organized structures finally disappear by separation.

Shuai Chang said: You have shown force distance curves in your presentation. What is the difference between your force measurement instrument and regular AFM?

Kazue Kurihara replied: There are several advantages of SFA when compared to AFM as follows.

(1) More defined geometry in the case of SFA, which employs a crossed cylindrical geometry. Thus, the measured force can be converted to the interaction between flat surfaces, so quantitative theoretical treatments are easy. In case of AFM, the shape of AFM tips usually varies, so this is not easy. There is a colloidal probe AFM method, but it may not be so easy to find a suitable probe and control their surface potential by applying the voltage.

(2) The size of the surface is relatively large, *i.e.* 30 μm in diameter at contact, so it is possible to monitor multiple molecules like real systems. If the ultimate goal of AFM is a single molecular measurement, this difference is important since a real system consists of many molecules.

(3) Because of the large surface area, it is possible to measure weaker forces.

(4) The large surface enabled us to combine the force measurement with other complementary techniques for surface characterization, including cyclic voltammetry, fluorescence spectroscopy, and X-ray diffraction.

(5) It is possible to employ the measurement for a wider variety of samples. We can modify surfaces using most surface modification techniques, including silane coupling agents, polymer coating and metal sputtering.

Shuai Chang enquired: In what kind of media can you perform your experiments? Liquid, air, organic solvents?

Kazue Kurihara responded: We usually perform the measurements in liquids (water and organic solvents), but sometimes in air (or gas).

Si-Min Lu opened the discussion of the paper by Shuai Chang: I am interested in the tunneling distance in the stochastic collision electrochemistry. Have you estimated the tunneling distances of the CBA molecule, FC molecule, and AFC molecule?

Shuai Chang answered: CB7 is a structurally rigid molecule with a molecular height of 0.91 nm between two portal rims. The tunneling distance between the gold NP and the gold nanoelectrode at collision can be estimated to be close to 0.91 nm. For experiments with guest molecules (FC and AFC) present, the guest molecules are complexed inside the CB7 cavity and the tunneling distance is still dominated by the CB7 height.

Alain Walcarius asked: Did you observe any effect of the NP size on the response time (as it might influence the extent of the host–guest interactions)?

Shuai Chang replied: We have performed the collision experiments with different GNPs, *e.g.*, 10 nm and 40 nm GNPs, in the study of a different molecule (4-aminothiophenol) using the same setup.[1] No electrochemical current changes were detected, but the event rate appears to be higher with the smaller sized GNPs. In this work (DOI: 10.1039/d1fd00054c), we did not carry out experiments with different sized NPs, and we expect a similar effect of the size of the gold NPs, which should have some effect on the response time but little effect on the electrochemical current amplitude.

1 J. Guo, J. Pan, S. Chang, X. Wang, N. Kong, W. Yang and J. He, *Small*, 2018, **14**, 1704164.

Alain Walcarius enquired: Perhaps you could also think about functionalizing the nanoparticles to tune their interactions with the macrocyclic ligand?

Shuai Chang answered: Thanks for the great suggestion. Functionalizing the nanoparticles with different ligands could definitely tune the interactions and allow more complex measurements.

Wenrong Yang said: The electrochemical sensing of host–guest chemistry triggered by nanoparticle collisions is an very interesting approach, which could be used to study the dynamics of other noncovalent (weak) interactions. Could the authors discuss if pH would play an important role in such noncovalent (weak) interactions?

Shuai Chang responded: Thanks for the comments. In the current study (DOI: 10.1039/d1fd00054c), the experiments were carried out at a neutral pH (pH = 7.4) and the pH effect on the NP collisions and the corresponding electrochemical current changes was not studied. But we do expect that pH would play an important role in affecting such noncovalent interactions because the electron rich portal of CB7 and the properties of the negatively charged gold nanoparticles can be greatly affected by the local environmental pH, thus the colliding interactions (distance and frequency) between the molecule and NP can be impacted.

Ben Slater commented: (1) Is the inclusion complex a 1 : 1 or 1 : 2 complex? In your methods, you describe it as a 1 : 2 complex, but the diagrams show a 1 : 1 complex. You also report a final concentration of 0.5 mM; how did you verify that you have 100% yield?

(2), did you separate excess guest molecules or do you have mixture of 0.5 mM of the inclusion complex and 0.5 mM of guest Fc? No method to separate the excess from inclusion complex is reported.

(3) Apart from the electrochemical measurements, do you have any evidence for the formation of the inclusion complex? Did you characterise the inclusion complex by other methods, for example NMR?

(4) Another report sonicates for 1 day; how can you be sure that the complex has been formed in around 30 minutes?[1]

(5) Please can you comment on the difference in the baseline conductivity of Fig. 2B of the paper (DOI: 10.1039/d1fd00054c)? CB[7]-Fc is around 3× higher than CB[7]AFc, why is this?

1 C. I. R. Magalhães, A. C. Gomes, A. D. Lopes, I. S. Gonçalves, M. Pillinger, E. Jin, I. Kim. Y. H. Ko, K. Kim, I. Nowik and R. H. Herber, *Phys. Chem. Chem. Phys.*, 2017, **19**, 21548–21555.

Shuai Chang replied: (1) Two guest molecules, FC and AFC, can bind to CB[7] in a 1 : 1 stoichiometry with a high binding constant. In this work, we prepared the host–guest solution at a 1 : 2 ratio to facilitate the complete occupation of the CB[7] cavity by guest molecules. The final concentration is 0.5 mM. It may not be a 100% yield because the host–guest complex is in a dynamic binding equilibrium. Based on the 0.5 mM and 1 mM mixture of CB7 and FC, together with a binding constant of 4×10^5 M, we calculated that the binding yield is ~99.5%. AFC has a higher binding affinity than FC toward CB7 and the yield will be even higher.

(2) We did not separate excess guest molecules from the inclusion complex, but used it directly for the electrode modification.

(3) Yes, we also characterized the inclusion complex using other methods, such as NMR[1] and surfaced enhanced Raman spectroscopy.[2]

(4) Both FC and AFC have a high binding affinity toward CB7, and we found that 30 minutes is enough for the complexed state to be formed, as characterized in the above referenced articles.

(5) The difference in the baseline conductivity of Fig. 2B of the paper (DOI: 10.1039/d1fd00054c) could be associated with the variation of the exposed electrode area.

1 M. Huang, M. Sun, X. Yu, S. He, S. Liu, W. M. Nau, Y. Li, T. Wu, Y. Wang, S. Chang and J. He, *J. Phys. Chem. C*, 2020, **124**(29), 16143–16148.
2 Q. Ai, J. Zhou, J. Guo, P. Pandey, S. Liu, Q. Fu, Y. Liu, C. Deng, S. Chang, F. Liang and J. He, *Nanoscale*, 2020, **12**(32), 17103–17112.

Steven Linfield commented: You showed that the two current levels observed during nanoparticle collisions with a CB7-Fc modified nanoelectrode were the result of the ferrocene molecule's rotational state. However, some L-type events are stable for tens of seconds, even ~100 s in one case. Do you ever see changes between the two rotational states (and therefore current level) during an L-type event? If the two rotational states are interchangeable, why are currents related to this observed more during the shorter S-type events than in the longer L-type events? And what could the ratio of the ferrocene configurations (horizontal and vertical) observed in the current tell us about the stability of ferrocene in the host?

Shuai Chang responded: In both S and L types of current switching signals, we could see two current states, which we attributed to the two rotational states of FC in a CB7 cavity. This can be seen in the histogram displayed in Fig. 2C and D of the paper (DOI: 10.1039/d1fd00054c). We do see some events showing the interchange of two states in an individual current switching spike, but these are quite rare. The S and L types of signals should be predominantly related to the properties of CB7 molecules on the nanoelectrode surface. We see both S and L types of signals in the NP impact experiments on CB7, CB7-FC and CB7-AFC, and find that the addition of guest molecules does not show an obvious impact on the appearance rate of S *versus* L signals. The rotation of FC inside the CB7 cavity is induced by the interaction between the gold tip and the FC.[1] Different rotational geometries could stay stable at different gap distances between two gold surfaces. In the current distribution, the two peaks for the S type events are not as well defined as those of the L type. We have reasoned that this is because in the L types of signals, FC is more stabilized because NP is transiently trapped on the molecular surface and this serves better to stabilize the geometry, as discussed in the manuscript. And for your last question, the ratio of the ferrocene configurations (horizontal and vertical) observed in the current may not be directly correlated to the stability of ferrocene in the host, because both horizontal and vertical states of FC can be stable depending on the gap distance between the two gold electrodes. Thus, if we dig further, the ratio of the ferrocene configurations could possibly reflect the statistically weighted gap conditions in various NP colliding events. One thing to note here is that the NP impact experiments are very different from the exact single molecule conductance measurements because the collision motion between NP and the electrode in the NP impact experiment is a random Brownian motion, where the tunneling distance at the collision can not be manipulated. While the single molecule conductance experiments could precisely control the tunneling distance between two electrodes.

1 M. Huang, M. Sun, X. Yu, S. He, S. Liu, W. M. Nau, Y. Li, T. Wu, Y. Wang, S. Chang and J. He, *J. Phys. Chem. C*, 2020, **124**(29), 16143–16148.

Yanfang Wu queried: Can you comment on the possible effect of the size of gold nanoparticles on the amplitude of the switching current signals (or electrochemistry activity) from ferrocyanide for the situation when gold nanoelectrode was modified with CB7 (or CB7 with a guest molecule) in your system? Also, how important do you think the homogeneity of gold nanoparticles may be in your study?

Shuai Chang answered: In a previous study of a different molecule (4-aminothiophenol) using the same setup,[1] we have performed the collision experiments with different GNPs, *e.g.*, 10 nm and 40 nm size GNPs. No electrochemical current changes were detected, but the event rate appears to be higher with the smaller sized GNPs. In this work (DOI: 10.1039/d1fd00054c), we did not carry out experiments with different sized NPs, and we expect a similar effect of the size of gold NPs. Similar for the homogeneity of gold NPs, we assume that it does not have a significant impact on the electrochemistry activity in our measuring system.

1 J. Guo, J. Pan, S. Chang, X. Wang, N. Kong, W. Yang and J. He, *Small*, 2018, **14**, 1704164.

Yi-Lun Ying opened the discussion of the paper by Weilin Xu: What about the homogeneity of the nanoparticles? Would you please comment on what is the major challenge in future nanoscale electrochemical measurement for studying nanoparticles in catalysis?

Weilin Xu responded: In the case, the homogeneity of the nanoparticle is not perfect, actually these particles are heterogeneous in both size and performance. The major challenge in future nanoscale electrochemical measurement for studying nanoparticles in catalysis can be summarized into the following two points: (i) how to monitor or analysis the electrocatalysis-induced structure variation of nanoparticles precisely; (ii) how to correlate the structure of a nanoparticle with its electrocatalytic performance precisely at single particle level.

Yi-Tao Long queried: About the platinum NPs, how did you characterize them? TEM or HRTEM?

Weilin Xu responded: As for the characterization of the Pt NPs, in this work, these nanoparticles were characterized by multiple types of tools, such as TEM, HRTEM, XRD, XPS, to reveal the physico-chemical properties of these particles, while for the electrocatalytic properties, CV, CO-stripping and electrochemical impedance spectroscopy (EIS) were adopted to characterize their electrocatalytic properties.

Ali Reza Kamali asked: Very nice work! The core/shell nanostructure that you suggest, with copper located at the center of the nanoparticles and Pt just on the surface, is very interesting, and may find many applications in different fields. However, it might be difficult to come to this conclusion based on Fig. 1 presented in the article (DOI: 10.1039/d1fd00047k). Have you conducted further characterization to further support this suggestion? Thank you.

Weilin Xu answered: Thanks for your comment here. Indeed, Fig. 1 of the paper (DOI: 10.1039/d1fd00047k) shows that the Cu may be surrounded by Pt skin, while the further XPS results shown in Fig. 2b of the paper, which show that the only metallic Cu(0) can be detected with no Cu oxide, can further support that the Cu atoms are in the core with a Pt skin outside. Of course, more direct evidence could be obtained with some other complex tools, such as STEM-EELS, through which we probably can directly see its core/shell structure. Thanks for your comment. We will try this in future if such tool is available.

Ali Reza Kamali commented: As a comment, further characterisation such as STEM-EELS analysis might be helpful. Thanks.

Weilin Xu responded: Yes, more complex tools, such as EEL, can be adopted to characterize the microstructure of the alloy nanoparticles. If it can be done deeply, more precise information can be revealed for a better understanding of its morphology control. Indeed, the precise structure of the Pt nanoparticles could be characterized more clearly *via* tools such as STEM-EELS. We will try this in future if such tools are available. While in the present, based on some other results obtained, such as those shown in Fig. 1 and 2 of the paper (DOI: 10.1039/d1fd00047k), we still can tell the possible structure of the particles we obtained in this work.

Hui Ma asked: Thank you for your wonderful talk. In this work, you provide a great method to fabricate Pt–Cu nanoparticles. Could this method be extended to other materials? And how can the thickness of the shell be controlled?

Weilin Xu replied: Thanks for your comment on our work. Basically, such a method can be extended to the synthesis of some other nanomaterials with different metals, for example, with the same protocol, we have synthesized PtFe and PtCo alloy nanoparticles with similar structures. For the thickness of the Pt shell, in the present, we found that it can be tuned or controlled simply by varying the mass ratio between Pt and Cu.

Yi-Tao Long opened a general discussion of the paper by Ayumi Hirano-Iwata: Have you tried different materials for supporting the bilayer? Could you please provide more details about the bilayer formation?

Ayumi Hirano-Iwata responded: We have tried SiN/Si, Teflon, and porous Al_2O_3 as materials for suspending BLMs. All of them were useful for BLM formation. BLMs in SiN/Si chips were highly stable, but channel incorporation was sometimes very difficult. BLMs in Teflon chips were fragile, and they were often broken by mechanical shocks, including solution exchanges. However, the incorporation of biological channels was more frequently observed. BLMs in porous Al_2O_3 exhibited an intermediate nature between the SiN/Si and Teflon systems. BLMs were formed by the monolayer-folding method as follows. The membrane supporting material was placed in a Teflon recording chamber so that the supporting material separated the two compartments. A buffer solution was added to each compartment. The level of the buffer surface was set to below the aperture. A lipid solution was then carefully spread on the buffer solution in each compartment. After evaporation of the solvent, a BLM was formed by gradually raising the buffer level until it surpassed the aperture. A more detailed procedure was described in our previous paper.[1] For the Teflon and porous Al_2O_3 systems, pretreatment of the aperture edge with *n*-hexadecane was necessary before adding the buffer solution.

1 D. Yamaura, D. Tadaki, S. Araki, M. Yoshida, K. Arata, T. Ohori, K. Ishibashi, M. Kato, T. Ma, R. Miyata, H. Yamamoto, R. Tero, M. Sakuraba, T. Ogino, M. Niwano and A. Hirano-Iwata, *Langmuir*, 2018, **34**, 5615.

Hui Ma commented: I notice that the noise of the chip at different potentials is quite different. I am curious about what caused this phenomenon?

Ayumi Hirano-Iwata answered: This is a difficult question to answer. As shown in Fig. 5 of the paper (DOI: 10.1039/d1fd00045d), the noise level varied with the lateral voltage in an unsystematic manner. We noticed that the noise currents at the lateral voltage of 1 V were larger than those at different lateral voltages. But we do not know the reason for the unsystematic noise level.

Hui Ma said: In Fig. 5 of your paper (DOI: 10.1039/d1fd00045d), the noise of your system under light is greater than that in dark. What caused this phenomenon? Is it because the lipid bilayer is unstable? Will this influence the current recording?

Ayumi Hirano-Iwata replied: There is still a lot we do not know about noise. As you commented, we often observed higher noise currents under the light than in the dark. However, we sometimes observed the opposite phenomena, for example, for PCBM-doped BLMs at lateral voltages of 0 and 1 V. At the moment, we have no clear answer to the unsystematic noise level. But the noise current was still lower than the photo-induced current, which enabled us to detect the photoresponse.

Yi-Lun Ying enquired: The structure of the electrode-equipped Si chip plays an important role in the formation of the bilayer. Is there any simulation model available for the design of the chip?

Ayumi Hirano-Iwata replied: Unfortunately, we do not have a good simulation model for the chip design. At the moment, we designed the shape of the SiN aperture edge based on the stability of the BLMs and SiN septa; the thinner the edge angle of the SiN aperture, the more stable the BLMs, but the SiN aperture itself becomes more fragile. As a result of this trade-off, the edge angle is currently 45°.

Shelley Minteer responded: Not that I am aware of.

Popular Pandey said: Thank you for the nice presentation. Why was an electrode-equipped Si chip used for PCBM-doped BLMs, and an electrode-equipped Teflon chip used for BLMs incorporated with voltage-gated sodium channels? Also, could you please comment on why the current noise level for the two BLMs systems are very different at applied AC/DC voltages?

Ayumi Hirano-Iwata replied: BLMs formed in the electrode-equipped Si chips were more stable with tolerance to a high transmembrane voltage of ±1 V, while BLMs in the electrode-equipped Teflon chips ruptured at ±1 V. For sensor applications like PCBM-doped BLMs as a light-sensitive membrane, the higher applied transmembrane voltage is preferable to obtain larger responses. Therefore, we used the Si chip platform for the PCBM-doped BLMs. On the other hand, for recording channel activities, the probability of ion channel incorporation was sometimes more important than the stability of the BLMs, especially for ion channels with low open probability, like voltage-gated sodium channels (Na_v channels). From our experience, incorporation of the $Na_v1.5$ channel was more frequently observed for BLMs in the electrode-equipped Teflon chips than those in the electrode-equipped Si chips. We chose the Teflon chip platform for recording ion channel activities.

Regarding the current noise level for the two BLM systems, the pronounced differences lie in the flexibility of the supporting material (Teflon *vs.* Si_3N_4) and the necessity of *n*-hexadecane to be present around the aperture, as discussed in the paper in detail (DOI: 10.1039/d1fd00045d). In brief, we tentatively assume that the BLMs formed in the flexible Teflon films are likely to vibrate mechanically due to the Coulomb force from the AC lateral voltage, generating a displacement current noise near the BLM surface. On the other hand, the noise would be suppressed in the Si chip due to its mechanical stability. At the same time, because it is necessary to apply *n*-hexadecane near the aperture to facilitate BLM formation on Teflon chips, the edge of the lateral electrodes around the aperture would be in indirect contact with the BLM. A DC lateral voltage would not induce an overly large electric field around the edge of the lateral electrodes. In the Si chip systems, *n*-hexadecane is not required to form BLMs. The BLM is in direct contact with the SiO_2-protected electrodes at the edge of the aperture. We speculate that a larger electric field would be generated at the sharp edges of the lateral electrodes under the lateral DC voltage, resulting in high noise.

Shane O'Neill asked: Do you know if the size of the lipid bilayer has an influence on the lateral voltage effect on the transmembrane voltage or current?

Ayumi Hirano-Iwata answered: Since we used *n*-hexadecane for the pretreatment of the aperture edge, the BLMs were surrounded by the *n*-hexadecane torus in the aperture. The size of the BLMs varied from membrane to membrane due to the *n*-hexadecane torus, even if they were formed in the same aperture. Nonetheless, from our experience, the lateral voltage was most efficiently applied to BLMs when the BLMs were formed in the apertures of 120 ± 20 μm. When we used apertures larger than 150 μm, the effect of the lateral voltage was rarely observed. With apertures smaller than 100 μm, *n*-hexadecane often clogged them. Based on these observations, we usually used an aperture size in the range of 100–140 μm.

Martin A. Edwards asked: Do you know the mechanism of the behavior you observed? Did you investigate how the behavior varies as a function of the PCBM concentration? Perhaps there is a critical concentration required to observe this behavior, or you will gain some other insight from doing this.

Ayumi Hirano-Iwata replied: The mechanism of the photo-induced current observed for BLMs doped with PCBM is still under investigation. We have examined the PCBM concentrations of 1, 2, and 4 mg mL^{-1} in the lipid solution to form the BLMs. Much higher responses were observed at the PCBM concentration of 2 mg mL^{-1} (present study, DOI: 10.1039/d1fd00045d) compared to at 1 mg mL^{-1}. However, the photoresponse was quite unstable at 4 mg mL^{-1} due to the formation of the PCBM aggregates.

Yi-Tao Long said: Nature uses ion channels for ion transports. The crystal size of ion channel protein is around or under 1 nm. Usually, the channel surface is made from one molecule. Grouping the amino acids together provides the confined chemical and physical environments for single entity sensing. Could you please comment on perspective of the artificial bilayer and channel system?

Ayumi Hirano-Iwata responded: There would be two possible directions for the artificial lipid bilayer systems; one is a platform for single-molecule sensing, including DNA sequencing, the other is a platform for the functional analysis of membrane proteins, including ion channels. So far, BLM systems with reconstituted ion channels have mainly been used for the single-channel analysis of bacterial channels and have provided insight into the channel gating mechanism. On the other hand, for human ion channel analysis, the whole-cell patch-clamp method has exclusively contributed. However, single-channel analysis is essential for further understanding human channel functions and the pharmaceutical and medicinal applications. For this purpose, new factors are necessary that can extend the regulation of the highly complex human channel gating mechanisms. We believe that the lateral voltage has a high potential as that new regulatory factor. Now that cell-free expression systems have made it possible to synthesize various ion channels, the combination of synthesized human channels and lateral voltages opens up a new frontier in ion channel science.

Conflicts of interest

Patrick Unwin is co-inventor of granted patent PCT/GB2011/051518 "Pipets containing Electrolyte and Electrodes", which describes dual-channel SECCM.

Faraday Discussions

ROYAL SOCIETY OF CHEMISTRY

PAPER

Potential-induced wetting and dewetting in pH-responsive block copolymer membranes for mass transport control†

Seung-Ryong Kwon, ‡[a] Seol Baek ‡[b] and Paul W. Bohn *[bc]

Received 6th August 2021, Accepted 1st September 2021

DOI: 10.1039/d1fd00048a

Wetting and dewetting behavior in channel-confined hydrophobic volumes is used in biological membranes to effect selective ion/molecular transport. Artificial biomimetic hydrophobic nanopores have been devised utilizing wetting and dewetting, however, tunable mass transport control utilizing multiple transport modes is required for applications such as controllable release/transport, water separation/purification and energy conversion. Here, we investigate the potential-induced wetting and dewetting behavior in a pH-responsive membrane composed of a polystyrene-*b*-poly(4-vinylpyridine) (PS-*b*-P4VP) block copolymer (BCP) when fabricated as a hierarchically-organized sandwich structure on a nanopore electrode array (NEA), *i.e.* BCP@NEA. At pH < pK_a(P4VP) (pK_a ~ 4.8), the BCP acts as an anion-exchange membrane due to the hydrophilic, protonated P4VP cylindrical nanodomains, but at pH > pK_a(P4VP), the P4VP domains exhibit charge-neutral, hydrophobic and collapsed structures, blocking mass transport *via* the hydrophobic membrane. However, when originally prepared in a dewetted condition, mass transport in the BCP membrane may be switched on if sufficiently negative potentials are applied to the BCP@NEA architecture. When the hydrophobic BCP membrane is introduced on top of 2-electrode-embedded nanopore arrays, electrolyte solution in the nanopores is introduced, then isolated, by exploiting the potential-induced wetting and dewetting transitions in the BCP membrane. The potential-induced wetting/dewetting transition and the effect on cyclic voltammetry in the BCP@NEA structures is characterized as a function of the potential, pH and ionic strength. In addition, chronoamperometry and redox cycling experiments are used to further characterize the potential response. The multi-modal mass transport system proposed in this work will be useful for ultrasensitive sensing and single-molecule

[a]*Department of Chemistry, Research Institute of Natural Science, Gyeongsang National University, Jinju, 52828, South Korea*

[b]*Department of Chemistry and Biochemistry, University of Notre Dame, Notre Dame, Indiana, 46556, USA*

[c]*Department of Chemical and Biomolecular Engineering, University of Notre Dame, Notre Dame, Indiana, 46556, USA. E-mail: pbohn@nd.edu; Fax: +1 574 631 8366; Tel: +1 574 631 1849*

† Electronic supplementary information (ESI) available. See DOI: 10.1039/d1fd00048a

‡ Seung-Ryong Kwon and Seol Baek contributed equally to this work.

studies, which require long-time monitoring to explore reaction dynamics as well as molecular heterogeneity in nanoconfined volumes.

Introduction

Living organisms achieve selective ion/molecular transport using actively-controlled protein channels in cell membranes, a function which serves many biological functions, such as the transmission of nerve signals, muscle contraction and substance transport.[1,2] In biological channels such as nicotinic acetylcholine receptor channels, potassium and magnesium channels, a hydrophobic cavity in the channel pathway plays a key role in selectively transporting ions or molecules, exhibiting a fast gating action.[3–5] The flux of ions/molecules with co-permeating water is regulated by the wettability of the hydrophobic domain in the channel, which is reversibly switched between the wetted and dewetted states in response to an external stimulus, such as a concentration gradient or an electric field. Not surprisingly and motivated by these characteristics, great effort has been expended on preparing artificial biomimetic nanopores for applications in water purification/collection, biochemical sensing and smart valves for controlled release and delivery.[6–12]

The use of wettability switching controlled by an external stimulus, *e.g.*, electric field, light, pH and pressure, to produce repeatable high and low ionic currents in hydrophobic nanopores has attracted significant interest. For example, Smirnov and co-workers investigated wetting and dewetting behavior in hydrophobic nanopores controlled by light,[13] pH,[14] pressure[15] and voltage.[16] They demonstrated the reversible switching of wettability in hydrophobic nanopores controlled by an electric field and demonstrated a 1000-fold difference in the conductance between the wetted (high conductivity) and dewetted (low conductivity) states.[16] Siwy and co-workers further developed this concept by employing single hydrophobic nanopores with high aspect ratios – a 12 μm-long conically shaped pore with a 16 nm-diameter opening – achieving on/off switching (non-conducting and conducting ionic current) based on electric-field-induced reversible wetting and dewetting.[17] Jiang and co-workers realized dual-stimuli-, *i.e.*, electrostatic-charge- and electric-field-, responsive wetting and dewetting behaviors using surface-charged nanochannels,[18] and more recently reported light- and electric-field-controlled wetting behavior in functionalized conical polymer nanopores by adapting a light-responsive host–guest reaction.[19]

Our laboratory has long been interested in mass transport behavior in confined geometries composed of nanoscale pores. Electrochemical transistors were demonstrated in 3-electrode embedded nanopore electrode arrays (NEAs) based on potential-controlled wetting and dewetting inside defect-mediated nanochannels.[20] Furthermore, surface charge on the nanopore wall has been shown to exhibit a significant effect on ion accumulation and exclusion at a low ionic strength. By exploiting this effect, a ~2000-fold current amplification was achieved.[21] In addition, ion-selective electrochemical detection was realized by tailoring the surface charge in nanopores.[22] Recently, hierarchically organized membrane-coated NEAs have been devised using either Nafion™ or polystyrene-*b*-poly-4-vinylpyridine (PS-*b*-P4VP) to achieve cation- and anion-selectivity, respectively. Coupling these block copolymer (BCP) structures to NEAs

(BCP@NEA) enables the current amplification characteristics of the NEA to be combined with the pH-dependent ion-permselectivity of the BCP membranes.[23,24]

To date, however, the regulation of mass transport by integrating inputs from multiple control signals has not been realized. In this regard, the PS-*b*-P4VP BCP@NEA structure is of interest, because the PS-*b*-P4VP membrane exhibits pH-dependent mass transport behavior above and below the pK_a of the P4VP block ($pK_a \sim 4.8$). At pH < pK_a, the P4VP cylindrical domains are protonated at the pyridine nitrogen atom, rendering them swollen and hydrophilic, and consequently the BCP membrane acts as an anion-exchange membrane. In contrast, at pH > pK_a, the P4VP domains exhibit charge-neutral, hydrophobic and collapsed structures. Thus, inspired by the previous mass transport control in hydrophobic nanopores, here we report the systematic investigation of the potential-induced wetting and dewetting behavior of BCP@NEA structures as a function of pH, membrane thickness and ionic strength to achieve a mass transport system with a multi-modal control paradigm. As shown in Fig. 1, the BCP@NEA architectures exhibit: (1) anion-selective transport at pH < pK_a(P4VP) and, at pH > pK_a(P4VP) (2a), no transport in the absence of external potential and (2b) strong, charge-independent ion transport upon switching the P4PV nanochannels to the wetted state at sufficiently negative potentials. These characteristics of the BCP membrane make it possible to reversibly confine solution in the nanopores and to

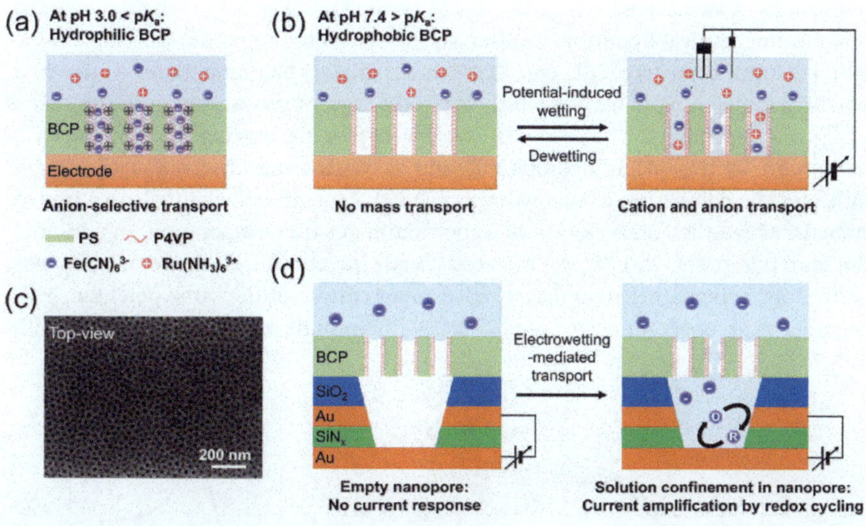

Fig. 1 Schematic illustration of the PS-*b*-P4VP BCP membrane showing 3 types of mass transport actions controlled by pH and potential. (a) Anion-selective transport at pH < pK_a(P4VP). (b) On/off mass transport switching between no transport (left) and charge-independent transport controlled by the applied potential (right) at pH > pK_a(P4VP). (c) Top-view SEM image of the BCP membrane spun onto a planar Au electrode showing the hexagonally packed cylindrical domains of P4VP. (d) Schematic illustration of a BCP@NEA structure at pH > pK_a(P4VP). Mass transport into the nanopore is initially blocked by the hydrophobic character of the membrane (left). It may be temporarily opened for mass transport across the membrane by the electric-field-induced wetting of the P4VP nanochannels, resulting in nanopore filling and subsequent redox cycling-based current amplification using BE and TE as the working and counter electrodes, respectively (right).

isolate it from the bulk solution. The multi-modal mass transport system developed in this work holds potential for ultrasensitive sensors, as well as single molecule studies, which require the long-time monitoring of reaction dynamics.

Results and discussion

Mass transport in PS-*b*-P4VP BCP membranes *vs.* pH

In order to study the mass transport behavior in the pH-responsive PS-*b*-P4VP membrane, a BCP membrane was prepared by spin-coating 2% (w/v) PS-*b*-P4VP in dioxane onto a planar Au electrode at 6000 rpm for 1 min, producing an 81 nm thick membrane, as shown in the SEM image in the inset of Fig. 2(a). BCP membranes were characterized electrochemically at different pH values. First, a pH 8.4 solution containing 5 mM $Fe(CN)_6^{3-}$ in 0.1 M KCl was used to render the P4VP domains charge-neutral, collapsed and hydrophobic. The hydrophobic BCP membranes obtained at pH > pK_a(P4VP) block the mass transport of aqueous solutions. This behavior has previously been observed and assigned to the evaporation of water being more favorable in the hydrophobic P4VP nanochannels than the condensation of water.[10,25]

Cyclic voltammetry was performed to investigate the effect of an external potential on the mass transport through the BCP membrane. At pH 8.4 > pK_a(P4VP), no current response is observed in the potential range from +0.5 V > E_{appl} > −0.4 V *vs.* Ag/AgCl, where a Nernstian response would be expected from $Fe(CN)_6^{3/4-}$, but a rapid increase in the current is observed at $E_{appl} \le -0.4$ V, as shown in Fig. 2(a). It should be noted that electrochemical hysteresis is observed in the voltammograms, especially in the first potential scan (see ESI, Fig. S1,† which shows voltammograms of the first potential cycle). A rapid increase in the cathodic current is observed in the negative-going scan starting at $E_{appl} \sim -0.85$ V. Given that the resting state of the BCP membrane at this pH is hydrophobic and dewetted, the observation of current indicates the conversion of the hydrophobic P4VP membrane channels to a wetted state, in which the condensation of water dominates over evaporation. In addition, the currents in Fig. 2(a) display hysteresis, with the current on the return positive-going scan being larger than the negative-going current at the same potential. This suggests that once the hydrophobic P4VP membrane channels are filled with

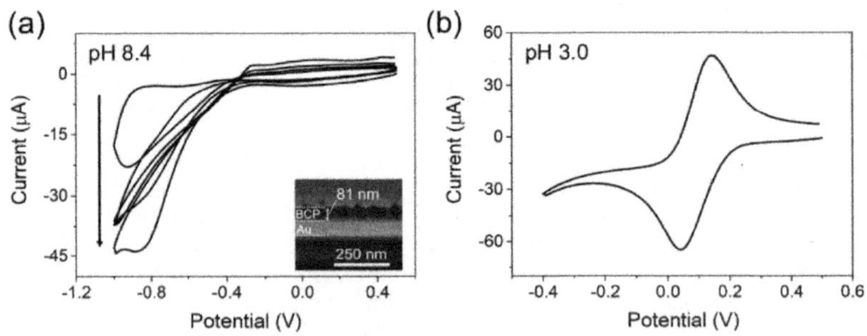

Fig. 2 Voltammograms of 5 mM $Fe(CN)_6^{3-}$ in 0.1 M KCl at pH 8.4 (a) and 3.0 (b) on the BCP-coated planar Au electrode. The arrow indicates the direction of the increasing scan number. The cross-sectional SEM image in panel (a) shows the thickness of the BCP membrane on the Au electrode obtained by spin-coating at 6000 rpm for 1 min.

electrolyte solution, they tend to maintain the wetted state. Upon reaching $E_{appl} > -0.4$ V, the P4VP channels recover the original dewetted state, as shown by the negligible current at $-0.4\ V < E_{appl} < 0.5$ V in the return scan. Furthermore, increasing limiting currents are obtained with successive scans, suggesting that more membrane channels are involved in the potential-dependent wetting and dewetting responses with successive scans. Together, the two distinct electrochemical responses, *i.e.*, (a) no current response in the potential range $+0.5\ V > E_{appl} > -0.4$ V and (b) increasing current response with electrochemical hysteresis in the potential range between $-0.4\ V > E_{appl} > -1.0$ V, suggest the primary role of potential-dependent wetting and dewetting in the hydrophobic BCP membrane at pH 8.4.

The behavior of the same BCP membrane was also characterized at pH 3.0, *i.e.*, pH < pK_a(P4VP). It has previously been shown that under these conditions, the P4VP nanodomains of the BCP membrane are hydrophilic and protonated, exhibiting approximately cylindrical structures. Under these conditions, a reversible current response of $Fe(CN)_6^{3/4-}$ is obtained, as expected from the anion-selective character of the BCP membrane at pH < pK_a(P4VP), Fig. 2(b). Notably, the cathodic peak current obtained at pH 3.0 is somewhat larger than the maximum current at pH 8.4 upon electric-field-induced wetting (*viz.* −45 and −65 μA at pH 8.4 and 3.0, respectively), implying that the sum total of the mass transport pathways under potential-induced wetting at pH 8.4 is less than those presented by the fully wetted P4VP domains at pH 3.0.

Cation transport through hydrophobic BCP membranes

Since the pH-responsive PS-*b*-P4VP functions as an anion-exchange membrane at pH < pK_a(P4VP), cyclic voltammetry was conducted with a positively charged redox species, *i.e.*, $Ru(NH_3)_6^{3+}$, to explore the charge-dependence of mass transport at pH > pK_a(P4VP) in the presence of an applied potential in order to define the conditions at which the hydrophobic BCP membrane switches to a wetted state. Thus, 5 mM $Ru(NH_3)_6^{3+}$ in 0.1 M KCl solution at pH 7.8 was characterized on two BCP membranes of different thicknesses on planar Au electrodes to probe the effects of charge selectivity and membrane thickness. On a 160 nm thick BCP membrane, gradually increasing current responses are observed at negative potentials with increasing scan numbers, reaching a quasi-steady state at *ca.* the 7th potential scan, as shown in Fig. 3(a). Electrochemical hysteresis was observed, similar to the experiments with $Fe(CN)_6^{3-}$, which is consistent with the potential-induced wetting interpretation. In contrast, an 89 nm thick BCP membrane produced a response which reaches a maximum after only 2 potential cycles, as shown in Fig. 3(b). Notably, the maximum current on the 89 nm membrane was *ca.* 2-fold higher than that on the 160 nm membrane, reflecting the enhanced transport characteristics of the thinner BCP membrane. The key observation is that even though the PS-*b*-P4VP BCP membranes are anion-permselective at pH < pK_a(P4VP), positively charged $Ru(NH_3)_6^{3+}$ is readily transported through the membrane at $E_{appl} < -0.4$ V, thus confirming that the potential-induced wetting transitions in the BCP membrane are independent of the charge of the test species.

Ionic strength-dependence of wetting/dewetting in the BCP membranes

The potential-induced wetting and dewetting were also examined as functions of the ionic strength (I) in the range 0.006 M < I < 1 M, corresponding to 0–1 M

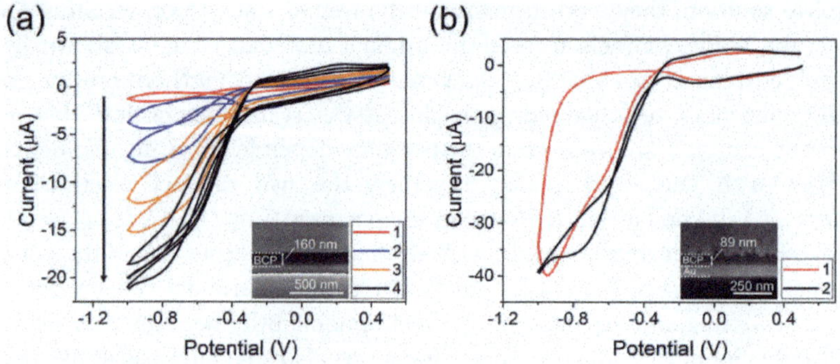

Fig. 3 Voltammograms of 5 mM $Ru(NH_3)_6^{3+}$ in 0.1 M KCl at pH 7.8 on (a) 160 nm thick and (b) 89 nm thick BCP membranes on Au. The arrow in panel (a) indicates the increasing scan number. The insets show cross-sectional SEM images of the BCP membranes spin-coated on Au.

supporting electrolyte (SE), *i.e.*, KNO_3, with 1 mM $K_3[Fe(CN)_6]$ at pH 8.4. Interestingly, a negligible voltammetric response was obtained from 1 mM $Fe(CN)_6^{3-}$ in the absence of SE over the potential range of $-1.0\ V \leq E_{appl} \leq +0.5\ V$ used for the previous measurements, as shown by the red curve of Fig. 4(a). However, extending the negative potential limit to −1.2 V produces electrochemical

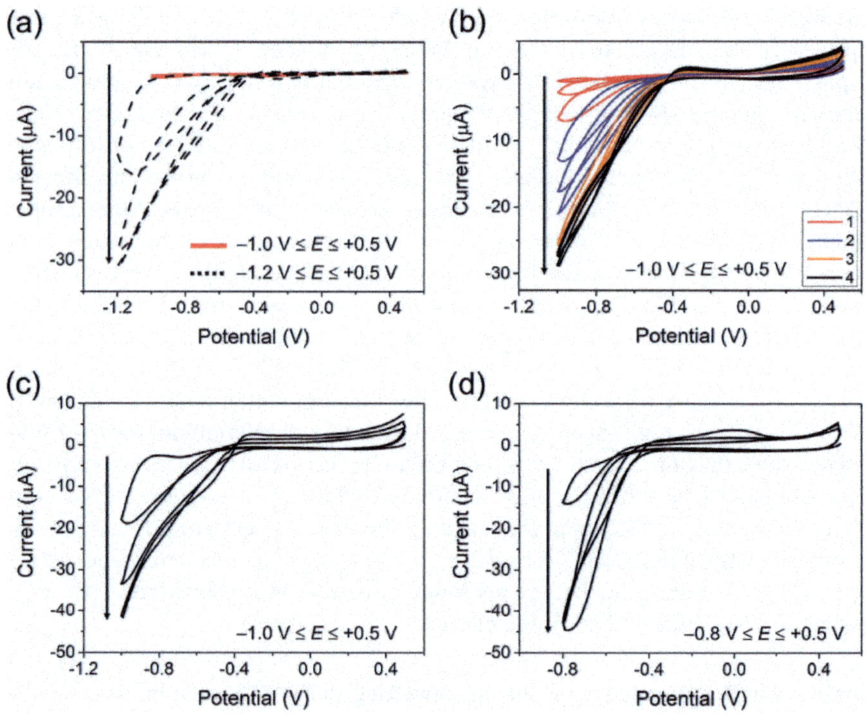

Fig. 4 Voltammetric responses of 1 mM $Fe(CN)_6^{3-}$ at varying concentrations of supporting electrolyte: (a) 0 mM, (b) 10 mM, (c) 100 mM and (d) 1000 mM KNO_3.

behavior similar to that observed in Fig. 2(a), *i.e.*, scan-dependent limiting currents and electrochemical hysteresis, consistent with wetting/dewetting. The BCP membrane behavior was subsequently characterized in a pH 3.1 electrolyte solution containing the same concentration of redox species, *i.e.*, 1 mM $Fe(CN)_6^{3-}$ without SE. As shown in Fig. S2,† the current gradually increased with the scan number and quasi-reversible voltammetric responses centered at the formal potential of $Fe(CN)_6^{3-}$ were obtained. The gradual recovery of the $Fe(CN)_6^{3/4-}$ current is consistent with regaining anion-selectivity with the protonated pyridine in P4VP nanocylinders and demonstrates the mechanical and chemical stability of the BCP membranes to changing potentials, pH and ionic strength conditions.

Next, the concentration of the supporting electrolyte was increased from 10 mM to 1000 mM. At 10 mM SE, as shown in Fig. 4(b), the BCP membrane exhibits potential-induced electrowetting over a narrower potential range, only requiring −1.0 V to induce transport. Once again, a gradually increasing current is observed with an increase in the scan number and a steady-state current response is obtained after 7 potential cycles. When the SE concentration is further increased to 100 mM, only 3 potential cycles are required to reach the maximum current value at −1.0 V, as shown in Fig. 4(c), and at 1000 mM SE, the maximum SE concentration used in these experiments, potential-induced wetting is further promoted, such that only −0.8 V is needed to induced the enhanced currents, as shown in Fig. 4(d). These experiments clearly demonstrate that it is easier to induce transport across the BCP membrane at high ionic strengths and suggests that the underlying electrowetting transitions are facilitated under these conditions.

Isolation of electrolyte solution in nanopores

Having demonstrated the ability to control transport across the PS-*b*-P4VP membrane by the external potential, as well as the pH, we next constructed architectures combining the BCP membrane with NEAs, *i.e.* BCP@NEAs. These structures were built to test the ability to isolate and confine an electrolyte solution apart from the bulk solution using the potential-controlled wetting and dewetting of PS-*b*-P4VP BCPs. NEAs were fabricated using a combination of nanosphere lithography and reactive ion etching applied to multilamellar metal–insulator–metal–insulator (MIMI, from bottom to top) structures. BCP membranes were transferred from Si substrates onto the NEAs using thermal release tape, as shown in Fig. S3.† Transferring the membrane onto the NEA produced no apparent damage, as shown in Fig. S3(b),† and cross-sectional SEM images revealed the good attachment of the membrane on the outer surface of the NEA with well-preserved nanopore structures, as shown in Fig. S3(c) and (d).† It should be noted that rather thick ∼1.2 μm BCP membranes were required to maintain the physical integrity with the NEA structures. However, these thicker membranes proved competent to control mass transport *via* potential-induced wetting and dewetting, as shown below. Moreover, these MIMI structures were terminated with a 500 nm top SiO_2 layer to facilitate the capture of the aqueous solution, as well as to reduce the influence of the membrane hydrophobicity on the electrochemical response.

When the BCP@NEA device was exposed to 50 mM $Fe(CN)_6^{3/4-}$ without SE at pH 7.6 in the absence of an applied potential, mass transport through the hydrophobic BCP membrane was blocked, as expected. This is evidenced by a negligible faradaic current being observed at either the bottom (BE) or top (TE) electrodes, collected in a 3-electrode configuration with an external Pt counter and Ag/AgCl reference electrode, as shown in Fig. S4(a).† When the BE and the TE were used as the working and reference/counter electrodes, respectively, in a 2-electrode configuration under the same solution conditions, a very small, but measurable, faradaic current due to $Fe(CN)_6^{3/4-}$ at 0 V was superimposed onto the capacitive current, as shown in Fig. S4(b),† likely resulting from a small amount of leakage.

Next, BE was used as working electrode in a 3-electrode configuration with an external Pt counter and Ag/AgCl reference electrode to perform chronoamperometry with multiple potential steps ranging from −0.5 V to −2.8 V in order to stimulate mass transport through the hydrophobic P4VP nanochannels upon potential-induced wetting. As shown in Fig. 5(a), negligible current is observed until −2.0 V, suggesting a threshold potential to induce wetting in the hydrophobic P4VP domains of the membrane. However, when E_{BE} is increased to −2.5 V, a current bust with fluctuations is observed, which is consistent with the P4VP nanochannels being switched to the wetted state. The current fluctuations are similar to the stochastic fluctuations reported by others, which have been assigned to rapid dynamics, reflecting the competition between the condensation and evaporation of water in the hydrophobic nanochannels.[16–20,26] At $E_{BE} = -2.8$ V, a second current bust is recorded with stronger fluctuations, which is consistent with a higher number of nanochannels participating in potential-induced wetting/dewetting.

After introducing electrolyte solution into the interior of the nanopores through these multiple potential steps, cyclic voltammetry was performed to confirm the successful introduction of the electrolyte solution, and its subsequent isolation from the bulk by the BCP membrane, by switching the potential back to

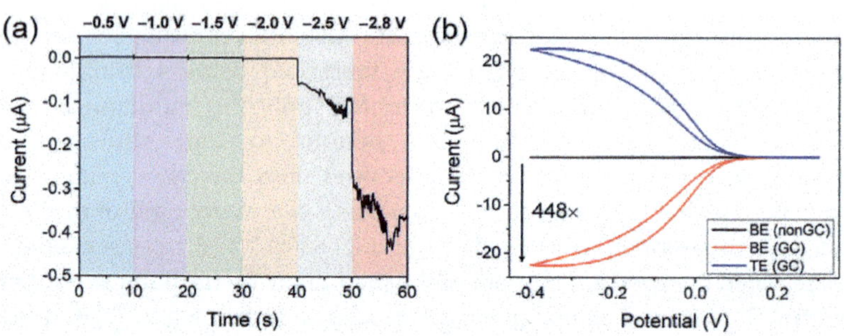

Fig. 5 (a) Chronoamperometric response of 50 mM $Fe(CN)_6^{3/4-}$ with no supporting electrolyte at pH 7.6 to multiple potential steps at a BCP@NEA structure. BE was used as the working electrode, with $E_{BE} = -0.5$ V, −1.0 V, −1.5 V, −2.0 V, −2.5 V and −2.8 V, in a 3-electrode configuration with an external Pt wire counter and Ag/AgCl reference electrode. The current sampling frequency was set at 0.5 kHz. (b) Voltammograms measured in the generator–collector (GC) mode and non-GC mode on a BCP@NEA structure. For the GC mode operation, E_{BE} was swept from +0.3 V to −0.4 V while E_{TE} was held at +0.3 V in a 4-electrode configuration with a Pt counter and Ag/AgCl reference electrode. The non-GC mode current was obtained with BE in a 3-electrode configuration with E_{TE} left at the open-circuit potential.

a value less negative than the threshold potential for wetting, *i.e.*, $E_{BE} \sim -2.0$ V. Fig. S4(c)† shows a faradaic response for 50 mM $Fe(CN)_6^{3/4-}$ (together with capacitive current), but the current is much smaller than the value obtained from a similar NEA without the BCP (*viz.* ~0.05 μA and ~0.9 μA with and without the BCP, respectively).[20,27] The observation of a stable *i*–*V* characteristic after switching the membrane to positive potentials suggests that most of the hydrophobic P4VP membrane channels recovered the original dewetted state, thereby blocking the egress of the redox species out of the NEA nanopores.

Finally, the redox cycling current was measured using BE and TE as the working and reference/counter electrodes, respectively, in a 2-electrode configuration. Unlike the current measured in the 3-electrode configuration, as shown in Fig. S4(c),† a significant current amplification is achieved with an enhancement factor of ~200, as shown in Fig. S4(d),† although non-canonical current behavior is obtained in the 2-electrode configuration, especially at low ionic strengths, due to the use of the Au TE as the reference electrode.[27] The difference between the non-GC (3-electrode with the counter and reference outside the NEA) and the GC-mode (2-electrode with BE and TE as the working and counter/reference) currents clearly reflects the difference between the blocked transport of the redox species across the BCP (3-electrode non-GC) and the facile mass transport between BE and TE in the 2-electrode GC-mode configuration. When the redox cycling (GC-mode) current was measured in a 4-electrode system, with E_{BE} being swept from +0.3 V to −0.4 V while holding E_{TE} at +0.3 V, a ~450-fold current amplification was achieved compared to the non-GC current, as shown in Fig. 5(b). Taken together, the significant difference in the current between the GC and non-GC modes clearly establishes the efficient isolation and confinement of the electrolyte solution in the NEA nanopores by utilizing the potential-induced control of mass transport across the hydrophobic BCP membrane of a BCP@NEA structure, a control concept that we attribute to the wetting/dewetting phenomena inside the membrane channels.

Conclusions

The mass transport behavior in PS-*b*-P4VP BCP membranes has been characterized as a function of the pH and applied potential in order to develop a controlled transport system, using both BCP on planar Au electrodes and BCP@NEA structures. The dominant features established are: (1) anion-selective transport at pH < pK_a(P4VP), (2a) negligible mass transport at pH > pK_a(P4VP) with no applied potential and (2b) charge-independent mass transport at pH > pK_a(P4VP) at sufficiently negative applied potentials. These effects are consistent with the potential-dependent wetting and dewetting of the membrane channels, as proposed by others.[16–20,26] Cyclic voltammetry performed on BCP membrane-coated Au electrodes demonstrates the pH-responsive anion-permselectivity of the BCP membranes and potential-controlled on/off mass transport switching. Solutions with pH > pK_a(P4VP) confirm that P4VP domains exist in a charge-neutral, collapsed and hydrophobic state, which basically stays in the dewetted state, blocking the transport of the aqueous solution in the absence of external stimuli, such as applied potentials. However, the P4VP membrane channels can be reversibly switched between the dewetted and wetted states by changing the applied potential between the positive and negative values of the threshold potential, respectively. The threshold values at which these phenomena occur are dependent on the environment, as shown by the ionic strength dependence. Furthermore, BCP@NEA structures have been employed to demonstrate intelligent sampling operations. The sequence used here consists of (a) the negative potential-induced opening of the PS-*b*-P4VP BCP membrane channels, which (b) enables the transport of the redox active probe species into the NEA nanopores, followed by (c) switching the applied potential back to a positive value or releasing the applied potential, rendering the membrane channels back to a dewetted state, which then (d) traps the probe species in the NEA nanopore. Importantly, these manipulations of the transport of the electrolyte solution are achieved at pH > pK_a(P4VP), conditions under which no transport occurs in the absence of an applied potential, and are thus attributed to the reversible switching between the wetted and dewetted states of the P4VP membrane channels.

The next phase of this research will be aimed at practical sensing applications, especially the use of these structures to achieve signal amplification from nanoconfined enzymes. In conventional electrochemical sensors, the catalytic products of enzymes immobilized on electrode surfaces can diffuse away from the electrode, diminishing the magnitude of the signal from the enzymatic products by reducing their effective concentration. By confining enzyme products in nanopores, the small volumes can support ultrasensitive sensing by accumulating enzyme products over time and generating an amplified current. Additionally, the intelligent mass transport control enabled by the PS-*b*-P4VP BCP membrane may also be useful in single-molecule studies, where long residence times are required to monitor reaction dynamics, as well as static and dynamic heterogeneity.

Experimental section

Chemicals and materials

Potassium nitrate (KNO_3), potassium ferricyanide(III) ($K_3Fe(CN)_6$), potassium ferrocyanide(II) ($K_4Fe(CN)_6$) and hexaammineruthenium(III) chloride ($Ru(NH_3)_6Cl_3$) were purchased from Sigma-Aldrich. Polystyrene-*b*-poly(4-vinyl pyridine) (PS_{48400}-*b*-$P4VP_{21300}$ $M_w/M_n = 1.09$) was obtained from Polymer Source Inc. The poly(dimethylsiloxane) (PDMS) monomer and curing agent were purchased from Dow Corning. All electrolyte solutions were prepared using deionized (DI) water for electrochemical measurements ($\rho \sim 18.2$ MΩ cm, Millipore Milli-Q system). Gold electrodes were prepared by depositing 100 nm Au (UNIVEX 450B, Oerlikon) on pre-cleaned glass slides (Glass D, Applied Microarrays, Inc.).

Preparation of the BCP membranes

Approximately 200 μL of a solution 2–3% (w/v) PS-*b*-P4VP in dioxane was drop cast onto a Au electrode (or a Si wafer) and then immediately spin-coated at various speeds (*typ.* 500–6000 rpm for 1 min). The as-spun BCP membranes were immersed in ethanol for 30 min then dried with N_2. To transfer the BCP membranes from a silicon carrier onto a NEA device, thermal release tape was first attached to the BCP membrane. When immersed in water, the BCP membrane was detached from the Si wafer and the remaining moisture on the BCP membrane was removed by N_2. Then, the BCP membrane was attached to the NEA surface, taking care to avoid trapping air at the interface. In the final step, the BCP membrane-coated NEA was placed on a hot plate heated at 100 °C to spontaneously detach the thermal release tape from the BCP.

Fabrication and characterization of the BCP@NEA membranes

NEAs were fabricated by following the previous methods reported elsewhere by this laboratory, except that in these experiments a much thicker layer of SiO_2 (500 nm) was used.[27–29] The BCP@NEA structures were characterized using scanning electron microscopy (FEI-Helios Dual-beam focused ion beam). In order to minimize the charging effects, 5 nm Ir was deposited onto the BCP membrane prior to SEM imaging. Typically, an accelerating voltage of 5 kV and an electron beam current of 0.1 nA were used for the clear structural observation of the BCP@NEA.

Electrochemical measurements

In order to perform electrochemical measurements, a PDMS well with a diameter of 6 mm was constructed on the BCP-coated Au electrode and then filled with *ca.* 100 μL electrolyte solution. Cyclic voltammetry (CV) and multiple potential step experiments were conducted in a 3-electrode system with a Pt wire counter electrode and a Ag/AgCl reference electrode (Harvard Apparatus, USA) using either a CHI842C or CHI760A electrochemical workstation (CH Instruments, USA). All voltammograms were obtained with a scan rate of 100 mV s^{-1}. For current–time measurements with multiple potential steps, the current sampling frequency was set at 0.5 kHz to faithfully trace the transient changes in the current.

Author contributions

S.-R. Kwon contributed to the NEA device fabrication, BCP@NEA construction, electrochemical measurements and analysis. S. Baek conducted the BCP membrane preparation, SEM image analysis, electrochemical measurements and analysis. P. W. Bohn contributed to planning the experiments and analyzing the data. All of the authors participated in scientific discussion for data analysis and the writing/revising of the manuscript.

Conflicts of interest

There are no conflicts to declare.

Acknowledgements

The development of the BCP@NEA structures was supported by the National Science Foundation through grant 1904196 and the wetting/dewetting experiments were supported by the Department of Energy Office of Science through grant DE FG02 07ER15851. The authors gratefully acknowledge the Notre Dame Nanofabrication Facility and the Integrated Imaging Facility for providing fabrication and characterization support.

References

1 D. A. Doyle, J. M. Cabral, R. A. Pfuetzner, A. Kuo, J. M. Gulbis, S. L. Cohen, B. T. Chait and R. MacKinnon, *Science*, 1998, **280**, 69–77.

2 E. Gouaux and R. MacKinnon, *Science*, 2005, **310**, 1461–1465.

3 O. Beckstein and M. S. P. Sansom, *Phys. Biol.*, 2006, **3**, 147–159.

4 M. Ø. Jensen, D. W. Borhani, K. Lindorff-Larsen, P. Maragakis, V. Jogini, M. P. Eastwood, R. O. Dror and D. E. Shaw, *Proc. Natl. Acad. Sci. U. S. A.*, 2010, **107**, 5833–5838.

5 C. Neale, N. Chakrabarti, P. Pomorski, E. F. Pai and R. Pomès, *PLoS Comput. Biol.*, 2015, **11**, e1004303.

6 S. Balme, F. Picaud, M. Manghi, J. Palmeri, M. Bechelany, S. Cabello-Aguilar, A. Abou-Chaaya, P. Miele, E. Balanzat and J. M. Janot, *Sci. Rep.*, 2015, **5**, 10135.

7 L. Chen, W. Wang, B. Su, Y. Wen, C. Li, Y. Zhou, M. Li, X. Shi, H. Du, Y. Song and L. Jiang, *ACS Nano*, 2014, **8**, 744–751.

8 Q. Zhang, J. Kang, Z. Xie, X. Diao, Z. Liu and J. Zhai, *Adv. Mater.*, 2018, **30**, 1703323.

9 Y. Cheng, X. Jiao, L. Zhao, Y. Liu, F. Wang, Y. Wen and X. Zhang, *J. Mater. Chem. B*, 2018, **6**, 6269–6277.

10 Z. Zhang, X.-Y. Kong, K. Xiao, Q. Liu, G. Xie, P. Li, J. Ma, Y. Tian, L. Wen and L. Jiang, *J. Am. Chem. Soc.*, 2015, **137**, 14765–14772.

11 E. D. Steinle, D. T. Mitchell, M. Wirtz, S. B. Lee, V. Y. Young and C. R. Martin, *Anal. Chem.*, 2002, **74**, 2416–2422.

12 G. Pérez-Mitta, A. G. Albesa, C. Trautmann, M. E. Toimil-Molares and O. Azzaroni, *Chem. Sci.*, 2017, **8**, 890–913.

13 I. Vlassiouk, C.-D. Park, S. A. Vail, D. Gust and S. Smirnov, *Nano Lett.*, 2006, **6**, 1013–1017.

14 F. Rios and S. N. Smirnov, *Chem. Mater.*, 2011, **23**, 3601–3605.

15 S. Smirnov, I. Vlassiouk, P. Takmakov and F. Rios, *ACS Nano*, 2010, **4**, 5069–5075.

16 S. N. Smirnov, I. V. Vlassiouk and N. V. Lavrik, *ACS Nano*, 2011, **5**, 7453–7461.

17 M. R. Powell, L. Cleary, M. Davenport, K. J. Shea and Z. S. Siwy, *Nat. Nanotechnol.*, 2011, **6**, 798–802.

18 K. Xiao, Y. Zhou, X.-Y. Kong, G. Xie, P. Li, Z. Zhang, L. Wen and L. Jiang, *ACS Nano*, 2016, **10**, 9703–9709.

19 G. Xie, P. Li, Z. Zhao, Z. Zhu, X.-Y. Kong, Z. Zhang, K. Xiao, L. Wen and L. Jiang, *J. Am. Chem. Soc.*, 2018, **140**, 4552–4559.

20 S.-R. Kwon, S. Baek, K. Fu and P. W. Bohn, *Small*, 2020, **16**, 1907249.

21 C. Ma, N. M. Contento and P. W. Bohn, *J. Am. Chem. Soc.*, 2014, **136**, 7225–7228.

22 K. Fu and P. W. Bohn, *ACS Appl. Mater. Interfaces*, 2017, **9**, 24908–24916.

23 K. Fu, D. Han, S.-R. Kwon and P. W. Bohn, *ACS Nano*, 2018, **12**, 9177–9185.

24 S. Baek, S.-R. Kwon, K. Fu and P. W. Bohn, *ACS Appl. Mater. Interfaces*, 2020, **12**, 55116–55124.

25 Y. Wu, K. Wang, H. Tan, J. Xu and J. Zhu, *Langmuir*, 2017, **33**, 9889–9896.

26 L. Innes, D. Gutierrez, W. Mann, S. F. Buchsbaum and Z. S. Siwy, *Analyst*, 2015, **140**, 4804–4812.

27 S.-R. Kwon, K. Fu, D. Han and P. W. Bohn, *ACS Nano*, 2018, **12**, 12923–12931.

28 K. Fu, D. Han, C. Ma and P. W. Bohn, *Nanoscale*, 2017, **9**, 5164–5171.

29 K. Fu, D. Han, C. Ma and P. W. Bohn, *Faraday Discuss.*, 2016, **193**, 51–64.

Faraday Discussions

ROYAL SOCIETY OF CHEMISTRY

PAPER

Nanopore-based measurement of the interaction of P450cam monooxygenase and putidaredoxin at the single-molecule level†

Hui Chen, ‡[a] Yao Lin,‡[ab] Yi-Tao Long, [b] Shelley D. Minteer *[a] and Yi-Lun Ying *[bc]

Received 10th July 2021, Accepted 5th August 2021
DOI: 10.1039/d1fd00042j

Protein–protein interactions occur in a wide range of biological processes and are of great significance to life function. Characterization of transient protein–protein interactions remains a significant barrier to our understanding of cellular processes. Nanopores provide unique nanoscale environments that accommodate single molecules from the surrounding bulk solution. This method permits label-free sensing at the single-molecule level with extremely high sensitivity. Herein, the interaction between a single P450cam monooxygenase and its redox partner putidaredoxin (Pdx) was monitored *via* transient ionic current by using functionalized glass nanopores. Results show that the volume of P450cam determines the blockage current while the interactions between the P450cam and Pdx give a long blockage duration. Our glass nanopore sensor with adjustable diameter could be applied for real-time sensing of protein–protein interactions between individual proteins with a wide range of molecular weight.

Introduction

Protein–protein interactions are ubiquitous in cellular processes ranging from antibody–antigen interactions to protein supercomplex formation for substrate channeling (*i.e.* electron transport chain supercomplex or the Krebs cycle metabolon) to molecular machines.[1–5] Protein–protein interactions frequently occur due to hydrophobicity effects, electrostatics, and/or hydrogen bonding. Due to their critical role in disease states, cellular function, and even biotechnology,

[a]*Department of Chemistry, University of Utah, USA. E-mail: minteer@chem.utah.edu*
[b]*State Key Laboratory of Analytical Chemistry for Life Science, School of Chemistry and Chemical Engineering, Nanjing University, China. E-mail: yilunying@nju.edu.cn*
[c]*Chemistry and Biomedicine Innovation Center, Nanjing University, China*

† Electronic supplementary information (ESI) available. See DOI: 10.1039/d1fd00042j

‡ These authors contributed equally to this work.

there is a great deal of interest in analytical tools for studying protein–protein interactions.

Over the last two decades, there have been a variety of *in vitro* analytical methods developed for studying protein–protein interactions, including affinity chromatography,[6] crosslinking mass spectrometry analysis,[7] Förster resonance energy transfer (FRET) assays,[8] surface plasmon resonance,[9] and isothermal titration calorimetry.[10] All of these methods look at an ensemble of protein–protein pairs. In an effort to better understand protein–protein interactions and their diversity, there is a desire for measurements at the single-molecule level. However, most of the single-molecule techniques are highly dependent on the labeling of biomolecules or limited by the vacuum conditions, which hardly achieves the direct and real-time monitoring of the protein–protein interactions in solution.

Nanopore-based techniques for mimicking the transportation processes of membrane protein channels were proposed for sensing at single-molecule level. Nanopores can be fabricated by a range of biological proteins and solid-state materials.[11–14] It utilizes nanoscale confinement to capture a label-free single molecule from the bulk solution. Currently, biological nanopores have been applied in analyzing protein–protein interactions, but it requires careful design of peptides as capturing probes or engineering of the membrane channel with the target proteins.[15–18] Moreover, the small diameter of the biological nanopore ($d <$ 10 nm) limits its application in large proteins. With the advantage of tunable diameter ($d \sim$ 10–100 nm), the glass nanopore is favorable to characterize a wide molecular range of single proteins for achieving ultrasensitive sensing in a mixture,[19,20] probing binding kinetics and aggregation states.[21,22] For example, an antibody functionalized glass nanopore sensor enables the analysis of the kinetic antigen–antibody interactions at the single-molecule level.[23] Moreover, the amyloidosis process of Aβ1-42 peptide was monitored *via* reading distinct current fluctuations from a glass nanopore.[22]

Cytochrome P450cam (CYP101A1) is a bacterial heme monooxygenase from *Pseudomonas putida* that catalyzes the regio and stereoselective C–H bond activation of camphor and converts camphor to 5-exohydroxycamphor.[24] For the catalytic process of P450cam, the electrons from reducing equivalents (NAD(P)H) first are transferred from the FAD of putidaredoxin reductase (PdR) to the Fe_2S_2 of putidaredoxin (Pdx). Finally, Pdx delivers electrons to heme of P450cam to support the C–H bond activation reaction.[25] The interaction between P450cam and Pdx, including recognition, binding and disassociation, is known to induce changes in structure and spectroscopic features of P450cam.[26] In view of this, the P450cam–Pdx system is an ideal model to investigate the protein–protein interaction.

In this paper, we detail a nanopore-based method for studying protein–protein interactions by utilizing the model system of P450cam monooxygenase and its redox partner Pdx at the single-molecule level. Our results demonstrate that the translocation of P450cam through a Pdx modified glass nanopore associates with the energy barrier from protein–protein interactions, leading to a long duration time.

Experimental

Reagents and materials

Sodium chloride was purchased from Sigma-Aldrich (St. Louis, MO, USA). Tris(hydroxymethyl)aminomethane was purchased from Fisher Scientific. 11-

Mercaptoundecanoic acid (MUA, 98%), *N*-hydroxysuccinimide (NHS, 98%), and 1-ethyl-3-(3-diethylaminopropyl)carbodiimide hydrochloride (EDC, ≥98%) were purchased from Sigma-Aldrich (St. Louis, MO, USA). All reagents were of analytical grade. All solutions were prepared by Milli-Q ultrapure water with a resistance of 18.2 MΩ cm at 25 °C (EMD Millipore, Billerica, USA) and were filtered with 0.22 μm pore-size filter.

Fabrication of glass nanopore

Quartz capillaries with 0.5 mm inner diameter and 1 mm outer diameter (QF100-50-10, Sutter Instrument Co., USA) were used for all experiments. Capillaries were cleaned by sonication in acetone and ethanol for 30 min, respectively, followed by drying under a nitrogen gas stream to remove residual ethanol. The glass nanopores were fabricated by using a P-2000 CO_2-laser puller (Sutter Instrument Co., USA) with the parameters: heat = 625, fil = 4, vel = 60, del = 150, pull = 192.

Modification of glass nanopore with Pdx

The quartz nanopipette was further coated with a gold layer using electron-beam evaporation. Then the gold-coated nanopore was filled with 10 mM MUA in ethanol for 12 h. After thorough washing with ethanol, 50 mM MES (pH 5.0) containing 100 mM EDC and 50 mM NHS was added for 1 h to activate the COOH-terminated group of the MUA. Afterwards, Pdx (0.1 mg mL^{-1}) in Tris–HCl (10 mM, pH = 7.5) was added and incubated for 4 h after careful removal of the EDC/NHS solution (Fig. 1).

Nanopore experiment and data acquisition

The gold-coated glass nanopore or Pdx modified nanopore was filled with 1 M NaCl containing 10 mM Tris–HCl (pH 7.5) using a microloader and then centrifuged at a speed of 3000 rpm for 3 min to remove the air bubbles at the tip. A pair of Ag/AgCl electrodes were employed to apply a potential to the nanopore. The working electrode is inserted inside the nanopore, and the other electrode, which is grounded, was immersed in the solution outside the nanopore. Current traces were measured using a home-developed nanopore system (Cube-D1, from Prof. Yi-Tao Long's group) at a sampling rate of 100 kHz with a 10 kHz low-pass Bessel filter. Data analysis was performed using a home-designed software (http://

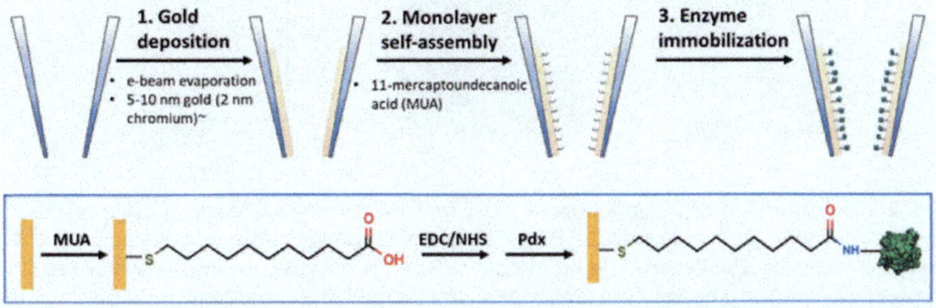

Fig. 1 The schematic illustration for Pdx immobilization on gold-coated glass nanopore.

hysz.nju.edu.cn/ytlong/Data/View/267) and Origin 9.2 (OriginLab Corporation, Northampton, USA). The final concentration of P450cam in the outside solution was 2.4 nM.

P450cam, Pdx, and FeSII expression and purification

The genes of P450cam (Uniport: P00183) and Pdx (Uniport: P00259) were synthesized by GenScript and inserted into pET-20b plasmid with a C-terminal histidine tag. The gene of FeSII protein (Uniport: C1DE01) was purchased as a synthetic gene from Integrated DNA Technologies, Inc. and cloned into a pre-linearized pET28a expression vector (Novagen) by Gibson assembly. The *E. coli* BL21(DE3) strains containing the protein expression plasmids of P450cam, Pdx, and FeSII were cultivated in the LB medium supplemented with 1.2% glycerol at 37 °C. When A_{600} reached about 0.8, 100 μM isopropyl β-D-thiogalactopyranoside (IPTG, a final concentration) was added and the cultivation temperature was decreased to 18 °C for ~16 hours. After induction with IPTG, cells were harvested by centrifugation. The cell pellets were resuspended in 100 mM MOPS buffer (pH 7.0) containing 50 mM NaCl, followed by sonication. After centrifugation, the supernatant was loaded onto a column packed with HisPur Ni-NTA Resin (Fisher Scientific, Pittsburgh, PA, USA) and eluted with 100 mM MOPS buffer (pH 7.0) containing 300 mM NaCl and 250 mM imidazole. The purified proteins (10 μL) were loaded into 12% SDS-PAGE to check the quality of purification.

Determination of protein concentration

Protein mass concentration was measured by the Bio-Rad Bradford protein–dye reagent method (Bio-Rad, Hercules, CA, USA) with bovine serum albumin as a reference protein.

Results and discussion

In order to produce the protein needed to study this protein–protein interaction, P450cam and Pdx were each cloned, expressed, and purified from BL21 *E. coli* cells. His-tag affinity protein purification was used to purify the protein from the lysed *E. coli* cells. The SDS-PAGE characterization of the proteins is shown in

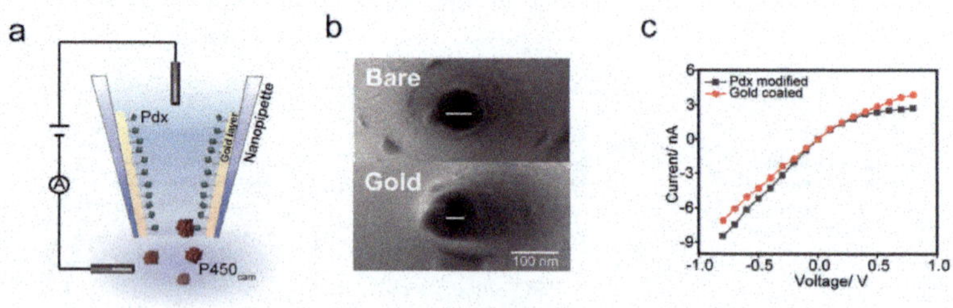

Fig. 2 (a) Schematic representation of label-free monitoring of single protein–protein interactions with glass nanopores. The green protein represents the Pdx protein, and the red one indicates P450cam. The electrode put in the outside solution is appointed as ground. (b) Scanning electron microscope (SEM) images of the bare glass nanopore and the gold-coated nanopore. (c) The representative I–V responses of gold-coated glass nanopores before and after Pdx modification measured in 1 M NaCl.

Fig. S1.† Glass nanopores were fabricated in quartz capillaries with an outer diameter of 1.00 mm and an inner diameter of 0.5 mm by using a laser puller. The diameter of fabricated glass nanopores after pulling was 50 nm. Then, e-beam evaporation was used to deposit a chromium adhesion layer followed by a gold layer along the inner surface of the nanopore. As a result, the diameter of the gold-coated nanopore decreases to about 30 nm (Fig. 2b). After gold deposition, a self-assembled monolayer was formed with 11-mercaptoundecanoic acid (MUA) as described in a previous study.[23] Due to the negatively charged carboxyl group, the MUA modified nanopore shows an obvious ionic current rectification. Then, that monolayer was used to attached Pdx *via* EDC/NHS coupling. After modification, the exposed carboxyl group on the Pdx leads to a change in the surface charge density of the nanopore. As a result, the rectification ratio of positive current value, I_+, to the negative current values, I_-, at ±0.8 V decreased from 0.54 to 0.32.

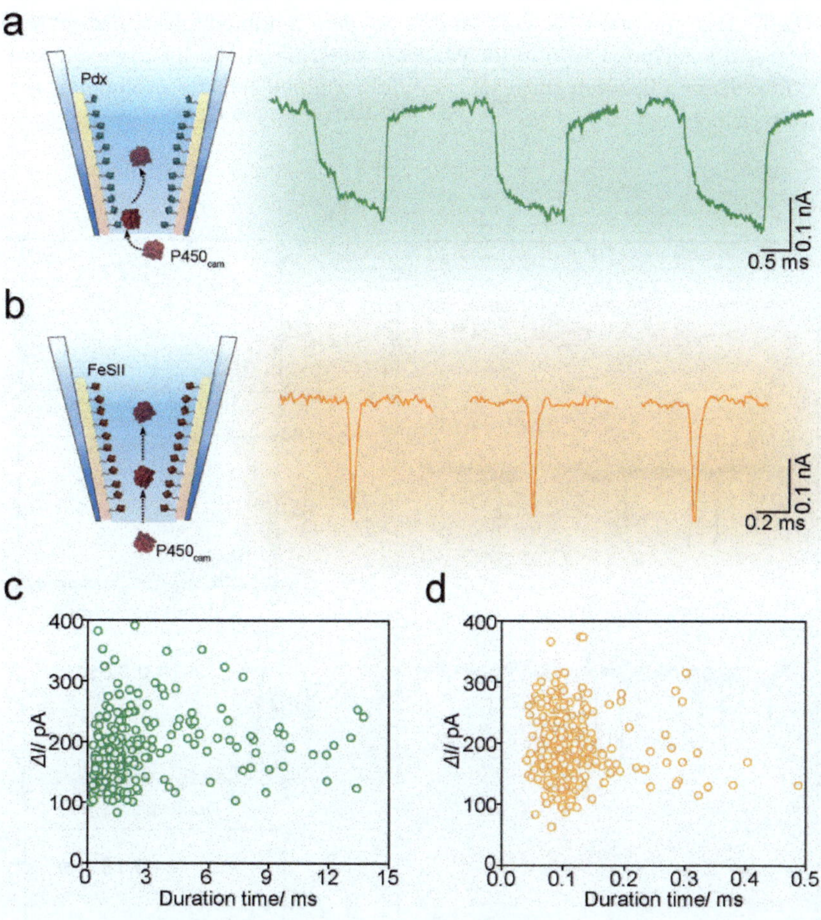

Fig. 3 (a) Schematic of P450cam translocation and representative translocation signals using Pdx-modified glass nanopores. (b) Schematic of P450cam translocation and representative translocation signals using FeSII-modified glass nanopores. (c) Scatter plots of current blockade events of Pdx and P450cam with duration time. (d) Scatter plots of current blockade events of FeSII and P450cam with duration time. The data were collected under +200 mV.

The stable time-series current trace from the Pdx-modified nanopore further confirms the functionalization of the glass nanopore, as shown in Fig. S2.†

After preparation of the Pdx modified glass nanopore, we performed electrochemical measurements with P450cam in solution to evaluate the current changes associated with protein–protein interactions between Pdx and P450cam. Since the P450cam carries a net charge of −18.1, the applied positive potential was used to drive the P450cam traversing through the nanopore. As shown in Fig. 3a, these protein–protein interactions lead to reproducible current signals with a long statistic duration of 1.1 ms. A control experiment was performed by replacing Pdx with the FeSII (Shethna) protein which is the same size as Pdx (Fig. S1†), but does not interact with P450cam. As shown in Fig. 3b, the duration of blockage is shorter with no intermolecular interactions. Statistical analysis revealed that a single P450cam molecule undergoes a rapid translocation through the FeSII modified glass nanopore with a statistic duration of 0.1 ms at 200 mV (Fig. S3†). The P450cam induced blockage current amplitude (ΔI) in FeSII modified glass nanopore is comparable to Pdx modified glass nanopores. Therefore, the excluded volume of P450cam determines the blockage current while interactions between the P450cam and modified Pdx are prolonging the duration of blockage.

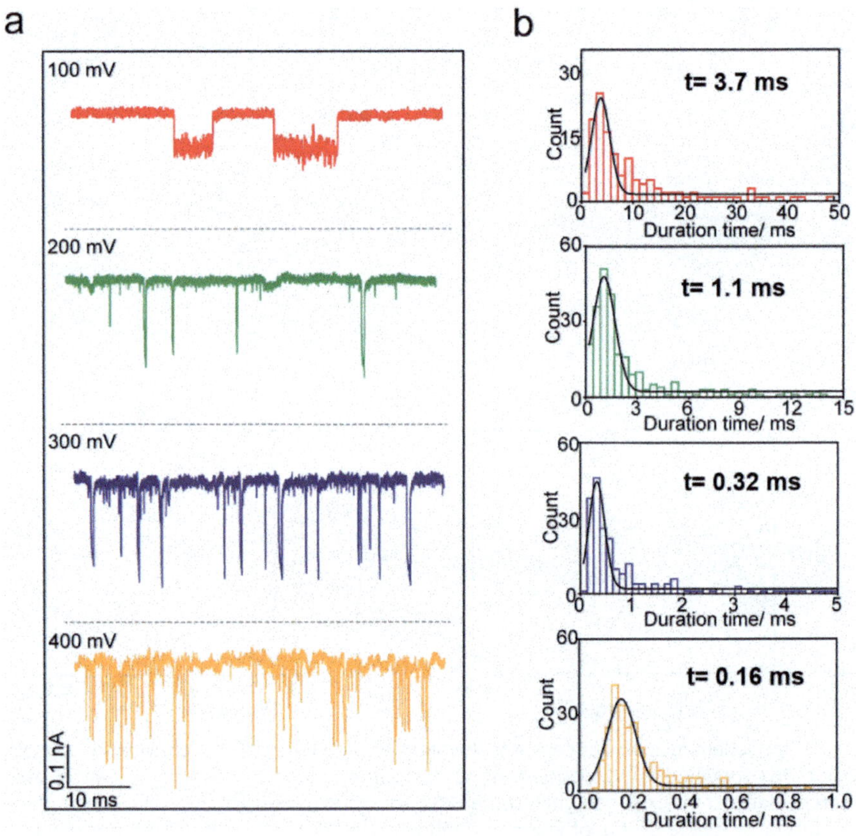

Fig. 4 (a) Current traces of P450cam translocations using Pdx-modified glass nanopores at different voltages. (b) Histograms of duration time at different voltages.

As the applied potential increased from 100 mV to 400 mV, the duration time of P450cam exponentially decreased (Fig. 4 and S4†). This result demonstrates that the translocation of P450cam is associated with the energy barrier from protein–protein interactions. The increased blockage numbers with applied potential further show the capture of P450cam is driven by the electrophoretic force (Fig. S5†).

Conclusions and future work

The use of a glass nanopore modified with a self-assembled monolayer of protein was utilized to study protein–protein interactions of a single protein (Pdx) with a single binding protein (P450cam). Control experiments were performed with a similar size protein with no structural affinity to P450cam and the results confirm that the nanopore is an appropriate method for investigating single protein–protein interactions. In the future, this technique could be used to study electrostatic, hydrogen bonding, and other intermolecular interactions controlling protein–protein complexation in electron transport chains, substrate channeling, molecular motors, and disease progression. By developing a suitable algorithm for analyzing the noise in the blockage current, the transient conformational changes of the protein could be further resolved during the protein–protein interactions inside a nanopore.

Meanwhile, the impact of the orientation of the immobilized Pdx on the interaction between P450cam and Pdx is a noteworthy issue. For Pdx, two key residues around the 2Fe–2S center, Asp38 and Trp106, are critical for the binding with P450cam. Pdx_{Asp38} forms ion pairs with $P450cam_{Arg112}$. Pdx_{Trp106} forms nonbonded interactions with $P450cam_{Ala113}$ and an H-bond with $P450cam_{Asn116}$.[27] Consequently, the orientation of these two residues on the surface of the immobilized Pdx is a key factor in the binding of P450cam and Pdx. In this study, Pdx was immobilized on the inner surface of the nanopore *via* EDC/NHS chemical modification. It is difficult with this method to ensure that the bonding area composed by Asp38 and Trp106 is in the optimal orientation. Based on the current experimental conditions, it is very difficult to estimate the impact degree of the orientation of Pdx on the spread in translocation times as it is almost impossible to estimate the percentage of protein in different orientations. To address this problem, the future work is to fuse a short gold binding peptide (GBP)[28] at the C-terminus of Pdx. The Pdx–GBP fusion protein could be directionally immobilized on the inner surface of the nanopore *via* the gold binding activity of the GBP to fully expose the P450cam binding area, thereby finally eliminating the influence of the different orientation of Pdx on the distribution of translocation time.

Conflicts of interest

There are no conflicts to declare.

Acknowledgements

This research was supported by the National Natural Science Foundation of China (No. 21922405) and China Scholarship Council (No. CSC201806740044). We thank Henry White and Martin Edwards for equipment use and Kaili Xin for the effort in data analysis.

References

1 I. M. Nooren and J. M. Thornton, *EMBO J.*, 2003, **22**, 3486–3492.
2 D. R. Davies and G. H. Cohen, *Proc. Natl. Acad. Sci. U. S. A.*, 1996, **93**, 7–12.
3 J. A. Letts and L. A. Sazanov, *Nat. Struct. Mol. Biol.*, 2017, **24**, 800–808.
4 J. De Las Rivas and C. Fontanillo, *Briefings Funct. Genomics*, 2012, **11**, 489–496.
5 F. Wu and S. Minteer, *Angew. Chem., Int. Ed.*, 2015, **54**, 1851–1854.
6 V. Le Brun, W. Friess, S. Bassarab and P. Garidel, *Pharm. Dev. Technol.*, 2010, **15**, 421–430.
7 J. Vasilescu, X. Guo and J. Kast, *Proteomics*, 2004, **4**, 3845–3854.
8 K. Truong and M. Ikura, *Curr. Opin. Struct. Biol.*, 2001, **11**, 573–578.
9 M. Kim, K. Park, E.-J. Jeong, Y.-B. Shin and B. H. Chung, *Anal. Biochem.*, 2006, **351**, 298–304.
10 M. M. Pierce, C. Raman and B. T. Nall, *Methods*, 1999, **19**, 213–221.
11 N. Varongchayakul, J. Song, A. Meller and M. W. Grinstaff, *Chem. Soc. Rev.*, 2018, **47**, 8512–8524.
12 Y.-L. Ying, C. Cao, Y.-X. Hu and Y.-T. Long, *Natl. Sci. Rev.*, 2018, **5**, 450–452.
13 L. Restrepo-Pérez, C. Joo and C. Dekker, *Nat. Nanotechnol.*, 2018, **13**, 786–796.
14 S. McGinn, D. Bauer, T. Brefort, L. Dong, A. El-Sagheer, A. Elsharawy, G. Evans, E. Falk-Sörqvist, M. Forster and S. Fredriksson, *Nat. Biotechnol.*, 2016, **33**, 311–330.
15 A. K. Thakur and L. Movileanu, *Nat. Biotechnol.*, 2019, **37**, 96–101.
16 C.-W. Ho, V. Van Meervelt, K.-C. Tsai, P.-J. De Temmerman, J. Mast and G. Maglia, *Sci. Adv.*, 2015, **1**, e1500905.
17 M. Fahie, C. Chisholm and M. Chen, *ACS Nano*, 2015, **9**, 1089–1098.
18 M. A. Fahie, B. Yang, B. Pham and M. Chen, *ACS Sens.*, 2016, **1**, 614–622.
19 M. Raveendran, A. R. Leach, T. Hopes, J. L. Aspden and P. Actis, *ACS Sens.*, 2020, **5**, 3533–3539.
20 S. Umehara, M. Karhanek, R. W. Davis and N. Pourmand, *Proc. Natl. Acad. Sci. U. S. A.*, 2009, **106**, 4611–4616.
21 X. Wang, M. D. Wilkinson, X. Lin, R. Ren, K. R. Willison, A. P. Ivanov, J. Baum and J. B. Edel, *Chem. Sci.*, 2020, **11**, 970–979.
22 R.-J. Yu, S.-M. Lu, S.-W. Xu, Y.-J. Li, Q. Xu, Y.-L. Ying and Y.-T. Long, *Chem. Sci.*, 2019, **10**, 10728–10732.
23 Y.-L. Ying, R.-J. Yu, Y.-X. Hu, R. Gao and Y.-T. Long, *Chem. Commun.*, 2017, **53**, 8620–8623.
24 T. L. Poulos, *Chem. Rev.*, 2014, **114**, 3919–3962.
25 A. H. Follmer, S. Tripathi and T. L. Poulos, *J. Am. Chem. Soc.*, 2019, **141**, 2678–2683.
26 S.-H. Liou, M. Mahomed, Y.-T. Lee and D. B. Goodin, *J. Am. Chem. Soc.*, 2016, **138**, 10163–10172.
27 S. Tripathi, H. Li and T. L. Poulos, *Science*, 2013, **340**, 1227–1230.
28 Y. S. Lee, S. Baek, H. Lee, S. S. Reginald, Y. Kim, H. Kang, I.-G. Choi and I. S. Chang, *ACS Appl. Mater. Interfaces*, 2018, **10**, 28615–28626.

Faraday Discussions

ROYAL SOCIETY OF CHEMISTRY

PAPER

Carbon nanospike coated nanoelectrodes for measurements of neurotransmitters†

Qun Cao,[a] Zijun Shao,[a] Dale Hensley[b] and B. Jill Venton*[a]

Received 16th August 2021, Accepted 23rd September 2021

DOI: 10.1039/d1fd00053e

Carbon nanoelectrodes enable the detection of neurotransmitters at the level of single cells, vesicles, synapses and small brain structures. Previously, the etching of carbon fibers and 3D printing based on direct laser writing have been used to fabricate carbon nanoelectrodes, but these methods lack the ability of mass manufacturing. In this paper, we mass fabricate carbon nanoelectrodes by growing carbon nanospikes (CNSs) on metal wires. CNSs have a short, dense and defect-rich surface that produces remarkable electrochemical properties, and they can be mass fabricated on almost any substrate without using catalysts. Tungsten wires and niobium wires were electrochemically etched in batch to form sub micrometer sized tips, and a layer of CNSs was grown on the metal wires using plasma-enhanced chemical vapor deposition (PE-CVD). The thickness of the CNS layer was controlled by the deposition time, and a thin layer of CNSs can effectively cover the entire metal surface while maintaining the tip size within the sub micrometer scale. The etched tungsten wires produced tapered conical nanotips, while the etched niobium wires were long and thin. Both showed excellent sensitivity for the detection of outer sphere ruthenium hexamine and the inner sphere test compound ferricyanide. The CNS nanosensors were used for the measurement of dopamine, serotonin, ascorbic acid and DOPAC with fast-scan cyclic voltammetry. The CNS nanoelectrodes had a large surface area and numerous defect sites, which improved the sensitivity, electron transfer kinetics and adsorption. Finally, the CNS nanoelectrodes were compared with other nanoelectrode fabrication methods, including flame etching, 3D printing, and nanopipettes, which are slower to make and more difficult for mass fabrication. Thus, CNS nanoelectrodes are a promising strategy for the mass fabrication of nanoelectrode sensors for neurotransmitters.

Introduction

Carbon-based electrodes are standard electrochemical sensors for neurotransmitter detection due to their conductivity, biocompatibility, and small size.[1,2]

[a]*Department of Chemistry, University of Virginia, Charlottesville, Virginia, 22904, USA. E-mail: jventon@virginia.edu*

[b]*Center for Nanophase Material Science, Oak Ridge National Lab, Oak Ridge, Tennessee, 37831, USA*

† Electronic supplementary information (ESI) available. See DOI: 10.1039/d1fd00053e

Nanoelectrodes are required to approach the site of neurotransmitter release in tiny spaces, such as single synapses, or in small brain structures, such as those in *Drosophila* or zebrafish, which have micrometer-scale brain structures.[3–5] In recent studies, several methods have been developed to fabricate carbon nanoelectrode sensors for neurotransmitter detection.[6–8] Carbon fibers are the standard electrode material for neurotransmitter detection,[9] and they can be flame etched or electrochemically etched to the sub micrometer scale.[10–16] For example, the Ewing group used flame etched carbon-fiber nanotip electrodes to study catecholamine released by single cells.[14,17,18] However, the manual etching of a carbon fiber lacks precise control over the electrode size and geometry, and the method is not suitable for batch manufacturing. Our group developed 3D printed carbon nanoelectrodes based on direct laser writing,[8,19] allowing customizable electrode sizes and geometries with perfect reproducibility. 3D printing requires a specialized printer and is not extremely rapid. Another approach is to make nanopipettes, where quartz capillaries are pulled to create a nanoscale orifice, and carbon can be selectively deposited onto the inner wall of the quartz capillary templates.[20–23] The opening of the carbon nanopipette can be correlated with vesicle size.[24] The Shen group developed nanopipette electrodes without carbon deposition, which facilitates the detection of non-redox-active neurotransmitters, such as acetylcholine.[25–27] The active surface areas of carbon nanopipette electrodes are located inside the capillary, so their surface areas are smaller and the analyte must diffuse into the nanopipette. Overall, there are still limited methods to fabricate carbon nanoelectrodes for neurotransmitter detection, and better approaches are needed to mass produce nanoelectrode sensors with high sensitivities.

Carbon nanospikes (CNSs) are a new electrode material and have been used for electrochemical energy and neurotransmitter applications.[28,29] They exhibit curled, spike-like structures which are approximately 50 nm in length, which densely overlap to increase the electrode surface area. CNSs are rich in edge-plane carbon sites, improving electron transfer and adsorption kinetics.[28,30–32] With their larger surface area and more defect carbon sites, CNSs outperform carbon nanotubes and carbon fibers as neurochemical sensors.[33] CNSs are produced by plasma-enhanced chemical vapor deposition (PE-CVD) without any catalyst; thus, they can be uniformly grown on almost any substrate.[34] The deposition time controls the layer thickness, and a thin layer of CNSs effectively covers the entire substrate surface. The fabrication of CNSs is simple and suitable for mass fabrication, and their electrochemical properties for neurotransmitter detection outperform those of carbon nanotubes and carbon fibers.[32] Although CNS electrodes have been grown on various substrates for neurotransmitter detection, they have not been grown on nanometer scale substrates to form carbon nanoelectrodes.

In this work, we have demonstrated the growth of CNSs on metal wires with nanotips to fabricate CNS nanoelectrodes. Tungsten and niobium wires were electrochemically etched to the nanoscale, and a thin layer of CNSs was grown onto the tungsten and niobium surfaces to form W/CNS and Nb/CNS nanoelectrodes, respectively. The CNS layers completely cover the metal wires while the electrode tip is still on the nanometer scale. The CNSs have a larger surface area to improve the sensitivity, and the numerous defect sites improve the electron transfer and adsorption kinetics. Cyclic voltammetry (CV) and electrochemical impedance spectroscopy (EIS) were used to characterize the CNS nanoelectrodes. The Nb/CNS nanoelectrodes have good sensitivity for dopamine and other neurochemical detection using fast-scan cyclic voltammetry (FSCV) detection. The growth of CNSs on the nanoscale substrates was compared with other carbon nanoelectrode fabrication methods, including the flame etching of carbon fibers, nano 3D printing and nanopipettes. The entire fabrication process of the CNS nanoelectrodes is done in batches; we fabricated 30 nanosensors on a 2.5 cm long substrate, but even more could be mounted on a larger conductive substrate. Therefore, the method has promising potential for the mass fabrication of carbon nanoelectrode sensors.

Methods

Etching of metal wires

Tungsten wires and niobium wires (50 μm diameter) were purchased from Advent Research Materials, Ltd (Oxford, England). The metal wires were cut to approximately 2 cm in length, and parallelly immobilized at the edge of a niobium foil. The extended length of the metal was about 1 cm, and metal wires were vertically inserted into the etching solution at about 0.5 cm depth. About thirty 50 μm diameter metal wires can be mounted onto a 2.5 cm × 2.5 cm niobium foil. An AC current supplier (Interworld Highway LLC, Long Branch, NJ) was used to etch the metal wires to a sub-micrometer diameter at the tip. One clamp of the AC supplier was connected to the niobium foil, and another was connected to a graphite electrode. The tungsten wires were etched at 2 V and 60 Hz in 2 M NaOH for about 5 min, and the niobium wires were etched at 2 V and 60 Hz in concentrated HCl for about 20 min.

Growth of the CNSs

Metal wires were immobilized on a stainless-steel stage and placed into a custom-built DC plasma-enhanced chemical vapor deposition (PE-CVD) chamber. The metal wires served as the cathode for the DC plasma. The pressure of the chamber was 6 Torr, and the wires were exposed to 100 sccm ammonia and 80 sccm acetylene for 6 min. The DC plasma discharge was operated at 250 mA and 480–550 V at 650 °C.

Surface and electrochemical characterizations

Scanning electron microscopy (SEM) images were taken on a Merlin field emission SEM (Zeiss, Thornwood, NY) and FEI Quanta 650 SEM (Thermo Fisher Scientific, Waltham, MA). A secondary electron detector was used at an accelerating voltage of 1.0–3.0 kV.

Cyclic voltammetry and electrochemical impedance spectroscopy experiments were performed with a Reference 600 potentiostat (Gamry Instruments, Warminster, PA). Prior to the measurements, the CNS coated wires were inserted into a pulled glass capillary and sealed with 5 min epoxy (J-B weld, Sulphur Springs, TX). A three electrode system was used. The working electrode was a CNS electrode, the reference electrode was a standard Ag/AgCl electrode, and the counter electrode was a Pt wire. EIS measurements were performed with a frequency range of 100 kHz to 0.1 Hz, and the amplitude of the AC voltage was 10 mV with a DC biased potential of 0.2 V.

FSCV experiments were performed with a ChemClamp potentiostat and head-stage (Dagan, Minneapolis, MN). The CNS coated wires were inserted into a pulled glass capillary and sealed with 5 min epoxy, and the glass capillary was filled with 1 M KCl solution to make an electrical connection. A two-electrode system was used. The working electrode was the CNS nanoelectrode and the reference electrode was Ag/AgCl. The PBS buffer (131.5 mM NaCl, 3.25 mM KCl, 1.2 mM $CaCl_2$, 1.25 mM NaH_2PO_4, 1.2 mM $MgCl_2$, and 2.0 mM Na_2SO_4, and the pH was adjusted to 7.4 by concentrated NaOH) and neurochemical solutions were injected through a flow cell at 2 mL min^{-1} by a syringe pump (Harvard Apparatus, Holliston, MA) and modulated by a six-port loop injector with an air actuator (VIVI Valco Instruments, Houston, TX). The data were analyzed with HDCV Analysis software (Department of Chemistry, University of North Carolina at Chapel Hill).

Results and discussion

We used electrochemically etched tungsten wires and niobium wires for the growth of the CNSs. Fig. 1a shows that the etched tungsten wire has a conical geometry with sharp nanotips, which were measured to be 286 ± 92 nm (n = 10, s.d), and the surface is smooth, which is beneficial for the growth of the CNSs. Tungsten wires are advantageous because they can be easily etched to sharp nanotips,[35] but they are not electrochemically inert at the potential of 1.3 V, which is the switching potential of the typical FSCV waveform for dopamine detection.[30] Therefore, the CNSs must cover the entire tungsten wire so that the electroactive surface is only carbon. Fig. 1b shows the SEM image of the CNSs grown on etched tungsten wires, and the CNSs are uniformly grown. The nanotip size of the W/CNS electrodes was measured to be 448 ± 90 nm (n = 10, s.d). Therefore, the thickness of the CNS layer is approximately 80 nm, which slightly blunts the nanotip, but

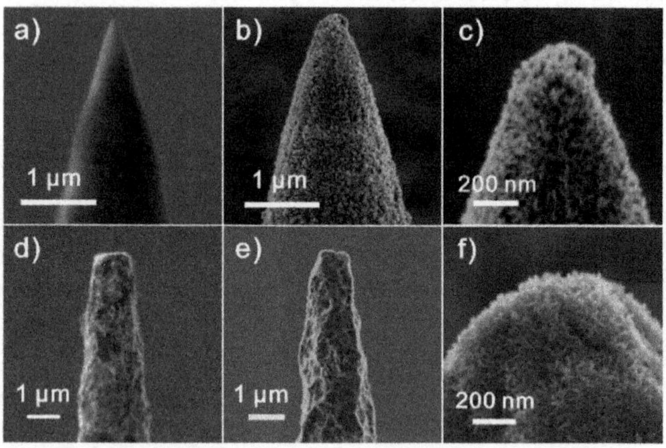

Fig. 1 SEM images of (a) the etched tungsten wire, (b) CNSs grown on the tungsten wire, (c) a more enlarged image of the CNSs on the tungsten wire, (d) the etched niobium wire, (e) CNSs grown on the etched niobium wire, and (f) a more enlarged image of the CNSs on the niobium wire.

the tip size is still at the nanoscale (Fig. 1c). Fig. 1d shows that the etched niobium wires are not as sharp and smooth as tungsten wires, because niobium is relatively inert and difficult to etch.[36] But the apex angle of the Nb/CNS electrode is smaller, yielding a long and thin tip. The nanotip size of the etched Nb wires is 908 ± 242 nm ($n = 6$, s.d). As shown in Fig. 1e, the growth of the CNSs on the niobium wire slightly increases the tip size to 1015 ± 133 nm ($n = 6$, s.d). The increase of the nanotip size is about 50 nm because the etched Nb wires have a rough surface. More enlarged SEM images of the CNSs are shown in Fig. 1c and f, which exhibit the dense spike-like features of the CNSs grown on the tungsten and niobium substrates. We calculated the roughness of the CNS nanoelectrodes by measuring the background double layer charging current.[33] We fabricated two batches of Nb/CNS nanoelectrodes; the first batch has a roughness of 4.9 ± 0.5, and the second batch has a roughness of 5.1 ± 0.8, indicating that the roughness of the electrode surfaces are nearly identical among batches.

As shown in Fig. 2, the W/CNS nanoelectrodes and Nb/CNS nanoelectrodes were characterized by cyclic voltammetry (CV) and electrochemical impedance spectroscopy (EIS). CV was performed with a surface insensitive redox system of $Ru(NH_3)_6{}^{3+/2+}$ (Fig. 2a) and a surface sensitive redox system of $Fe(CN)_6{}^{3-/4-}$ (Fig. 2b).[37] For both redox complexes, the Nb/CNS and W/CNS electrodes exhibit similar shapes and current densities, indicating that the CNSs on the two types of wires exhibit similar electrochemical behavior. Fig. 2c shows the Nyquist plot from EIS obtained in $K_3Fe(CN)_6$ solution. The Nb/CNS and W/CNS electrodes have similar semicircles in size, demonstrating that the CNSs grown on them have similar electron transfer kinetics. The Nyquist plot shows only one semicircle for both the W/CNS and Nb/CNS nanoelectrodes, indicating that no other capacitance exists other than the double layer capacitance. These results further demonstrate that the CNSs effectively cover the metal wires, and the electrochemical responses only come from the CNSs and not the metal wires.

Fast-scan cyclic voltammetry (FSCV) is the standard technique for the real-time measurement of neurotransmitters, and we used FSCV to characterize the detection of dopamine by the Nb/CNS nanoelectrodes.[38] Fig. 3a and b show the background current and the background-subtracted dopamine response of the Nb/CNS nanoelectrodes, which have characteristic shapes similar to the cyclic voltammograms of carbon-fiber electrodes. Dopamine is oxidized to dopamine-*o*-quinone at around 0.6 V and the reduction happens at −0.2 V. The oxidation peak

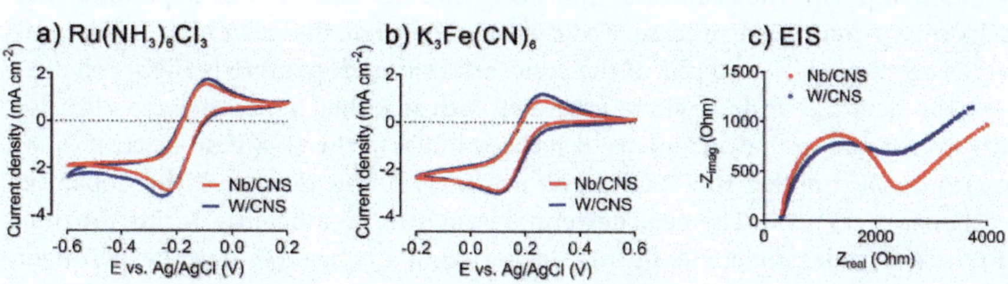

Fig. 2 Cyclic voltammogram and electrochemical impedance measurements of the W/CNS and Nb/CNS electrodes. (a) Cyclic voltammogram in 10 mM $Ru(NH_3)_6Cl_3$ and (b) cyclic voltammogram in 10 mM $K_3Fe(CN)_6$ (scan rate = 100 mV s^{-1}). (c) Nyquist plots of the Nb/CNS and W/CNS electrodes.

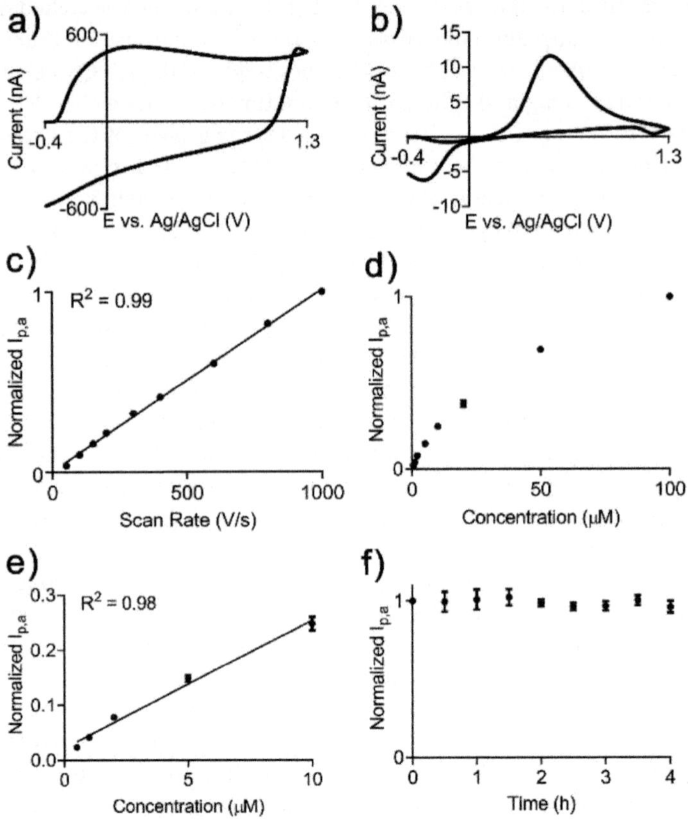

Fig. 3 FSCV study of the Nb/CNS nanoelectrodes. (a) Background current. (b) Background-subtracted current of 1 μM dopamine. (c) The dopamine current is linear with respect to the scan rate ($R^2 = 0.99$, $n = 3$). (d) Dopamine response at concentrations of 1 μM to 100 μM ($n = 3$). (e) The dopamine current is linear with respect to the concentration up to 10 μM ($R^2 = 0.98$). (f) Stability test. 1 μM dopamine was measured every 0.5 h for 4 h.

is larger than the reduction peak because dopamine has tighter adsorption than dopamine-*o*-quinone.[39] Fig. S1† shows the background signal and the detection of dopamine by the W/CNS nanoelectrodes, which have similar behavior to that of the Nb/CNS electrodes. Fig. 3c shows that the anodic peak current has a linear relationship with the scan rate, indicating that the detection of dopamine is an adsorption-controlled process.[39] Fig. 3d–e show that the current is linear with concentrations up to 10 μM. If the concentration of dopamine is higher than 10 μM, the electrode surface will be saturated with adsorbed dopamine and diffusion will contribute to the current. To test the stability of the CNS nanoelectrodes, we continuously applied the FSCV waveform for 4 h and measured the dopamine response every 0.5 h. The peak current did not drop significantly during the time, so the electrodes are stable during the period of a typical biological experiment.

FSCV detection was tested for other neurochemicals, including serotonin, DOPAC, adenosine and ascorbic acid, using the Nb/CNS nanoelectrodes. Serotonin is a monoamine neurotransmitter,[40] and its reduction peak is more positive than that of dopamine (Fig. 4a). Here, using the dopamine waveform for

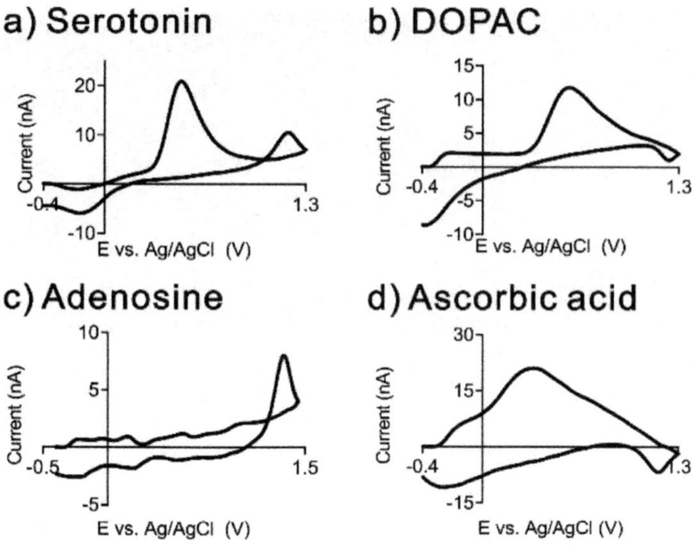

Fig. 4 Detection of other neurochemicals using CNS nanoelectrodes. (a) 1 μM serotonin, (b) 20 μM DOPAC, (c) 1 μM adenosine and (d) 200 μM ascorbic acid.

detection, the peaks for serotonin are very characteristic, including a peak at the switching potential.[40] Compared with carbon fiber and carbon nanotube electrodes, CNS electrodes are less sensitive to serotonin, probably due to serotonin fouling which blocks more active sites for the CNS electrodes.[33] DOPAC is a catechol that is present in the brain in higher concentrations and is a major metabolite of the neurotransmitter dopamine.[41] At physiological pH, DOPAC is an anion, and the molecules are electrostatically repelled from adsorption, so its CV reduction peak is more negative than that of dopamine (Fig. 4b).[42] Adenosine has a higher oxidation potential, and the oxidation peak of adenosine shows up on the reverse scan with a 1.45 V switching potential (Fig. 4c).[43] Interestingly, a secondary peak is observed for adenosine on carbon-fibers, but is less evident at the CNS electrodes.[44] A broad peak is observed for ascorbic acid, which is a common interferent in the brain at concentrations 100 to 1000-fold higher than those of dopamine (Fig. 4d).[45] One major advantage of CNSs nanoelectrodes is that they exhibit a very high surface area to volume ratio. Compared with smooth carbon fiber electrodes, the surface area of CNS electrodes is 5-fold larger given the same geometric area. Therefore, the background signal and the current responses are larger at CNS nanoelectrodes, and they exhibit enhanced sensitivity for neurotransmitter sensing.[32,46] Also, CNS electrodes have greater selectivity for cationic neurochemicals over anionic neurochemicals than those of other carbon materials,[32] which may be explored in future works.

Compared with other carbon nanoelectrode fabrication methods, growing CNSs on etched metal wires is promising for mass fabrication. Fig. 5 and Table 1 compare the advantages and disadvantages of different fabrication techniques, including CNS nanoelectrodes, flame etched carbon fiber nanoelectrodes, 3D printed nanoelectrodes, and carbon nanopipettes. Flame etching of carbon fibers is simple and it is easy to obtain small nanotips, so it is the most used method for

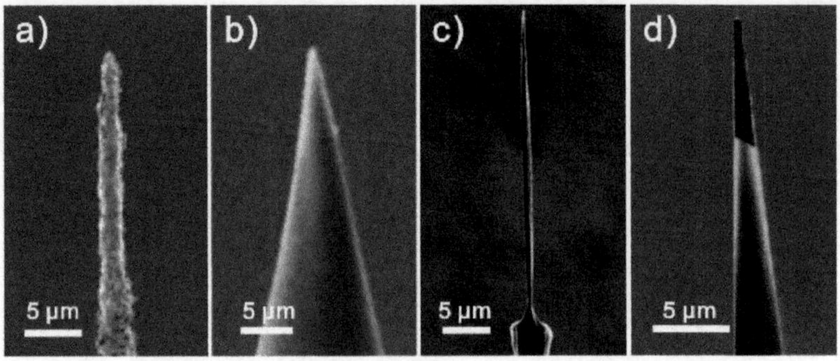

Fig. 5 SEM images comparing different carbon nanoelectrodes. (a) Nb/CNS nanoelectrode, (b) 3D printed nanoelectrode and (c) flame etched carbon fiber nanoelectrode, adapted with permission from ref. 14, copyright (2015) Wiley-VCH, and (d) carbon nanopipette, reprinted with permission from ref. 22, copyright (2015) American Chemical Society.

carbon nanoelectrode fabrication.[7,47] Both CNS nanoelectrodes and carbon fiber nanoelectrodes have conical nanotips, which is the most common morphology yielded by an etching process. Flame etching is easier than growing CNSs, but it is a hand fabrication process and not amenable for batch fabrication. Nano 3D printing enables fully customized electrode geometries and surface structures, but the method is not very fast and it requires specialized equipment.[8,19] Therefore, nano 3D printing is most useful for the fabrication of rationally designed carbon nanoelectrode sensors with high reproducibility. Carbon nanopipettes are fabricated by chemical vapor deposition, similar to CNS electrodes, but for carbon nanopipettes, carbon is selectively deposited on the inner surface of the quartz capillary. Both methods are suitable for batch manufacturing. Carbon nanopipettes have the cavity structure which confines the analytes, so thin layer electrochemistry is the main electrochemical process when using FSCV[46,48] Also,

Table 1 Comparisons of CNS nanoelectrodes, carbon fiber nanoelectrodes, 3D printed nanoelectrodes and carbon nanopipettes

Nano-electrode	Tip size	Electrode geometry	Surface morphology	Fabrication method	Features	Limitations	Ref.
CNS	300 nm to 1 μm	Conical	Rough	PE-CVD	Suitable for mass fabrication	Relatively large nanotip	30, 32 and 34
Carbon fiber	50–300 nm	Conical	Smooth	Flame etching	Simple and small size	Manual fabrication	10–14, 17, 18 and 50–52
3D printing	100–600 nm	Custom	Custom	Direct laser writing	Customized geometry	Require specific equipment	8, 19 and 53–58
Carbon nano-pipettes	20–600 nm	Conical	Cavity	CVD	Cavity structure	Small surface area	22–24, 48 and 59–61

the nanoscale cavity of carbon nanopipettes can be used for unique applications such as resistive-pulse sensors and rectification sensors.[21,24,49] Thus, there are different applications for the different types of carbon nanoelectrodes, but CNS electrodes would be suitable for direct implantation into small organisms or near synapses.

CNS nanoelectrodes require the deposition of a thin layer of CNSs, so the nanotip areas are larger than those of other types of nanoelectrodes, even though the geometric areas may be similar. CNSs are short and dense, so they increase the electroactive surface area to volume ratio to improve the sensitivity. Thus, the currents measured here are larger than those for other nanoelectrodes of a similar size, because CNS electrodes have rough surfaces with greater surface areas. CNSs are also rich in edge-plane carbon sites, which enhance the detection of neurotransmitters by improving adsorption. One issue with carbon nanoelectrodes is that the currents are small, and these CNS electrodes overcome that challenge. The CNS electrodes also show more resistance to anion neurochemicals, such as ascorbic acid; and they are less sensitive to neurotransmitters which can induce fouling, such as serotonin. The benefits of CNS electrodes for selectivity should be emphasized in future studies. Overall, CNSs have large surface areas and enhanced electrochemical performance. We fabricated thirty nanoelectrode sensors on a 2.5 cm wide substrate for etching in one batch, and then put all the wires into the PECVD simultaneously. Therefore, the fabrication of CNS nanoelectrodes is a promising strategy for mass manufacturing of carbon nanoelectrode sensors.

Conclusion

In summary, we have developed a new carbon nanoelectrode batch fabrication method by growing CNSs onto etched metal wires. In one batch, we mounted approximately 30 tungsten or niobium wires on a 2.5 cm wide substrate and electrochemically etched them to form nanotips. For mass manufacturing, more than one hundred metal wires may be mounted on a larger conductive substrate. A thin layer of CNSs was uniformly grown on the metal surface using PE-CVD, which is also a mass fabrication process. The CNS nanoelectrodes exhibit a large surface to volume ratio and good electrochemical properties. The CNS nanoelectrodes were characterized by FSCV detection for dopamine, and they are useful for detecting a variety of neurochemicals. Compared with other fabrication methods of carbon nanoelectrodes, CNS nanoelectrodes have a larger surface area to volume ratio, and thus larger currents. All of the fabrication processes of CNS nanoelectrodes are operated in batch, therefore, the development of CNS nanoelectrodes exhibits promising potential for mass fabricating carbon nanoelectrode sensors.

Conflicts of interest

There are no conflicts of interest.

Acknowledgements

This work was funded by NIH R01EB026497 and NIH R01MH085159. The growth and SEM imaging of the carbon nanospikes were conducted at the Center for Nanophase Materials Sciences, Oak Ridge National Laboratory, which is a DOE

Office of Science User Facility. This manuscript has been authored by UT-Battelle, LLC under contract No. DE-AC05-00OR22725 with the U.S. Department of Energy.

References

1 Q. Cao, P. Puthongkham and B. J. Venton, *Anal. Methods*, 2019, **11**, 247–261.
2 C. Yang, M. E. Denno, P. Pyakurel and B. J. Venton, *Anal. Chim. Acta*, 2015, **887**, 17–37.
3 X. Zhang, A. Hatamie and A. G. Ewing, *Curr. Opin. Electrochem.*, 2020, **22**, 94–101.
4 M. Shin, Y. Wang, J. R. Borgus and B. J. Venton, *Annu. Rev. Anal. Chem.*, 2019, **12**, 297–321.
5 M. Shin, J. M. Copeland and B. J. Venton, *ACS Chem. Neurosci.*, 2018, **9**, 1872–1883.
6 R. Chen, K. Alanis, T. M. Welle and M. Shen, *Anal. Bioanal. Chem.*, 2020, 6121–6132.
7 M. Shen and M. L. Colombo, *Anal. Methods*, 2015, **7**, 7095–7105.
8 Q. Cao, M. Shin, N. V. Lavrik and B. J. Venton, *Nano Lett.*, 2020, **20**, 6831–6836.
9 M. L. Huffman and B. J. Venton, *Analyst*, 2009, **134**, 18–24.
10 W. Z. Wu, W. H. Huang, W. Wang, Z. L. Wang, J. K. Cheng, T. Xu, R. Y. Zhang, Y. Chen and J. Liu, *J. Am. Chem. Soc.*, 2005, **127**, 8914–8915.
11 T. G. Strein and A. G. Ewing, *Anal. Chem.*, 1992, **64**, 1368–1373.
12 X. Zhang, W. Zhang, X. Zhou and B. Ogorevc, *Anal. Chem.*, 1996, **68**, 3338–3343.
13 Y.-T. Li, S.-H. Zhang, L. Wang, R.-R. Xiao, W. Liu, X.-W. Zhang, Z. Zhou, C. Amatore and W.-H. Huang, *Angew. Chem., Int. Ed.*, 2014, **53**, 12456–12460.
14 X. Li, S. Majdi, J. Dunevall, H. Fathali and A. G. Ewing, *Angew. Chem., Int. Ed.*, 2015, **54**, 11978–11982.
15 A. M. Strand and B. J. Venton, *Anal. Chem.*, 2008, **80**, 3708–3715.
16 Q. Cao, J. Lucktong, Z. Shao, Y. Chang and B. J. Venton, *Anal. Bioanal. Chem.*, 2021, **413**, 6737–6746.
17 C. Gu, X. Zhang and A. G. Ewing, *Anal. Chem.*, 2020, **92**, 10268–10273.
18 Y. Wang and A. Ewing, *ChemBioChem*, 2021, **22**, 807–813.
19 C. Yang, Q. Cao, P. Puthongkham, S. T. Lee, M. Ganesana, N. V. Lavrik and B. J. Venton, *Angew. Chem., Int. Ed.*, 2018, **57**, 14255–14259.
20 B. M. Kim, T. Murray and H. H. Bau, *Nanotechnology*, 2005, **16**, 1317–1320.
21 K. Hu, Y. Wang, H. Cai, M. V Mirkin, Y. Gao, G. Friedman and Y. Gogotsi, *Anal. Chem.*, 2014, **86**, 8897–8901.
22 H. R. Rees, S. E. Anderson, E. Privman, H. H. Bau and B. J. Venton, *Anal. Chem.*, 2015, **87**, 3849–3855.
23 C. Yang, K. Hu, D. Wang, Y. Zubi, S. T. Lee, P. Puthongkham, M. V. Mirkin and B. J. Venton, *Anal. Chem.*, 2019, **91**, 4618–1624.

24 A. G. Ewing, K. Hu, R. Jia, A. Hatamie, K. L. Le Vo and M. V. Mirkin, *J. Am. Chem. Soc.*, 2020, **142**, 16910–16914.
25 M. L. Colombo, J. V. Sweedler and M. Shen, *Anal. Chem.*, 2015, **87**, 5095–5100.
26 N. T. Iwai, M. Kramaric, D. Crabbe, Y. Wei, R. Chen and M. Shen, *Anal. Chem.*, 2018, **90**, 3067–3072.
27 M. L. Colombo, S. McNeil, N. Iwai, A. Chang and M. Shen, *J. Electrochem. Soc.*, 2016, **163**, H3072–H3076.
28 L. B. Sheridan, D. K. Hensley, N. V. Lavrik, S. C. Smith, V. Schwartz, C. Liang, Z. Wu, H. M. Meyer and A. J. Rondinone, *J. Electrochem. Soc.*, 2014, **161**, H558–H563.
29 Y. Song, R. Peng, D. K. Hensley, P. V. Bonnesen, L. Liang, Z. Wu, H. M. Meyer, M. Chi, C. Ma, B. G. Sumpter and A. J. Rondinone, *ChemistrySelect*, 2016, **1**, 1–8.
30 A. G. Zestos, C. Yang, C. B. Jacobs, D. Hensley and B. J. Venton, *Analyst*, 2015, **140**, 7283–7292.
31 A. S. Shanta, S. Shamsir, Y. Song, D. K. Hensley, A. J. Rondinone, S. K. Islam and N. McFarlane, *2018 40th Annu. Int. Conf. IEEE Eng. Med. Biol. Soc.*, 2018, pp. 4281–4284.
32 Q. Cao, D. K. Hensley, N. V. Lavrik and B. J. Venton, *Carbon*, 2019, **155**, 250–257.
33 Q. Cao, D. K. Hensley, N. V Lavrik and B. J. Venton, *Carbon*, 2019, **155**, 250–257.
34 A. S. Shanta, K. A. Al Mamun, S. K. Islam, N. McFarlane and D. K. Hensley, *Int. J. High Speed Electron. Syst.*, 2017, **26**, 1740008.
35 S. Kerfriden, A. H. Nahlé, S. A. Campbell, F. C. Walsh and J. R. Smith, *Electrochim. Acta*, 1998, **43**, 1939–1944.
36 Y. Uehara, T. Fujita, M. Iwami and S. Ushioda, *Rev. Sci. Instrum.*, 2001, **72**, 2097–2099.
37 R. L. McCreery, *Chem. Rev.*, 2008, **108**, 2646–2687.
38 B. J. Venton and Q. Cao, *Analyst*, 2020, **145**, 1158–1168.
39 B. D. Bath, D. J. Michael, B. J. Trafton, J. D. Joseph, P. L. Runnels and R. M. Wightman, *Anal. Chem.*, 2000, **72**, 5994–6002.
40 K. E. Dunham and B. J. Venton, *Analyst*, 2020, **145**, 7437–7446.
41 P. Takmakov, M. K. Zachek, R. B. Keithley, E. S. Bucher, G. S. McCarty and R. M. Wightman, *Anal. Chem.*, 2010, **82**, 9892–9900.
42 M. L. Huffman and B. J. Venton, *Electroanalysis*, 2008, **20**, 2422–2428.
43 M. D. Nguyen and B. J. Venton, *Comput. Struct. Biotechnol. J.*, 2015, **13**, 47–54.
44 B. E. K. Swamy and B. J. Venton, *Anal. Chem.*, 2007, **79**, 744–750.
45 H. Cheng, L. Li, M. Zhang, Y. Jiang, P. Yu, F. Ma and L. Mao, *TrAC, Trends Anal. Chem.*, 2018, **109**, 247–259.
46 Q. Cao, Z. Shao, D. K. Hensley, N. V. Lavrik and B. J. Venton, *Langmuir*, 2021, **37**, 2667–2676.
47 J. Clausmeyer and W. Schuhmann, *TrAC, Trends Anal. Chem.*, 2016, **79**, 46–59.
48 Z. Shao, P. Puthongkham, K. K. Hu, R. Jia, M. V. Mirkin and B. J. Venton, *Electrochim. Acta*, 2020, **361**, 137032.
49 Y. Yu, J.-M. Noël, M. V Mirkin, Y. Gao, O. Mashtalir, G. Friedman and Y. Gogotsi, *Anal. Chem.*, 2014, **86**, 3365–3372.
50 Y.-T. Li, S.-H. Zhang, X.-Y. Wang, X.-W. Zhang, A. I. Oleinick, I. Svir, C. Amatore and W.-H. Huang, *Angew. Chem., Int. Ed.*, 2015, **54**, 9313–9318.
51 X. Li, J. Dunevall and A. G. Ewing, *Acc. Chem. Res.*, 2016, **49**, 2347–2354.

52 J. G. Roberts, E. C. Mitchell, L. E. Dunaway, G. S. McCarty and L. A. Sombers, *ACS Nano*, 2020, **14**, 2917–2926.
53 A. Abdalla and B. A. Patel, *Curr. Opin. Electrochem.*, 2020, **20**, 78–81.
54 X. Zhang, A. Vyatskikh, H. Gao, J. R. Greer and X. Li, *Proc. Natl. Acad. Sci. U. S. A.*, 2019, **116**, 6665–6672.
55 C. M. Portela, B. W. Edwards, D. Veysset, Y. Sun, K. A. Nelson, D. M. Kochmann and J. R. Greer, *Nat. Mater.*, 2021, **20**, 1491–1497.
56 G. Seniutinas, A. Weber, C. Padeste, I. Sakellari, M. Farsari and C. David, *Microelectron. Eng.*, 2018, **191**, 25–31.
57 J. Bauer, A. Schroer, R. Schwaiger and O. Kraft, *Nat. Mater.*, 2016, **15**, 438–443.
58 B. Cardenas-Benitez, C. Eschenbaum, D. Mager, J. G. Korvink, M. J. Madou, U. Lemmer, I. De Leon and S. O. Martinez-Chapa, *Microsyst. Nanoeng.*, 2019, **5**, 1–13.
59 Y. Yu, Y. Gao, K. Hu, P. Y. Blanchard, J. M. Noël, T. Nareshkumar, K. L. Phani, G. Friedman, Y. Gogotsi and M. V. Mirkin, *ChemElectroChem*, 2015, **2**, 58–63.
60 S. E. Anderson and H. H. Bau, *Nanotechnology*, 2014, **25**, 245102.
61 K. Hu, D. Wang, M. Zhou, J. H. Bae, Y. Yu, H. Xin and M. V. Mirkin, *Anal. Chem.*, 2019, **91**, 12935–12941.

Faraday Discussions

ROYAL SOCIETY OF CHEMISTRY

PAPER

Development of multifunctional nanopipettes for controlled intracellular delivery and single-entity detection

Popular Pandey,[a] Alberto Sesena-Rubfiaro,[a] Santosh Khatri[a] and Jin He *[ab]

Received 20th August 2021, Accepted 4th October 2021
DOI: 10.1039/d1fd00057h

The intracellular delivery of biomolecules and nanoscale materials to individual cells has gained remarkable attention in recent years owing to its wide applications in drug delivery, clinical diagnostics, bio-imaging and single-cell analysis. It remains a challenge to control and measure the delivered amount in one cell. In this work, we developed a multifunctional nanopipette – containing both a nanopore and nanoelectrode (pyrolytic carbon) at the apex – as a facile, minimally invasive and effective platform for both controllable single-cell intracellular delivery and single-entity counting. While controlled by a micromanipulator, the baseline changes of the nanopore ionic current (*I*) and nanoelectrode open circuit potential (*V*) help to guide the nanopipette tip insertion and positioning processes. The delivery from the nanopore barrel can be facilely controlled by the applied nanopore bias. To optimize the intracellular single-entity detection during delivery, we studied the effects of the nanopipette tip geometry and solution salt concentration in controlled experiments. We have successfully delivered gold nanoparticles and biomolecules into the cell, as confirmed by the increased scattering and fluorescence signals, respectively. The delivered entities have also been detected at the single-entity level using either one or both transient *I* and *V* signals. We found that the sensitivity of the single-entity electrochemical measurement was greatly affected by the local environment of the cell and varied between cell lines.

Introduction

Controlled cell manipulation and investigation *via* the localized intracellular delivery of biomolecules are becoming game-changing approaches to understand molecular biology at the single-cell level, manipulate cell function, reprogram cell behavior and treat diseases.[1,2] The controlled delivery of functional nano-entities, such as synthetic nanoparticles and nanostructures, RNA, plasmid DNA, proteins and peptides, into living cells has many applications, including gene expression

[a]*Physics Department, Florida International University, Miami, Florida, 33199, USA. E-mail: jinhe@fiu.edu*
[b]*Biomolecular Sciences Institute, Florida International University, Miami, Florida, 33199, USA*

and regulation, bio-imaging, cell based-therapies, drug delivery, tissue engineering and wound healing.[2–12] In the last two decades, various delivery techniques have been developed.[4,13] The most common techniques are viral and chemical transfections at the bulk cell level. Viral transfection is reliable with high efficiency, but it is expensive and limited to gene delivery.[14] Chemical transfection techniques in general have lower transfection efficiencies, with problems of endosomal entrapment and the degradation of the transfected entities.[4] Both techniques lack control over the dosage and cannot deliver genes to a single cell.

Physical delivery approaches, such as microinjection, electroporation,[15] fluidic force microscopy, carbon nanotube endoscopes, nanostraws, nanowires and nanoinjection, have been introduced to achieve intracellular delivery.[4,16] These techniques can deliver a wide range of exogenous materials to a wide range of cell lines. They are normally of low throughput but can deliver to individual cells and enable single-cell studies.[17–19] Glass or quartz micropipettes are commonly used for pressure-controlled microinjection. However, the microinjection also induces low cell viability due to the large tip diameters from 0.5 to 5 μm.[20–22] In recent years, various electroporation methods, such as nanofountain probes, nanostraws, flow-through electroporation, microchip electroporation, fluidic force microscopy and 2D nanopore-based electroporation platforms, have been developed for intracellular delivery.[23–34] However, they also suffer problems of irreversible damage to the cell membrane and cell death due to high electric fields, expensive device fabrication and poor dosage control during delivery.[29,35–37] Carbon nanotube endoscopes and fluidic force microscopes have also been used for pressure-controlled cellular delivery.[38–42] These platforms have high cell viability post-injection however suffer limitations such as expensive device fabrication, and the requirement of the specialized probe.

Recently, nanopipettes with pore diameters below 300 nm have been utilized for single-cell delivery.[43–47] Different from other nanoscale delivery devices, such as nanostraws and nanowires, nanopipettes can be easily fabricated from quartz/glass capillaries using pipette pullers. The reduction of the nanopipette tip size greatly improves the cell viability after injection.[21,38] The nanopipette delivery can also be controlled by the voltage, which enables the quick, facile and precise control of the delivered entities.[21,43,44] Pourmand's group have demonstrated a voltage-driven delivery system using single/double barrel nanopipettes *via* a scanning ion conductance microscope.[43] Mirkin's group have used a SECM platform as a high precision micromanipulator for volume-controlled intracellular delivery *via* a nanopipette.[47] A metal coated nanopipette was also used as an electrowetting nanoinjector with controlled dosage delivery and high transfection efficiency.[48]

The accurate assessment and control of the delivery dosage is the main challenge in single-cell transfection. Currently, fluorescence intensity, injection volume and injection duration-based approaches have been widely used in an attempt to quantify the delivered entities.[35,36,48,49] However, these approaches are unable to quantify the delivered number of molecules with high precision. Quantification of delivered entities at the single-molecule level has been achieved during delivery *via* nanopore devices.[49,50] The Timp group utilized silicon nitride nanopores for the quantitative intracellular delivery of DNA plasmids *via* electroporation.[51] The Angelis group implemented a 3D hollow nanoelectrode-based system to detect translocation signals during the delivery of gold nanorods into cells.[52] While both works demonstrate the capability of nanopores as a quantitative delivery platform, both approaches demand expensive nanofabrication steps and often require a high threshold voltage to deliver entities.

Because of the rapid developments of nanopipette-based single-entity electrochemistry (SEE) measurement techniques,[45] we are interested in achieving nanopipette-based intracellular delivery and SEE detection at the same time. Here, we developed a multifunctional nanopipette-based intracellular delivery method that allows for the SEE detecting of delivered molecules and monitoring of the delivery process. The multifunctional nanopipette consists of a nanopore and a carbon nanoelectrode (CNE) at the apex of a nanopipette.[53] We first investigated and optimized the conditions to achieve intracellular SEE detection using both the nanopore based ionic current detection and CNE based open-circuit-potential (OCP) detection during the voltage-controlled intracellular delivery of various nanoentities. Recent studies have suggested that the molecular crowding effect can be a beneficial factor for nanopore sensing.[54,55] Indeed, we found that SEE detection inside the cell not only can be realized but often improved. We also found that *in situ* SEE detection is very useful in monitoring the delivery rate and the dissipation of delivered entities, and probing the local cellular environment.

Experimental methods

Materials

FITC dye was purchased from Sigma-Aldrich. Thiol-PEG 5k (MW-5000, purity >95%) for the surface modification of 40 nm gold nanoparticles (GNPs) was purchased from Nanocs. The citrate stabilized 40 nm GNPs were bought from Ted Pella, Inc. Other chemicals in ACS grade were purchased from Fisher Scientific. The purchased chemicals were used without further purification. 130 mM PBS was prepared by diluting 1× PBS in DI water. The anion concentration of the final solution was 0.13 M and it had an ionic strength of 0.147 M. The intracellular solution was prepared with the following composition: NaCl 8 mM, KCl 132.5 mM, $MgCl_2$ 2 mM, HEPES 10 mM, 7.2 pH adjusted with NaOH. The extracellular solution contained: NaCl 130 mM, KCl 3 mM, $CaCl_2$ 2 mM, $MgCl_2$ 1.2 mM, HEPES 10 mM, glucose 5 mM, 7.4 pH adjusted with NaOH. Tyrode's solution was prepared with: NaCl 140 mM, KCl 5 mM, $CaCl_2$ 1.8 mM, $MgCl_2$ 1 mM, HEPES 5 mM, NaH_2PO_4 1 mM, glucose 10 mM, 7.4 pH adjusted with NaOH. FITC stock solution was prepared in DMSO. 20 μM of the FITC solution was prepared by diluting the stock solution in intracellular medium which contains only 0.2% DMSO. All solutions were prepared using deionized water (∼18 MΩ) from a water purification system (Ultra Purelab system, ELGA/Siemens).

Quartz nanopipette fabrication

The fabrication of multifunctional nanopipettes from quartz theta capillary tubes (FG-G QT120-90-7.5, Sutter Instrument) has been described in detail previously.[53] In brief, the dual-barrel nanopipettes were fabricated from cleaned capillary tubes using a laser pipette puller (P-2000, Sutter Instrument) with the following parameters: HEAT = 825, FIL = 3, VEL = 40, DEL = 220, PUL = 190. Then, one barrel was filled with pyrolytic carbon to form a CNE at the tip.[53] Single barrel quartz nanopipettes were also fabricated from cleaned capillary tubes (QF-100-50-7.5, Sutter Instrument) with the following pipette puller parameters: HEAT = 795, FIL = 3, VEL = 40, DEL = 145, PUL = 180. Optical microscopy, current–voltage (*I*–*V*) measurements and/or cyclic voltammetry (CV), and scanning electron microscopy (SEM, JEOL JSM 6335F) were used to characterize the fabricated nanopipettes.

For the delivery experiments, the nanopore barrel was backfilled with nanoscale entities (molecules or GNPs) suspended in 130 mM PBS solution (for bath delivery) or in intracellular solution (for intracellular delivery). Then, the nanopipette was centrifuged at 2000 rpm for 2 min. All intracellular delivery experiments were performed in an extracellular medium. The loaded nanopipette tips were imaged by a dark-field microscope for the GNPs and a fluorescence microscope for the fluorescent entities to evaluate the filling before the delivery experiments.

Preparation of PEGylated GNPs

The PEGylated GNPs were prepared using a previously reported method[56] with small modifications. In brief, 6 μL of aqueous 10 mM thiol-PEG 5k was added to 1 mL of 40 nm citrate-coated GNPs (150 pM stock concentration), producing a molar ratio of thiol-PEG to GNPs of $10^5 : 1$. The mixture was incubated overnight at room temperature to achieve full replacement of the citrate ligands by thiol-PEG on the GNP surface. The mixture was then purified by centrifugation twice, each at 7000 rpm for 10 min. The supernatant was removed, and the pellet was re-suspended in 1 mL of DI water. The prepared PEGylated GNPs were re-suspended in the intracellular medium or PBS solution. The PEG modification and the stability of the PEGylated GNPs in the intracellular medium or PBS solution were confirmed *via* UV–Vis spectroscopy.

Electrical measurements

The current–time (*i*–*t*) and potential–time (*V*–*t*) traces were recorded using the experimental setups illustrated in Fig. 1a and 3a, respectively. We used freshly prepared Ag/AgCl wire electrodes and an Axopatch 200B amplifier (Molecular Devices Inc., CA) in the voltage-clamp mode to measure the ionic current through the nanopore. A battery-powered high-input impedance differential amplifier was used to measure the local potential of the CNE. The bandwidth of the low-pass filter was 5 kHz for the current. The data were recorded with a sampling rate of 50 kHz using Digidata 1440A and the software Axoscope 10.5 (Molecular Devices).

HEK293 cell culture

The HEK293H cell line was obtained from ThermoFisher Scientific Invitrogen (Waltham, MA) and cultured as previously described.[57] Briefly, HEK293H cells were maintained with DMEM/high glucose medium (Corning, 10-013-CV) supplemented with 10% fetal bovine serum (Gibco, 16140-071) and 1% Pen Strep (Gibco, 15140-122) in a humidified incubator with 5% CO_2 at 37 °C. HEK293H cells were seeded in a 35 mm Petri dish at a density of 2×10^5 cells per dish. The HEK293H cells were rinsed two times with the extracellular solution before experiments.

Cell culture and differentiation of hiPSCs

Following the previously reported protocol,[58] hiPSCs from reprogrammed fibroblasts (GM23338, Coriell Institute for Medical Research, NJ, USA) were cultured on Matrigel-coated plates (BD Biosciences) with stem basal medium (mTeSR1, STEMCELL Technologies) until they reached 80–90% of confluence. Cardiac differentiation of the hiPSCs was initiated by treatment with 12 μM CHIR99021 (Tocris, 4423) diluted in RPMI/B27-insulin during the initial 24 h; the medium was changed every day in succession for 3 days. Subsequently, the hiPSCs were treated with 5 μM IWP4 (Tocris, 5214) that was diluted by RPMI/B27-insulin, which was removed after 24 h. From day 4, the hiPSCs were maintained in RPMI/B27-insulin with daily medium change. Spontaneous contractions of the cardiomyocytes (CMs) were observed around 8–10 days after the initiation of the differentiation process. According to the previous protocol description,[59] the CMs were purified by 5 mM L-lactate (Sigma, 71718) diluted in RPMI-glucose (Gibco, 11879-020), which was replaced daily in succession for 4 days. After purification, the CMs were dissociated with TrypLE Express (Gibco, 12605010) at 37 °C for 15 min. Thereafter, the cell suspension was centrifuged for 5 min at 1500 rpm to remove the enzyme. The pellet of cells was mixed with RPMI/B27-insulin supplemented with fetal bovine serum 10% (Gibco, 16140-071), 1% Pen Strep (Gibco, 15140-122), and 5 μM Y-27632 (Tocris 1254). For nanopipette injection, the CMs were replanted in a 35 mm Matrigel coated dish at a density of 6×10^5 cells per dish. Spontaneous beating was observed around 5–6 days after replanting.

Nuclear staining of HEK cells

The staining of the live HEK nuclei was performed using the medium exchange method. Briefly, the original cell culture medium was removed and immediately replaced with a medium containing Hoechst33258 dye with ~1 μg mL^{-1} concentration. The cells were incubated at 37 °C for 2 hours. Then, the cells were washed with an extracellular medium twice before experiments.

Trypan blue (TB) exclusion test

The TB test was performed to evaluate the cell membrane damage due to nanopipette injection into the cell membrane.[60] Trypan blue was added to the extracellular medium in a 1 : 1 ratio to check the membrane integrity. After 10–15 min of incubation, optical bright field (BF) images were taken. The cells with intact cell membranes appear transparent as trypan blue cannot enter the cytosol. While cells with a blue color represent cell membrane damage and hence are dead cells.

Fluorescence imaging

All fluorescence images were taken using a home-built inverted fluorescence microscope. A LED light source (X-cite 120 LED Boost, USA) was used to illuminate the sample. The emitted fluorescence light was collected by a 20× objective lens (Nikon, CFI S Plan Fluor ELWD) and recorded by a sCMOS camera (ThorLabs, USA).

Dark-field microscopy imaging

The dark-field (DF) microscopy images were taken by an inverted optical microscope (Nikon Eclipse Ti-U) equipped with a dark-field condenser (Nikon, Ti-DF, NA ~0.8–

0.95) and a 40× objective lens (Nikon CFI Super Plan Fluor ELWD, NA = 0.6). A CCD camera (Point Gray Grasshopper 3) was used to capture the dark-field images.

Data analysis and simulation

The electrical current and potential data were analyzed using Clampfit programs and OriginPro 2018b. The moving average smoothing method with a 2 ms time window was typically applied to the data before statistical analysis. The d*V*/d*t* curves were smoothed by the moving average method using a 5 ms time window. ImageJ was used to analyze the fluorescence images. We used finite element method (FEM) simulation to solve the coupled Poisson–Nernst–Planck (PNP) partial differential equations. To simplify the simulation, the fluidic flow term was not included, and the system was simulated at a steady state. COMSOL Multiphysics 5.2 with AC/DC and Chemical Reaction Engineering modules was used for the FEM simulation.

Results and discussions

Optimizing delivery and single-entity detection in a controlled environment

Before the live-cell experiments, we first conducted the delivery and single-entity detection experiments using the nanopore-CNE multifunctional nanopipettes in a bath solution to optimize the experimental conditions and better understand the detection mechanism. The schematic of the experimental setup is given in Fig. 1a. To minimize the cellular damage induced by post intracellular delivery,[61] we used relatively long tapered (LT) nanopipettes with a shank length of about 6.2 ± 0.4 mm, as shown in Fig. 1b. Based on the SEM image, the half cone angle of the nanopipette tip is ~3.5°. To better match the size of nanoentities, the nanopore diameter is in the range of 10–30 nm for biomolecules and 40–90 nm for GNPs. The average effective surface area of the CNEs is estimated to be 0.24 μm^2. The exiting NPs have to pass through a long nanochannel before reaching the orifice of the LT nanopipette. The nanopipette tip geometry has been previously shown to affect greatly on resistive pulse signals.[37] Here, we mainly focus on the OCP detection by the CNE.

The nanopore barrel is backfilled with 130 mM PBS solution (see the composition in the methods section). Before adding proteins and/or NPs inside, both the *i*–*t* and *V*–*t* traces acquired through the nanopore and CNE, respectively, were stable and featureless at different nanopore biases, V_{pore}. Fig. 1c shows that the potential baseline remains the same value at different V_{pore}. Therefore, the long channel of the LT nanopipette can effectively screen the changes of V_{pore} on the potential changes near the nanopore. After introducing nanoentities into the nanopore barrel, transient *i* and *V* changes often appear in the time traces at the appropriate V_{pore}. These characteristic currents and potential changes are caused by the individual exiting entities *via* the orifice of the nanopore barrel. The nanoentities can be driven out by the applied V_{pore} through the electrokinetic effect. We typically applied the V_{pore} with a magnitude of 0.2 V from the beginning and optimized the V_{pore} in experiments based on the observed event rate. The polarity of V_{pore} normally depends on the charge polarity of the nanoentities.

Fig. 1d shows the typical *i*–*t* and *V*–*t* traces at −0.2 V V_{pore} during the delivery of 40 nm PEGylated GNPs in 10 mM (i) and 130 mM (ii) PBS baths. The salt

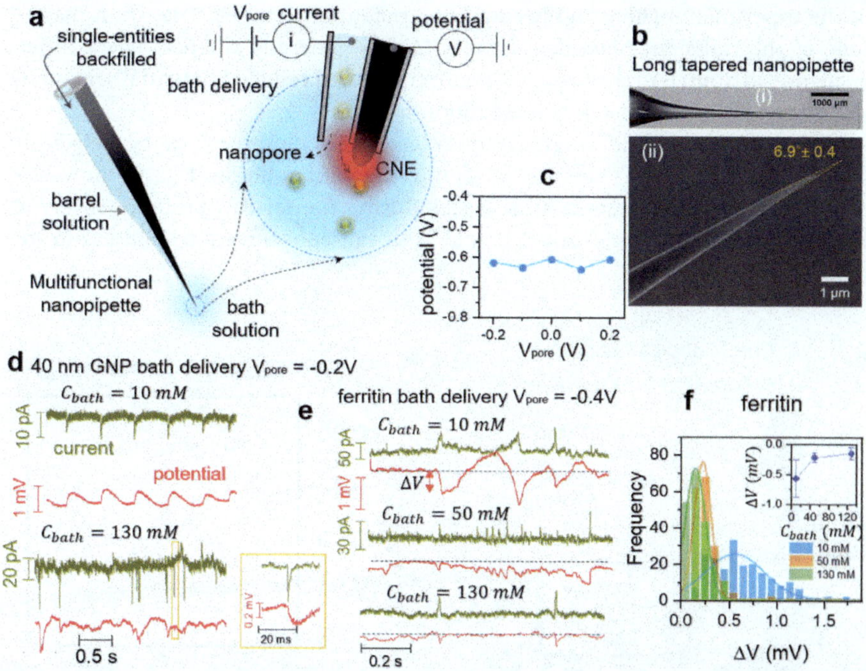

Fig. 1 (a) Schematic of the experimental setup of the bath delivery and single-entity detection *via* a multifunctional long tapered nanopipette at $V_{pore} = -0.2$ V and with a 130 mM PBS bath. The zoomed in image of the nanopipette apex is also shown (not to scale). The V and V_{pore} are the potential on the floating CNE and the applied nanopore bias, respectively. The sensing zone of the CNE is shown by the red color gradient. (b) Optical microscopy (i) and SEM images (ii) of the side view of a LT nanopipette tip. (c) The baseline V *vs.* V_{pore} plot for a LT nanopipette in a 130 mM PBS bath. (d) The i–t (olive) and V–t (red) time traces during the delivery of PEGylated 40 nm GNPs in (i) 10 mM (ii) 130 mM PBS baths. The zoomed-in current and potential changes in the 130 mM PBS bath are also shown. (e) The I–t (olive) and V–t (red) time traces during the bath delivery of ferritin with bath solution concentrations of 10, 50 and 130 mM *via* a nanopipette at −0.4 V V_{pore}. The black dashed lines denote potential baselines. (f) The histograms showing the ΔV distributions at different bath salt concentrations. $N = 208$, 148 and 200 for the 10 mM, 50 mM and 130 mM PBS baths, respectively. The solid lines in the histogram are Gaussian fits. The inset shows the plot of mean ΔV as a function of the bath salt concentration. The error bars represent the standard deviation from the mean.

concentration is an important factor that influences the sensitivity of the current and potential detections. In the 10 mM PBS bath, distinct and evenly spaced current spikes are observed. The uniformly distributed transient signals indicate the buildup of GNPs inside the nanopore barrel near the orifice.[62] Due to the strong electrostatic repulsion between the negatively charged GNPs and the silica surface near the nanopore mouth, the motion of the exiting GNPs is slowed down, which is beneficial for the single-entity detection. In general, ∼75% (7 of 9 attempts) of the GNP delivery in the 10 mM PBS bath show highly correlated current and potential changes. However, delivery in the 130 mM PBS bath features stochastic current spikes with shorter time durations, t_D, and bigger magnitudes compared with the results in the 10 mM PBS bath. Meanwhile, the potential changes fluctuate a lot in shape and magnitude. Less than 20% (2 of 6 attempts) of the experiments show highly correlated potential changes. The mean magnitude of the transient potential changes (ΔV) is generally smaller. These differences are attributed to the faster GNP movement and reduced potential sensitivity when the bath solution salt concentration is high.

We also studied the single-entity detection of biomolecules by delivering ferritin proteins (d = 11 nm, pI = 5.4) into the bath solution. Fig. 1e shows the typical current and potential time traces during the ferritin protein delivery at $V_{pore} = -0.4$ V with 10, 50 and 130 mM bath salt concentrations. Similar to the

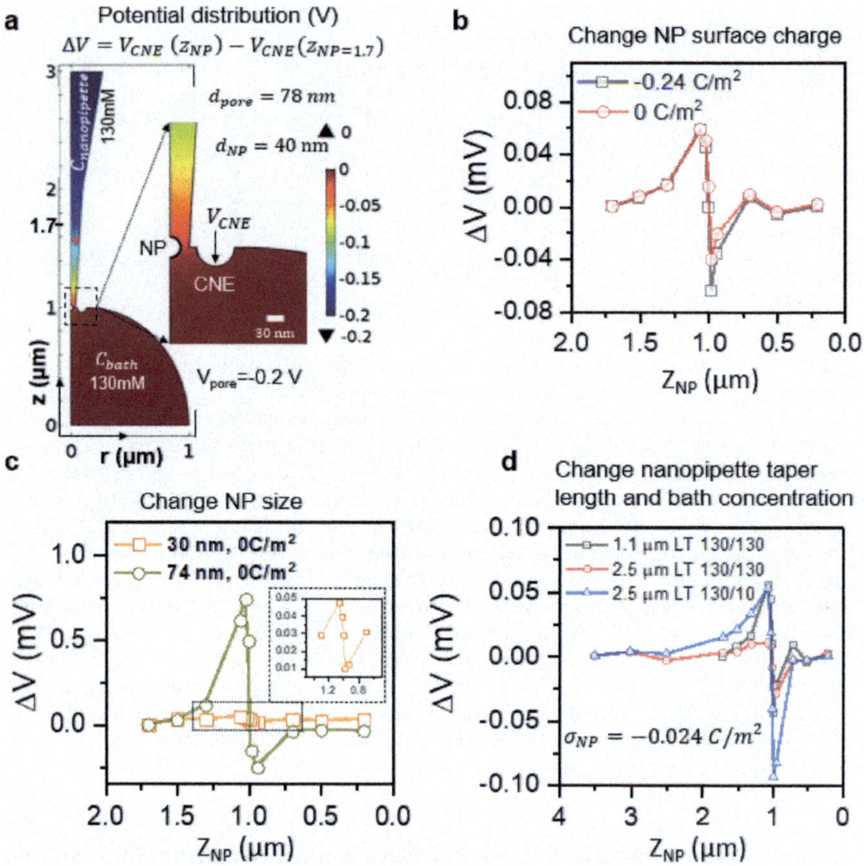

Fig. 2 (a) The simulation model and the electric potential distribution when a NP is at the nanopore ($r = 0$, $z = 1$) in a 130 mM KCl bath. The concentration inside the nanopipette is always 130 mM. A 3 nm thick insulating layer is placed on top of a 40 nm GNP to mimic the structure of a PEGylated GNP. The GNP surface charge density is -0.024 C m^{-2}. The nanopipette surface charge density is -0.005 C m^{-2}. There is no charge on the floating CNE surface. The downward red arrow denotes the NP motion direction. The nanopipette tip region is zoomed in on the right. The scale bar is 30 nm. (b) The simulated ΔV *vs.* Z_{NP} plots for the 40 nm NP with -0.24 C m^{-2} and 0 C m^{-2} surface charge densities. (c) The ΔV *vs.* Z_{NP} plots for neutral dielectric NPs with 30 nm and 74 nm sizes. The plot in the dashed rectangle is zoomed in the inset. (d) ΔV *vs.* Z_{NP} plots for a 40 nm NP delivery from nanopipettes with 1.1 μm and 2.5 μm taper lengths and at different bath salt concentrations. 130/10 denote nanopore barrel/bath salt concentrations. We used the V_{CNE} (Z_{NP} = 3.5 μm) as the reference point here.

GNP results, the transient potential change is bigger and the current and potential correspondence is better at a low (10 mM) salt concentration. The mean ΔV as a function of the bath concentration is shown in Fig. 1f. The mean ΔV decreases with the increase of the bath concentration.

To better understand the potential sensing mechanism for intracellular delivery, FEM simulations were performed at the steady state. Fig. 2a shows the simulation model and the potential distribution with 130 mM KCl solution in both the bath and barrel when a NP is at the orifice. To compare with the experimental results, we monitor the change of the CNE potential (V_{CNE}) when the NP is positioned at different z locations along the symmetry axis $r = 0$. To quantify the change of V_{CNE} (ΔV) induced by the NP position change, we define a reference point ($r = 0$, $z = 1.7$ μm) far away from the CNE. Near and beyond this point, the presence of a NP has no impact on the V_{CNE}.

We first investigate the effect of the surface charge of the NP on the ΔV change. As shown in Fig. 2b, the ΔV magnitude is at its maximum when the NP is at the orifice and zero when the NP is far away ($z = \pm 1.7$ μm). The shapes of the ΔV–Z_{NP} plots are always biphasic. When the NP is negatively charged (*i.e.* -0.24 C m^{-2}), the downward peak becomes slightly higher but retains the biphasic shape. This is different from the mainly downward peak shape for the negatively charged NPs at a low salt concentration.[53,62] Here, the biphasic shape is mainly due to the resistive changes induced by the NP translocation motion. The charge sensing mechanism-based potential detection becomes minor at a high salt concentration.

Because the ΔV during delivery is mainly due to the resistive change at the high salt concentration, the size difference between the NP and the nanopore should impact the ΔV. As shown in Fig. 2c, when the NP is only 30 nm while the nanopore size is 78 nm, the ΔV *vs.* Z_{NP} plot has a biphasic shape but the change is very small (see the inset of Fig. 2c). When the NP size is increased to 74 nm, the ΔV change is about 10 times bigger. Therefore, the size match between the NP and the nanopore plays a significant role in the ΔV. Both the nanopore current sensing and CNE OCP sensing can achieve a higher sensitivity when the nanopore size is comparable to the NP size.

The taper length of the nanopipette can also impact the measured ΔV. As shown in Fig. 2d, two nanopipette geometries were simulated with charged glass wall lengths of 1.1 μm and 2.5 μm. The ΔV change as a function of the Z_{NP} induced by the NP through the 1.1 μm nanopipette in a 130 mM KCl bath is biphasic (gray) while the ΔV change with the 2.5 μm nanopipette is more of a downward dip (red). We also simulated the ΔV change with the 2.5 μm nanopipette in a 10 mM KCl bath. A much larger biphasic ΔV change is observed (blue), demonstrating the effect of the bath salt concentration.

The FEM simulation results suggest that the potential sensing at a high salt concentration is quite different from the case at low salt concentration. In general, the ΔV change is smaller and the resistance change induced potential redistribution is the main cause of the potential change at the CNE. Because of the same resistive change origin, the potential change and current change are expected to be highly correlated. Meanwhile, the ΔV can be affected by several other factors, including the nanopipette tip geometry, NP surface charge and size match, and local salt concentration, which may lead to highly variable ΔV shapes. It should also be noted that the simulation is at a steady state. The contribution of

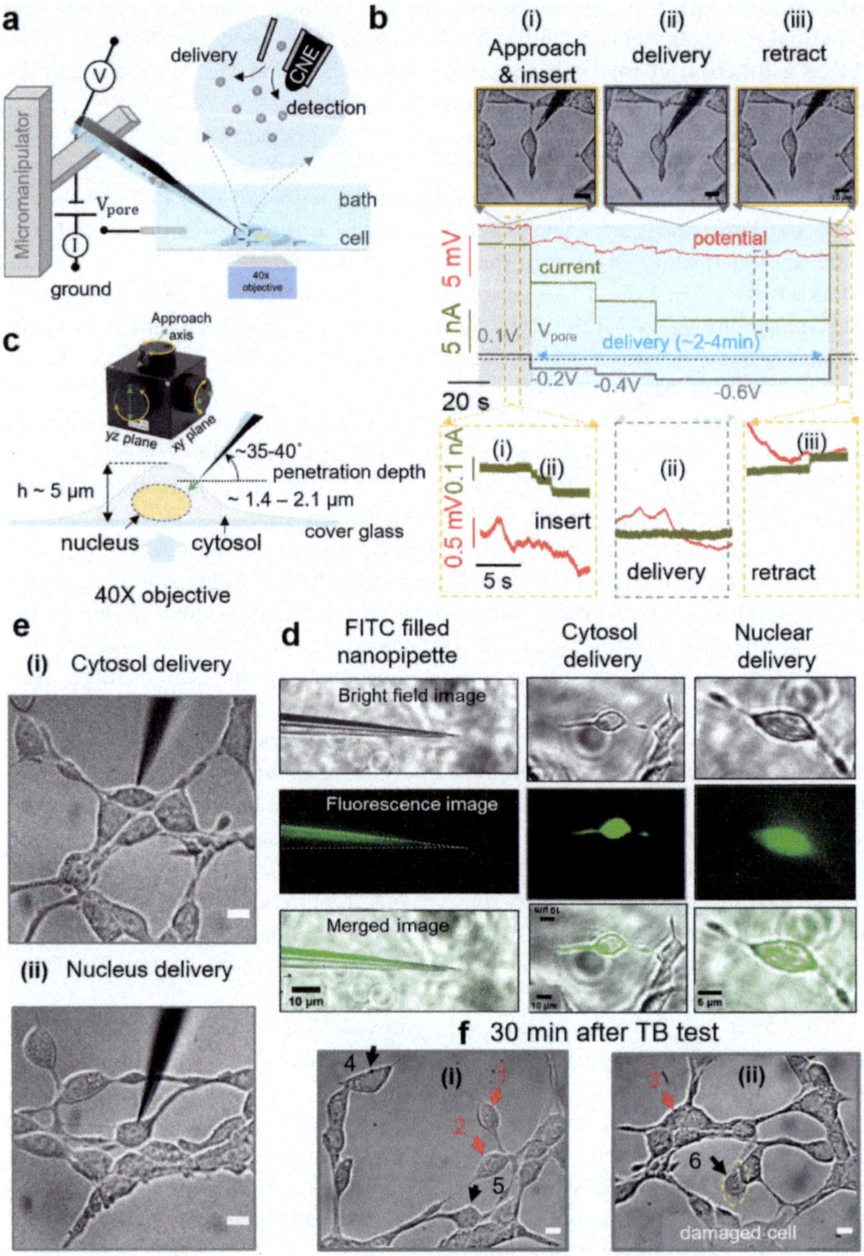

Fig. 3 (a) The schematic of the experimental setup of the intracellular delivery of single-entities *via* a multifunctional nanopipette. The extracellular bath solution is grounded and V_{pore} is the bias applied to the nanopore barrel filled with intracellular medium. A high impedance differential amplifier is connected to the CNE barrel to measure the local potential change at the floating CNE. (b) Typical *i*–*t* (olive) and *V*–*t* (red) traces during the approach and insertion (i), 2–4 min delivery (ii) and retraction (iii) of a multifunctional nanopipette. (c) The schematic of a controlled cytosol *vs.* nucleus injection into the cell *via* a multifunctional nanopipette using a high precision micromanipulator. The top inset in (c) shows the three fine orthogonal control knobs with 60 nm resolution. (d) Optical images of (i) cytosol and (ii) nucleus delivery in HEK cells. (e) The fluorescence images of the FITC

NP motion cannot be simulated. As shown in Fig. 1d, at a low bath salt concentration, the GNP motion is much slower, which also helps to achieve a better SEE signal. Due to the molecular crowding effect inside the cell, we also expect that the nanoentities will come out of the nanopore mouth slower.

Multifunctional nanopipette for intracellular injection

The experimental setup for the voltage-controlled intracellular delivery of single-entities *via* a multifunctional nanopipette is illustrated in Fig. 3a. The nanopore barrel is filled with an intracellular medium containing the nanoentities. The nanopipette tip position is controlled by a micromanipulator and monitored using an optical microscope. The micromanipulator has three fine orthogonal control knobs, each with a spatial resolution of 60 nm. The intracellular delivery of the nanoentities is accomplished by applying an appropriate V_{pore} to the nanopore barrel. Meanwhile, we record the current and potential changes, which helps to guide the delivery process. The delivery efficiency, as well as the distribution of the delivered biomolecules within a cell, are then characterized by fluorescence microscopy.

Fig. 3b shows the typical current (olive) and potential (red) time traces during the (i) approach and insertion, (ii) delivery of the FITC dye and (iii) retraction of a multifunctional nanopipette tip. The corresponding optical images for each step are also shown at the top. In step (i), before the nanopipette tip reaches the cell membrane, the current baseline at $V_{pore} = 0.1$ V is stable and flat. Since the FITC dye has a net negative charge of $-1.02e$,[63] a positive V_{pore} bias is applied to prevent the leakage of the dye into the bath. When the nanopipette tip makes contact and punctures the cell membrane, the characteristic ionic current drop in steps is observed (see the zoomed-in trace). In step (ii), with the tip inside the cell, the V_{pore} is changed to negative to begin the delivery of the FITC dye molecules. Most of the delivery is finished in 2–4 minutes. When the delivery is finished, V_{pore} is switched back to 0.1 V. Finally, in step (iii), the nanopipette tip is retracted from the cell and detached from the cell membrane. Similar to step (i), we can observe a step-wise increase of the baseline current (see the zoomed-in trace). It is worth noting that we can also observe the potential changes during the tip insertion and retraction. The potential changes normally happen earlier than the corresponding current changes, which is attributed to the longer potential detection range. However, the potential baseline is also noisier.

As illustrated in Fig. 3c and d, the inserted nanopipette tip can be controlled by the fine approach axis control knob of the manipulator to stay in the cytosol or further insert into the nucleus of the HEK293 cell in step (ii). Fig. 3e presents the fluorescence images of the FITC filled nanopipette tip (i) and HEK cells after cytosol (ii) and nucleus (iii) deliveries. During the cytosol delivery, the nanopipette tip is inserted ~1 μm deep inside the cell. The fluorescent dye molecules

filled multifunctional nanopipette and cytosol and nucleus delivery of 20 μM of the FITC dye. The merged fluorescence image shows the cytosol and nucleus delivery. (f) The optical images of cells 30 min after the TB exclusion test to evaluate cell viability post-delivery. (i, ii) Cells labelled with 1, 2, 3 and 4, 5, 6 arrows denote cytosol and nucleus intracellular medium delivered cells respectively. The arrows denote the injection points. The dotted encircled cell 6 was intentionally damaged. Scale bars = 10 μm.

can diffuse quickly to be evenly distributed in the entire cytosol and even diffuse to the neighboring cells through intercellular connections. If the nanopipette tip penetrates more than 2 μm, the tip can often get into the nucleus to enable nucleus delivery. The fluorescent image in Fig. 3e(iii) reveals that the majority of the dye molecules are concentrated around the nucleus.

To evaluate the cell membrane damage due to the delivery of nanoentities *via* the nanopipette, a TB exclusion test was performed after the delivery process. The delivery experiments were conducted under the typical conditions for 3–5 min. The cytosol delivery cells (labeled 1 through 3 in Fig. 3f) and the nucleus delivery cells (labeled 4 and 5 in Fig. 3f) did not change color after treating the cells with TB for more than 30 min, confirming the integrity of the cell membrane. In contrast, when the membrane was intentionally damaged by inserting the nanopipette tip much deeper (*i.e.* >3 μm) into the cell (cell 6 in Fig. 3f(ii)), the cell only appeared slightly darker in the grayscale BF image taken by the monochromatic CCD camera (slightly blue by eye).

The delivery efficiency of this method is high. Typically, one nanopipette can conduct up to 5–8 intracellular deliveries within 20–30 min with a delivery efficiency of 80–90%.

The ionic current based single-molecule detection of delivered molecules in live cells

Though the delivery of small fluorescent dye molecules into HEK cells is successful, the transient current and potential changes induced by the exiting of individual dye molecules cannot be observed due to the relatively large average pore diameter (20 nm). Owing to the poor sensitivity of potential detection for the very small dye molecules, we used single barrel quartz nanopipettes with a small pore diameter (<15 nm) to investigate the intracellular single-molecule detection of the delivered molecules by the nanopore current.

We first inserted the nanopipette tip in the cytosol of the HEK cells, as shown in the optical microscopy image in Fig. 4a. The typical *i–t* trace during the cytosol delivery of FITC dye molecules is also shown. No clear current spikes can be resolved. This is common in ~70% of the cytosol deliveries of FITC that we conducted. Therefore, the nanopore current sensitivity is still poor for the delivered FITC molecules in the cytosol. To confirm whether the delivery is successful, we took fluorescence images after the dye delivery. As shown in Fig. 4b, the green fluorescence intensity of FITC is uniform inside the delivered cell, confirming the successful cytosol delivery of the dye molecules.

We then inserted the nanopipette tip into the nucleus, as shown in the optical image in Fig. 4c. A typical *i–t* trace during the nucleus delivery of FITC dye molecules is shown in Fig. 4c. Interestingly, the *i–t* trace features numerous distinguishable current spikes, corresponding to individual delivered dye molecules. The increase in the nanopore sensitivity to enable single dye molecule detection can be attributed to the increased molecular crowding effect inside the nucleus. The fluorescence intensity image post-nucleus delivery is presented in Fig. 4d. The green fluorescence intensity is obviously non-uniform inside the cell and is higher at the nucleus, confirming the nucleus delivery.

We further compared the *i–t* traces of the cytosol and nucleus delivery, as shown in Fig. 4a and c, respectively. When the intracellular delivery began by

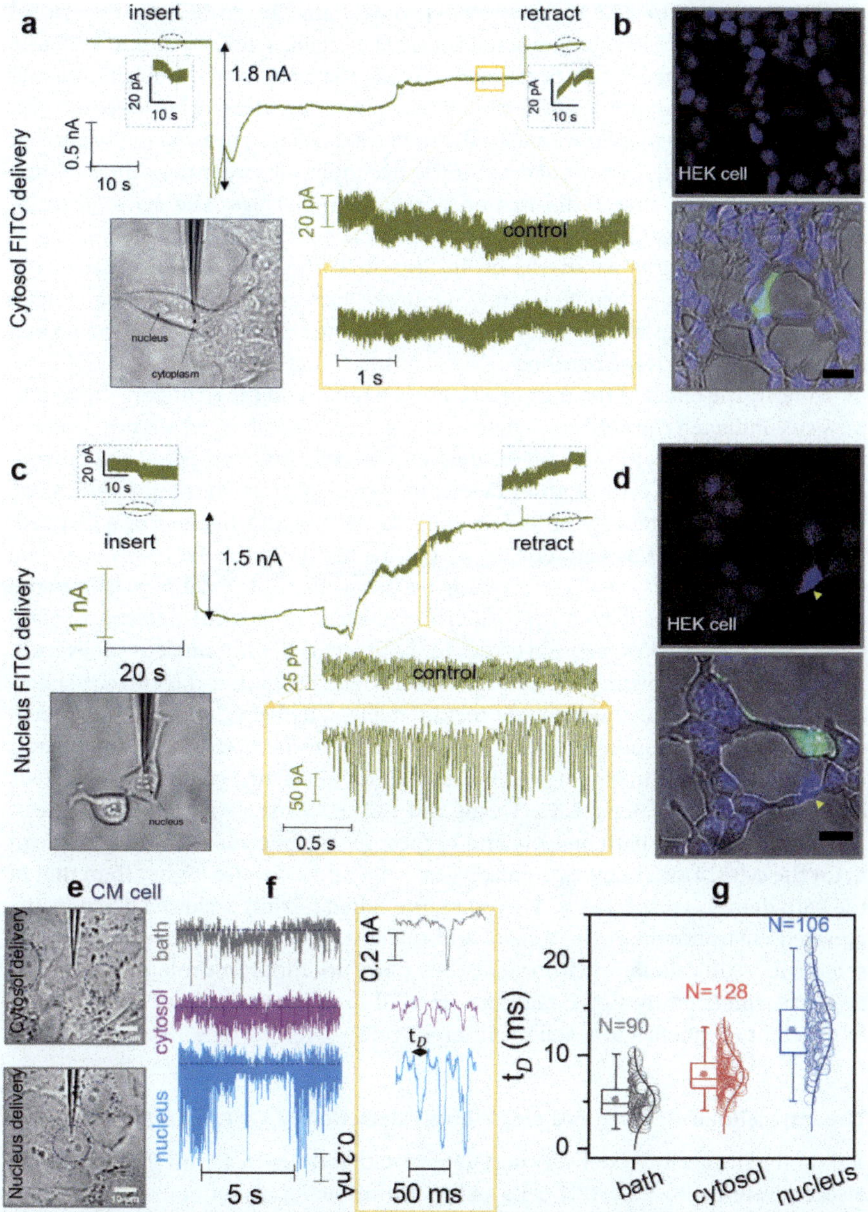

Fig. 4 (a and b) The i–t trace (a) of cytosol delivery of the fluorescent FITC molecules in HEK cells at $V_{pore} = -0.5$ V and the corresponding fluorescence and the merged (fluorescence and bright-field) images (b). Hoechst33258 dye (violet) was used to stain the nucleus and the green fluorescence is from the FITC dye. Scale bars = 10 μm. (c and d) The i–t trace (c) of nucleus delivery of the fluorescent FITC molecules in HEK cells and the corresponding fluorescence and merged images (d). The bright blue fluorescence in (d) is due to a floating dead cell. The current change due to the insertion and retraction of the nanopipette tip is shown in the zoomed in image. The highlighted sections in (a) and (c) are zoomed in below. A single barrel quartz nanopipette with a pore diameter of ~10 nm was used in the experiment. (e) The bright field images of the CM cytosol and nucleus ferritin delivery. The dashed circles denote the nuclei of the CM cells. (f) Current time traces of the switching the V_{pore} to −0.5 V, the overall current change magnitude for the cytosol delivery (1.8 nA) is ~17% larger than that for the nucleus delivery (1.5 nA). This is attributed to the bigger access resistance when the tip is in the nucleus. During the delivery, the magnitude of the current baseline keeps decreasing towards zero. The current drop is more obvious for the nucleus delivery. The drop in the current baseline indicates the accumulation of the dye molecules near the orifice, which is attributed to the slowed dissipation of the dye molecules. The dense environment of the nucleus further retarded the motion of the dye molecules, leading to the clogging effect. As we discussed in Section 1, the slowed movement of the molecule is actually beneficial for nanopore detection. Indeed, the single-molecule current spikes in the nucleus delivery often appeared after the partial reduction of the current baseline.

To verify the effect of the intracellular environment on the nanopore detection, we also conducted the delivery experiments of negatively charged ferritin proteins (about 11 nm in size) to the hiPSC derived CM cells and compared the current signals between the cytosol and nucleus deliveries. Fig. 4e shows the optical BF images during the delivery of ferritin into the cytosol and nucleus of a CM cell. The nucleus of the CM cell, circled by the dashed line, can be resolved in the images. Fig. 4f shows the corresponding *i–t* traces during the delivery. Compared with the *i–t* trace of the control experiment with no ferritin molecules, high-density current spikes are observed in both cytosol and nucleus deliveries. Between the traces during the cytosol and nucleus deliveries, the current spikes are generally more distinguishable for the nucleus delivery. The statistical analysis of the current spike duration time (t_D) for the bath, cytosol and nucleus deliveries is shown in Fig. 4g. The mean t_D for the bath, cytosol and nucleus deliveries are 4.7 ± 1.6 ms, 7.5 ± 1.9 ms and 12.5 ± 3.7 ms, respectively. The mean t_D is smallest for the bath delivery and highest for the nucleus delivery. The mean t_D for the cytosol and nucleus deliveries are ~1.6 and 2.6 times higher than that of the bath delivery, respectively. Therefore, the exited ferritin is slowed down by the crowded environment of the cytosol and nucleus, which enhances the nanopore sensitivity. The cellular environment thus plays an important role in SEE detection. The single-entity signal may also be used to probe the cellular environment of different organelles in a cell or different cells.

The intracellular delivery and single-entity detection of GNPs in HEK cells

To test the intracellular delivery and detection of large NPs, 40 nm PEG–GNPs are also delivered into the HEK cells using multifunctional nanopipette tips. The nanopipettes for GNP delivery have a mean nanopore diameter of 64 ± 7 nm. Fig. 5a shows the BF and DF images of a PEG–GNP filled multifunctional nanopipette tip. The CNE barrel is dark and the nanopore barrel is bright in the BF image. While, in the DF image, the nanopore barrel is much brighter due to the enhanced scattering light from the loaded GNPs.

Fig. 5b shows the *i–t* (olive) and *V–t* (red) traces during the intracellular medium delivery into the HEK cytosol at V_{pore} = −0.3 V. The *i–t* trace is featureless

bath (gray), cytosol (purple) and nucleus (blue) delivery of ferritin in the CM cells and the respective zoomed in traces are shown on the right. (g) The bar chart for the t_D results in bath delivery, and CM cell cytosol and nucleus deliveries of ferritin.

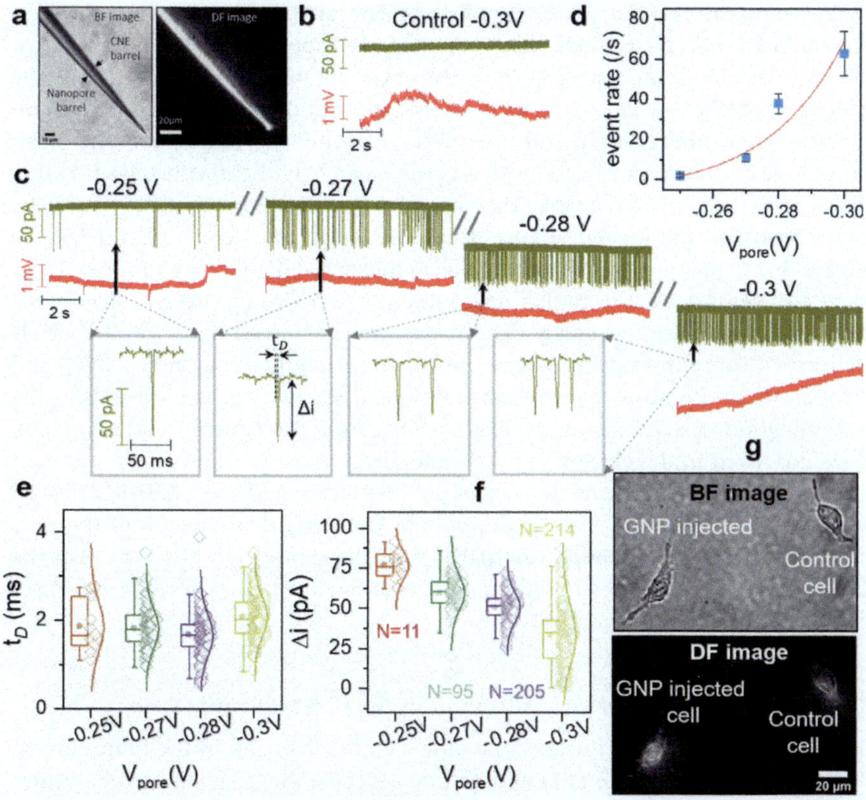

Fig. 5 (a) BF and DF images of the 40 nm PEGylated GNP filled multifunctional nanopipette. (b) *I*–*t* (olive) and *V*–*t* (red) traces during the intracellular medium (b) and GNP (c) delivery into the HEK cell *via* a nanopipette. Zoomed in current spikes at all V_{pore} values are also shown inside the gray rectangles. (d) The V_{pore} effect on the event rate (/s) of intracellular GNP delivery. The error bars represent the standard deviation from three measurements, each of 10 s duration. The solid red line is the exponential fit. The bar chart of current t_D at different V_{pore}. (e) The bar plot of current spike t_D at different V_{pore}. (f) The bar plot of current spike amplitude (Δi) at different V_{pore}. (g) The BF and DF images after the delivery of GNPs into the HEK cell. The DF image of the HEK cell without GNP delivery is also shown as a control.

during intracellular medium delivery. Fig. 5c shows the typical *i*–*t* (olive) and *V*–*t* (red) traces during the delivery of PEG–GNPs into the HEK cytosol at different negative V_{pore} biases. Compared with the small dye molecules and ferritin, the intracellular current detection of larger sized GNPs is relatively easy. Many current spikes appeared in the *i*–*t* trace, confirming the successful intracellular delivery of GNPs. Compared with the current spikes in the 130 mM bath solution (Fig. 1c), the current spikes here are denser with bigger magnitudes, though a slightly higher negative V_{pore} is needed to drive the GNPs out. This is attributed to the more crowded cytosol. Representative individual current spikes at different V_{pore} are also shown in the insets. Different from the current signal, the *V*–*t* trace shows rare corresponding transient changes during the GNP delivery. Therefore, intracellular potential sensing is still challenging in the cytosol of the HEK cells.

Based on the current spikes, the plot of event rate (/s) as a function of V_{pore} is shown in Fig. 5d. The GNP delivery rate increases exponentially with the V_{pore}. At -0.25 V, the event rate for the GNP delivery is about 0.7 s^{-1}. While at -0.3 V, the event rate reaches $\sim$61 s^{-1}. Statistical analysis of t_D and Δi is shown in Fig. 5e and f, respectively. Interestingly, with the increase of the V_{pore} magnitude, the mean amplitude of Δi is obviously reduced and the mean t_D is slightly increased. This is opposite to the general results of the nanopore translocation events. We attribute these results to the increased GNP accumulation at the vicinity of the delivery region due to the quickly increased delivery rate at the higher magnitude of V_{pore}. Due to the crowded intracellular environment, the delivered GNPs in the cytosol cannot dissipate away quickly. They generate extra resistance for the newly delivered GNPs and slow down their movement, leading to the increased t_D and decreased Δi. Because the current baseline does not change obviously, the accumulation of GNPs should be slightly away from the nanopore orifice, which does not result in the clogging of the nanopore.

To confirm the successful delivery of the GNPs into the cytosol of HEK cells, DF imaging was performed. Fig. 5g shows both the BF and DF images of the cells after $\sim$3 min of GNP delivery. Compared with the control cell, the cytosol of the GNP-delivered cell appeared brighter post-injection due to the strong scattering of the GNPs.

The delivery and detection of GNPs in individual hiPSC derived CMs

So far, we have achieved intracellular single-entity detection by the ionic current but not the potential. The FEM simulations revealed that the potential detection is impacted by several factors. We have found that it is sensitive to the changes in the local cell environment, such as ion concentration and crowding. To understand the effect of different cell environments on the potential detection sensitivity, PEG–GNPs are also delivered into the hiPSC-CMs using multifunctional nanopipettes. The nanopipettes used for the intracellular delivery into the HEK and CMs cells have similar nanopore diameters.

Fig. 6a shows the i–t (olive) and V–t (red) traces of PEG–GNP delivery into the cytosol of hiPSC-CMs under different V_{pore}. The current and potential spikes are very clear with high signal-to-noise ratios. Similar to the case of HEK delivery, the delivery rate is very sensitive to the magnitude of the applied V_{pore}. As shown in Fig. 6b, the delivery rate of GNPs at -0.3 V is $\sim$0.8 s^{-1}. While at -0.9 V, the delivery rate reaches $\sim$24 s^{-1}. Compared with the case of HEK delivery, the delivery rate is about 2.5 times lower, while the V_{pore} magnitude is 3 times higher, reflecting the bigger delivery resistance in the CM cytosol.

Statistical analyses of Δi and ΔV are shown in Fig. 6c and d, respectively. Compared with the results for HEK in Fig. 5, the magnitude of the mean Δi is much higher and the ΔV can be detected. In addition, the Δi and ΔV are gradually increased, while t_D (not shown) is gradually decreased upon increasing the V_{pore}. The gradual increase of Δi and ΔV and decrease of t_D suggest the typical electrophoretic driven delivery of PEGylated GNPs. Therefore, unlike the delivery of GNPs into HEK, the effect of GNP accumulation near the tip is not obvious. The different results in different cells can be attributed to the lower delivery rate in CMs, which allows more time for the delivered GNPs to dissipate away from the tip. The GNPs may also dissipate better once inside the CMs under the higher V_{pore}.

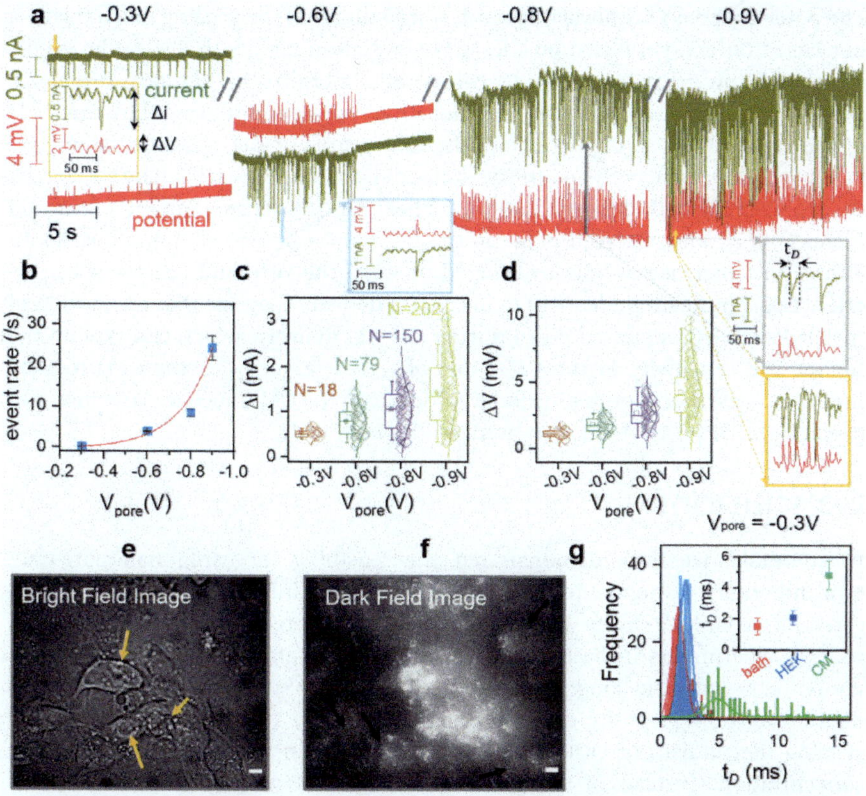

Fig. 6 Controlled intracellular delivery of PEGylated GNP into CM cells using a multifunctional nanopipette. (a) Representative *i*–*t* (olive) and *V*–*t* (red) traces of PEGylated GNP delivery into CMs at various V_{pore}. The zoomed in current spikes corresponding to each V_{pore} are shown below. (b) The delivery rate (/s) as a function of V_{pore}. The solid red line is the exponential fit. 15 s of PEG–GNP delivery data was considered to estimate the event rate. The error bars represent the standard deviation from three measurements, each of 15 s duration. (c and d) The bar plots of Δi *vs.* V_{pore} (c) and ΔV *vs.* V_{pore} (d). (e) BF and (f) DF images of PEGylated GNP delivered CM cells. The yellow arrows in (e) denote the nanopipette insertion locations and the black arrows in (f) show the DF image of CMs with no PEG–GNP delivery. Scale bars = 20 μm. (g) Histograms showing the t_D comparison during bath and PEGylated GNP delivery into HEK (Fig. 5c) and CM cells at $V_{pore} = -0.3$ V. $N = 221$, 214 and 112 for bath, HEK and CM delivery, respectively. The inset shows the corresponding t_D plot. The error bars denote the standard deviation from the mean. The solid lines on the histogram denote the Gaussian fits.

To confirm the successful delivery of PEG–GNPs into the cytosol of CMs, DF imaging was performed. Fig. 6e and f show the BF and DF images, respectively, taken about 20 min after the delivery. Compared to the control cells where no PEG–GNPs were injected (cells indicated by the black arrows in Fig. 6d), the cytosol of the PEG–GNP delivered CM cells (cells indicated by the yellow arrows in Fig. 6c) appeared brighter due to the scattering of the GNPs.

The major difference between the results in HEK and CM cells is the appearance of highly correlated current and potential signals in the time traces. The individual current and potential peaks can be clearly seen in the insets of Fig. 6a.

The potential peaks are always upward, suggesting that the sensing mechanism is not based on charge. Based on the FEM simulation results in Fig. 2, the highly correlated upward potential peaks can be attributed to the big local resistance changes induced during the GNP translocation. We also conducted the t_D analysis of the current spikes. Fig. 6g shows the histograms of the t_D at the same V_{pore} (−0.3 V) for both the bath and intracellular delivery of PEG–GNPs. The mean t_D is the smallest for bath delivery and the highest for CM delivery. About a four-fold increase of t_D was observed between the bath and CM cytosol deliveries. About a three-fold increase of t_D was observed between the HEK and CM cytosol deliveries. The significant increase of t_D in the CM delivery suggests that the motion of the GNPs is further slowed down during the CM delivery, which can explain the higher SEE sensitivity in the CMs. The observed larger t_D in the CMs may be induced by the more complex cellular structure of the CMs, such as sarcomeres, *t*-tubules and SERCA, which are absent in the HEK cells.

Conclusions

In summary, we have demonstrated the feasibility of using nanopore-CNE multifunctional nanopipettes for voltage-controlled, minimally-invasive and targeted intracellular delivery with *in situ* single-entity detection capability. Through control experiments in bath solution and FEM simulations, we show that the nanopipette with the long taper geometry is suitable to conduct intracellular delivery and single-entity detection simultaneously. We also investigated the sensing mechanism of OCP detection by the CNE in a near physiological salt concentration. Instead of charge sensing, the CNE mainly detects the local potential change induced by the drastic resistance change during the translocation of nanoentities. We have shown that various nanoentities, including dye molecules, ferritin protein and 40 nm GNPs, can be delivered into the HEK and CM cells in a controlled fashion with high efficiency. Meanwhile, the SEE detection of these nanoentities during delivery is achieved.

Interestingly, the current and potential SEE signals are strongly affected by the local environment near the nanopipette tip. Compared with the results in the bath solution, the current single-entity signals of the nanopore show improved signal-to-noise ratios. For example, we can detect individual small dye molecules delivered in the cytosol of HEK cells by the nanopore but not in the bath solution with a similar salt concentration. By using a micromanipulator, we selectively deliver the nanoentities to the cytosol or nuclei of the cells. The SEE sensitivity is always higher during the nucleus delivery. For the cytosol delivery results between HEK and CM cells, we always observe a higher SEE sensitivity in the CMs. The single-entity potential sensing by CNE can be achieved in the CMs but not in the HEK cells. These results are attributed to the molecular crowding effect. As suggested by the duration time of the current spikes, the movement of the nanoentities *via* a nanopore is slower with the increase of the crowding effect. The slower movement of the exiting nanoentities can effectively enhance the SEE detection.

With the assistance of the SEE signals, we can improve the accuracy of the delivery and monitor the delivery process and the cellular response in greater detail, which cannot be achieved by the fluorescence imaging. For example, when the entities are delivered very fast, we can detect the accumulation of delivered entities near the nanopipette tip. With the instant feedback by the SEE signals, we can control the delivery rate, thus reducing the aggregation induced cellular damage and nanopipette clogging. We believe that this developed platform is ideal for single-cell studies with both cellular manipulation and detection functions.

Conflicts of interest

There are no conflicts to declare.

Acknowledgements

This work was partially supported by the National Science Foundation (NSF) grants CBET-1454544 and the NSF Engineering Research Centers Program under the NSF Cooperative Agreement EEC-1647837. The authors want to thank Prof. Fenfei Leng for the help with the nucleus staining method and Govinda Ghimire for preparing the PEGylated GNPs. P. P. also acknowledges the support of the FIU CASE distinguished postdoc fellowship.

References

1 W. Poon, B. R. Kingston, B. Ouyang, W. Ngo and W. C. W. Chan, *Nat. Nanotechnol.*, 2020, **15**, 819–829.

2 D. Kim, C.-H. Kim, J.-I. Moon, Y.-G. Chung, M.-Y. Chang, B.-S. Han, S. Ko, E. Yang, K. Y. Cha, R. Lanza and K.-S. Kim, *Cell Stem Cell*, 2009, **4**, 472–476.

3 M. P. Stewart, A. Sharei, X. Ding, G. Sahay, R. Langer and K. F. Jensen, *Nature*, 2016, **538**, 183–192.

4 M. P. Stewart, R. Langer and K. F. Jensen, *Chem. Rev.*, 2018, **118**, 7409–7531.

5 N. P. Restifo, M. E. Dudley and S. A. Rosenberg, *Nat. Rev. Immunol.*, 2012, **12**, 269–281.

6 R. Duangrat, A. Udomprasert and T. Kangsamaksin, *Cancer Sci.*, 2020, **111**, 3164–3173.

7 A. D. Fesnak, C. H. June and B. L. Levine, *Nat. Rev. Cancer*, 2016, **16**, 566–581.

8 L. Warren, P. D. Manos, T. Ahfeldt, Y.-H. Loh, H. Li, F. Lau, W. Ebina, P. K. Mandal, Z. D. Smith, A. Meissner, G. Q. Daley, A. S. Brack, J. J. Collins, C. Cowan, T. M. Schlaeger and D. J. Rossi, *Cell Stem Cell*, 2010, **7**, 618–630.

9 L. Rohani, C. Fabian, H. Holland, Y. Naaldijk, R. Dressel, H. Löffler-Wirth, H. Binder, A. Arnold and A. Stolzing, *Stem Cell Res.*, 2016, **16**, 662–672.

10 C. J. Bishop, J. Kim and J. J. Green, *Ann. Biomed. Eng.*, 2014, **42**, 1557–1572.

11 S. Patel, A. Athirasala, P. P. Menezes, N. Ashwanikumar, T. Zou, G. Sahay and L. E. Bertassoni, *Tissue Eng., Part A*, 2019, **25**, 91–112.

12 U. Rothbauer, K. Zolghadr, S. Tillib, D. Nowak, L. Schermelleh, A. Gahl, N. Backmann, K. Conrath, S. Muyldermans, M. C. Cardoso and H. Leonhardt, *Nat. Methods*, 2006, **3**, 887–889.

13 A. K. Fajrial, Q. Q. He, N. I. Wirusanti, J. E. Slansky and X. Ding, *Theranostics*, 2020, **10**, 5532–5549.

14 T. K. Kim and J. H. Eberwine, *Anal. Bioanal. Chem.*, 2010, **397**, 3173–3178.

15 E. Neumann, M. Schaefer-Ridder, Y. Wang and P. H. Hofschneider, *EMBO J.*, 1982, **1**, 841–845.

16 P. Actis, *Small Methods*, 2018, **2**, 1700300.
17 F. Tang, K. Lao and M. A. Surani, *Nat. Methods*, 2011, **8**, S6–S11.
18 S. S. Rubakhin, E. V. Romanova, P. Nemes and J. V. Sweedler, *Nat. Methods*, 2011, **8**, S20–S29.
19 J. R. Heath, A. Ribas and P. S. Mischel, *Nat. Rev. Drug Discovery*, 2016, **15**, 204–216.
20 Y. Zhang, C. B. Ballas and M. P. Rao, *Annu. Int. Conf. IEEE Eng. Med. Biol. Soc.*, 2012, **2012**, 594–597.
21 M. Simonis, W. Hübner, A. Wilking, T. Huser and S. Hennig, *Sci. Rep.*, 2017, **7**, 41277.
22 Z. Nan, Q. Xu, Y. Zhang and W. Ge, *IEEE Access*, 2019, **7**, 55543–55553.
23 S. S. P. Nathamgari, N. Pathak, V. Lemaitre, P. Mukherjee, J. J. Muldoon, C.-Y. Peng, T. McGuire, J. N. Leonard, J. A. Kessler and H. D. Espinosa, *Small*, 2020, **16**, 2002616.
24 G. He, J. Feng, A. Zhang, L. Zhou, R. Wen, J. Wu, C. Yang, J. Yang, C. Li, D. Chen, J. Wang, N. Hu and X. Xie, *Nano Lett.*, 2019, **19**, 7201–7209.
25 Y. Cao, E. Ma, S. Cestellos-Blanco, B. Zhang, R. Qiu, Y. Su, J. A. Doudna and P. Yang, *Proc. Natl. Acad. Sci. U. S. A.*, 2019, **116**, 7899.
26 J. Shi, Y. Ma, J. Zhu, Y. Chen, Y. Sun, Y. Yao, Z. Yang and J. Xie, *Molecules*, 2018, **23**, 3044.
27 P. Mukherjee, S. S. P. Nathamgari, J. A. Kessler and H. D. Espinosa, *ACS Nano*, 2018, **12**, 12118–12128.
28 D. Zhao, D. Huang, Y. Li, M. Wu, W. Zhong, Q. Cheng, X. Wang, Y. Wu, X. Zhou, Z. Wei, Z. Li and Z. Liang, *Sci. Rep.*, 2016, **6**, 18469.
29 Y. Li, M. Wu, D. Zhao, Z. Wei, W. Zhong, X. Wang, Z. Liang and Z. Li, *Sci. Rep.*, 2016, **5**, 17817.
30 X. Xie, A. M. Xu, S. Leal-Ortiz, Y. Cao, C. C. Garner and N. A. Melosh, *ACS Nano*, 2013, **7**, 4351–4358.
31 W. Kang, F. Yavari, M. Minary-Jolandan, J. P. Giraldo-Vela, A. Safi, R. L. McNaughton, V. Parpoil and H. D. Espinosa, *Nano Lett.*, 2013, **13**, 2448–2457.
32 J. E. Bestman, R. C. Ewald, S.-L. Chiu and H. T. Cline, *Nat. Protoc.*, 2006, **1**, 1267–1272.
33 M. L. Yarmush, A. Golberg, G. Serša, T. Kotnik and D. Miklavčič, *Annu. Rev. Biomed. Eng.*, 2014, **16**, 295–320.
34 M. S. Venslauskas and S. Šatkauskas, *Eur. Biophys. J.*, 2015, **44**, 277–289.
35 R.-C. Qian, J. Lv and Y.-T. Long, *Anal. Chem.*, 2018, **90**, 13744–13750.
36 Y. T. Chow, S. Chen, R. Wang, C. Liu, C.-w. Kong, R. A. Li, S. H. Cheng and D. Sun, *Sci. Rep.*, 2016, **6**, 24127.
37 M. Bakouei, S. Abdorahimzadeh and M. Taghipoor, *Phys. Chem. Chem. Phys.*, 2020, **22**, 25306–25314.
38 Z. Orynbayeva, R. Singhal, E. A. Vitol, M. G. Schrlau, E. Papazoglou, G. Friedman and Y. Gogotsi, *Nanomedicine*, 2012, **8**, 590–598.
39 R. Singhal, Z. Orynbayeva, R. V. Kalyana Sundaram, J. J. Niu, S. Bhattacharyya, E. A. Vitol, M. G. Schrlau, E. S. Papazoglou, G. Friedman and Y. Gogotsi, *Nat. Nanotechnol.*, 2011, **6**, 57–64.
40 O. Guillaume-Gentil, E. Potthoff, D. Ossola, P. Dörig, T. Zambelli and J. A. Vorholt, *Small*, 2013, **9**, 1904–1907.

41 A. Meister, M. Gabi, P. Behr, P. Studer, J. Vörös, P. Niedermann, J. Bitterli, J. Polesel-Maris, M. Liley, H. Heinzelmann and T. Zambelli, *Nano Lett.*, 2009, **9**, 2501–2507.
42 P. Stiefel, F. I. Schmidt, P. Dörig, P. Behr, T. Zambelli, J. A. Vorholt and J. Mercer, *Nano Lett.*, 2012, **12**, 4219–4227.
43 R. Adam Seger, P. Actis, C. Penfold, M. Maalouf, B. Vilozny and N. Pourmand, *Nanoscale*, 2012, **4**, 5843–5846.
44 J. Lv, R.-C. Qian, Y.-X. Hu, S.-C. Liu, Y. Cao, Y.-J. Zheng and Y.-T. Long, *Chem. Commun.*, 2016, **52**, 13909–13911.
45 R.-J. Yu, Y.-L. Ying, R. Gao and Y.-T. Long, *Angew. Chem., Int. Ed.*, 2019, **58**, 3706–3714.
46 B. Babakinejad, P. Jönsson, A. López Córdoba, P. Actis, P. Novak, Y. Takahashi, A. Shevchuk, U. Anand, P. Anand, A. Drews, A. Ferrer-Montiel, D. Klenerman and Y. E. Korchev, *Anal. Chem.*, 2013, **85**, 9333–9342.
47 F. O. Laforge, J. Carpino, S. A. Rotenberg and M. V. Mirkin, *Proc. Natl. Acad. Sci. U. S. A.*, 2007, **104**, 11895.
48 E. Shekaramiz, G. Varadarajalu, P. J. Day and H. K. Wickramasinghe, *Sci. Rep.*, 2016, **6**, 29051.
49 A. P. Ivanov, P. Actis, P. Jönsson, D. Klenerman, Y. Korchev and J. B. Edel, *ACS Nano*, 2015, **9**, 3587–3595.
50 N. Varongchayakul, J. Song, A. Meller and M. W. Grinstaff, *Chem. Soc. Rev.*, 2018, **47**, 8512–8524.
51 V. Kurz, T. Tanaka and G. Timp, *Nano Lett.*, 2014, **14**, 604–611.
52 J.-A. Huang, V. Caprettini, Y. Zhao, G. Melle, N. Maccaferri, L. Deleye, X. Zambrana-Puyalto, M. Ardini, F. Tantussi, M. Dipalo and F. De Angelis, *Nano Lett.*, 2019, **19**, 722–731.
53 N. Panday, G. Qian, X. Wang, S. Chang, P. Pandey and J. He, *ACS Nano*, 2016, **10**, 11237–11248.
54 C. C. Chau, S. E. Radford, E. W. Hewitt and P. Actis, *Nano Lett.*, 2020, **20**, 5553–5561.
55 C. C. Chau, E. W. Hewitt and P. Actis, *Curr. Opin. Electrochem.*, 2021, **25**, 100654.
56 Y. Liu, M. K. Shipton, J. Ryan, E. D. Kaufman, S. Franzen and D. L. Feldheim, *Anal. Chem.*, 2007, **79**, 2221–2229.
57 A. S. Rubfiaro, P. S. Tsegay, Y. Lai, E. Cabello, M. Shaver, J. Hutcheson, Y. Liu and J. He, *ACS Appl. Bio Mater.*, 2021, **4**, 1632–1639.
58 X. Lian, J. Zhang, S. M. Azarin, K. Zhu, L. B. Hazeltine, X. Bao, C. Hsiao, T. J. Kamp and S. P. Palecek, *Nat. Protoc.*, 2013, **8**, 162–175.
59 S. Tohyama, F. Hattori, M. Sano, T. Hishiki, Y. Nagahata, T. Matsuura, H. Hashimoto, T. Suzuki, H. Yamashita, Y. Satoh, T. Egashira, T. Seki, N. Muraoka, H. Yamakawa, Y. Ohgino, T. Tanaka, M. Yoichi, S. Yuasa, M. Murata, M. Suematsu and K. Fukuda, *Cell Stem Cell*, 2013, **12**, 127–137.
60 W. Strober, *Curr. Protoc. Immunol.*, 1997, **21**, A.3B.1–A.3B.2.
61 J. Guo, A. Sesena Rubfiaro, Y. Lai, J. Moscoso, F. Chen, Y. Liu, X. Wang and J. He, *Analyst*, 2020, **145**, 4852–4859.
62 P. Pandey, N. Panday, S. Chang, P. Pang, J. Garcia, X. Wang, Q. Fu and J. He, *ChemElectroChem*, 2018, **5**, 3102–3112.
63 L. Yin, W. Wang, S. Wang, F. Zhang, S. Zhang and N. Tao, *Biosens. Bioelectron.*, 2015, **66**, 412–416.

Faraday Discussions

PAPER

Nanostructured carbon-fiber surfaces for improved neurochemical detection†

Ayah J. Syeed,‡ Yuxin Li, ‡ Blaise J. Ostertag, Jared W. Brown and Ashley E. Ross *

Received 10th August 2021, Accepted 13th September 2021
DOI: 10.1039/d1fd00049g

Fundamental insight into the extent to which the nanostructured surface and geometry impacts neurochemical interactions at electrode surfaces could provide significant advances in our ability to design and fabricate ultrasensitive neurochemical detection probes. Here, we investigate the extent to which the nanostructure of the carbon-fiber surface impacts detection of catecholamines and purines with fast-scan cyclic voltammetry (FSCV). Carbon-fibers were treated with argon (Ar) plasma to induce variations in the nano- and micro-structure without changing the functionalization of the surface. We tested variations in topology by measuring the extent to which the flow rate, RF power, and treatment time affect the surface roughness. Flow rates from 50–100 sccm, plasma power from 20–100 W, and treatment times from 30 s to 5 min were compared. Two Ar-treatments were chosen from the optimization studies for comparison, and the surface roughness was evaluated using atomic force microscopy (AFM). To ensure no changes in chemical composition, fibers were analyzed with X-ray photoelectron spectroscopy (XPS). On average, at the optimized Ar-plasma treatment procedure, oxidative current for adenosine and ATP increased by 3.5 ± 1.4-fold and 3.2 ± 0.6-fold, and guanosine and GTP by 1.7 ± 0.3-fold and 1.8 ± 0.3-fold, respectively ($n = 9$). Dopamine increased by 1.7 ± 0.3-fold. The extent to which changes in the electrode structure impact adsorption, sensitivity, and electron transfer rates were measured. A COMSOL Multiphysics simulation was developed to enable the modeling of mass transport of electroactive species at varying electrode geometries. Overall, this study provides critical insight into the extent to which the nanostructure of the surface impacts the electrochemical detection of neurochemicals.

Introduction

Insight into the extent to which nanostructured surface topology and geometry impact neurochemical interactions at electrode surfaces is important for tailoring

University of Cincinnati, Department of Chemistry, 312 College Dr 404 Crosley Tower, Cincinnati, OH 45221-0172, USA. E-mail: Ashley.ross@uc.edu; Tel: +1-513-556-9314

† Electronic supplementary information (ESI) available. See DOI: 10.1039/d1fd00049g

‡ Co-first author.

electrode surfaces for optimal detection. Fast-scan cyclic voltammetry (FSCV)[1–3] at carbon-fiber microelectrodes (CFMEs) is a well-established electroanalytical method used to detect neurotransmitters. Several reports have investigated the extent to which surface geometries impact dopamine detection with FSCV,[4–7] and a recent review highlighted the necessity to consider the 3D structure of the electrode when investigating surface interactions.[4] Cavity carbon-nanopipette electrodes have been developed to increase dopamine current due to local redox cycling in the 3D cavity.[5] This paper demonstrated that detection sensitivity to dopamine could change based on the 3D structure of the electrode; however, to date, it is unclear how 3D structure impacts neurochemical detection beyond catecholamines. Likewise, it remains difficult to delineate changes in the electrochemistry to surface geometry *vs.* surface functionalization or the carbon allotrope. Purines are important extracellular signaling molecules in the body whose functions range from modulating neurotransmission, providing neuroprotection,[3,8,9] to regulating inflammatory signaling.[10] Extensive studies using FSCV to monitor purine signaling exist;[1,11–14] however, only a few reports have explored the impact of the electrode surface for purine detection.[2,15,16] To further understand purine interactions at electrode surfaces, it is critical to assess how changes in the surface topology impact interaction. Additionally, detection of the purine nucleotides, GTP and ATP, remain challenging at bare carbon-fiber electrodes due to limited interaction at the electrode surface.[2,15,17] Here, we investigate the extent to which small changes in surface topology, *via* argon plasma treatment of the carbon-fiber surface, impacts detection of catecholamines and purines with FSCV. This work provides a useful approach to study how geometrical changes alone, while keeping the carbon allotrope and functionalization unchanged, impact neurochemical detection.

Varying the geometries and nanostructure of the carbon surface can significantly increase the surface area resulting in higher sensitivity, induce thin-layer cell-like behavior, and manipulate the ratio of edge plane *vs.* basal plane sites.[7,18–20] The Venton lab has demonstrated that carbon nanomaterial-based electrodes exhibit nanocavities and crevices at the surface, which can facilitate higher temporal resolution measurements of dopamine with FSCV.[6,7,21] This work has provided a foundation for understanding that electrode surface geometry is important for detection with FSCV. To advance our fundamental understanding of electrode structure–analyte interactions, it is important to develop methods in which you can compare analyte interactions prior to and after changes in the surface roughness. Several methods exist to roughen and manipulate the surface topology of carbon, including strong acid treatments, plasma treatments, and even electrochemical scanning at high potentials.[22–26] However, to correlate changes in electrochemical behavior to surface geometry, it is imperative to choose methods that do not manipulate surface functionalization.

We have shown that N_2 and O_2 plasma-treated CFMEs improve purine detection due to changes in functionalization and topology.[16] This foundational work provided information on the extent to which purine detection is dependent on surface functionalization and topology. Despite this, these plasma treatments induce both surface functionalization and topology changes.[24,27] Ar-plasma initiates bombardment of inert Ar^+ ions on the surface to physically etch carbon while causing no chemical reactions on the surface.[28,29] Controlling the surface roughness with Ar-plasma can ensure that changes in the electrochemistry are a result of geometrical changes at the surface.

Here, we show Ar plasma-treated CFMEs significantly improve detection of the purines adenosine, ATP, guanosine, GTP, and the catecholamine dopamine with FSCV providing evidence that changes in surface roughness alone greatly impacts detection. Ar-plasma treatment was optimized by controlling the flow rate, power, and time of treatment. Ar plasma-treated CFMEs were characterized by scanning electron microscopy (SEM), atomic force microscopy (AFM), Raman spectroscopy, and X-ray photoelectron spectroscopy (XPS) to quantify the degree of roughness, defects, and to validate limited changes in surface functionalization. A COMSOL multiphysics model was developed to enable simulation of how changes in surface geometry impact dopamine interaction at carbon-based surfaces during FSCV. Overall, we provide a new and simple method to fundamentally study neurochemical interactions at variously roughened carbon surfaces.

Methods

Chemicals/reagents

Dopamine (DA), adenosine (AD), adenosine triphosphate (ATP), guanosine (GN), and guanosine-5′-triphosphate (GTP) were purchased from Sigma-Aldrich (St. Louis, MO, USA). All neurochemical stock solutions in this study were 10 mM, dissolved in 0.1 M HCl, and stored at 4 °C. Tris buffer at a pH of 7.4 was used to dilute stock solutions to produce daily working solutions (1 μM DA, 5 μM AD, ATP, GN, and GTP) for testing electrodes. Tris buffer includes 15 mM Tris, 1.25 mM NaH_2PO_4, 2.0 mM Na_2SO_4, 3.25 mM KCl, 140 mM NaCl, 1.2 mM $CaCl_2$ dihydrate, and 1.2 mM $MgCl_2$. All solutions were made from deionized water (Milli-Q, Millipore, Billerica MA).

Disk carbon-fiber microelectrode fabrication and plasma treatment

Disk carbon-fiber microelectrodes were fabricated as previously described.[16] Briefly, carbon-fiber microelectrodes were fabricated from 7 μm T-650 carbon fibers (a gift from Mitsubishi Chemical Carbon Fiber and Composites Inc., Sacramento CA) by vacuum aspirating a single carbon-fiber into a capillary glass tube (1.2 × 0.68 mm, A&M Systems, Sequim WA). The glass capillary was pulled into two using a vertical Narishige PE-22 electrode puller (Tokyo, Japan). The carbon-fiber was trimmed under a microscope near the glass seal. Electrodes were sealed using epoxy. Epoxy Resin 828 was mixed with 14% (w/w) 1,3-phenylenediamine (Sigma Aldrich) in a 80–85 °C water bath, and electrodes were dipped momentarily in the heated epoxy/hardener mixture followed by a brief rinse in acetone.[16] Electrodes were cured at 150 °C overnight. All electrodes were polished by a fine diamond abrasive plate set to 45° to create a polished disk electrode. To plasma-treat the electrodes, a CS-1701 RIE plasma etcher (Nordson MARCH, Westlake, Ohio, USA) with a 13.56 MHz RF generator was used. Argon plasma treatments were kept at a constant pressure of 75 torr and the plasma gas flow rate, power, and treatment time were varied and compared to optimize the carbon-fiber modification.

Fast-scan cyclic voltammetry (FSCV)

All FSCV data was collected using a WaveNeuro potentiostat with a 5 megaohm head stage (Pine Instruments, Durham NC) with HDCV software (UNC-Chapel

Hill, Mark Wightman) and a National Instruments PC1e-6363 computer interface board (Austin, TX). In this study, an electrochemical waveform scanning from −0.4 to 1.45 V (*vs.* Ag/AgCl) and back with a 400 V s^{-1} scan rate and a repetition rate of 10 Hz was used due to the fact that purines oxidize at higher potentials.[1,15] All electrodes were equilibrated for 10 min prior to testing. Electrodes were calibrated using a flow injection analysis set-up using a flow rate of 1 mL min^{-1} with a Fusion 200 Two-Channel Chemyx Syringe pump (Stafford, TX).

Surface characterization

To compare different properties of the bare carbon-fiber and Ar plasma-treated carbon-fiber, a combination of surface analysis techniques were used. Scanning electron micrographs (SEM) were used to qualitatively assess the surface. SEM images were collected with an FEI XL30 SEM at an accelerating voltage of 5.00 kV, 610 mm away. Surface roughness was measured using atomic force microscopy (AFM, Veeco Dimension 3100). AFM imaging was collected in the *X*-axis direction. The arithmetic average of the absolute values of surface height deviations measured from the mean plane $\left(R_a = \frac{1}{L}\int_0^L |Z(x)|dx\right)$ was used to determine surface roughness where $Z(x)$ is a function that describes the surface profile in terms of height (Z) and position (x) over the evaluation length (L). X-ray photoelectron spectroscopy (XPS, Axis HSi 165 Ultra by Kratos) was used to measure and compare the elemental composition of the surface. The X-ray source was operated at 12 kV and 10 mA emission.

COMSOL multiphysics modeling

COMSOL Multiphysics (Burlington, MA) was used to model the redox couple dopamine (DA) and dopamine-*o*-quinone (DOQ) at the electrode surface (1) of varying electrode geometries, similar to previous work.[29] Purines were not modelled with COMSOL due to the complexity of their redox reactions.

$$\text{DOQ} + 2e^- \rightleftharpoons \text{DA} \tag{1}$$

Using the electroanalysis module, a 2D-axisymmetric model was simulated for simplicity of the electrode/solution environment, in which the boundary of the model was a 90° semi-circle with a radius of 105 μm. The electrode surface was drawn as a horizontal line with a radius of 3.5 μm (to mimic the 7 μm in diameter carbon-fiber). This model allows for a two-dimensional environment integrated over 360° to encompass the entire model in which the electrochemical reaction occurs only at the surface of the electrode, and the distribution of current over the electrode's surface is fully uniform. Changes in the electrode geometry after Ar-plasma treatment were modelled to match the average topographical changes induced by the plasma (50 nm features – see AFM data).

The electroanalysis module has various nodes to define what features and constraints to apply to the electrochemical study. The electrolyte node signifies mass transfer of the analyte within the electrolyte by diffusion, dictated by Fick's 2nd law (2):

$$\frac{\partial c_i}{\partial t} + \nabla J_i + u\nabla c_i = R_i, \quad J_i = -D_i\nabla c_i \tag{2}$$

where D_i is the diffusion coefficient of the analyte in $m^2\ s^{-1}$, R_i is the expression for the reaction rate in mol $m^{-3}\ s^{-1}$, and c_i is the analyte concentration in mol m^{-3}. In this case, convection (u) is not activated. Therefore the equation becomes (3):

$$\frac{\partial c_i}{\partial t} + \nabla J_i = R_i,\ \ J_i = -D_i \nabla c_i \tag{3}$$

The second node is axial symmetry, in which the y-axis of the electrochemical environment in the two-dimensional model is selected to signify the simulation's rotation about this axis. A default boundary condition in the electroanalysis module is set for the exterior boundary of the electrochemical environment, and is denoted by the third node, no flux, (4), as the cell consists of no mass flux across the exterior walls or through the electrode surface and is applied for all species:

$$-nJ_i = 0 \tag{4}$$

and all boundaries that do not face a conductor are dictated by the insulation node (5):

$$-ni_s = 0 \tag{5}$$

The initial value node allows you to define the initial concentrations of the redox couple. Here, the initial values for "*cRed*" correspond to [DA] and were set to 20 μM, and "*cOx*" corresponds to [DOQ] and was set to an initial value of 0 μM.

The electrode surface node is the key node of the electroanalysis module which enables the user to set the waveform to apply the cyclic voltammetry study, and model the interface of the electrochemical cell between the electrode surface and electrolyte domain (6, 7 and 8).

$$ni_l = i_{total} \tag{6}$$

$$i_{total} = \sum_m i_{loc,m} + i_{dl} \tag{7}$$

$$-nJ_i = R_{i,tot},\ \ R_{i,tot} = \sum_m R_{i,m} + R_{dl,i} \tag{8}$$

Within the electrode surface node is the subnode, electrode reaction. For charge transfer reactions occurring at an electrode–electrolyte interface, the electrode reaction defines the electrode kinetics by the Butler–Volmer equation for local current density (9):

$$i_{loc,expr} = i_{0,ref}(T)\left[\prod_{i:v_i>0}\left(\frac{c_i}{c_{i,ref}}\right)^{v_i}\exp\left(\frac{\alpha_a F\eta_{ref}}{RT}\right) - \prod_{i:v_i<0}\left(\frac{c_i}{c_{i,ref}}\right)^{-v_i}\exp\left(\frac{\alpha_c F\eta_{ref}}{RT}\right)\right] \tag{9}$$

Lastly, the double layer capacitance subnode is used to indicate the electrical double layer capacitance (C_{dl}) based on the electrode material; in this case, the electrode material is carbon fiber. The value we used for C_{dl} is 0.2 Fm^{-2}, based on prior work.[20]

A similar approach was used to model dopamine's interaction at an Ar-plasma modified surface. To model and account for the average topographical changes induced by Ar-plasma, 50 nm-tall vertical line segments were added to the simulated flat carbon-fiber electrode surface. This enabled a simplified model of the average surface height deviations induced by the Ar-plasma (R_a ~50 nm). The electrode surface node only accounts for exterior boundary–electrolyte interfaces, so the perforated electrode surface node was used to account for the additional electrode surface created by plasma enhancements. The perforated electrode surface node operates similar to the electrode surface node in that the electrode reaction subnode is still dictated by electrode kinetics following the Butler–Volmer equation for local current density, and the double layer capacitance subnode still defines the double layer capacitance.

Fast-scan cyclic voltammograms were simulated using the submodule cyclic voltammetry. A traditional FSCV DA waveform was used for these simulations, scanning from a starting potential of −0.4 V to +1.3 V at a scan rate of 400 V s^{-1}. A formal equilibrium potential of 0.2 V was used for the redox reactions due to DA's standard oxidation potential. The diffusion coefficients for DA and DOQ were defined as 2.6×10^{-10} m^2 s^{-1}, and the number of electrons transferred was set to 2.[30] A reference exchange current density of 1500 A m^{-2} was used to best mimic the conditions for simulating a carbon-fiber microelectrode.[29] Parameters of this study were maintained for both bare carbon-fiber and surface roughened carbon-fiber to analyze the impact surface roughness changes have on DA's interaction at the surface. To plot the cyclic voltammograms of faradaic oxidation current, manual background subtraction of the capacitive current was done to remove non-faradaic current, similar to experimental data analysis. To generate the capacitive current, the model was run for an initial [DA] of 0 μM.

Statistics

All statistics were conducted with GraphPad Prism V. 9.0 (GraphPad Software Inc., La Jolla, CA, USA). Statistical p-values were significant at 95% confidence level ($p < 0.05$). Values are reported as the mean ± standard error of the mean (SEM), and n represents the number of electrodes. AFM data were analyzed using NanoScope Analysis 1.5 software (Bruker Corporation, Santa Barbara, CA, USA) to determine the surface roughness. The XPS data were analyzed by peak fitting in OriginLab 2020 (OriginLab, Northampton, Massachusetts, USA).

Results and discussion

Argon plasma-treated carbon-fiber microelectrodes procedure

Plasma-treated carbon induces changes to the structural topology and functionalization of the surface. Previously, our lab showed that O_2 and N_2 plasma-treated carbon-fiber microelectrodes significantly improved purine and catecholamine detection by changing the surface functionalization and roughness.[16] Here, argon (Ar) plasma was used to measure the impact of changes in the surface nanostructure, with no change in surface chemistry, on purine and catecholamine detection exclusively. Ar plasma power, flow rate, and treatment time can introduce different levels of roughness.[29] We compare the extent to which the power, flow rate, and treatment time impact the electrochemical detection of dopamine, adenosine, ATP, guanosine, and GTP. Dopamine serves as a control analyte because there exists strong evidence that increased surface roughness improves dopamine detection with FSCV.[2,23,30] An optimal procedure was chosen for carbon-fiber modification, and a detailed surface characterization was done to understand correlations in surface changes to changes in electrochemical detection.

The parameters of Ar-plasma treatment were optimized to determine the ideal surface topology changes for electrochemical detection of purines and dopamine. All electrodes were pretested at the "adenosine waveform" (see Methods) with 1 μM dopamine (DA) and 5 μM each adenosine (AD), ATP, guanosine (GN), and GTP. Next, electrodes were treated with Ar-plasma under varying conditions followed by immediate post-testing with the same concentrations of analyte (Fig. S1–S3,† $n = 9$). The ratio of post-tested current (i_{post}) to pretested current (i_{pre}) was plotted as a function of the analyte. Ratios above 1.0 indicate an increase in oxidative current, and ratios below 1.0 indicate a decrease in oxidative current. Briefly, treatment time was varied from 30 s to 5 min while keeping the gas flow rate and power constant (100 Watts, 100 sccm, Fig. S1†). Short, 30 s, treatment times were not sufficient to etch the carbon surface (Fig. S1A†) and induced no change in oxidative current for purines and dopamine. Longer treatment times (>3 min) visibly damaged the electrode surface and decreased the measured oxidative current; therefore, we chose a 1.0 min treatment time as optimal (Fig. S1B†). Treatment power was varied from 20 Watts (W) to 100 W while keeping the treatment time and flow rate constant (Fig. S2†). Although 100 W treatment adequately caused increases in oxidative current for all analytes, 70 W produced larger increases in oxidative current for the purines AD and ATP. Additionally, data collected at 100 W was more variable; therefore, we chose 70 W as optimal. Lastly, we varied the gas flow rate from 50 sccm to 100 sccm (Fig. S3†) while keeping the treatment time and power constant. A flow rate of 100 sccm was chosen as optimal (Fig. S3C†) due to larger increases in oxidative current for all analytes. Combining all of this together, the optimized procedure for Ar-plasma treating carbon-fibers is a 1 min treatment time, at 70 W, and a gas flow rate of 100 sccm. The extent to which this procedure impacts the surface is further explored in the sections below.

The optimized Ar-plasma treatment impacts purine detection more than dopamine. The cyclic voltammograms (CVs) of each analyte are compared before and after the optimized treatment (Fig. 1), and the ratio of oxidative peak current before treatment to after treatment was calculated and plotted as a function of analyte (Fig. 2, $n = 9$). On average, the background current for Ar-plasma treated electrodes (Fig. 1A) increased by 1.5 ± 0.1-fold (Fig. 1A, $n = 9$), likely due to the increase in surface area from changes in topology. On average, the oxidative peak current for dopamine only increased by 1.7 ± 0.3-fold (Fig. 1 and 2, $n = 9$). Previously, we showed that dopamine oxidative peak current increased by 2.2-fold after O_2-plasma treatment on carbon fiber electrodes due to increases in both surface roughness and surface oxide groups.[16] This may indicate that the majority of the changes observed after plasma treatment are due to the changes in the surface roughness, with minimal effect from the increase in oxide functionality. On average, Ar-plasma treatment increased AD and ATP oxidative current by 3.5 ± 1.4-fold and 3.2 ± 0.6-fold, respectively, which suggests that changes in surface topology impact adenine-based purine detection more than DA. Interestingly, only 1.7 ± 0.3-fold and 1.8 ± 0.3-fold changes for GN and GTP,

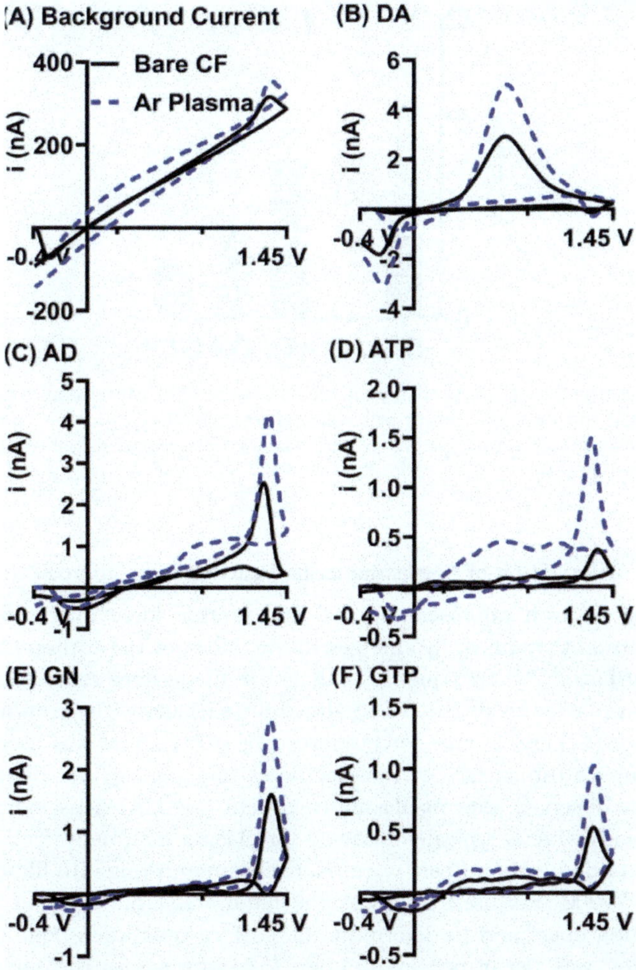

Fig. 1 Ar-plasma treatment increases oxidative peak current for purines and dopamine with observable changes in the background charging current. The CVs for the pre-treated carbon-fiber electrodes (solid line) are compared to the post-treatment with Ar-plasma (dashed line) for all analytes and the background current. (A) A change in the background current was observed after treatment. The oxidative peak current increased for all analytes after plasma treatment: (B) dopamine, DA, (C) adenosine, AD, (D) adenosine triphosphate, ATP, (E) guanosine, GN, (F) and guanosine triphosphate, GTP.

respectively, were measured (Fig. 1 and 2, $n = 9$). Similar discrepancies between adenine- and guanine-based purines were observed previously for O_2 and N_2 plasma-treated electrodes, suggesting that although these analytes are very similar in structure, they interact quite differently at the electrode surface.[16] Lim *et al.* described that the purine structure has significant impact on the electrode–analyte interface on the carbon-fiber surface.[15] Overall, this provides additional evidence that careful analysis of specific analyte–electrode interactions is needed in order to develop ultrasensitive electrode surfaces for neurochemical detection.

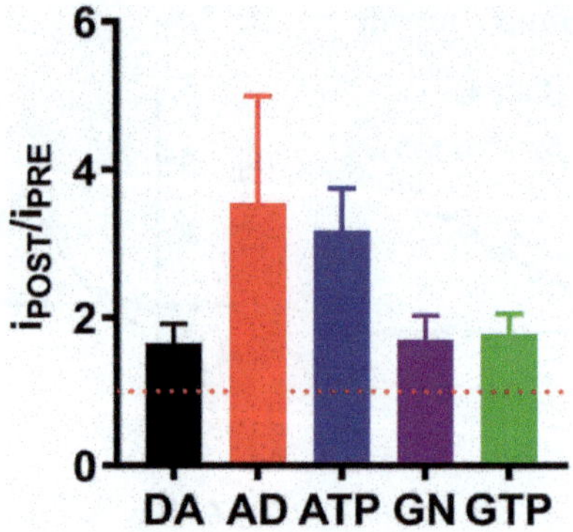

Fig. 2 Ar-plasma increases oxidative peak current for purines more than dopamine. The plots represent the ratio of current post-plasma treatment (i_{post}) to the current before plasma treatment (i_{pre}). A ratio above 1.0 indicates an increase in oxidative peak current after treatment ($n = 9$).

Surface characterization of Ar-plasma treated carbon-fiber microelectrodes

Scanning electron microscopy (SEM) and atomic force microscopy (AFM). Plasma treatment significantly changes the topology of the carbon-fiber microelectrode surface.[24,27–29,31] To quantify changes in the surface roughness after the optimized Ar-plasma treatment, we analyzed the surfaces with scanning electron microscopy (SEM) and atomic force microscopy (AFM). SEM was used to qualitatively compare the surface of the electrodes (Fig. 3). Visible changes in the surface were observed after Ar-plasma treatment (Fig. 3A). This morphology is consistent with what is typically observed for plasma treatment.[28,29,32] Comparatively, unmodified carbon-fibers were relatively smooth (Fig. 3B) like previously reported.[31,33] AFM was used to assess the large area uniformity of Ar-plasma treated carbon-fibers and to determine the surface roughness (Fig. 3B and D). AFM imaging was performed at multiple locations across the sample. Clear changes in surface topology were observed for Ar-plasma treated fibers compared to unmodified surfaces (Fig. 3B and D). To more accurately capture the changes in surface roughness, we calculated the average roughness across the entire image (R_a). On average, the R_a value for bare carbon fiber was 13.5 ± 0.1 nm ($n = 4$), similar to previous reports.[34] Ar-plasma treated carbon-fibers (optimized method) had an average R_a value of 50.0 ± 0.3 nm ($n = 4$). Roughness values between bare and Ar-plasma treated fibers were significantly different (unpaired t-test, $p <$ 0.0001). Comparatively, we assessed the changes in surface roughness for carbon-fibers that were treated with 100 W Ar-plasma. The 100 W Ar-plasma treatment increased the surface roughness more than the 70 W treatment (Fig. S4,† R_a was 79.1 nm) indicating that there is a balance between changes in surface roughness and optimal electrochemical detection. We speculate that increasing roughness too much could be destroying the conductivity of the surface, similar to prior reports,[4] which could negatively impact the changes in oxidative current after treatment. An interesting observation is that the changes in oxidative current for purines at Ar-plasma treated fibers are similar to what was observed for O_2 plasma-treated fibers from our previously published work, with almost identical

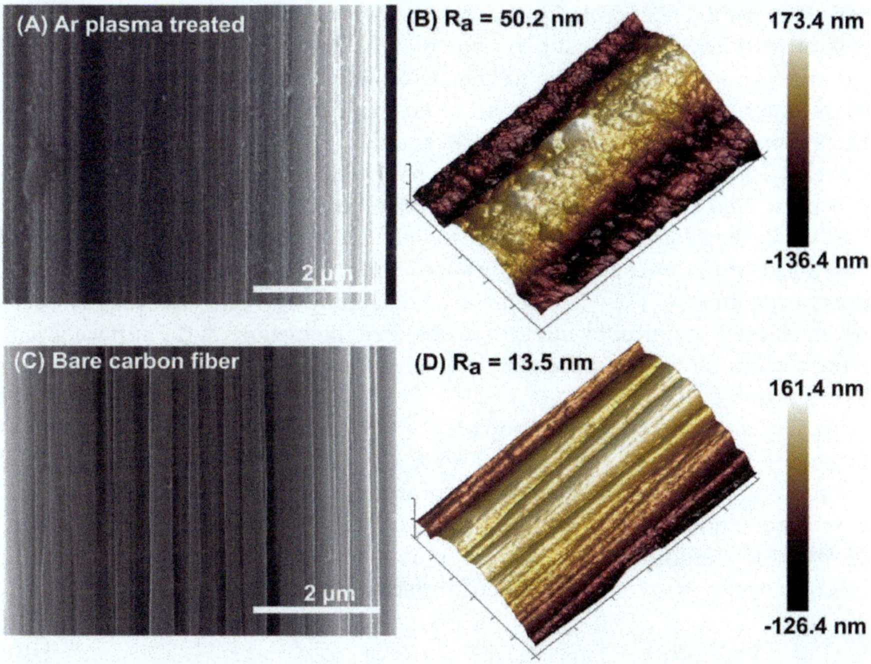

Fig. 3 Surface roughness changes after Ar-plasma treatment. (A) SEM of Ar-plasma treated carbon-fiber, (B) AFM of Ar-plasma treated carbon-fiber. Scale bars shown on the graph. (C) SEM of bare carbon-fiber, (D) AFM of bare carbon-fiber. A large increase in the surface roughness was observed after Ar-plasma treatment compared to the bare carbon-fiber.

changes in surface roughness. This suggests that the surface topology is a major driving factor for improving purine detection.

Raman spectroscopy. Raman spectroscopy was used to assess the changes in surface defects after Ar-plasma treatment. The ratio of surface defects (D band, $\sim$1350 cm^{-1}) to graphitic carbon (G band, $\sim$1590 cm^{-1}) is used to evaluate the amount of sp^3 and sp^2 hybridized carbon. The bare carbon-fiber D/G ratio is 1.86 $\pm$ 0.03 (Fig. S5A,† $n = 12$), similar to prior reports.[35,36] The D/G ratio increases to 1.92 $\pm$ 0.01 after treating carbon-fibers with the optimized Ar-plasma procedure (Fig. S5B†). We also compared the D/G ratio at the 100 W procedure and show that higher Ar-plasma powers increase the D/G ratio further (1.96 $\pm$ 0.03, Fig. S5C†). The changes in surface defects are significantly different between the 100 W Ar plasma treatment and bare carbon-fiber (One-way ANOVA, Bonferroni post-test, $p = 0.0125$, $n = 12$), yet the changes in D/G between the bare carbon-fiber and the optimized Ar-plasma treatment (70 W) are not significantly different ($p > 0.05$). These results indicate that changes in the surface roughness can impact the degree of defects on the surface, but are not likely the primary factor in impacting the electrochemical detection of purines and dopamine.

X-Ray photoelectron spectroscopy (XPS). Ar-plasma treatment is not expected to induce noticeable changes in surface functionalization. We used XPS to demonstrate that our optimized Ar-plasma treatment is not significantly changing the chemical composition of the surface. The overall XPS survey for both bare carbon-fibers (black) and carbon-fibers treated with the optimized Ar-plasma treatment (red) are superimposed and plotted in Fig. 4 for comparison. On average, no changes in the weight percentages of the three main elements (C, N, and O) were observed (Fig. 4). More specifically, the oxygen content was 22.39% on bare carbon-fiber, and 21.79% on the Ar-plasma treated fiber. Likewise, the carbon content was 77.61% on bare carbon fibers, and 78.27% on Ar-plasma treated carbon fibers. No measurable peak for nitrogen was observed. Further deconvolution of the peaks to quantitate the percentage of specific functional groups revealed no significant differences in surface functionalization. Comparing this to the electrochemical data provides evidence that surface roughness can significantly impact the observed interactions at the surface. Even more interesting is that these effects seem to be hugely impactful on adenine-based purines.

Surface roughness changes impact the surface coverage and sensitivity. A Langmuir isotherm was used to quantitate and compare the surface coverage of purines and dopamine on the optimized Ar-plasma-treated electrode surface to bare carbon-fiber microelectrodes at equilibrium.[15,16] The surface coverage provides a platform to compare the monolayer electrode surface saturation coverage by adsorbate under different surface conditions:

$$\Gamma_A = \frac{\Gamma_s \beta[A]}{1 + \beta[A]} \tag{10}$$

Γ_A is surface concentration, Γ_s is the saturation surface coverage, and β is the thermodynamic equilibrium constant for adsorption. The [A] is the concentration

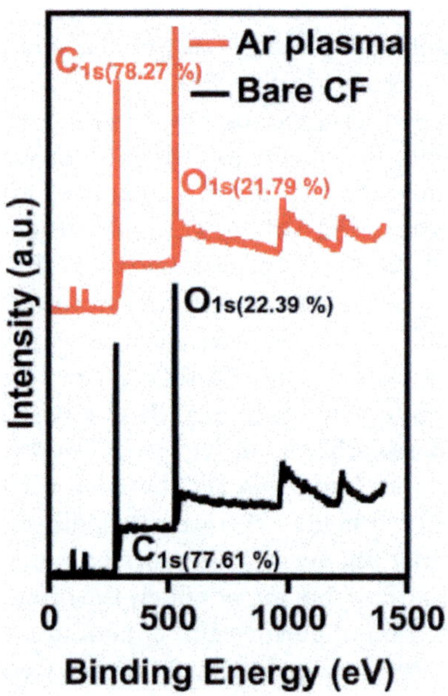

Fig. 4 Ar-plasma does not significantly impact the functionalization of the carbon surface. XPS survey spectra of untreated carbon-fiber (black) and Ar-plasma treated (red) carbon-fiber. The oxide functionality is not impacted by Ar-plasma treatment.

of the analyte. The surface coverage for purines was determined using eqn (3), since both reversible and irreversible steps are involved:[15,37]

$$i_p = \frac{n'^2 \theta F^2}{2.718RT} \nu A \Gamma \tag{11}$$

i_p is the peak oxidative current, T is the temperature (25 °C), R is the ideal gas constant, ν scan rate, F is Faraday's constant, n' is the number of electrons transferred prior to the chemically irreversible steps, θ is the number of electrons transferred in the redox reaction divided by n', and A is the area of the electrode surface (4×10^{-7} cm^2 for 45° angled disk electrode).[38] For adenosine and ATP, the $n' = 2$ and $\theta = 3$. For guanine and GTP, the $n' = 2$ and $\theta = 2$.[15] Dopamine undergoes a one-step reversible redox reaction; therefore eqn (12) was used to calculate its surface coverage:

$$i_p = \frac{n^2 F^2}{4RT} \nu A \Gamma \tag{12}$$

In eqn (12) for dopamine, the $n = 2$ because n is the number of electrons in the redox reaction.

A linearized form of eqn (10) can be used to calculate b (eqn (13) and (14)), the equilibrium coefficient which dictates adsorption strength,[39] because at low concentrations, Γ_A is much smaller than $\Gamma_{s:}$

$$\Gamma_A = b[A] \tag{13}$$

$$b = \Gamma_s \beta \tag{14}$$

Eqn (14) can be used to determine the thermodynamic equilibrium constant (β).

The oxidative current for all analytes was measured as a function of concentrations ranging from 1 to 200 μM. The surface coverage was calculated as described above. The average surface coverage for DA increased from 60.2 pmol cm^{-2} to 100.5 pmol cm^{-2} after Ar-plasma treatment, equating to an approximate 66.9% increase (Fig. 5, Table 1, n = 5–11). Likewise, the sensitivity (calibration slope, see Fig. 6 and Table 1) improved approximately 1.7-fold after the Ar-plasma treatment. On average, the AD and ATP surface coverage increased from 3.88 pmol cm^{-2} and 1.68 pmol cm^{-2} to 5.18 pmol cm^{-2} and 2.68 pmol cm^{-2}, respectively, after Ar plasma treatment, equating to an approximate 33.5% and 59.5% increase (Fig. 5, Tables 1 and S1†). The GN and GTP surface coverage increased from 11.1 pmol cm^{-2} and 3.68 pmol cm^{-2} to 14.5 pmol cm^{-2} and 6.69 pmol cm^{-2} , respectively, after Ar-plasma treatment equating to an approximate 30.6% and 81.8% increase (Fig. 5, Tables 1 and S1†). Similarly, the adsorption strength for AD and ATP increased 27.3% and 57.1% after Ar-plasma treatment, and GN and GTP increased by 60.7% and 100%, respectively. The sensitivity for ATP increased 4.2-fold but only increased 1.4-fold for GTP after Ar-plasma treatment. For AD and GN, the sensitivity increased 1.6-fold and 1.4-fold. After Ar plasma treatment, significant differences between the surface coverage of dopamine and purines were observed (One-Way ANOVA, Bonferroni post-tests, DA compared to AD $p = 0.0002$***; compared to ATP $p < 0.0001$****; compared to GN

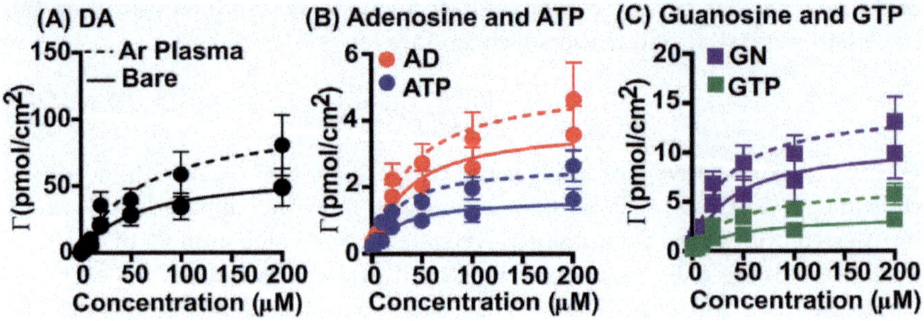

Fig. 5 Ar-plasma treatment impacts the surface coverage and adsorption strength for purines and dopamine. Langmuir isotherms for (A) DA, (B) AD and ATP, and (C) GN and GTP at the bare carbon fiber (CF, solid line) and optimized Ar-plasma treatment (dashed line). Concentrations tested range from 1 μM to 200 μM. (n = 5–11).

$p = 0.0019$**; compared to GTP $p = 0.0002$***). From these results, it appears that changes in surface topology produce varying effects on adsorption of purines with various structures. More specifically, disubstituted purines (GN and GTP) appear to increase in adsorption strength significantly more than monosubstituted purines (AD and ATP); likely due to having more "sites" for adsorption to the surface. These changes equate to similar increases in surface coverage for mono- *vs.* disubstituted purines. Overall, these results demonstrate the importance of not only looking at changes in current before and after a modification, but critically assessing the extent to which modifications affect surface adsorption and strength, is important for providing a more in-depth understanding of how analytes interact at the surface.

Changes in the difference between the oxidative peak potential and reductive peak potential (ΔE_p) are indicative of the electron transfer slowing down or speeding up at the surface. Purines are irreversible; therefore, a ΔE_p cannot be calculated. For dopamine, we show that after Ar-plasma treatment, the ΔE_p is significantly smaller compared to unmodified carbon-fibers (Table 1, paired *t*-test, $p = 0.0091$, $n = 11$). This shows that slightly faster electron transfer is observed at surfaces roughened with Ar-plasma, likely due to momentary trapping in the crevices on the surface, similar to prior work on roughened surfaces.[21,23]

COMSOL multiphysics simulation of dopamine at roughened surfaces and fast-scan rates. COMSOL Multiphysics software provides a modeling platform to simulate the interaction of dopamine at a carbon-fiber electrode with FSCV parameters. Here, we used COMSOL to simulate dopamine at traditional carbon electrodes, and electrodes which increased significantly in surface topology, to model the changes that are induced by Ar-plasma. The electroanalysis module was used to simulate traditional FSCV conditions (−0.4 V to 1.3 V, 400 V s^{-1}) of 20 μM DA as the test analyte at a carbon-fiber disk electrode. FSCV is a specialized, background subtracted technique, so analyzing data from the simulation requires a methodical approach. Once the parameters and conditions of the model were set (see Methods), the cyclic voltammetry study was used to simulate the voltammetric study, producing the CV for both non-faradaic background current and faradaic current produced by the redox reaction of interest. After acquiring the CV for 20 μM DA, the [DA] was set to 0 μM to simulate the CV in the absence of analyte

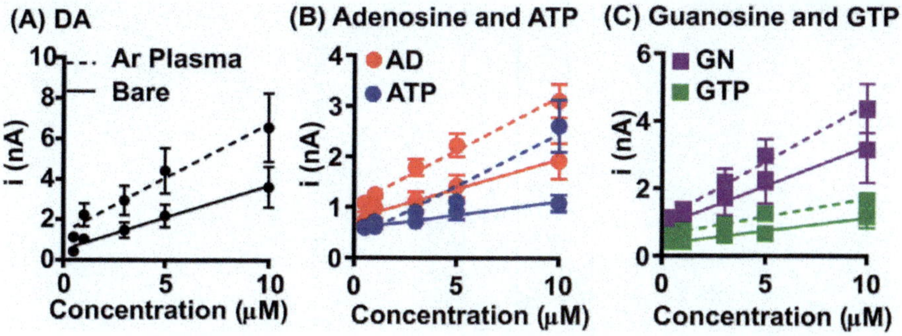

Fig. 6 Ar-plasma treatment increases sensitivity of carbon-fibers for purines and dopamine. Concentrations tested from 1 μM to 10 μM for (A) DA, (B) AD and ATP, and (C) GN and GTP at the bare carbon fiber (CF, solid line) and Ar-plasma (dashed line) treated electrodes. (n = 5–11).

giving only the non-faradaic signal (Fig. 7C and D). Once both were simulated, we manually subtracted the non-faradaic current to give the background-subtracted CV we used to analyze with FSCV (Fig. 7E).

COMSOL was used to predict changes in faradaic current for DA at a surface which best mimicked Ar-plasma treated surfaces (Fig. 7B). Bare carbon-fiber disk electrodes were modeled as a horizontal line in the 2D-axisymmetric model allowing for a flat, circular electrode surface following integration (Fig. 7A). Fig. 7A represents the concentration gradient for DOQ at the surface. The simulation was run with analyte present and analyte removed, acquiring a CV for both cases (Fig. 7C). The average surface roughness calculated from the experimental AFM data for the optimized Ar-plasma treated electrodes was used to inform the design of the surface-roughened electrode in COMSOL. The value for surface roughness with AFM was determined to be 50.2 nm, so vertical line segments of 50 nm were added to the bare carbon-fiber disk electrode to simulate nanoelectrodes which would be created by changes in the surface topology (Fig. 7B). Fig. 7B demonstrates changes in the concentration gradient of DOQ at the modified surface geometry. The perforated electrode surface node was used here to account for the surface roughness additions as the boundaries are now interior electrode–electrolyte interfaces, different from the solely exterior boundary of a bare CF disk electrode. CVs with and without 20 μM DA present at the roughened electrode surface, as described above, were simulated (Fig. 7D). A near 2-fold increase in oxidative current for DA was simulated (Fig. 7E), similar to our experimental data (1.7-fold) suggesting that changes in surface roughness are a primary source for increases in oxidative peak current at Ar-plasma treated electrodes (Fig. 7E). Importantly, we observe the expected shift in oxidative peak for DA at fast scan rates. The standard potential was modelled as 0.2 V; however, at 400 V s^{-1} we observe that the simulated oxidative peak shifts to 0.5 V. This result is similar to what we observe experimentally, where DA's oxidation peak is typically observed at 0.5–0.6 V. This validates that the model is adequately accounting for shifts in oxidative peak potential as a function of scan rate. We would like to note that the shape of the generated CVs differ slightly from expected FSCV CVs because COMSOL does not account for adsorption of DA to the

Table 1 Ar-plasma treatment effects the interaction at carbon-fiber surfaces ($n = 5$–11)

	Surface coverage (Γ, pmol cm^{-2})		Adsorption strength (b, cm)		Sensitivity (nA µM^{-1})		$\Delta E_{p,DA}$	
	Bare	Ar	Bare	Ar	Bare	Ar	Bare	Ar
AD	3.88 ± 1.2	5.18 ± 1.6	0.11	0.14	0.12 ± 0.01	0.21 ± 0.01	—	—
ATP	1.68 ± 0.44	2.68 ± 0.63	0.07	0.11	0.05 ± 0.01	0.21 ± 0.03	—	—
GN	11.1 ± 3.45	14.5 ± 2.91	0.28	0.45	0.28 ± 0.02	0.38 ± 0.02	—	—
GTP	3.68 ± 0.75	6.69 ± 1.23	0.08	0.16	0.07 ± 0.01	0.10 ± 0.02	—	—
DA	60.2 ± 13	100.5 ± 19	1.09	1.67	0.32 ± 0.03	0.54 ± 0.05	0.93 ± 0.004	0.90 ± 0.01**

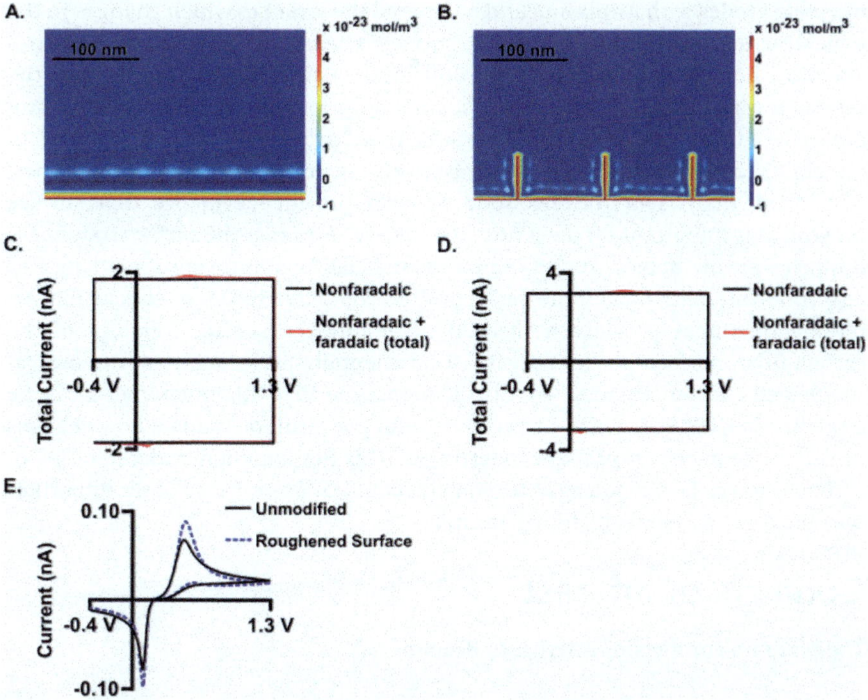

Fig. 7 A COMSOL Multiphysics model of 20 μM dopamine interactions demonstrate changes in faradaic current at roughened electrode surfaces. A waveform scanning from −0.4 to 1.3 V at 400 V s^{-1} was used in the simulation. Manual background subtraction to remove non-faradaic current was done to reveal simulated background-subtracted CVs for dopamine. (A) Concentration gradient color plot of dopamine-*o*-quinone (DOQ) at a bare carbon-fiber electrode of radius 3.5 μm and at a carbon-fiber electrode of radius 3.5 μm with, (B) the addition of 50 nm-tall surface segments to simulate the changes in topography. (C) Overlayed CV of simulated capacitive current (black) and capacitive plus faradaic current (red) for bare carbon-fiber electrodes, and for (D) a surface roughened carbon-fiber electrode. (E) Manually subtracted background CV of bare carbon-fiber (black) and surface roughened carbon-fiber (blue dashed line).

electrode's surface. This results in changes in the peak shape and redox cycling behavior at the electrode surface showing that COMSOL simulations for adsorption-limited processes with FSCV are not perfectly captured in this 2D-axisymmetric model. We also note that the reduction potential for DA is greater than 0 V in the simulated CVs, which is much greater than the expected value of −0.2 V observed with FSCV. We speculate that this is due to the COMSOL model not adequately capturing that DA is actually quasi-reversible. Overall, this data demonstrates that COMSOL can be used to easily simulate interactions at varying electrode geometries with FSCV, which could impact our understanding of surface interactions at fast-scan rates in the future.

Conclusion

Plasma-treated carbon fiber can be used to enhance topology and sensitivity to neurochemicals on the carbon surface. Here, we treated carbon-fiber microelectrodes with Ar-plasma and compared the extent to which changes in the nanostructure, without changes in surface chemistry, impacts the electrochemical detection of purines and dopamine. Fundamental insight into the extent to which nanostructured surface topology and geometry impacts neurochemical interactions at electrode surfaces could provide significant advances in our ability to design and fabricate ultrasensitive neurochemical detection probes. A detailed analysis of the electrode surfaces provided evidence that surface roughness is important for detection. Overall, we show that the surface roughness improves purine detection and is more beneficial for improving sensitivity and adsorption strength of adenine-based purines compared to DA. In conclusion, Ar-plasma treatment provides a method to modify the surface topology of the carbon-fiber surface to investigate fundamentally how analytes interact at roughened carbon surfaces. We also demonstrate that this phenomena can be modelled in COMSOL which provides a useful platform to simulate how changes in surface geometry impact FSCV detection. This fundamental understanding of carbon fiber surface topology and geometry could be extended to other important biological neurochemicals in the future.

Conflicts of interest

There are no conflicts of interest to declare.

Acknowledgements

The authors would like to thank the Department of Chemistry at the University of Cincinnati, the UC Women in Science and Engineering (WISE) program, and the National Institute of Allergy and Infectious Diseases of the National Institutes of Health under Award Number R01AI151552 for support for this research. The content is solely the responsibility of the authors and does not necessarily represent the official views of the NIH. We would like to thank the Advanced Materials and Characterization Core at the University of Cincinnati for assistance with collecting the SEM and AFM data. We would also like to thank the Surface Analysis Core at the Ohio State University for assistance with collecting the XPS data.

References

1 M. T. Cryan and A. E. Ross, *Anal. Chem.*, 2019, **91**, 5987–5993.
2 A. E. Ross and B. J. Venton, *The Analyst*, 2012, **137**, 3045.
3 A. E. Ross and B. J. Venton, *J. Neurochem.*, 2015, **132**, 51–60.
4 Q. Cao, P. Puthongkham and B. Jill Venton, *Anal. Methods*, 2019, **11**, 247–261.
5 C. Yang, K. Hu, D. Wang, Y. Zubi, S. T. Lee, P. Puthongkham, M. V. Mirkin and B. J. Venton, *Anal. Chem.*, 2019, **91**, 4618–4624.
6 C. B. Jacobs, I. N. Ivanov, M. D. Nguyen, A. G. Zestos and B. J. Venton, *Anal. Chem.*, 2014, **86**, 5721–5727.
7 Z. Shao, P. Puthongkham, K. Hu, R. Jia, M. V. Mirkin and B. J. Venton, *Electrochim. Acta*, 2020, **361**, 137032.
8 E. K. Jackson, D. Boison, M. A. Schwarzschild and P. M. Kochanek, *J. Neurochem.*, 2016, **137**, 142–153.

9 M. Garcia-Gil, M. Camici, S. Allegrini, R. Pesi, E. Petrotto and M. G. Tozzi, *Int. J. Mol. Sci.*, 2018, **19**, E3598.
10 G. Burnstock and J.-M. Boeynaems, *Purinergic Signal.*, 2014, **10**, 529–564.
11 S. T. Lee and B. J. Venton, *ACS Chem. Neurosci.*, 2018, **9**, 505–513.
12 M. Ganesana and B. J. Venton, *PloS One*, 2018, **13**, e0196932.
13 M. T. Cryan and A. E. Ross, *Analyst*, 2018, **144**, 249–257.
14 Y. Li, M. E. Weese, M. T. Cryan and A. E. Ross, *Anal. Methods*, 2021, **13**, 2320–2330.
15 G. N. Lim and A. E. Ross, *ACS Sens.*, 2019, **4**, 479–487.
16 Y. Li and A. E. Ross, *Analyst*, 2020, **145**, 805–815.
17 B. E. K. Swamy and B. J. Venton, *Anal. Chem.*, 2007, **79**, 744–750.
18 R. R. Kamath and M. J. Madou, *Anal. Chem.*, 2014, **86**, 2963–2971.
19 Z. Zhou, T. Liu, A. U. Khan and G. Liu, *Sci. Adv.*, 2019, **5**, eaau6852.
20 E. O. Barnes, G. E. M. Lewis, S. E. C. Dale, F. Marken and R. G. Compton, *Analyst*, 2012, **137**, 1068–1081.
21 Q. Cao, D. K. Hensley, N. V. Lavrik and B. J. Venton, *Carbon*, 2019, **155**, 250–257.
22 J. Li, L. Ma, X. Li, C. Lu and H. Liu, *Ind. Eng. Chem. Res.*, 2005, **44**, 5478–5482.
23 C. Yang, Y. Wang, C. B. Jacobs, I. N. Ivanov and B. J. Venton, *Anal. Chem.*, 2017, **89**, 5605–5611.
24 H. Bubert, X. Ai, S. Haiber, M. Heintze, V. Brüser, E. Pasch, W. Brandl and G. Marginean, *Spectrochim. Acta Part B At. Spectrosc.*, 2002, **57**, 1601–1610.
25 C. Jones, *Surf. Interface Anal.*, 1993, **20**, 357–367.
26 P. Takmakov, M. K. Zachek, R. B. Keithley, P. L. Walsh, C. Donley, G. S. McCarty and R. M. Wightman, *Anal. Chem.*, 2010, **82**, 2020–2028.
27 S. Corujeira-Gallo and H. Dong, *J. Compos. Mater.*, 2017, **51**, 3239–3256.
28 F. Ghamouss, E. Luais, C. Thobie-Gautier, P.-Y. Tessier and M. Boujtita, *Electrochim. Acta*, 2009, **54**, 3026–3032.
29 D. Tashima, A. Sakamoto, M. Taniguchi, T. Sakoda and M. Otsubo, *Vacuum*, 2008, **83**, 695–698.
30 J. A. Johnson and R. M. Wightman, *Electrochem. Soc. Interface*, 2017, **26**, 53–57.
31 N. Dilsiz, *J. Adhes. Sci. Technol.*, 2000, **14**, 975–987.
32 F. Ghamouss, E. Luais, C. Thobie-Gautier, P.-Y. Tessier and M. Boujtita, *Electrochim. Acta*, 2009, **54**, 3026–3032.
33 D. T. C. Chang, *J. Ind. Technol.*, 1998, **15**, 7.
34 A. C. Schmidt, X. Wang, Y. Zhu and L. A. Sombers, *ACS Nano*, 2013, **7**, 7864–7873.
35 M. E. Weese, R. A. Krevh, Y. Li, N. T. Alvarez and A. E. Ross, *ACS Sens.*, 2019, **4**, 1001–1007.
36 Y. Li, C. M. Fleischer and A. E. Ross, *Chem. Commun.*, 2020, **56**, 8023–8026.
37 L. M. Gonçalves, C. Batchelor-McAuley, A. A. Barros and R. G. Compton, *J. Phys. Chem. C*, 2010, **114**, 14213–14219.
38 J. G. Roberts, B. P. Moody, G. S. McCarty and L. A. Sombers, *Langmuir*, 2010, **26**, 9116–9122.
39 C. W. Atcherley, N. D. Laude, K. L. Parent and M. L. Heien, *Langmuir*, 2013, **29**, 14885–14892.

Faraday Discussions

ROYAL SOCIETY OF CHEMISTRY

DISCUSSIONS

Advanced nanoelectrochemistry implementation: from concept to application: general discussion

Paul W. Bohn, Xiangkun Elvis Cao, Shuai Chang, Dongfei Chen, Samuel Confederat, Dominik Duleba, Peisan E, Martin A. Edwards, Andrew Ewing, Luke Gundry, Jin He, Ali Reza Kamali, Frédéric Kanoufi, Seung-Ryong Kwon, Ndrina Limani, Steven Linfield, Xu Liu, Yi-Tao Long, Si-Min Lu, Bing-Wei Mao, Shelley Minteer, Popular Pandey, Hang Ren, Ashley Ross, Ben Slater, Patrick Unwin, Swathi Naidu Vakamulla Raghu, Jill Venton, Alain Walcarius, Hui Wei, Yanfang Wu, Li Xiao, Weilin Xu, Yi-Lun Ying, Ping Yu and Zhu Zhang

DOI: 10.1039/d2fd90004a

Yi-Tao Long opened a general discussion of the paper by Paul W. Bohn: The electrochemical nanoconfinement makes a major impact on controlling the mass transport. Could you please comment on the multi-phase effects on the mass transport under nanoconfinement. What's the next breakthrough application we could expected in controlling the mass transport in nanoconfinement.

Paul W. Bohn responded: You are correct that we see evidence of altered mass transport under conditions of confinement. Although we have not implemented a comprehensive study, we can say some general things about the factors that influence mass transport under these conditions. Across many different experiments, it appears that permselectivity is a dominant factor in determining transport behaviour in nanopores. Beyond this, ion migration appears to play an important role. This is especially true and accentuated by the alteration in ion shielding at small dimensions, which impacts the magnitude of ion migration currents. Lastly, diffusion is an important mass transport mechanism at the small scale, principally because transport between electrodes that are only a few tens to hundreds of nm apart can occur on the micro second time scale and thus be quite efficient. A final thought in this regard concerns a potential role for surface hopping processes. It is widely appreciated that the micro channels can be dominated by surface production processes when the size of the channel becomes small enough, and this becomes a player at the nanoscale as well, but it is a mechanism that we have not yet tried to exploit explicitly.

Yi-Tao Long added: In the last two *Faraday Discussions*, we discussed the well-defined nanointerface. We could see that there is progression, achieving good development. But still, developing a well-defined nanointerface is a challenge. Could you please comment on the future of nanointerfaces?

Paul W. Bohn answered: I think there are two important parts to answering your excellent question. First, is the question of being able to reproducibly fabricate architectures with sub-nanometer control over spatial positioning and orientation of functional components. For the parts of the structure that are amenable to traditional methods of nanofabrication, we are actually very close, as can be ascertained from some of the cross-sectional SEM images from our laboratory and others. When we move from fabrication strategies developed to support microelectronics to those that are more chemically oriented, for example the self-organization of block copolymers utilized in the structures from our laboratory, then the control is less precise. As an example, although we draw cartoons in which the P4VP nanodomains are all oriented perpendicular to the planer interface, in fact we have good cross-sectional SEM images illustrating that these cylindrical domains are oriented over a range of angles relative to the interface. This is just one example where the molecular approaches to bottom-up fabrication present real challenges. The last point with respect to this part of the question has to do with the incorporation of molecules as functional elements. For example something as well studied and straightforward as enzyme immobilization still produces layers with a large degree of structural heterogeneity, both within the enzyme itself and among neighboring enzyme molecules in a layer.

The second, and perhaps even more challenging, part of the question concerns the ability to translate the high precision nanoscale construction capabilities addressed above to very large macroscopic areas. The tools that we have at our disposal are well-designed for μm-scale architectures with nm-scale precision. However, we will need to develop new approaches that allow us to develop structures at the millimetre and centimetre scale in order to see the full implementation of nano-interface-enabled applications. This is a major challenge for the nanoscience field as a whole, not just nanoelectrochemistry.

Frederic Kanoufi continued: As mentioned by Yi-Tao Long, you have given this nanoelectrode array very interesting developments. In earlier work, you also proposed that those nanoelectrode arrays could be monitored optically. Could it be a complementary way to dynamically visualize the wetting/dewetting of the vestibule, or to probe single molecule activity within the vestibule? The strategy presented by Valavanis *et al.* (DOI:10.1039/d1fd00063b) in session 3, monitoring change in the reflecting properties of the bottom electrode could also be interesting.

Paul W. Bohn answered: Your question is right on the mark. In these experiments, we have constructed hierarchically organized membrane systems in which the BCP is located 0.5–1.0 μm from the floor of the nanopore, which places it sufficiently far away so that we could not use our electrochemical zero mode waveguide strategy directly. However, it should not be too hard to reduce the thicknesses of the NEA nanopore metal–insulator–metal structures such that the BCP membrane would be optically accessible through the NEA nanopore. With these altered (thinner) structures, it should be possible to capture fluorescence from the probe species within the nanocylindrical P4VP domains of the BCP under conditions in which these pores are wetted.

Si-Min Lu added: To continue on the comment from Yi-Tao Long, what do you think the next avenue should be for nanopore electrochemistry in terms of single entity measurements? Especially for the precise measurements of the transient electron transport of single entities in Faradaic processes.

Paul W. Bohn replied: In order to address single entities in nanopore electrochemistry, it is first important to define the nature of the single entity – they could be molecules, nanoparticles, viruses, or even bacteria (although at that point one is really talking about microscale structures not nanostructures). The reason it is important to define the problem is that we are still in a regime where addressing 1 single electron transfer event in single molecules by purely electrochemical means is extraordinarily difficult. At this level, the two major routes to measurement, both of which are inherently nanoscale processes, involve some type of amplification process, such as redox cycling, or the conversion of electrochemical events into spectroscopic processes, such as those that occur in electrofluorogenic reactions. If one confines the question of transient electron transport, then spectroscopic approaches have many advantages. Specifically, a fluorescent redox molecule can be cycled through the excited state many times, thus creating an amplified signal for a single redox turnover event. In addition spectroscopic processes can be studied under conditions that are not always limited, and detectors are routinely capable of measuring single photons. The trick, of course, is defining a redox system that has the requisite electrofluorogenic character. If the single entities of interest are larger, such as nanoparticles, viruses, and other microorganisms, then direct electrochemical study becomes more feasible, since the entities can be capable of transferring a large number of electrons in a single interaction. However, for the most demanding experiments, such as the transient electron transport processes that you have in mind, I think spectroelectrochemical approaches will for the near future, be a superior choice to purely electrochemical studies.

Popular Pandey asked: Thank you for the great talk. Of the 160 nm and 89 nm BCP membranes, the 89 nm BCP membrane exhibited a 2-fold increase in transport properties (main text Fig. 3 in your article, DOI: 10.1039/d1fd00048a). Is this due to the thinner BCP membrane? Also, how stable are the BCP membranes over the range of applied potential?

Paul W. Bohn replied: Yes, we assign the increase in transport rate to the membrane thickness being smaller, which should decrease the transport resistance in a general sense. As for the stability of the BCP membranes, we have not observed any overt signs of mechanical or chemical degradation over the range of potentials used in our experiments. Presumably there are limits to the potentials which can be applied without degradation, but we have not explored that question in detail.

Yanfang Wu said: Thanks a lot for the inspiring talk. Could you please provide some more detail/insight into the change in pore size of the BCP membrane from hydrophilic to hydrophobic? What is the rationale in terms of the pore size of the BCP membrane to perform potential-induced wetting and dewetting? This is a very interesting study and one day the system may be able to fulfil operations with liquid samples like loading, filtering and analytical analysis afterwards.

Paul W. Bohn responded: This is an interesting, but still open, question. There have been a large number of publications from other laboratories addressing the nature of the pH-dependent change in properties of the nanocylindrical P4VP domains, but I do not think there is at present a definitive answer to the question of the change in pore size above and below the pK_a.

The rationale for exploring potential-induced wetting and de-wetting came from observations of similar phenomena in other polymer nanopore systems. In particular, I would point to the work of Smirnov and coworkers, for example.[1] We certainly agree with your assessment that these phenomena may be capable of enabling operations like liquid sampling, loading, and analytical determinations within the nanopores of our structures.

1 S. N. Smirnov, I. V. Vlassiouk and N. V. Lavrik, *ACS Nano*, 2011, **5**, 7453–7461.

Frederic Kanoufi asked: To continue on the suggestion made by Serge Lemay yesterday (DOI: 10.1039/d2fd90003c): one way to capture and analyze single molecules is to isolate, concentrate then analyse them. You showed a promising way to isolate and analyze molecules into these nanosized vestibules, maybe soon at the single entity level. What about concentrating them into the vestibule? Could you use the polymer layer as an affinity preconcentration layer for this purpose, or use the electric field, *etc.*...?

Paul W. Bohn replied: This is an excellent observation. In fact, this is exactly the strategy we had in mind when preparing Fig. 1(d) in our manuscript and the data presented at the end of the oral presentation. It should be possible to manipulate potential as the BCP membrane can be used as a potential-gated valve to allow the capture and trapping of analyte species within the vestibule of the NEA nanopores. Once trapped they can be subjected to a variety of chemical and electrochemical manipulations in order to enhance their utility in chemical sensing. For example, one can immobilize enzymes within the NEA nanopores and utilize these to convert non-redox active substrates into redox-active products that can then be detected with high sensitivity using the redox cycling capabilities of the two-electrode nanopores discussed in the paper. In fact, we just published an early version of this concept.[1]

1 J. Jia, S.-R. Kwon, S. Baek, V. Sundaresan, T. Cao, A. R. Cutri, K. Fu, B. Roberts, J. D. Shrout and P. W. Bohn, *Anal. Chem.*, 2021, **93**(43), 14481–14488.

Zhu Zhang commented: The insulating layer between two Au layers is SiN_x and the top insulating layer is SiO_2, is the reason that the SiN_x is better than SiO_2 in terms of insulation? Could you give a explanation on that?

Paul W. Bohn responded: In the early days of these experiments, we tried both silicon nitride and silicon oxide insulating layers between the two metallic electrode layers. In our hands, we had better yield employing plasma enhanced CVD deposited silicon nitride. For future applications where we want to reduce the size of the insulating layer – for example to enhance mass transport between the two electrodes in redox cycling experiments – it may be advantageous to utilize atomic layer deposition (ALD) of alternative dielectrics. We have had success in depositing both HfO_2 and TiO_2 by this approach, but we have not yet done a careful comparison of these materials to silicon nitride.

Zhu Zhang continued: Do you see any etching of the insulation layer during the experiments?

Paul W. Bohn replied: We have not obtained any evidence of etching of the insulating layers, either during fabrication or the experiments themselves. If etching were to occur, in the silicon nitride layer, it would produce under cutting of the top electrode, while etching of the silicon oxide top layer would likely just reduce the overall dimensions of that layer.

Zhu Zhang further questioned: How do you remove the Ga ions from your nanopore?

Paul W. Bohn responded: We use a nitric acid post-milling etch procedure described previously.[1]

1 S. P. Branagan, N. M. Contento and P. W. Bohn, *J. Am. Chem. Soc.*, 2012, **134**, 8617–8624.

Zhu Zhang commented: The membrane transport from another substrate to nanopores array – how do you check the membranes have good contact?

Paul W. Bohn answered: This is an important and experimentally challenging part of our procedure. First, we do a visual check by optical microscopy, then when we immerse the structure in the target solution, we look for the presence of any non-specific transport into the NEA nanopores. Although these procedures are not fool proof, they certainly identify samples where there is large scale failure, and make it possible to move forward with structures in which we have confidence of their structural integrity.

Zhu Zhang added: You introduced some fluorescent proteins in your nanopores, how do you introduce the singly fluorescent proteins?

Paul W. Bohn answered: Since we do not have a fool proof procedure to introduce exactly one protein molecule into every nanopore, we rely on a statistical distribution (Poisson) of enzyme molecules produced by carrying out the reactions under low loading conditions. By biasing the distribution so that the majority of pores have zero enzyme molecules, the number which exhibit a single enzyme molecule is maximized relative to pores that have multiple enzymes. Determining the conditions under which this is true is done experimentally, and painstaking measurements are typically required to identify the optimum conditions, *i.e.* enzyme concentration, buffer concentration, reaction time, nanopore pre-treatment, *etc.*

Alain Walcarius queried: How do you interpret the ionic strength-dependence of wetting/dewetting in the BCP membranes? Is there something to do with the thickness of the electrical double layer (on charged membrane) relative to the membrane pore size?

Paul W. Bohn replied: There are likely two major contributions to the ionic strength dependence. The first, as you suggest, likely has to do with the altered structure of the electrical double layer relative to the membrane nanopore at different ionic strengths. The average pore size of the block copolymer membranes was ~12 nm, placing them exactly in the range of Debye lengths applicable at the lower ionic strengths. The other likely contributor is the detailed spatial potential distribution, which would be altered at different ionic strengths, specifically the potential dropped across the thickness of the block copolymer membrane. These two factors are likely not completely separate, and we are currently performing detailed simulations to try to understand the effects we observe experimentally.

Dominik Duleba opened a general discussion of the paper by Shelley Minteer: How are you able to fill these long conical nanochannels with the surface modification reagents?

Shelley Minteer responded: The solution was injected into the glass nanopore with a microloader. Then, we centrifuged the glass nanopore for 5 min at around 1878 *g* for the removal of air bubbles in the nanopipette. For this step, we placed the nanopipette with the tip facing down into a homemade holder within a 2 mL centrifuge tube. For more detail refer to our previous papers.[1,2]

1 R. Gao, Yao Lin, Y.-L. Ying, Y.-X. Hu, S.-W. Xu, L.-Q. Ruan, R.-J. Yu, Y.-J. Li, H.-W. Li, L.-F. Cui and Y.-T. Long, *Nat. Protoc.*, 2019, 2015–2035.
2 R. Gao, L.-F. Cui, L.-Q. Ruan, Y.-L. Ying and Y.-T. Long, *J. Vis. Exp.*, 2019, **145**, e59003.

Steven Linfield asked: Control experiments with FeSII modified nanopores seem more like typical translocations through a conically shaped pore; a sharp decrease in current as the protein travels through the pore mouth (greatest sensing zone), followed by a relatively gradual return to background current. Translocations through a Pdx modified nanopore have a stranger shape; a sharp decrease in current followed by a further decline during the translocation event. What is the explanation for the shape of the P450cam translocations through Pdx modified nanopores in the context of protein–protein interactions? Also, did you record rectification in FeSII modified nanopores? Was it similar to Pdx modified nanopores or could some change in surface charges/ion exclusion zones contribute to differences in translocation speed?

Shelley Minteer replied: The further decline during the translocation of P450cam in the nanopore is because the P450cam undergoes multiple and frequent associations and disassociations with Pdx on the surface of the nanopore. We did not record rectification in FeSII modified nanopores.

Ping Yu questioned: how could you sort out the information on protein–protein interaction from the nanopore results?

Shelley Minteer responded: In our experiment, the Pdx is modified on the inner surface of the glass nanopore while the P450cam was driven into the glass nanopore under bias potential. The interaction between P450cam and Pdx produced a characteristic signal with a long statistic duration of 1.1 ms. In order to confirm that the protein–protein interaction indeed gives the long duration, we performed control experiments by replacing Pdx with FeSII (Shethna) protein. Note that FeSII is the same size as Pdx (Fig. S1 in the ESI of our article, DOI: 10.1039/d1fd00042j), but does not interact with P450cam. The results revealed that a single P450cam molecule undergoes a rapid translocation through the FeSII modified glass nanopore with a statistic duration of 0.1 ms at 200 mV. Therefore, the interaction between P450cam and Pdx prolongs the duration of the blockage.

Ndrina Limani commented: I was wondering if there is a way to check before the measurement if the protein was indeed immobilized in the nanopore? Also, is it stable enough over time so it does not impact the measurement?

Shelley Minteer answered: In our experiments, we compared the *I–V* responses before and after modification to confirm the immobilization of Pdx. As shown in Fig. 2c in our article, the rectification ratio at ±0.8 V decreased from 0.54 to 0.32 after modification. The exposed carboxyl group on the Pdx leads to a change in the surface charge density of the nanopore, giving the low rectification ratio. The open pore current traces were stable and consistent (Fig. S2 in the ESI of our article), which suggest the good stability of our modified nanopore during the measurement.

Hui Wei said: Since your proteins have His tags, have you tried to attach your protein to the Au with the His tag? Would the dissolved O_2 in protein solutions affect the measurement?

Shelley Minteer responded: The proteins with a His tag can not be immobilized on the Au surface directly. The Au surface would need to be modified by a nickel and copper chelating agent. So, we did not immobilize Pdx on the Au surface with the His tag. We did not investigate the effect of dissolved O_2 on the measurement, but we would not expect it to make a difference in the P450cam/Pdx protein system. However, if we were looking at protein–protein interactions where O_2 causes a conformational change in the protein (*i.e.* the Shethna protein and nitrogenase), then we would expect to see an effect of dissolved O_2 on the nanopore measurement.

Hui Wei added: It seems that when you measure/monitor the protein–protein interaction, you apply different voltages. Therefore, I am wondering whether there is possible electron transfer between your proteins and the gold layer?

Shelley Minteer replied: The potentials applied in these systems are small compared to the potentials needed for the bioelectrochemistry of the protein or bioelectrocatalysis catalyzed by the protein.

Ben Slater highlighted: (1) The method you refer to in ref. 23,[1] for the thiol modification, uses a different thiol, mercaptoacetic acid, to that in your work. Was there a reason for the change to the undecanoic acid in your work?

(2) On page 5 you state "Due to the negatively charged carboxyl group, the MUA modified nanopore shows an obvious ionic current rectification" where is this result? There are only *I–V* curves for the gold and Pdx nanopipettes (Fig. 2c in your article).

(3) On page 5 you state "After modification, the exposed carboxyl group on the Pdx leads to a change in the surface charge density of the nanopore" – what is the difference in carboxylate density between the MUA modified nanopipette and the PDX modified nanopipette? Can you show a comparison of all of the current rectifications please? Adding to the two shown in 2c.

1 Y.-L. Ying, R.-J. Yu, Y.-X. Hu, R. Gao and Y.-T. Long, *Chem. Commun.*, 2017, **53**, 8620–8623.

Shelley Minteer replied: (1) The modification procedure of mercaptoundecanoic acid (MUA) is the same as mercaptoacetic acid. According to a previous report, long-chain ($n > 10$) can form stable, close-packed, and well-ordered self-assembled monolayers on gold surface (DOI: 10.1021/jp054290n) therefore may act as a more friendly substrate for protein immobilization. That is one reason we chose MUA in this work. Another reason is that we had MUA in the lab and did not have to purchase mercaptoacetic acid.

(2) The modification was carried out according to a previous protocol, which included the step-by-step procedure and characterization, except mercaptoacetic acid was replaced MUA.[1] We agree that further characterization is needed and we will compare the modification difference between mercaptoacetic acid and mercaptoundecanoic acid on the nanopipette in our future work.

(3) In this experiment, we focused on the nanopore recognition of protein–protein interactions. More details about the modification can be found in ref. 1 and 2. In our future studies, we will do the suggested experiment to compare the modification difference between mercaptoacetic acid and mercaptoundecanoic acid on the nanopipette.

1 Y.-L. Ying, R.-J. Yu, Y.-X. Hu, R. Gao and Y.-T. Long, *Chem. Commun.*, 2017, **53**, 8620–8623.
2 Q. Li, Y.-L. Ying, S.-C. Liu, Y. Lin and Y.-T. Long, *ACS Sens.*, 2019, **4**(5), 1185–1189.

Seung-Ryong Kwon said: Could you differentiate current signals between a single binding event of P450cam to Pdx and multiple binding events of P450cam to several Pdx while P450cam translocates through the nanopore?

Shelley Minteer answered: In our stochastic analysis, the P450cam induced blockage current amplitude (ΔI) in FeSII modified glass nanopore is comparable to Pdx modified glass nanopores (Fig. 3c and d). This result demonstrates that each event represents a single P450cam interacting with Pdx modified glass nanopores. The current fluctuation during the translocation of P450cam is mainly due to the multiple and frequent associations and disassociation between P450cam and Pdx on the surface of the nanopore.

Hui Wei commented: It seems that both reduced and oxidized forms of Pdx proteins are used. Are both forms stable and easily prepared?

Shelley Minteer commented in reply: Only oxidized Pdx was used in this study. The oxidized Pdx is stable and easily prepared. As for the reduced Pdx, it can be reduced using NADH as the electron donor and ferredoxin reductase (Pdr) as an electron transport protein to reduce the iron–sulfur center of Pdx. It is very difficult to evaluate the stability of the reduced Pdx as there are many factors that affect its stability, such as the presence of O_2 and the surrounding electric field environment, *etc.*

Weilin Xu suggested: Is it possible to use nanopores to study your cascade enzyme catalysis? Thanks.

Shelley Minteer replied: This depends on the enzyme cascade system. Some of our cascades form metabolons (supercomplexes of the enzymes in the cascade – *i.e.* Kreb's cycle cascade) and they could be studied with the nanopore. However, some of our enzyme cascades involve enzymes that have no intermolecular interactions with each other (*i.e.* alcohol dehydrogenase, aldehyde dehydrogenase, and formate dehydrogenase) and, therefore, would not be easily studied using nanopores.

Yi-Tao Long asked: Could you please give some comments on the future of single enzyme electrochemical analysis? Should we go for biophysics-like studies or for sensors? Thanks!

Shelley Minteer answered: In the future, this technique could be used to study electrostatic, hydrogen bonding, and other intermolecular interactions controlling protein–protein complexation in electron transport chains, substrate channelling, molecular motors, and disease progression. By developing a suitable algorithm for analyzing the noise in the blockage current, the transient conformational changes of the protein could be further resolved during the protein–protein interactions inside a nanopore.

Popular Pandey remarked: How do you ensure the formation of monolayer Pdx on the inner wall of the nanopipette? What is the electric field effect on the stability of the Pdx enzyme? Is there any influence of the P450cam protein as well as Pdx enzyme orientation on the observed current blockade characteristics? Additionally, in Fig. 3a in your article, why does the blockade current have a slow drop and fast increase shape as P450cam translocates from the bath into the nanopipette? While the usual translocation signal features a fast current drop and slowly increases as it traverses through the narrow channel of the pipette.

Shelley Minteer answered: After modification, the exposed carboxyl group on the Pdx leads to a change in the surface charge density of the nanopore. As a result, the rectification ratio of the positive current value, $I+$, to the negative current values $I-$, at ±0.8 V decreased from 0.54 to 0.32 (Fig. 2c in our article). Moreover, the stable time-series current trace from the Pdx-modified nanopore further confirms the stability of the Pdx layer under bias potential, as shown in Fig. S2 in the ESI of our article. During the association between P450cam protein and Pdx layer, the orientation of the complex could affect the current fluctuations. The further decreases in the diameter of the nanopore would be beneficial to revealing the orientation of the complex. Comparing the outcome from P450cam and FeSII, the current fluctuation during the translocation of P450cam in the nanopore is mainly due to the multiple and rapid associations and disassociation between P450cam and Pdx on the surface of the nanopore.

Samuel Confederat said: Is the stability of the Pdx-modified nanopore limited by the range of the potential applied?

Shelley Minteer replied: Pdx is a small protein and is very stable at room temperature. Moreover, the Pdx is immobilized on the surface of the nanopore which means the local concentration of Pdx is very high in this nanoscale space. The high local concentration helps maintain protein stability. It is also important to note that the protein is more stable on the surface of the nanopore than it would be at the surface of an electrode because we are not applying a potential directly to the gold surface that the protein is modified on.

Bing-Wei Mao commented: The nanopore electrode technique has played important roles in a variety of studies, including single molecule detection. Such detection is based on the blockage of the current flow, *i.e.* with detection of the reduced current. Could there be any possibilities in the future development of nanopore techniques that detections are based on current enhancement mechanism for further improving detection sensitivity?

Shelley Minteer replied: Thank you very much for your comments and suggestions. The ionic current flow through the glass nanopore is regulated by a series of competing effects including volume exclusion of transported molecules and current enhancement from the charged analyte at the tip. In the low salt concentration, counterions screen the glass surface charge within a region further from the pore. Therefore, the introduction of mobile counterions carrying the analyte outweighs the current reduction from the volume exclusion effects. Then, the enhanced signals could be observed upon translocation of a charged analyte. This mechanism has been applied in glass nanopore sensing including for DNA and peptides.[1,2] Especially, the nanobubble-induced current enhancement was applied in the wireless nanopore electrode for the detection of single redox molecules/ions, analyzing the metabolism of single cells, and discrimination of single nanoparticles in a mixture.[3–6] In the future, the further combination of volume exclusion and electrochemical charge enhancement at the nanopore confinement could benefit to produce a fingerprint current pattern for single-molecule sensing.

1 V. Wang, N. Ermann and U. F. Keyser, *Nano Lett.*, 2019, **19**(8), 5661–5666.
2 R.-J. Yu, S.-M. Lu, S.-W. Xu, Y.-J. Li, Q. Xu, Y.-L. Ying and Y.-T. Long, *Chem. Sci.*, 2019, **10**, 10728–10732.
3 Y.-L. Ying, Y.-X. Hu, R. Gao, R.-J. Yu, Z. Gu, L. P. Lee, and Y.-T. Long, *J. Am. Chem. Soc.*, 2018, **140**(16), 5385–5392.
4 R. Gao, Y.-L. Ying, Y.-X. Hu, Y.-J. Li and Y.-T. Long, *Anal. Chem.*, 2017, **89**(14), 7382–7387.
5 R. Gao, Y. Lin, Y.-L. Ying, X.-Y. Liu, X. Shi, Y.-X. Hu, Y.-T. Long and H. Tian, *Small*, 2017, 13(25), 1700234.
6 R. Gao, Y.-L. Ying, Y.-J. Li, Y.-X. Hu, R.-J. Yu, Y. Lin and Y.-T. Long, *Angew. Chem., Int. Ed.*, 2018, **57**(4), 1011–1015.

Dongfei Chen asked: What if the translocation of the proteins occurs from the inside of the nanopore to the outside (*i.e.*, bulk solution)? Will we expect to see any changes from that?

Shelley Minteer replied: If the P450cam adds to the inside of the glass nanopore, the multiple binding between several P450cam proteins and modified Pdx layer is expected. To avoid multiple binding interference, the P450cam was driven from outside of a glass nanopore into the side lumen. Therefore, each event represents a single P450cam interaction with the Pdx.

Ping Yu opened a general discussion of the paper by Jill Venton: How about the stability of the CNS electrode when we continuously cycle the electrode.

Jill Venton answered: CNSs are very stable in PBS buffer conditions. In Fig. 4F in our article (DOI: 10.1039/d1fd00053e), we continuously tested the electrodes for 4 hours, and we don't see an obvious current drop. They have a very long shelf life as well, as we tested them many weeks after being made and they are still good.

Andrew Ewing said: Why do you not see DA electrochemistry on Nb? That seems interesting.

Jill Venton answered: Yes that is interesting. It is known that gold and platinum are sensitive to dopamine, but in our previous work, we found some metals or alloys are not sensitive to dopamine, including niobium, tantalum, stainless steel, *etc.* This is probably due to the adsorption kinetics of dopamine on the metal surface.

Andrew Ewing said: Can you take something that's normally fast, for example ruthenium hexamine?

Jill Venton replied: I don't exactly remember where this comment came from in the discussion. We use ruthenium hexamine as an outer sphere test compound in our research to test surface insensitive processes. Most of the improvement in electrocatalytic effects we see are with inner sphere compounds that are surface sensitive.

Andrew Ewing remarked: I will bet the carbon layer is peeling off at high potentials for W?

Jill Venton responded: That may be true. However, the tungsten wire itself has electrochemical activity for dopamine and higher background charging capacitance. Thus, we need the wires to be fully covered or the exposed tungsten may interfere with the measurement. We have not checked the wires to see if the carbon is peeling off but it may be seen with long term measurements. They are stable for shorter term measurements if they are made fully coated.

Andrew Ewing added: I find it fascinating that Nb has no effect.

Weilin Xu asked: Did you try any other carbon materials with a porous structure as the electrode materials?

Jill Venton responded: We have been studying carbon nanotube yarn electrodes. These electrodes are porous so they can trap analytes. We have also used simulations to study the electrochemical process at the rough carbon electrode surface.

Ping Yu queried: Does dopamine still adsorb onto CNS? How could you conclude this? Thanks.

Jill Venton replied: Yes, dopamine adsorbs onto the CNS surface. We have demonstrated this in Fig. 3C in our article, where the dopamine current is linear to the scan rate. Edge-plane carbon can adsorb dopamine, and CNSs have rich edge-plane sites.

Peisan E asked: Have you looked at how the neurotransmitters may foul the CNS surface? Is there any functional groups at the CNS surface that may reduce the fouling effect?

Jill Venton replied: We haven't studied the antifouling properties in this work. CNSs are slightly doped by nitrogen, and they have dense nanostructures, which may reduce the fouling effect.

Ashley Ross commented: I was just curious about why enhanced redox cycling for DA isn't observed on the CNS coated electrodes (because of the apparent porosity of the surface)? And also, I was just curious if you can control the surface functionality of the CNS surface during the plasma enhanced CVD method when growing on metal wires?

Jill Venton commented in reply: Redox cycling is not observed because the surface roughness isn't high enough to trap the molecules. CNSs have a surface roughness of a few hundred nanometers, but roughness of over a micron is needed to see the redox cycling. With plasma-enhanced CVD we do not see enhanced surface functionality on the CNSs that we know of. Other plasma treatments in oxygen plasma can enhance oxygen groups, but we don't see that here because of the low oxygen.

Xu Liu remarked: Could you please explain a bit about the potential applications of carbon nanoelectrodes in other fields, in addition to measurements of neurotransmitters?

Jill Venton responded: As other application examples, the CNS nanoelectrodes would be useful as scanning electrochemical microscopy (SECM) tips in electrochemical imaging or for investigating catalytic responses of atom clusters.

Xiangkun Elvis Cao commented: Nice work on using carbon nanospike coated nanoelectrodes for neurotransmitter measurements. Your paper stated that CNS nanoelectrodes are a promising strategy for the mass fabrication of nanoelectrode sensors for neurotransmitters. Would you be able to share more insights over challenges your group encountered or envisioned for the scale-up process for mass fabrication and the potential pathways to solve these challenges?

Jill Venton responded: The most challenging thing we encountered is to uniformly control the size of the nanoelectrodes. Although we mounted a batch of metal wires onto the same conductive substrate, so that the same potential and time is applied to each sample, the size of the etched tips still varied, especially for niobium wires. In future work, we will continue to optimize the methods to etch metal wires. Finally, while our process is a few steps, we have to put the electrodes into several instruments. True mass manufacturing would integrate all these steps into a process line.

Bing-Wei Mao mentioned: It is not very easy to distinguish processes having a slightly different time constant from Nyquist plots. A more sensitive way could be to plot the EIS data in a complex capacitance plane (CCP or Cole–Cole plot) representation that it is easier to judge whether there is one or more kinds of capacitance.

Jill Venton replied: Thank you for the suggestions. The complex capacitance plane would better show the capacity difference. Here we used a more common complex impedance plot, because we also wanted to compare with the impedance plots of other carbon electrodes in the literature, such as carbon fiber microelectrodes and carbon nanotube electrodes.

Ali Reza Kamali commented: This is a very nice work; have you characterised carbon nanospikes in terms crystallinity, crystalline structure and surface area? Also, can you comment on the nature of the bonding between CNTs and the metal substrate, please? Thank you.

Jill Venton communicated in reply: The surface area and crystal structures can be found in previous work.[1,2] CNSs are grown layer by layer onto the metal substrates, so the carbon atoms will fit into the vacancies of the metal and form stable bonds.[3]

1 L. B. Sheridan, D. K. Hensley, N. V Lavrik, S. C. Smith, V. Schwartz, C. Liang, Z. Wu, H. M. Meyer and A. J. Rondinone, *J. Electrochem. Soc.*, 2014, **161**, H558–H563.
2 Q. Cao, D. K. Hensley, N. V. Lavrik and B. J. Venton, *Carbon*, 2019, **155**, 250–257.
3 H. L. Zhuang, G. P. Zheng and A. K. Soh, *Comput. Mater. Sci.*, 2008, **43**, 823–828.

Yi-Lun Ying addressed Jill Venton, Paul W. Bohn and Shelley Minteer: The topic of this session is "Advanced nanoelectrochemistry implementation: from concept to application". Could you please comment on the future application of your new techniques? How to bring the nanoelectrochemistry into the reality? What's the major gap?

Jill Venton responded: The future application of the electrodes will be the mass-manufacturing of carbon nanoelectrode sensors. Compared with other fabrication methods of carbon nanoelectrodes, CNS nanoelectrodes have a larger surface area to volume ratio, so they show larger currents for sensing. Both the etching of metal wires, and the growth of CNSs, are done in batch. The major gap would be controlling the quality of the nanoelectrodes in mass fabrication. For *in vivo* applications, these will enable localization in the synapse and better synaptic measurements.

Paul W. Bohn added: Thank you for this forward-looking question. I believe that work in our laboratory and many others have made significant progress in fleshing out the concepts that can be imported from nanoscience to enable advanced applications, such as high sensitivity chemical sensors. The single most important roadblock to further progress, in my opinion, lies in developing the fabrication and chemical manipulation strategies that can produce architectures with nanoscale functional properties over large, macroscopic distances. While there are tools capable of producing structures which are many square μm's, these need to be extended to produce structures which are square mm–cm's in area. Perhaps there are opportunities to develop reel-to-reel processes for direct printing of such structures. This would do more than anything else I can think of to drive the implementation of the structures in real-world applications.

Shelley Minteer said: In the future, our technique could be used to study intermolecular interactions controlling protein–protein complexation in supercomplexes, substrate channelling, molecular motors, and disease progression caused by protein–protein interactions. However, for optimal quantitative evaluation of binding constants for a single protein–protein interaction, the gap is in making a nanopore that only contains one immobilized protein (*i.e.* one binding site).

Ben Slater opened a general discussion of the paper by Jin He: I am interested in the centrifugation method for the nanopipettes (assuming it is to remove bubbles), are there limitations to this method, *e.g.*, maximum speed before it breaks? Can the centrifugation remove surface modifications? Do you think this would also successfully remove bubbles in solid state nanopores such as SiN?

Jin He replied: The centrifugation method was used previously to enrich the nonentities at the tip of the nanopipette.[1] For the experiments in our work here, the nanopipette was centrifuged at 2000 rpm for 2 min. Higher centrifugation speed and longer centrifugation time either damage or cause clogging issues during delivery. If you can wait, a lot of nanoentities we tested can also diffuse to the nanopipette tip overnight without using the centrifugation method.

To effectively remove the air bubbles at the nanopipette tip, we used the heating method. The nanopipettes were first filled with solution and then placed in the 120 °C oven for 30 s. With heating, we do not have air bubble issues. Both centrifugation and heating may not work with the SiN nanopore because there is no long fluidic channel for the SiN thin membrane based nanopore. We worked with SiN nanopores before; you may first fill the nanopore with pure ethanol, then mix it gradually by increasing the concentration of water.

1 R.-C. Qian, J. Lv and Y.-T. Long, *Anal. Chem.*, 2018, **90**(22), 13744–13750.

Samuel Confederat asked: In terms of the fabrication of these multi-functional probes, how do you control the pyrolytic carbon deposition on one barrel in such a way that it does not constrain the pore size of the other barrel (the open barrel used for translocations)? Furthermore, is there a large variation between the fabricated probes with regards to the carbon-deposited barrel?

Jin He answered: Thanks for the questions! The details of the fabrication process can be found in our previous publication.[1] In brief, we carefully controlled the pressure of the carbon source, butane gas, as well as the protection gas, N_2. The deposition time is also very short, only about 20 s. Based on the ionic current measurements, the pore size is still reduced after the carbon deposition. For nanopipettes with smaller pores, below 20 nm, the size reduction is generally small and no more than a few nm. The size reduction is sometimes beneficial for protein measurements. It should be mentioned that the nanopore barrel is still easy to fill after the carbon deposition, suggesting the carbon deposition at the inside surface of the nanopore barrel is very limited. Otherwise, the nanopipette tip inner surface becomes more hydrophobic, preventing the filling of aqueous solution. Based on the steady-state CV measurements of the fabricated CNE, there are relatively large variations of the effective surface area, which can be more than several times. However, in the low-resolution SEM images, we did not find such large variations in the CNE geometries. We believe a thin layer of carbon may deposit to the outer surface of the nanopipette tip, which contributes to the larger variation in the CV derived effective surface area. For each experiment, the students typically prepared 5–10 nanopipettes, which are enough to obtain several nanopipettes with similar geometries (based on ionic current and electrochemical current). We are fine with this yield. The potential signals are also less sensitive to the surface area variation of the CNE. However, the large variation still limits us in conducting high precision quantitative measurements.

1 N. Panday, G. Qian, X. Wang, S. Chang, P. Pandey and J. He, *ACS Nano*, 2016, **10**(12), 11237–11248.

Dominik Duleba said: How do you set up the centrifuging of the pipettes considering how extremely fragile they are? Secondly, according to your manuscript, Navier–Stokes was not included in the FEM simulations, do you think the electroosmotic flow has a negligible effect on the particle delivery?

Jin He responded: Please also see the reply to Ben Slater's opening question above. First, the quartz nanopipette is actually not that fragile, as shown in our recent papers.[1,2] Second, we still need to control the centrifuging speed and time to avoid tip damage. The nanopipette is placed in a typical 5 ml centrifuge tube. The backend of the nanopipette is secured by a PDMS slab, which cannot move in the centrifuge tube. To simplify the calculations, we did not consider the fluid velocity and ignore the electroosmotic flow. Indeed, the electroosmotic flow may play a non-negligible effect for particle delivery. It depends on the applied bias, size and surface charges of nanopore and nanoparticles. Also, large local ion and nanoparticle concentration change becomes more common at the nanopipette tip in bias regulated nanopipette delivery. The depletion of ion or the accumulation of charged entities may also induce stronger electroosmotic flows. In our experiments, at the higher applied nanopore bias, we often observed both upward and downward current spikes, suggesting the presence of electroosmotic flow. We will investigate these phenomena in follow-up work.

1 J. Guo, A. S. Rubfiaro, Y. Lai, J. Moscoso, F. Chen, Y. Liu, X. Wang and J. He, *Analyst*, 2020, **145**(14), 4852–4859.

2 L. Yu, Z. Wang, H. Chen, J. Guo, M. Zhang, Y. Liu, J. He and S. Chang, *ACS Appl. Nano Mater.*, 2020, **3**(4), 3410–3416.

Jill Venton queried: What do you plan to use the carbon nanoelectrode for in the electrode? Will you detect the electrochemical payload you deliver or another change in something redox active in the cell?

Jin He replied: Thanks for the question! In general, we are interested in using the multifunctional nanopipette for single-cell analysis. We are still learning about applications to cellular biology and conducting literature surveys to better decide the direction. One initial direction is to use the carbon nanoelectrode to conduct intracellular electrophysiology, such as the recording of action potentials of excitable cells. Also, we are interested in using the carbon nanoelectrode to manipulate the delivered particles. Your suggestion is certainly another interesting direction.

Martin A. Edwards said: Thanks for the interesting talk and paper. In the work, you repeatedly mentioned 'single-entity delivery'; however, the data you present seem to show 'countable delivery' – that is, while you identify each single entity, you don't deliver just one and then stop. Have you considered triggering off the resistive pulses and then reversing the driving force? We have demonstrated this to be a feasible approach to control nanoparticles and nanorods exiting/entering a nanopipette using either mechanical pressure[1] or electrokinetic phenomena[2,3] and are even able to manipulate a nanoparticle so it is passed back and forth many times. If you are able to implement this within your system, it would offer the opportunity to deliver either one single entity or a defined number of entities.

I would also like to draw your attention to other similar work using nanopipettes to deliver single entities to interfaces.[4–6] The results of these papers broadly support the interesting results you presented in your work.

1 S. R. German, T. S. Hurd, H. S. White and T. L. Mega, *ACS Nano*, 2015, **9**(7), 7186–7194.
2 M. A. Edwards, S. R. German, J. E. Dick, A. J. Bard and H. S. White, *ACS Nano*, 2015, **9**(12), 12274–12282.
3 Y. Zhang, M. A. Edwards, S. R. German and H. S. White, *J. Phys. Chem. C*, 2016, **120**, 20781–20788.
4 X.-W. Zhang, A. Hatamie and A. G. Ewing, *J. Am. Chem. Soc.*, 2020, **142**(9), 4093–4097.
5 K. McKelvey, M. A. Edwards and H. S. White, *J. Phys. Chem. Lett.*, 2016, 7, 3920–3924.
6 Y. Wang, H. Cai and M. V. Mirkin, *ChemElectroChem*, **2**(3), 343–347.

Jin He replied: Thanks for the great suggestion! Indeed, it is very interesting to conduct repeat measurements of the same nanoparticle inside the cell. In this way, we can better and more accurately probe the changes of the nanoparticle and also the surrounding intracellular environment. Also, thanks for the references. We will look at them carefully.

Hang Ren said: Is there any cross-talk between ionic current and open circuit potential (OCP)? Since the current is passed through the quasi-counter/reference Ag/AgCl electrode, its potential will change. Is this change what is being observed in the OCP measurement?

Also, what is the effect of the cell membrane junction for the ionic current and potential measurement? Since the ionic current is measured to be significant, does that mean the cell membrane is leaking? The current has to pass between the Ag/AgCl inside the pipette to the outside of the reference electrode. Does the current spike you saw come from the opening and closing of the ion channel? Can you put the reference electrode in the second barrel of the pipette so that there is no effect from the cell membrane potential?

Jin He answered: Thanks for the great questions. We have conducted a lot of control experiments. The change of Ag/AgCl electrode only induces the overall drift of the potential baseline. The fast transient potential changes are induced by the particle movements. These fast changes are also clearer in the first derivative of the potential time traces. The nanopore resistance we used for this work is typically in the giga or 10 s of giga Ohms range. Compared with the high resistance of the nanopore, the cell membrane is leaky. Therefore, the current of the equivalent circuit is dominated by the pore resistance. As shown in our current traces (Fig. 3b and 4c in our article, DOI: 10.1039/d1fd00057h), the current change is very small during the nanopipette tip cellular penetration process. For the smaller nanopipette in Fig. 4c in our article, the current change is less than 10 pA. In the control experiments of nanopipettes without loading nanoparticles, we did not observe the current spikes. The chance to observe current changes from individual membrane ion channels is expected to be low. We have measured the potential changes in the self-beating cardiac cell using these nanopipettes. The signals are also different. We will conduct the suggested experiment using a double-barrel nanopipette (without filling the second barrel with carbon).

Shuai Chang commented: You have mentioned that each ionic current spike corresponds to the delivery of a single NP, how did you quantify that? Are there any clustered NPs? And does the size of NPs matter in the quantification of the ionic current signals?

Jin He responded: The spike is generally determined by the shape, magnitude and time duration of the transient current spikes (or potential spikes). For larger entities such as GNPs (whose size is comparable to the nanopore size), the magnitude of the current spikes is usually quite uniform (see Fig. 5 in our article). The clustered NPs appeared as closely-spaced current spikes. So we know they are from individual NPs. For the same size nanopore, the size change of the NP will obviously affect the ionic current signals. The smaller NP moves faster and induces smaller current changes.

Jill Venton opened the discussion of the paper by Ashley Ross: Why do you think there are differences between adenine and guanine purines? Is it shape with the roughness – because adenine is bigger? Chemistry?

Ashley Ross responded: That's a great question. I do think it has to do with the shape in a sense. These differences could be due to the steric effect from the ribose group. The steric effect is thought to be different for guanine-based purines compared to adenine-based purines.

The literature suggests that where the intramolecular hydrogen-bonding between the NH_2 group of the guanine moiety and the OH group of the ribose at the 5′-position occurs, changes the shape of the nucleoside. Possibly these subtle differences in shape have an impact on how they approach and interact at the electrode surface.

Luke Gundry asked: My question is on the COMSOL and simulation, is it fair to assume that the 2E electron transfer is fully reversible with such a fast scan rate and how does the simulation compare to experimental results?

Ashley Ross answered: The simulation results are a bit different than what we observe experimentally. Experimentally, at such fast scan rates, the 2E transfer is not fully reversible; however, COMSOL simulations don't capture this in the model.

Andrew Ewing asked: Did you look to see if the type of C–O functional groups change on the carbon surface with the Ar etching? (Can they change the adsorption?)

Ashley Ross answered: Yes, we did further evaluation of the XPS data to analyze specific oxide functionality and did not observe any differences between the bare carbon fiber and the argon plasma-treated carbon fiber.

Andrew Ewing commented: It wouldn't take much, 1/2% might be enough to affect adsorption – molecular models may help to figure this out.

Martin A. Edwards remarked: It is possible and relatively straightforward to use COMSOL Multiphysics to model the adsorption/desorption and surface electrochemical processes. Whether this would capture any remaining behaviour that is not captured by your current simulations, or is a worthwhile endeavor, is something I don't know. However, the mechanics of setting up the simulations is not too onerous.

Ashley Ross responded: Thank you Martin!

Popular Pandey asked: Why does Ar-plasma treatment of the electrode impact purine detection more than dopamine? Does the carbon fiber disc microelectrode size also affect detection sensitivity? Have you applied this system to intracellular neurochemical recordings?

Ashley Ross answered: We have not used these Ar-plasma treated electrodes yet for neurochemical recording *in vivo*. In terms of comparing purines *vs.* dopamine, we only really see significantly higher improvement for the adenine-based purines. The guanine-based purines increase similarly to dopamine after Ar-plasma treatment. We speculate that it's probably a complicated combination of size, analyte structure and chemistry. But it is really hard to know for sure without more in-depth analysis of these specific analyte–electrode interactions.

Swathi Naidu Vakamulla Raghu said: After the plasma treatment, the surfaces investigated had varied functional groups. So do you think the type of functional group as a result of (if any) substitution that was measured, results in a difference in signal intensity for A and G purine. And on the purines, is one functional group more favorable for detection?

Ashley Ross responded: Absolutely, we actually published a paper a few years ago investigating the extent to which small changes in functional group type and placement on the purine ring impacts their interaction at the carbon surface. We found that amine-functionalized purines, especially disubstituted purines, interact more favorably than carbonyl-functionalized purines and even the placement of the functional group on the purine ring has an impact. So yes, this does influence the interaction at the carbon surface which ultimately leads to changes in detection.

Paul W. Bohn commented: There are likely a very large number of types/geometries/compositions of local sites on the etched electrode surface. Do you have any sense of how the reactivity changes among these – in analogy to SERS do you think there are "hot spots" of unusually high activity that dominate the behaviour for the remainder of the electrode?

Ashley Ross answered: You are correct, that these surfaces are quite heterogeneous. It would be interesting to do SECCM at these electrodes to truly look at these "hot spots" of reactivity. I don't think we would see the same level of reactivity across the surface at all.

Paul W. Bohn addressed Jin He and Ashley Ross: Both Dr He and Dr Ross present complex nanostructures obtained by challenging fabrication protocols. How tightly can you control the structural properties of your materials/devices. What is the dispersion in properties from one run to the next – either in tip shape (Jin He) or surface topography (Ashley Ross).

Jin He replied: Please also see the response to Samuel Confederat above. We have tried our best to control the structural variation in the fabrication process. We have had some problems (charging, tip structure is not stable under higher energy electron beam) to obtain the high-resolution images of the nanopipette tip by transmission electron microscopy. Based on the steady-state cyclic voltammetry measurements, we have relatively large variations in the effective surface area of the carbon nanoelectrode. Obviously, the dispersion in the nanopipette tip prevents us from being more quantitative in our data analysis. So far, most of our studies are still qualitative. For these qualitative studies, we can get reproducible results from one run to another with high yield. However, we cannot use the nanopipette to carry out in-depth studies like other single-molecule biophysics methods, such as protein nanopore, single-molecule force measurements or single-molecule FRET. In the long-term, nanofabricated on-chip nanopore/nanoelectrode by mass-production should help to solve this issue.

Ashley Ross added: The Ar-plasma procedure can be controlled by controlling the flow rate of gas, power, and time of treatment and we have shown that manipulating these parameters produces differences in the average change in surface roughness. However, I think moving to even more controlled methods of controlling the nanostructure of the surface would be beneficial in the future for studying precise analyte–electrode interactions.

Jill Venton asked Jin He and Ashley Ross: Can the speakers comment on what applications these new techniques will enable and where they think their fields will be going for biological measurements.

Jin He responded: Due to the limitation in the geometry and size control of glass/quartz nanopipette tip, the accuracy of the nanopipette nanopore based measurements cannot match with other solid state nanopores, such as ion or electron beam drilled silicon nitrite nanopore. Not to mention the atomic level resolution achieved by protein nanopores. However, the nanopipette is extremely versatile, easy and cheap to make and use, and can be conveniently integrated with other techniques. We try to seek the applications of nanopipette based single-entity techniques in cellular biology, especially single-cell analysis. There are already many works and applications of glass micropipettes/microelectrodes in single-cell studies. The nanopipette can extend the earlier work. The nanoscale size and high flexibility of the nanopipette tip induces minimized cellular damage. Therefore, the nanopipette can conduct intracellular measurements for a long time. Also, we are interested in integrating nanoelectrochemistry methods with optical methods for biological measurements.

Ashley Ross replied: I foresee these Ar-plasma treated surfaces being used to: (1) more sensitively detect purines in tissue, but also (2) enable improved fundamental studies on how changes in the nanostructure of the carbon-fiber surface impacts the interaction and detection of various neurochemical analytes. I do foresee the field though moving to more precise methods of controlling the nanostructure of the surface. I think moving beyond carbon-fiber, and to more controllable surfaces will significantly advance not only fundamental electrochemical studies but also push the field to be able to detect traditionally more difficult electroactive biomolecules.

Li Xiao opened a general discussion of the paper by Patrick Unwin: I usually study macro scale catalysts and have not done single particle reaction research before. I want to know, if we gradually let two single particles with different composition come close together, can it help us better understand synergistic effects? For example, PtRu for the hydrogen oxidation reaction, and NiFe for the oxygen evolution reaction, *etc.*

Patrick Unwin responded: There are considerable prospects for understanding and rationalising electrocatalytic processes by studying a spectrum of entity populations, from single entities (particles), to pairs, small aggregates and upwards, including interactions between different entities and groups of particles. A technique such as scanning electrochemical cell microscopy (SECCM) should allow a large number of measurements within these populations, and the structure and arrangement of entities can be determined by complementary microscopy techniques, so that detailed structure–activity datasets are constructed. SECCM is also amenable to multiscale measurements, by changing the tip size. The examples you suggest would be very interesting to look at with this type of approach.

Conflicts of interest

Patrick Unwin is co-inventor of granted patent PCT/GB2011/051518 "Pipets containing Electrolyte and Electrodes", which describes dual-channel SECCM.

Faraday Discussions

PAPER

Concluding remarks: next generation nanoelectrochemistry – next generation nanoelectrochemists

Patrick Unwin *

Received 28th January 2022, Accepted 7th February 2022
DOI: 10.1039/d2fd00020b

The aim of this paper is to describe the scientific journey taken to arrive at present-day nanoelectrochemistry and consider how the area might develop in the future, particularly in light of papers presented at this *Faraday Discussion*. By adopting a generational approach, this brief contribution traces the story of the nanoelectrochemistry family within the broader electrochemistry field, with a focus on scientific capability and themes that were important to each generation. I shall consider research questions and the impact of technology that was developed or available in each period. Nanoelectrochemistry is still somewhat niche, but is attracting increasing numbers of researchers. It is set to become a major part of electrochemistry and interfacial science. It is studied by people with a fairly unique skillset, and I shall speculate on the skills and expertise that will be needed by nanoelectrochemists to address the challenges and opportunities that lie ahead. I conclude by asking: who will be the nanoelectrochemists of the future and what will they do?

Historical origins of nanoelectrochemistry

We start our journey in a time when there was huge interest in batteries, and this is not the present day! Volta's development of the pile, in 1799, reported to the Royal Society in 1800, was called an 'alarm bell for experimenters' by Humphry Davy[1] and led to a race by scientists across Europe to construct bigger and more powerful batteries for use in 'electrochemical incandescence', the isolation of elements and chemical analysis. In 1813, John George Children reported 'the greatest galvanic battery that has ever been constructed',[2] in the purpose-built laboratory in the grounds of his home, Ferox Hall, Tonbridge, Kent, England. The battery consisted of '20 pairs of copper and zinc plates, [and] each plate was 6 feet [*ca.* 1.83 m] in length, 2 feet 8 inches [*ca.* 0.81 m] in breadth'. It was filled with 4000 l of acid! On the first weekend of July of that year, a group of 38 eminent chemists and natural philosophers assembled at Ferox Hall to witness the great battery and perform a series of experiments. The resulting article by Children[2] was

Department of Chemistry, University of Warwick, Coventry, CV4 7AL, UK. E-mail: p.r.unwin@warwick.ac.uk

published under the heading: 'Scientific intelligence and notices of subjects connected with science'; but was the construction of ever larger batteries the most intelligent trend for electrochemical analysis at this time? William Hyde Wollaston (who was among the chemists at Ferox Hall and made suggestions to Children about sample size) did not think so. Davy's biographer, John Parris, recollected:[3,4] 'shortly after he [Wollaston] had inspected the grand galvanic battery constructed by Mr Children… he accidentally met a brother chemist in the street, and seizing his button, (his constant habit when speaking on any subject of interest) he led him into a secluded corner; when taking from his waistcoat pocket a tailor's thimble, which contained a galvanic arrangement, and pouring into it the contents of a small vial, he instantly heated the platinum wire to white heat'. By miniaturising the experiment, and passing current through a Pt wire of just *ca.* 4 μm diameter, Wollaston was able to perform the same experiments using a pocket device for which others were using giant batteries.

Wollaston recognised the importance of *current density* in electrochemistry as early as 1801, reporting:[5]

'It has long been thought necessary to employ powerful machines and large Leyden jars, for the decomposition of water; but, when I considered that the decomposition must depend on duly proportioning the strength of the charge of electricity to the quantity of water, and that the quantity exposed to its action at the surface of communication [electrode] *depends on the extent of that surface*, I hoped that, by reducing the surface of communication, the decomposition of water might be effected by smaller machines, than have hitherto been used for that purpose; and, *in this hope, I have not been disappointed.*' (italics my emphasis).

To decrease the 'surface of communication', Wollaston set about constructing increasingly smaller electrodes and, in fact, fabricated microelectrodes (or, equivalently, ultramicroelectrodes)[6] in the same way as they were introduced separately by Fleischmann and Wightman from the late 1970s onwards (more on microelectrodes later). He wrote:[5]

'Having procured a small wire of fine gold, and given it as fine a point as I could, I inserted it into a capillary glass tube; and, after heating the tube, so as to make it adhere to the point and cover it in every part, I gradually ground it down, till, with a pocket lens, I could discern that the point of the gold was exposed.'

We might even reasonably speculate that Wollaston made the first nanoscale electrode:[5]

'In order to try how far the strength of the electric spark might be reduced by proportional diminution of the extremity of the wire, I passed a solution of gold in *aqua regia* through a capillary tube, and, by heating the tube, expelled the acid. There remained a thin film of gold, lining the inner surface of the tube, which, by melting the tube, was converted into a very fine thread of gold, through the substance of the glass.'

Although Faraday referred to the significance of Wollaston's work in his electrochemistry research in the 1830s,[7] it seems to have largely been forgotten. What is true is that Wollaston's work was distinct and well ahead of its time. Incredibly, some of the steps in the process described above resemble those used in the present-day fabrication of pipette-based gold nanopore devices, which are powerful for nanoscale analysis.[8] Nanoelectrochemistry has long relied on the crafting of small scale electrodes and devices.

Our journey recommences over a century later, after the Second World War, which marked a turning point for science and society. I shall describe developments generationally, not in terms of the birth years of scientists, but rather the themes, influences and trends in each generation.

Boomer electrochemistry (1946–64)

The 1947 *Discussion of the Faraday Society* on *Electrode Processes* provides a good indication of the key topics at the start of this period. Almost 200 people gathered in Manchester, England[9] to discuss papers under the headings general and theoretical, the deposition of hydrogen and metals, and anodic and other electrodic processes. It was anticipated that advances in "physical chemistry in the next decade will lie in the application of the science of the techniques of electronic engineering, developed particularly during the war, and the field of electrode processes may be one which will yield readily to attack by the new weapons".[9] Many of the papers from the 1947 *Discussion* are grounded in firm mathematics and theory, and some of the comments are 'robust' (even on paper!). Among those contributing in person or by communication were scientists who have made major contributions to the field and on whose work we continue to rely today, including Randles, Butler, Hickling, Mott, Levich, Bockris, Fleischmann, Parsons, Heyrovsky and Lingane. Perusing the volume, one is struck by the fact that a number of key problems to be solved in electrochemistry are longstanding. In the present *Discussion* the need to better understand ion and concentration distributions near electrochemical interfaces arose in several papers, and this was the case 75 years ago; for example, a communication from W. F. Berg at the 1947 *Discussion* proposed that an interferometric method, used to study the ion distribution around a growing crystal, could be applied to "electrode chemistry, since there is much speculation and little detailed knowledge of the concentration distribution of ions".[10]

This period also saw the birth of the Coulter Counter: there is a very nice description of this device, and many other historical developments in a 'Perspective and Prospectus on Single-Entity Electrochemistry' written by Lane Baker in 2018,[11] which is highly recommended. This period was naturally dominated by home-built instrumentation and electronics. While mainstream electrochemistry was to become (over)reliant on commercial instrumentation in later generations, the nanoelectrochemistry community is presently heavily engaged in instrumentation development, particularly in pushing the capability of low noise and high bandwidth current amplifiers for single entity analysis, as evident from the paper from Zhong *et al.* (DOI: 10.1039/D1FD00055A) in this *Discussion* on high throughput single entity analysis. Such developments will continue to be important in our field moving forward. In addition, the Boomer period was concerned about solutions and salt purity, particularly with a switch in focus away from polarography to the electrochemistry of solid electrodes, as described in Adams' classic text.[12] There was considerable interest in understanding the electrical double layer, electrode kinetics and mass transport, including increasing attention towards hydrodynamic electrodes, all of which are timeless themes in electrochemistry. Consequently, the work by this generation has a long influence.

Generation X (1965–80)

This period saw a significant diversification in the scope and application of electrochemistry. At the core, there was a focus on kinetics and mechanisms, especially electron transfer coupled to homogeneous processes (*e.g.* the Savéant school, and others),[13] which extended into subsequent generations. In this *Discussion*, Gundry *et al.* (DOI: 10.1039/D1FD00050K) showed that such processes are a useful testbed for the deep neural network classification of electrochemical processes. A further major advance in this period was the seminal work from Allen Hill's group on the electrochemistry of redox proteins, which ultimately led to the development of sensors in the 1980s that revolutionised blood glucose analysis and provided an inspiration for much other work on electrochemical sensors in subsequent generations.[14]

Chemically-modified electrodes,[15,16] particularly polymer-modified electrodes,[17] came to the fore in the 1970s and one gets a good sense of the status of this burgeoning field in the next generation from *Faraday Discussion 88* (1988) on *Charge Transfer in Polymeric Systems*. In the papers from Walcarius and co-workers on polyaniline nanowire arrays (DOI: 10.1039/D1FD00034A) and Bohn *et al.* on nanopore-based sensors (DOI: 10.1039/D1FD00048A), we see that it is now possible to control and characterise the architecture of complex electrode structures at the nanoscale. This ability to design functional electrodes with exquisite control will lead to a better understanding of mass transport at small scales, open up new sensing paradigms and lead to improved material performance.

This era marked the birth of neuro-electroanalysis through the pioneering work of Adams,[18] together with Wightman[19] and the scientists who came through these groups and those of their descendants, several of whom have contributed to this *Discussion*. Carbon materials of various types have proven to be the electrodes of choice for the electrochemical detection of neurotransmitters, and this *Discussion* highlighted the improved detection capabilities of carbon electrodes when the nanoscale architecture is engineered, controlled and characterised: (i) nanospike electrodes that could also be batch fabricated with good reproducibility, which is an important development, as described by Venton and co-workers (DOI: 10.1039/D1FD00053E) and (ii) the use of different Ar plasma treatments of carbon-fiber electrodes to create nanostructured electrodes without affecting the surface chemistry, by Ross *et al.* (DOI: 10.1039/D1FD00049G). These fundamental advances in sensor design are important steps towards the rational design of nanostructured electrodes for *in vivo* and *in vitro* measurements.

The patch clamp technique was invented in this period by Neher and Sakmann, making use of micropipettes to measure ion currents in individual living cells, cell membrane patches and tissue sections.[20,21] As we have seen in this *Discussion*, nanopipettes have become among the most important tools for various measurements in nanoelectrochemistry. Artificial bilayers are also useful model systems for understanding various physicochemical processes related to cell membranes and in this *Discussion*, Hirano-Iwata *et al.* (DOI: 10.1039/D1FD00045D) described the application of a lateral voltage to such devices. Bohn's paper (DOI: 10.1039/D1FD00048A) also took inspiration from biology in the design of a pH-sensitive nanopore membrane with potential-control of wetting/dewetting.

Spectroelectrochemistry assumed prominence, especially through leading developments in several areas by the Southampton Electrochemistry Group, where Fleischmann and colleagues first observed surface enhanced Raman scattering (SERS).[22] In this *Discussion*, we saw the importance of *in situ* SERS for revealing molecular details at electrode/polyelectrolyte interfaces in the paper by Xiao *et al.* (DOI: 10.1039/D1FD00051A).

In this era, there were significant efforts to improve the signal to background (noise) and concentration detection limits in electrochemistry, which are key themes for nanoelectrochemistry today. Although pulse polarography was pioneered by Barker in the Boomer period,[23] Osteryoung's work on pulse voltammetry in the 1970s[24] rekindled interest in this field. Alternative approaches to improve the detection of electrochemical signals include the work of Bruckenstein and Miller on hydrodynamic modulation voltammetry with lock-in detection.[25] There is certainly scope for the greater use of such concepts in nanoelectrochemistry, for example, in self-referencing hopping-mode scanning electrochemical probe microscopy (SEPM) methods, where the signal at the probe is measured in bulk and near the surface at each pixel (position) in a scan.[26]

Approaching the end of this period, there was the rediscovery of microelectrodes independently by Fleischmann and Wightman,[27,28] and they were perhaps unaware that they were fabricating small scale electrodes in a essentially identical way to Wollaston almost two centuries earlier (see above). The impact of voltammetric microelectrodes in electrochemistry[29] cannot be underestimated; they have revolutionised the spatiotemporal scale of electrochemical measurements, opened up many possibilities for making electrochemical measurements in a diversity of unusual environments, expanded quantitative knowledge of diffusion and mass transport in electrochemical systems, and enabled subsequent developments in SEPM, providing the foundation for much of what we do in nanoelectrochemistry today.

Bard and Faulkner's *Electrochemical Methods*,[30] published in 1980, brings this period to a close. This textbook signalled a shift in emphasis in electrochemistry towards techniques and methodology, and brought the ways in which we study electrochemical processes to the front and centre. When reviewing the developments in methodology in this period, one can see why this book was so timely and has had a significant impact on generations of electrochemists. The updated 3rd edition of this classic text, with Henry White as a new coathor, is imminent and is eagerly anticipated.

Millennial (1981–95)

We live in an age of imaging science, surrounded by impressive atomic-scale images of materials, surfaces and interfaces. The invention of scanned probe microscopy (SPM), originating with scanning tunnelling microscopy (STM) and atomic force microscopy (AFM), was a watershed in surface and interfacial science and heralded new ways to visualise electrochemical interfaces *in situ* with unprecedented resolution. These methods continue to be hugely important in understanding dynamic electrode surfaces at the nanoscale, as evident in the paper of Mao on the visualisation of lithium electrodeposition under battery electrode conditions using *in situ* AFM in a glovebox (DOI: 10.1039/D1FD00051A).

Following from the previous generation, there was an explosion of interest in microelectrodes, micropipettes and microelectrode arrays, leading to the use of microelectrodes for probing various interfaces and in scanning electrochemical microscopy (SECM). The spatial resolution was at the several micron scale. Hitherto unprecedented electrochemical detection limits were reported in beautiful studies of single cell exocytosis from Wightman's group[31] and the inspirational work of Bard and Fan on nanoelectrochemical cells based on SECM.[32] As electrochemical devices became smaller, the importance of being able to characterise hand-crafted nanoscale electrodes became paramount, although methods were lacking at that time.

This was the era of high-speed electrochemistry, with record-breaking scan rates employed in fast scan cyclic voltammetry. The digital revolution impacted electrochemistry with a transition from analog (chart recorders, oscilloscopes) to PC-based potentiostats and data acquisition systems. These developments, enabling in some ways, have not been without their drawbacks. Potentiostats available commercially were increasingly regarded as black boxes in some quarters and in subsequent generations, leading to a loss of knowledge and knowhow in the community. Likewise, while research on electrochemical sensors developed in this period, as with many large fields, it became something of a bandwagon, where sensor performance and metrics eroded fundamental science. More positively, digital simulation (home-coding) enabled increasingly complex and realistic electrochemical models to be developed. Efforts in this area were greatly aided by excellent texts from Feldberg[33] and Britz.[34] Code-development tended to be confined to individual labs and there are not many examples of code sharing at this time. The increasing use of COMSOL Multiphysics (especially) and other packages in subsequent generations has given many more groups access to numerical methods for analysing nanoelectrochemistry data (mass transport, kinetics) and also for the design of experiments (however, see also Generation Alpha below).

Generation Z (1996–2010)

Nanoelectrochemistry came into its own in this period, with various methods developed not only for the fabrication of nanoelectrodes and nanoscale devices, but also for their robust characterisation, visualisation and modelling. We see these key trends throughout the papers presented at this *Discussion*. Nanolithography strategies were used for the batch fabrication of multifunctional electrochemical imaging probes and nanogap cells for detecting and analysing single molecules. Where nanopores or nanopipettes were still crafted, they could be characterised with a high degree of confidence, especially using electron microscopy. Indeed, scanning electron microscopy, transmission electron microscopy, and variants thereof, arguably became as important as the electrochemical measurement. This was particularly true for nanomaterials electrochemistry, with the Brust and Schiffrin method[35] for the synthesis of gold nanoparticles proving seminal and leading to a nanomaterials electrochemistry revolution, driven in the present generation by the search for energy storage and electrocatalysis materials.

The rise of the nanomaterials field also brought about an identity crisis for electrochemical studies of such materials: how should the electrochemistry of nanomaterials be studied most informatively and what types of studies constitute nanoelectrochemistry? As an illustration, studies published at about the same time on carbon nanotubes could consider, on the one hand, complex samples of commercial materials, with various contaminants, deposited on a carbon support electrode (many examples of such "carbon-on-carbon" electrodes); on the other hand, considerable effort was put into fabricating and characterising devices in which the electrode was only an individual single walled carbon nanotube on an otherwise inert support.[36] In the opening to the Introductory Lecture, Gooding *et al.* (DOI: 10.1039/D1FD00088H) defined nanoelectrochemistry as "where electrochemical properties and/or electrode materials are investigated in the nanoscale regime" aided by measurement tools that can address nanoscale features. Thus, the latter measurements of a carbon nanotube, an example of *single entity electrochemistry*, conform to an accurate descriptor of nanoelectrochemistry. Macroscopic electrochemical measurements of a complex and under-characterised sample of nanomaterial spread on an electrode, by contrast, do not fall within the definition of nanoelectrochemistry.

Generation Alpha (2011–)

This is the generation of single entity electrochemistry, which was defined at the 2016 *Faraday Discussion* (vol. 193) as 'a new way of viewing electrochemical processes at the nanoscale, and provide a bottom-up approach for understanding electrochemical processes in complex systems'. The field of single entity electrochemistry, or the 'electrochemistry of things',[11] connects disparate problems that have common underpinning challenges, *e.g.* in measurement, instrumentation, data acquisition, analysis, modelling and so on. Single entity electrochemistry covers the study of: (i) individual nanoparticles, nanotubes and nanowires and (ii) nanopores and nanofluidics, which can be treated with a common language. But single entity electrochemistry also includes the study of complex surfaces and reactions at the nanoscale, where the idea is to break down surfaces into a set of more elementary structural or other features and try to understand the electrochemistry of the features. In the realm of molecular electroanalysis, single entity electrochemistry investigations vary from the detection of single molecules to single cells.

The new era of single entity electrochemistry, and nanoelectrochemistry more broadly, is connected strongly to imaging and visualization. Taking single particle electrochemistry as an example, studies can now be made with a range of different, but complementary, techniques, which provides a much more holistic view of structure–activity. SEPM[37] and optical imaging techniques[38] can be used to target individual nanoparticles, or groups of nanoparticles within an array or ensemble, and assess their activity and other properties. In this *Discussion*, Lemay and co-workers reported the innovative use of a CMOS-based nanocapacitor array device for local measurements of electrolyte impedance (DOI: 10.1039/D1FD00044F). The detection of the sedimentation of individual oil droplets in a water phase in real time, with sub-micron spatial resolution, illustrates the considerable potential of this approach. While nanoparticle impact studies do not usually involve imaging methods, if they are run for long times with high rate data acquisition, the resulting large datasets present similar challenges to those of imaging experiments, with regard to the extraction of signals, and data handling, archiving, analysis and presentation.

The use of optical microscopy methods and SEPM, and combinations thereof, for nanoscale electrochemical measurements is developing apace and there will be significant advances in understanding from studies where both optical (including spectroscopy) and electrochemical signatures of single entities are tracked in real time. Within the SEPM field, there are interesting hybrid techniques with dual or multifunctional probes: AFM-SECM,[39,40] SICM (scanning ion conductance microscopy)-SECM,[41] SECCM (scanning electrochemical cell microscopy)-SECM,[42] AFM-SICM[43] (FluidFM), *etc.* It is also possible to use these probes to exert considerable control over nanoscale systems, for example trapping, sampling and analysing single entities. These techniques, and others, are opening up exciting prospects for *operando* and correlative imaging at the nanoscale, which I shall address further below. As a consequence, huge datasets are being generated from multiple techniques applied in parallel or series that would not have been envisaged even a decade ago, and this presents a challenge to the experimentalist now and in the future.

Amber or red lights for Gen-Alpha are that COMSOL Multiphysics and similar programs should not simply be used as a black box. Furthermore, as nanoscale systems of interest become ever smaller, including nanoconfined systems, identified as a key topic in Gooding *et al.*'s Introductory Lecture (DOI: 10.1039/D1FD00088H), there will also be a need to transition from continuum models, including a reliance on the Gouy–Chapman–Stern model for the electrical double layer, to simulations that include atomistic detail.[44,45]

There is a healthy trend towards innovation in home-built electronic instrumentation in nanoelectrochemistry (mirroring what happened in the Boomer period for electrochemistry generally; see above), including reports at this *Discussion*. Furthermore, the use of patch clamp amplifiers from electrophysiology and the emergence of companies that are developing instruments that push the capability of current amplifiers and other instrumentation is to be welcomed.

Many electrochemical materials papers contain beautiful images of nanomaterials, for example, from aberration-corrected (scanning) transmission electron microscopy, (S)TEM, and other atomic-resolution microscopes, but the quality of the electrochemical measurements does not always match. We have to ensure that structural microscopy does not overshadow the electrochemistry itself, or compensate for inadequate electrochemical measurements. As I have described, developments in single entity electrochemistry techniques mean that structure and electrochemical activity/property measurements can be on at least an equal footing. Liquid cell TEM offers excellent opportunities for *in situ* structure–activity measurements, particularly if the nanomaterial can be presented for electrochemistry in the TEM cell in a well-defined and reproducible way.[46] That (S)TEM can be used routinely to characterise nanoscale electrodes and devices after measurements is a positive advance for nanoelectrochemistry, but it should not detract from the need to design reproducible batch fabrication methods.

Emerging general themes and challenges

Some of the key themes arising at this meeting and from developments in the wider nanoelectrochemistry field are summarised in Table 1, grouped around

Table 1 Emerging themes, challenges and opportunities in nanoelectrochemistry

Data	Simulations	Instrumentation	Complexity	Education & environment	Materials
Open data/ open access	Linking FEM & atomistic simulations	Open instrument architectures	Nanoscale spectroscopy	Discipline hopping & multidisciplinary teams	Rational design
Curation & mining		"Intelligent" experiments & instrumentation	Detection of reaction intermediates	Consortia and networks	Multiscale aspects
Big data, but sometimes sparse or missing		High throughput and parallelisation	Peering inside the double layer	Curriculum	
Image analysis and processing		Active control of nanosystems & nanoscale reactions	Interacting entities	Integration of fundamentals & applications	
Presentation			Single atoms		
			"Connections" between particles and particle/ support		

data, simulations, instrumentation, complexity, education and environment, and materials. This is not an exhaustive list, but rather is meant to be a starting point for discussion and debate. I expand on some of the topics in the sections that follow.

Data needs to be open access in a way that can be understood by other researchers, especially as datasets become ever bigger. This means that well-organised data curation and archiving is paramount. Some of the data in an experiment might be sparse, corrupted or missing and we can consider methods such as in-painting, *e.g.* as recently used in electrochemical imaging,[47] alongside the use of machine learning algorithms to deal with these situations. With these approaches, it should also be possible to speed up imaging and nano-electrochemistry measurements by collecting sparser data sets. The way in which large electrochemical datasets are presented is a major consideration. Movies are an attractive way to present large data and highlight key features,[48] in a spatio-temporal or other fashion, but interesting subtle (and rare) events within a large dataset need other methods of detection, display and presentation. We need to bring innovations in data mining to the nanoelectrochemistry field.

As mentioned earlier, nanoscale modelling largely relies on continuum models, but atomistic simulations, hand-in-hand with experiments, are really needed to understand what is happening at the nanoscale, particularly in the electrical double layer. Multiscale modelling, for example, through the integration of finite element method models and atomistic simulations also present great opportunities for understanding nanoscale electrochemical phenomena.

Under the instrumentation heading in Table 1, the idea behind open instrument architectures would be to make software and hardware designs more readily available, so that researchers from different laboratories work collaboratively and share instrumental features and design as an instrument evolves. We have tried to achieve this with the Warwick electrochemical scanned probe microscopy platform,[49] especially through the efforts of the originator, Kim McKelvey (Victoria University of Wellington, New Zealand), along with Martin Edwards (Arkansas, USA), Minkyung Kang (now at Deakin, Australia), and James Edmondson, Enrico Daviddi and Dimitrios Valavanis (Warwick) amongst others, who have helped other groups set up this platform. A revolution in intelligent experiments and instrumentation for nanoscale electrochemistry can be envisaged. A number of nanoscale techniques are excellent candidates for the implementation of AI protocols, aiding the high throughput and parallelisation of experiments, and the active control of nanoscale processes (*e.g.* feedback between synthesis and analysis).

Now to some aspects for innovation in the future. A major challenge is to be able to implement powerful spectroscopies at the nanoscale and preferably *in operando*. There has been significant progress in SERS and tip enhanced Raman spectroscopy,[50] but what about the challenge of small-scale mass spectrometry in electrochemistry? This would be incredibly enabling in the study of structure–activity in complex (*e.g.* multiple product) reactions. The other topics I highlight under the Complexity heading in Table 1 should be self-evident and are among some of the most interesting challenges for the community looking ahead. I shall explore some of them below.

Considering the scientific and wider environment in which we 'do' nanoelectrochemistry, we need mechanisms that encourage greater discipline hopping and the formation of multidisciplinary teams, both within individual groups and, more importantly, in larger networks and collaborations. Alongside, the curriculum and education is important, as discussed further below. And under the Education and environment heading, I return to the relationship between fundamentals and applications at the nanoscale: they go hand-in-hand, and are of mutual benefit, rather than there being a linear trajectory from fundamentals to applications.

The intelligent synthesis of functional materials and their use in various devices will have a dominant role in the scientific landscape this century, particularly for energy applications. The nanoelectrochemistry community has an amazing opportunity to lead in this area by providing the information that will unlock the rational design of functional materials and provide the underpinning science that will allow the prediction of nanomaterial behaviour in devices across lengthscales and timescales.

Multiple interacting entities at the nanoscale

As our investigations transition from single particles to ensembles and aggregates, we shall increasingly need complementary views from a range of techniques and perspectives to ensure an holistic understanding. I made this point in my concluding remarks in the *Discussion* with two pictures of a group of nanoelectrochemists having dinner together, but from different angles and at different times, to demonstrate that one picture alone leads to one interpretation, but two pictures provide a different conclusion! Moreover, interacting entities, and the nature of those interactions, can result in something much more beautiful and important than from single entities, as I illustrated with the example of The Beatles! These analogies were to draw attention to the importance of understanding interacting entities in electrochemistry, for example, in a very nice paper at the 140th *Faraday Discussion* (2008) from Behm's group[51] on the oxygen reduction reaction (ORR) at a model electrocatalyst comprising active platinum nanodiscs on a glassy-carbon support, where the spacing between the discs and electrolyte flow rate were shown to have a significant impact on the yield of H_2O_2, rationalised through the use of a reaction model. We now have the capability to perform similar types of measurements, but with platinum clusters, which have mobility on the electrode surface as a consequence of the ORR and the resulting heterogeneous fluxes and mass transport at the nanoscale.[52] In examining the transformation of reagent to product in electrocatalytic processes at ensembles of particles, we need to address the following questions: to what extent is the net transformation at a single entity or distributed across entities *via* inter-particle transport? In other words, what is the inter-entity diffusion of reaction intermediates? What is the reaction/interaction of reaction intermediates with the support? How do we go about trying to study these kind of processes?

Importance of the nano-entity electrode support

An aspect of considerable importance in the electrochemistry of nanomaterials (as a key example of single entity electrochemistry) is to understand the effect of the interaction between the support (current collector) and the entity being studied. The support–entity interaction can be tuned chemically, as illustrated in the paper from Ruixia Wang *et al.* (DOI: 10.1039/D1FD00054C) on collisions of functionalised nanoparticles at a functionalised nanoelectrode, where the interaction (and current) was controlled by host–guest chemistry, and the nature of the interaction was revealed through the electrochemical measurement. Recent work from Wei Wang's group has highlighted the importance of the eliminating effects of heterogeneous electrical contacts in order to measure the true activity of single nano-entities.[53] These studies serve to highlight the general importance of accounting for particle–electrode interactions in studies of particulate electrodes at all lengthscales, and is an aspect of single entity electrochemistry that needs further consideration.

The support used for the study of 2D materials is a particularly important consideration. In fact, the electrochemical and electrocatalytic activity of graphene heterostructures can be tuned *via* the support on which the graphene is placed.[54] In this *Discussion*, Tolbert and Hill described an innovative "carrier generation-tip collection SECCM" method for generating carriers locally in a 2D semiconductor and collecting them at an array of confined sites *via* a redox reaction at the SECCM meniscus (DOI: 10.1039/D1FD00052G). This extends electrochemical methodology to the visualization of key processes in solid state materials and is a particularly exciting development of SECCM.

"Well-defined" particles

In the paper from Yao *et al.* (DOI: 10.1039/D1FD00047K), control over the structure of PtCu nanoparticles was demonstrated. In the Introductory Lecture, Gooding *et al.* (DOI: 10.1039/D1FD00088H) suggested that effort should be put into the synthesis of well-defined nanoparticles that could be studied as an array to reveal structure–activity relationships. However, it must be recognised that even if nanoparticles are closely identical structurally, they can have very different activities, as demonstrated by Baker's group, who used SECCM to assess the electrocatalytic activity of gold nanoparticles with well-defined surface facets, showing that the activity of nominally identical nanocubes (under SEM) had very different activity.[55] This can be explained generally by subtle distributions in the atomic arrangements and surface capping agents (if used) from one nanoparticle to another, which affects the surface free energy.[56] This is an important message for the community and makes single entity (particle) measurements imperative: "well-defined" nanoparticles may not be as well-defined as we might think they are.

Developments in SECCM

SECCM is finding an ever diverse range of applications and there are new innovations from an increasing number of groups. At the *Discussion*, SECCM was the subject of three papers and four posters. I have already mentioned Tolbert and Hill's experiment above, also described in Poster 17. The paper from Ren's group introduced several methods for assessing and compensating for the *iR* drop in SECCM, which will be useful for electrocatalytic and electrochemical measurements under some conditions (DOI: 10.1039/D1FD00046B). To advance the analysis of SECCM data, Anderson and Edwards also presented simple and robust analytical equations for the current response, which will be very beneficial to SECCM users (Poster 5). SECCM is a powerful method for targeting individual particles within an ensemble, and there were examples of two such studies from Chen (Poster 9) and Choi (Poster 11), who also demonstrated environmental control of SECCM. Our group at Warwick has been collaborating closely with Kanoufi's group in Paris to integrate SECCM and interference reflection microscopy (IRM); in the paper from Valavanis *et al.*, it was shown that this capability allowed the status of the SECCM meniscus to be tracked *in situ* during voltammetric measurements and it was also possible to peer inside the meniscus to detect and visualise phase formation processes (DOI: 10.1039/D1FD00063B).

Nanopipettes

Nanopipettes are established tools in an increasingly wide range of nanoelectrochemistry measurements.[57] These devices can be employed in single channel, dual channel and multichannel formats, providing a range of delivery and detection possibilities, as witnessed in this *Discussion*. Dual channel nanopipettes with an electrolyte filled channel, back contacted with a quasi-reference counter electrode and a solid-carbon electrode as the second channel were introduced by Takahashi *et al.* for combined SECM-SICM,[41] and such probes have been used for quantitative molecular delivery (electrolyte channel)–detection (carbon electrode) external to cells.[58] An exciting development reported at the *Discussion* from He's group is that such probes can be used for intracellular injection and detection (DOI: 10.1039/D1FD00057H). The detection sensitivity afforded by nanopipettes was exemplified by Chen *et al.* who pushed the time limit of single channel functionalised nanopipette measurements to detect protein–protein interactions at the single molecule level (DOI: 10.1039/D1FD00042J). These advances, coupled with other attributes of nanopipettes, from nanoparticle delivery and sizing with exquisite control[59] to nanoscale mixing, synthesis and analysis,[60] make nanopipettes an outstanding platform for nanoelectrochemistry.

Peering inside the electrical double layer

The surface force apparatus (SFA) is a powerful means of examining electrostatic forces between two surfaces in electrolyte solution.[61,62] The developments reported in this *Discussion* by Kurihara's group (DOI: 10.1039/D1FD00060H) open up exciting prospects for performing electrochemical measurements in the confined space offered by the SFA, and generated considerable interest, as evidenced from the questions and comments on this work. Such measurements, alongside local surface charge measurements and new molecular dynamics models, point towards an improved understanding of the electrical double layer.

Single atoms in electrochemical processes

Single atom (electro)catalysis is a "hot" but established topic, concerned with macroscopic studies averaged over billions of atom catalysts dispersed in, or on, a support material. An important emerging topic in single entity nanoelectrochemistry, not touched upon in the *Discussion*, is the role of single atoms in electrochemical processes and the study of electrochemistry at *individual* single atoms. Ustarroz and co-workers identified the operation of non-classical routes to electrocrystallisation, involving the aggregation of small clusters to make larger particles,[63,64] and recent work from Macpherson's group has shown the involvement of single atoms in such processes.[65] These studies were enabled by the use of (identical-location) aberration-corrected TEM, where the electrode is a TEM grid. The use of SECCM to make electrodepositions on a TEM grid further makes possible the analysis of many different electrodeposition experiments in a combinatorial fashion.[66]

Inspirational work from Bard's group has proposed that a *single atom* can be deposited on a nanoelectrode and used for electrocatalysis.[67] This work is entirely electrochemical and uses classical models to infer the atom (or cluster) size from limiting current measurements of a redox reaction. Such work will be greatly advanced when complementary microscopy methods can be applied to prove unequivocally that single (individual) atom electrodes can be made.

Looking ahead: local to global

A wide range of single entity electrochemistry measurements are possible and many such experiments are now relatively straightforward. More challenging are studies in which the electrochemical activity and surface structure at the nanoscale are correlated, although this is increasingly possible through the use of a wide range of multi-microscopy strategies, with electrochemical imaging at the centre.[68] Detailed analyses of the resulting datasets, in tandem with modelling, is revealing structure–activity relationships with considerable detail, especially when complementary modelling is applied. This direction is crucial for both fundamental advances and the rational design of next generation electrodes and electrochemical devices.

Scanning the horizon, I believe that nanoscale electrochemistry will increasingly be connected to measurements at longer timescales and larger lengthscales so that electrochemical systems, and kinetic limitations, can ultimately be understood holistically at any scale (local to global). Thus, the future of nanoelectrochemistry is not simply better measurements of single entities or groups of entities, but will be concerned with enabling science that ultimately helps to connect diverse aspects of electrochemistry.

As nanoelectrochemistry develops, the analysis and visualisation of increasingly larger datasets will become critical. This transition will be facilitated by experiments that are intelligent and autonomous. Nanoelectrochemistry experiments are not only excellent candidates to be automated, but are ideally suited to become adaptive and autonomous, which is a very promising direction for materials discovery generally.[69] So, I expect that the human nanoelectrochemist of the future will spend much less of their time in the laboratory doing experiments and more time as a data scientist, analyst, engineer and archivist/steward, and communicating their work to a diversity of audiences. And, of course, interacting with their robot labmates (and other robot analysts)!

Curriculum and education

Electrochemistry has traditionally been chemistry-centric, and most students entering the field have have a chemistry background. Yet, many of the challenges and opportunities presented by nanoelectrochemistry, now and in the future, might be better suited to scientists from other disciplines and multidisciplinary teams. We saw that in the Boomer period, advances in electronics heralded new opportunities for electrochemistry, and throughout its history, major developments in electrochemistry have resulted from the cross-fertilisation of different fields: chemistry, electronics, physics, engineering, life sciences, computer science and mathematics. To thrive, nanoelectrochemistry needs a diverse student body, attracted from a diverse range of scientific, educational and cultural backgrounds and who can learn from each other.

So, we need mechanisms and programmes that are truly multidisciplinary. This was achieved by the Warwick University, EPSRC-funded MOAC Doctoral Training Centre (2003–15), led by Alison Rodger, which sought to catalyse research and training across the physical/life sciences interface. By running as a 3 year PhD, preceded by a one-year multidisciplinary MSc, involving taught courses and projects from across the science, engineering and medical faculties, this programme was able to attract a similarly scientifically diverse student body, and particularly students with a strong mathematics background. Individual modules taught by academics from different departments also led to positive new interactions at the research level. Research in the Warwick Electrochemistry & Interfaces Group benefitted enormously, particularly from the contributions of students with degrees in mathematics and computer science, for example, Kim McKelvey and Martin Edwards (both of whom are well known in the nanoelectrochemistry community), along with Kate Meadows, David Perry, Ashley Page, Hayley Powell, and Jenny Webb, among many other talented PhD students who are authors on papers from this period.

We have drawn on some of the elements of MOAC in the EU-funded SENTINEL (Single Entity Nanoelectrochemistry) Innovative Training Network (ITN), led by Paolo Actis (University of Leeds). Although some of the aims, including extensive globally disperse secondments for the participating PhD students, have been severely impacted by the COVID-19 pandemic, it has been successful in building a cohort of early career researchers from different backgrounds who share knowhow and work together to tackle frontier problems in nanoelectrochemistry.

An innovative curriculum needs good teaching materials, an aspect of which is tutorial-style articles. Over the past year, there have been some superb articles that are highly educational, from a discussion of shot noise[70] and stochasticity in nanoscale and single entity electrochemistry[71] to an overview of the working of the potentiostat.[72] These efforts are to be applauded and augur well for the future.

Conclusion

Historically, electrochemistry was dominated by individuals, as in the story of big science from 1813 and work from the Boomer generation, which I introduced by name-checking scientists from 75 years ago whose work we still use and link to individuals. For next generation (nano)electrochemistry, the emphasis will surely be on teams, centres and schools. Next generation problems in electrochemistry are too large and diverse to be tackled by individuals alone and will require increasing cooperation within large teams (which already happens extensively) and between teams based in different countries and continents that work together and help each other (a developing trend). By coming together, these teams will be able to address the challenges and opportunities presented by big data and complex instrumentation that will characterise next generation electrochemistry and nanoelectrochemistry.

Although there are many facets to nanoelectrochemistry, as illustrated by this *Discussion*, I believe that a major aspect of next generation electrochemistry will to *understand local* and connect this understanding to *global*. This will be facilitated by treating "electrochemical systems" through a "systems electrochemistry" approach, in which nanoelectrochemistry is a key enabling component, and is not treated in isolation. As described here, next generation electrochemistry will present a continuum of opportunities in fundamental science and applications.

Conflicts of interest

The author is co-inventor of granted patent PCT/GB2011/051518 "Pipets containing Electrolyte and Electrodes", which describes dual-channel SECCM.

Acknowledgements

The author dedicates this work to the memory of his late father, the historian Robert 'Bob' Unwin, who would have greatly enjoyed the theme of this article, especially the historical context of, and forward-look at, electrochemistry. In his later years, Bob was engaged in research on Humphry Davy and associated natural philosophers, upon which this article draws. The author thanks the Royal Society for a Wolfson Reseach Merit Award and the EPSRC for funding nanoelectrochemistry-related research (EP/F500378/1, EP/H023909/1, EP/V047981/1,

EP/L015307/1). The author is grateful to many generations of researchers in the Warwick Electrochemistry & Interfaces Group who have been inspirational and contributed to our nanoelectrochemistry research. Finally, the author thanks Dr Daniel Martín-Yerga for the TOC graphic for this article.

References

1 P. Unwin and R. Unwin, *Br. J. Hist. Sci.*, 2007, **40**, 181–203.
2 J. G. Children, *Ann. Philos.*, 1813, **2**, 32.
3 J. A. Parris, *Sir Humphry Davy*, Cambridge University Press, 1831, vol. 1, pp. 97–98.
4 M. C. Usselman, *Pure Intelligence: the Life of William Hyde Wollaston*, University of Chicago Press, 2015.
5 W. H. Wollaston, *Philos. Trans. R. Soc. London*, 1801, **91**, 427–434.
6 J. M. Pingarrón, J. Labuda, J. Barek, C. M. A. Brett, M. F. Camões, M. Fojta and D. B. Hibbert, *Pure Appl. Chem.*, 2020, **92**, 641–694.
7 M. Faraday, *Experimental Researches in Electricity*, University of London, Taylor, 1839, vol. 1.
8 Y.-L. Ying, Y.-X. Hu, R. Gao, R.-J. Yu, Z. Gu, L. P. Lee and Y.-T. Long, *J. Am. Chem. Soc.*, 2018, **140**, 5385–5392.
9 V. Gold, *Nature*, 1947, **160**, 306–307.
10 W. F. Berg, *Discuss. Faraday Soc.*, 1947, **1**, 44–45, general discussion.
11 L. A. Baker, *J. Am. Chem. Soc.*, 2018, **140**, 15549–15559.
12 R. N. Adams, *Electrochemistry at Solid Electrodes*, Marcel Dekker, NY, 1969.
13 J.-M. Savéant, *Elements of Molecular and Biomolecular Electrochemistry*, John Wiley, NJ, 2006.
14 J. E. Frew and H. A. O. Hill, *Anal. Chem.*, 1987, **59**, 933A–944A.
15 R. W. Murray, *Electroanal. Chem.*, 1984, **13**, 191–368.
16 R. F. Lane and A. T. Hubbard, *J. Phys. Chem.*, 1973, **77**, 1401–1410.
17 C. P. Andrieux and J.-M. Savéant, *J. Electroanal. Chem.*, 1978, **93**, 163–168.
18 R. N. Adams, *Anal. Chem.*, 1976, **48**, 1126A–1138A.
19 R. M. Wightman, E. Strope, P. M. Plotsky and R. N. Adams, *Nature*, 1976, **262**, 145–146.
20 E. Neher and B. Sakmann, *Proc. Natl. Acad. Sci. U. S. A.*, 1975, **72**, 2140–2144.
21 E. Neher and B. Sakmann, *Nature*, 1976, **260**, 799–802.
22 A. J. McQuillan, *Notes Rec. R. Soc.*, 2009, **63**, 105–109.
23 R. A. Osteryoung and J. Osteryoung, *Philos. Trans. R. Soc., A*, 1981, **302**, 315–326.
24 D. J. Myers and J. Osteryoung, *Anal. Chem.*, 1973, **45**, 381–271.
25 S. Bruckenstein and B. Miller, *Acc. Chem. Res.*, 1977, **10**, 54–61.
26 A. Page, D. Perry and P. R. Unwin, *Proc. R. Soc. London, Ser. A*, 2017, **473**, 20160889.
27 R. M. Wightman, *Anal. Chem.*, 1981, **53**, 1125A–1134A.
28 *Ultramicroelectrodes*, M. Fleischmann, S. Pons, D. R. Rolison and P. P. Schmidt, Datatech Systems, Morganton, NC, 1987.
29 R. M. Wightman and D. O. Wipf, *Electroanal. Chem.*, 1989, **15**, 267–353.
30 A. J. Bard and L. R. Faulkner, *Electrochemical Methods*, Wiley, NY, 1980.
31 R. M. Wightman, J. A. Jankowski, R. T. Kennedy, K. T. Kawagoe, T. J. Schroeder, D. J. Leszczyszyn, J. A. Near, E. J. Diliberto Jr and O. H. Viveros, *Proc. Natl. Acad. Sci. U. S. A.*, 1991, **88**, 10754–10758.

32 F.-R. F. Fan and A. J. Bard, *Science*, 1995, **267**, 871–874.

33 S. W. Feldberg, *Electroanal. Chem.*, 1969, **3**, 199–296.

34 D. Britz, *Digital Simulation in Electrochemistry*, Springer-Verlag, Berlin, 1988.

35 M. Brust, M. Walker, D. Bethell, D. J. Schiffrin and R. Whyman, *Chem. Commun.*, 1994, 801–802.

36 I. Heller, J. Kong, H. A. Heering, K. A. Williams, S. G. Lemay and C. Dekker, *Nano Lett.*, 2005, **5**, 137–142.

37 C. L. Bentley, J. Edmondson, G. N. Meloni, D. Perry, V. Shkirskiy and P. R. Unwin, *Anal. Chem.*, 2019, **91**, 84–108.

38 V. Brasiliense, A. N. Patel, A. Martinez-Marrades, J. Shi, Y. Chen, C. Combellas, G. Tessier and F. Kanoufi, *J. Am. Chem. Soc.*, 2016, **138**, 3478–3483.

39 J. V. Macpherson and P. R. Unwin, *Anal. Chem.*, 2000, **72**, 276–285.

40 C. Kranz, G. Friedbacher, B. Mizaikoff, A. Lugstein, J. Smoliner and E. Bertagnolli, *Anal. Chem.*, 2001, **73**, 2491–2500.

41 Y. Takahashi, A. I. Shevchuk, P. Novak, Y. Zhang, N. Ebejer, J. V. Macpherson, P. R. Unwin, A. J. Pollard, D. Roy, C. A. Clifford, H. Shiku, T. Matsue, D. Klenerman and Y. E. Korchev, *Angew. Chem., Int. Ed.*, 2011, **50**, 9638–9642.

42 B. P. Nadappuram, K. McKelvey, J. C. Byers, A. G. Güell, A. W. Colburn, R. A. Lazenby and P. R. Unwin, *Anal. Chem.*, 2015, **87**, 3566–3573.

43 L. Dorwling-Carter, M. Aramesh, H. Han, T. Zambelli and D. Momotenko, *Anal. Chem.*, 2018, **90**, 11453–11460.

44 A. R. Finney, I. J. McPherson, P. R. Unwin and M. Salvalaglio, *Chem. Sci.*, 2021, **12**, 11166–11180.

45 S.-J. Shin, D. H. Kim, G. Bae, S. Ringe, H. Choi, H.-K. Lim, C. H. Choi and H. Kim, *Nat. Commun.*, 2022, **13**, 174.

46 T. Tarnev, S. Cychy, C. Andronescu, M. Muhler, W. Schuhmann and Y.-T. Chen, *Angew. Chem., Int. Ed.*, 2020, **59**, 5586–5590.

47 J. F. Edmondson, G. N. Meloni, G. Costantini and P. R. Unwin, *ChemElectroChem*, 2020, **7**, 697–706.

48 M. Kang, D. Momotenko, A. Page, D. Perry and P. R. Unwin, *Langmuir*, 2016, **32**, 7993–8008.

49 https://warwick.ac.uk/fac/sci/chemistry/research/unwin/electrochemistry/wec-spm/, accessed 27 Jan 2022.

50 Z.-C. Zeng, S.-C. Huang, D.-Y. Wu, L.-Y. Meng, M.-H. Li, T.-X. Huang, J.-H. Zhong, X. Wang, Z.-L. Yang and B. Ren, *J. Am. Chem. Soc.*, 2015, **137**(37), 11928–11931.

51 Y. E. Seidel, A. Schneider, Z. Jusys, B. Wickman, B. Kasemo and R. J. Behm, *Faraday Discuss.*, 2009, **140**, 167–184.

52 J. Ustarroz, I. M. Ornelas, G. Zhang, D. Perry, M. Kang, C. L. Bentley, M. Walker and P. R. Unwin, *ACS Catal.*, 2018, **8**, 6775–6790.

53 W. Wei, T. Yuan, W. Jiang, J. Gao, H.-Y. Chen and W. Wang, *J. Am. Chem. Soc.*, 2020, **142**, 14307–14313.

54 J. Hui, S. Pakhira, R. Bhargava, Z. J. Barton, X. Zhou, A. J. Chinderle, J. L. Mendoza-Cortes and J. Rodríguez-López, *ACS Nano*, 2018, **12**, 2980–2990.

55 M. Choi, N. P. Siepser, S. Jeong, Y. Wang, G. Jagdale, X. Ye and L. A. Baker, *Nano Lett.*, 2020, **20**, 1233–1239.

56 B. H. Kim, *et al.*, *Science*, 2020, **368**, 60–67.

57 L. Sun, S. Watanabe and T. Ando, *Anal. Chem.*, 2022, **94**, 324–335.

58 A. Page, M. Kang, A. Armitstead, D. Perry and P. R. Unwin, *Anal. Chem.*, 2017, **89**, 3021–3028.
59 M. A. Edwards, S. R. German, J. E. Dick, A. J. Bard and H. S. White, *ACS Nano*, 2015, **9**, 12274–12282.
60 P. D. Morris, I. J. McPherson, M. A. Edwards, R. I. Walton and P. R. Unwin, *Angew. Chem., Int. Ed.*, 2020, **59**, 19696–19701.
61 T. S. Groves, C. S. Perez-Martinez, R. Lhermerout and S. Perkin, *J. Phys. Chem. Lett.*, 2021, **12**, 1702–1707.
62 A. M. Smith, A. A. Lee and S. Perkin, *J. Phys. Chem. Lett.*, 2016, **7**, 2157–2163.
63 J. Ustarroz, X. Ke, A. Hubin, S. Bals and H. Terryn, *J. Phys. Chem. C*, 2012, **116**, 2322–2329.
64 J. Ustarroz, J. A. Hammons, T. Altantzis, A. Hubin, S. Bals and H. Terryn, *J. Am. Chem. Soc.*, 2013, **135**, 11550–11561.
65 H. E. M. Hussein, R. J. Maurer, H. Amari, J. J. P. Peters, L. Meng, R. Beanland, M. E. Newton and J. V. Macpherson, *ACS Nano*, 2018, **12**, 7388–7396.
66 I. M. Ornelas, P. R. Unwin and C. L. Bentley, *Anal. Chem.*, 2019, **91**, 14854–14859.
67 M. Zhou, J. Dick and A. J. Bard, *J. Am. Chem. Soc.*, 2017, **139**, 17677–17682.
68 C. L. Bentley, M. Kang and P. R. Unwin, *J. Am. Chem. Soc.*, 2019, **141**, 2179–2193.
69 B. Burger, *et al.*, *Nature*, 2020, **583**, 237–241.
70 R. Gao, M. A. Edwards, J. M. Harris and H. S. White, *Curr. Opin. Electrochem.*, 2020, **22**, 170–177.
71 H. Ren and M. A. Edwards, *Curr. Opin. Electrochem.*, 2021, 100632.
72 A. W. Colburn, K. J. Levey, D. O'Hare and J. Macpherson, *Phys. Chem. Chem. Phys.*, 2021, **23**, 8100–8117.

Poster Presentations

Graphene quantum dots@Benzoquinone@β-cyclodextrin system for a dual mode sensing, **W. Aidli**, *Università degli studi di Milano*

High-throughput nanoelectrochemistry: individual nanoelectrodes investigated via the array microcell method (AMCM), **S. Alden, L. Baker, N. Lavrik, Y. Wang, L. Zhang**, *Indiana University, Center for Nanophase Materials Sciences, Oakridge National Laboratory, Oakridge, TN*

Development of new electrode materials for hybrid battery/supercapacitor devices, **S. Alshehri, G. G. Wildgoose, J. Fielden**, *University of East Anglia*

Nanoelectrode ensembles made of carbon nanotubes, **G. Dangel, H. Kumakli, C. Rahm, R. White, N. Alvarez**, *University of Cincinnati*

Evaluating analytical expressions for scanning electrochemical cell microscopy, **K. Anderson, M. A. Edwards**, *University of Arkansas*

Electrochemistry at the edge of graphene, **A. Yadav, M. Wehrhold, T. J. Neubert, R. M. Lost, K. Balasubramanian**, *Humboldt-Universitaet zu Berlin*

Micro-spectroscopic analysis of the SEI formed on Si nano-wire electrodes in ionic liquid electrolytes, **N. Carboni, S. Brutti, O. Palumbo, G. B. Appetecchi, G. Maresca, H. Geaney, K. M. Ryan, F. Capitani, A. Paolone, S. Fantini, R. Lin, P. A. Martin**, *ISC-CNR*

Rapid and ultrasensitive electrochemical detection of circulating tumor DNA by hybridization on the network of gold-coated magnetic nanoparticles, **D. Chen, Y. Wu, S. Hoque, R. D. Tilley, J. J. Gooding**, *The University of New South Wales*

Direct probing of the oxygen evolution reaction at single $NiFe_2O_4$ nanocrystal superparticles with tunable structures, **X. Lu, Q. Chen**, *Donghua University*

Functionalization of CMOS based nanocapacitor arrays for biomolecule detection, **S. Chinnathambi, M. Saghafi, and S. Lemay**, *University of Twente*

Probing CO_2 reduction activity of single Au-nanocrystal, truncated ditetragonal prisms (TDPs) by gas chamber-scanning electrochemical cell microscopy (SECCM), **M-H Choi, S. Jeong, N. P. Siepser, G. Jagdale, X. Ye, L. A. Baker**, *Indiana University, Chemistry*

Study of the aging of TiO_2 sol for electrochemical sensors for the determination of emerging contaminants, **S. Comis, L. Nicolucci, V. Pifferi, M. Longhi, L. Falciola**, *Università degli studi di Milano*

Differential Tafel analysis applied to OER catalysts – towards unambiguous extraction of Tafel slope data from electrochemical ensemble measurements, **M. Corva, J. K. Fuchs, N. Blanc, P. Cignoni, Z. Liu and K. Tschulik**, *Ruhr University Bochum*

Thermodynamic description of the formation of SEI: theoretical investigation of the hydrolysis of the lithium salt components of electrolytes, **S. Di Muzio, A. Paolone and S. Brutti**, *Istituto dei Sistemi Complessi- Consiglio Nazionale delle Ricerche*

The effect of electrolyte concentration and pore size on ion current rectification inversion, **D. Duleba, P. Dutta, S. Denuga, R. Johnson**, *University College Dublin*

Mechanochemical formation of electroactive graphite nanoplatelets, **C. H. Lucas, J. M. Vila-Fungueiriño, M. del Carmen Giménez-López**, *Universidade de Santiago de Compostela*

Electrochemically probing exciton transport in monolayers of two-dimensional semiconductors, **C. Tolbert, C. Hill**, *University of Wyoming*

Silicon based photoelectrochemical imaging and monitoring of cardiomyocyte organoids, **R. Jacques, B. Zhou, E. Marhuenda, M. Zhong, T. Iskratsch, S. Krause**, *Queen Mary University of London*

Designing low cost and durable Na-ion batteries, **Y. Li, Y. Lu, Y-S Hu**, *Institute of Physics, Chinese Academy of Sciences*

Approaching sub-detection limit nanoelectrochemistry with fluorescence-enabled electrochemical microscopy, **S. Linfield, S. Gawinkowski and W. Nogala**, *Institute of Physical Chemistry, Polish Academy of Sciences*

Controllable stochastic electron transfer at a confined interface, **S-M Lu, M. Chen, Y-T Long**, *Nanjing University*

Electrochemically confined effects on single enzyme detection with nanopipettes, **H. Ma, Y-T Long**, *Nanjing University*

How the tailoring of metal oxide-based nanocomposites can influence the chemiresistive sensing: the role of graphene oxide versus porphyrin, **E. Pargoletti**, *University of Milan*

Single virus detection model using lipid bilayer covered CMOS-based nanocapacitor arrays, **M. Saghafi, S. Chinnathambi, N. Overeem, J. Huskens, S. Lemay**, *University of Twente*

Automated resolution of stoichiometric changes during electrochemical cycling by density functional theory assisted electrochemical quartz crystal microbalance, **I. Scivetti, G. Teobaldi**, *The Science and Technology Facilities Council*

Plasmonic properties of AuPd nanoparticles for nitrite reduction reaction, **K. Nascimento da Silva, S. Shetty, P. Brasseur, S. Belhout, E. Sitta, P. H. C. Camargo**, *University of Helsinki, Universidade Federal de São Carlos*

Improving sulfur cathode for highly efficient and stable full cell magnesium-sulfur battery, **G. Singh, T. Keene**, *University College Dublin*

Nanostructured zirconia (ZrO_2): from anodization to potential bio-medical implants, **S. N. Vakamulla Raghu, K. Chuluunbandi, G. Onyenso, S. Mohajernia, M. S. Killian**, *University of Siegen, FAU-Erlangen-Nuremberg*

Single-molecule electrochemistry trigged by silver nanoparticle collisions, **J. Wang, J. He and W. Yang**, *Deakin university*

Biological Interfaces, **M. Wood**, *University of Cambridge*

Selectively detecting attomolar concentrations of proteins using gold lined nanopores in a nanopore blockade sensor, **Y. Wu, Y. Yao, S. Cheong, R. D. Tilley, J. Gooding**, *UNSW Sydney*

The deactivation mechanism of palladium nanoparticles in acidic system, **W. Nan, D. Zhan**, *Department of Chemistry, Xiamen University*

Potentiodynamic scattering microscopy of a tungsten ultramicroelectrode: "seeing" ionic currents under an optical microscope, **Z. Zhang, S. Faez**, *Utrecht University*

The Faraday Division poster prize went to Myunghoon Choi of Indiana University, United States, for their poster on 'Probing CO_2 reduction activity of single Au-nanocrystal, truncated ditetragonal prisms (TDPs) by gas chamber-scanning electrochemical cell microscopy (SECCM)'. The Faraday Discussions poster prize went to Kamsey Anderson of University of Arkansas with their poster on 'Evaluating analytical expressions for scanning electrochemical cell microscopy'.

List of participants

Ms Wafa Aidli, *Università degli studi di Milano*, *Italy*
Ms Kristen Alanis, *Indiana University Bloomington*, *USA*
Professor Tim Albrecht, *University of Birmingham*, *United Kingdom*
Miss Sasha Alden, *Indiana University*, *USA*
Ms Sarah Alshehri, *University of East Anglia*, *United Kingdom*
Dr Noe Alvarez, *University of Cincinnati*, *USA*
Ms Einav Amit, *The Hebrew University of Jerusalem*, *Israel*
Mrs Kamsy Anderson, *University of Arkansas*, *USA*
Mr Masamichi Asano, *Mitsubishi Heavy Industries America*, *USA*
Dr Albert Assonkeng, *Cape Breton University*, *Canada*
Ms Seol Baek, *University of Notre Dame*, *USA*
Professor Lane Baker, *Indiana University*, *USA*
Professor Dr Kannan Balasubramanian, *Humboldt-Universitaet zu Berlin*, *Germany*
Ms Iris Berg, *The Hebrew University of Jerusalem*, *Israel*
Professor Renata Bilewicz, *University of Warsaw*, *Poland*
Dr Paul Bohn, *University of Notre Dame*, *USA*
Mr Uriel Bruno Mota, *Institut National de la Recherche Scientifique*, *Canada*
Mr Mark Buckingham, *King's College London*, *United Kingdom*
Mr Xiangkun Elvis Cao, *Cornell University*, *USA*
Dr Georgina Carbajal De la Torre, *Universidad Michoacana de San Nicolas de Hidalgo*, *Mexico*
Dr Nicholas Carboni, *Istituto dei Sistemi Complessi - Consiglio Nazionale delle Ricerche*, *Italy*
Dr Shuai Chang, *Wuhan University of Science and Technology*, *China*
Miss Dongfei Chen, *The University of New South Wales*, *Australia*
Professor Qianjin Chen, *Donghua University*, *China*
Dr Selvaraj Chinnathambi, *University of Twente*, *Netherlands*
Mr Myunghoon Choi, *Indiana University, Chemistry*, *USA*
Professor Fabio Cicoira, *Polytechnique Montreal*, *Canada*
Mx Silvia Comis, *Università degli Studi di Milano*, *Italy*
Mr Samuel Confederat, *University of Leeds*, *United Kingdom*
Mr Manuel Corva, *Ruhr University Bochum*, *Germany*
Ms Allison R. Cutri, *University of Notre Dame*, *USA*
Ms Shekemi Denuga, *University College Dublin*, *Ireland*
Mr Shahar Dery, *The Hebrew University of Jerusalem*, *Israel*
Ms Linoy Dery, *The Hebrew University of Jerusalem*, *Israel*
Dr Simone Di Muzio, *Istituto dei Sistemi Complessi - Consiglio Nazionale delle Ricerche*, *Italy*
Dr Huan Doan, *University of Bristol*, *United Kingdom*
Mr Dominik Duleba, *University College Dublin*, *Ireland*
Ms Pallavi Dutta, *University College Dublin*, *Ireland*
Dr Peisan E, *Teesside University*, *United Kingdom*
Professor Martin A. Edwards, *University of Arkansas*, *USA*

Professor Andrew Ewing, *University of Gothenburg*, *Sweden*
Dr Sanli Faez, *Utrecht University*, *Netherlands*
Professor Dr Luigi Falciola, *Università degli Studi di Milano*, *Italy*
Ms Emer Farrell, *University College Dublin*, *Ireland*
Dr Stefano Fornasaro, *University of Trieste - Department of Engineering and Architecture*, *Italy*
Miss Meigeng Gao, *Xi'an Jiaotong-Liverpool University*, *China*
Professor Dr Pengcheng Gao, *China University of Geosciences*, *China*
Professor Justin Gooding, *The University of New South Wales*, *Australia*
Mr Luke Gundry, *Monash University*, *Australia*
Professor Jin He, *Florida International University*, *USA*
Mrs Hanieh Helli, *Università degli Studi di Milano*, *Italy*
Dr Magdalena Helmer, *Nature*, *United Kingdom*
Dr Carlos Herreros Lucas, *El Centro Singular de Investigación en Química Biológica y Materiales Moleculares (CiQUS), Universidade de Santiago de Compostela*, *Spain*
Professor Caleb Hill, *University of Wyoming*, *USA*
Professor Ayumi Hirano-Iwata, *Tohoku University*, *Japan*
Professor Shuo Huang, *Nanjing University*, *China*
Miss Rachel Jacques, *Queen Mary University of London*, *United Kingdom*
Mr Grant Jeffcoat, *University of Arkansas*, *USA*
Dr Robert Johnson, *University College Dublin*, *Ireland*
Professor Ali Reza Kamali, *Northeastern University*, *China*
Dr Frédéric Kanoufi, *Universite de Paris / Centre national de la recherche scientifique (CNRS)*, *France*
Professor Dr Manuela Sonja Killian, *University of Siegen*, *Germany*
Professor Steffi Krause, *Queen Mary University of London*, *United Kingdom*
Professor Kazue Kurihara, *Tohoku University*, *Japan*
Professor Dr Seung-Ryong Kwon, *Gyeongsang National University*, *South Korea*
Mr Cody Leasor, *Indiana University*, *USA*
Professor Serge Lemay, *University of Twente*, *Netherlands*
Miss Ndrina Limani, *CEA-Saclay*, *France*
Dr Steven Linfield, *Institute of Physical Chemistry, Polish Academy of Sciences*, *Poland*
Dr Yifan Liu, *Harvard University*, *USA*
Ms Xu Liu, *Cornell University*, *USA*
Professor Yi-Tao long, *Nanjing University*, *China*
Dr Mariangela Longhi, *Università degli Studi di Milano, Dipartimento di Chimica*, *Italy*
Dr Si-Min Lu, *Nanjing University*, *China*
Dr Hui Ma, *Nanjing University*, *China*
Professor Daniel Mandler, *Hebrew University of Jerusalem*, *Israel*
Professor Bing-Wei Mao, *Xiamen University*, *China*
Mr Fabio Marcuccio, *University of Leeds*, *United Kingdom*
Dr Atsuko Masuya-Suzuki, *Tohoku University*, *Japan*
Dr Ian McPherson, *University of Warwick*, *United Kingdom*
Dr Gabriel Meloni, *University of Warwick, Department of Chemistry*, *United Kingdom*
Dr Svetlana Menkin, *University of Cambridge*, *United Kingdom*
Professor Shelley Minteer, *University of Utah*, *USA*
Miss Parisa Moazzam, *The Australian National University*, *Australia*
Dr Pranati Nayak, *University of Oxford*, *United Kingdom*
Mr Shane O' Neill, *University College Dublin*, *Ireland*
Dr Chandana Pal, *Brunel University*, *United Kingdom*
Dr Popular Pandey, *Florida International University*, *USA*

Dr Eleonora Pargoletti, *University of Milan*, *Italy*
Mr Soumyadeep Paul, *The University of Tokyo*, *Japan*
Dr Neil Rees, *University of Birmingham*, *United Kingdom*
Professor Hang Ren, *The University of Texas at Austin*, *USA*
Dr Ashley Ross, *University of Cincinnati*, *USA*
Mr Mohammad Saghafi, *University of Twente*, *Netherlands*
Ms Samia Said, *University College London*, *United Kingdom*
Dr Ivan Scivetti, *The Science and Technology Facilities Council*, *United Kingdom*
Ms Natasha Siepser, *Indiana University*, *USA*
Ms Kaline Nascimento da Silva, *University of Helsinki*, *Finland*
Dr Gita Singh, *University College Dublin*, *Ireland*
Mr Sebastian Skaanvik, *University of Aarhus*, *Denmark*
Dr Ben Slater, *Imperial College*, *United Kingdom*
Mr Maxwell Sparey, *Keysight Technologies - SENTINEL ITN*, *Austria*
Professor Zhong-Qun Tian, *Xiamen University*, *China*
Ms Chloe Tolbert, *University of Wyoming*, *USA*
Professor Kristina Tschulik, *Ruhr-University Bochum*, *Germany*
Professor Patrick Unwin, *University of Warwick, UK*, *United Kingdom*
Ms Swathi Naidu Vakamulla Raghu, *University of Siegen*, *Germany*
Mr Dimitrios Valavanis, *University of Warwick*, *United Kingdom*
Professor Jill Venton, *University of Virginia*, *USA*
Dr Alain Walcarius, *Laboratoire de Chimie Physique et Microbiologie pour les Matériaux et l'Environnement - le Centre National de la Recherche Scientifique - University of Lorraine (LCPME - CNRS - Univ. Lorraine)*, *France*
Miss Wei-Wei Wang, *Xiamen University*, *China*
Mr Yunong Wang, *Indiana University*, *USA*
Dr Jianmei Wang, *Deakin University*, *Australia*
Professor Dr Hui Wei, *Nanjing University*, *China*
Mr Donal Whelan, *University College Dublin*, *Ireland*
Dr Mary Wood, *University of Cambridge*, *United Kingdom*
Dr Yanfang Wu, *The University of New South Wales, Sydney*, *Australia*
Professor De-Yin Wu, *Xiamen University*, *China*
Mrs Li Xiao, *Wuhan University*, *China*
Mr Xiangdong Xu, *University of Warwick*, *United Kingdom*
Professor Weilin Xu, *Changchun Institute of Applied Chemistry*, *China*
Professor Wenrong Yang, *Deakin University*, *Australia*
Professor Yi-Lun Ying, *Nanjing University*, *China*
Professor Dr Ping Yu, *Institute of Chemistry, Chinese Academy of Sciences*, *China*
Mr Chonglin Yuan, *Shenzhen University*, *China*
PhD Lucia Zanetti, *University of Padova*, *Italy*
Dr Xu Zhang, *Cape Breton University*, *Canada*
Mr Zhu Zhang, *Utrecht University*, *Netherlands*
Mr Xinlian Zhang, *Fudan University*, *China*
Professor Shelley Minteer, *University of Utah*, *USA*
Professor Hong-Yuan Chen, *Nanjing University*, *China*
Dr Kallie Willets, *Temple University*, *USA*